Automatic control systems have become essential features in virtually every area of technology, from machine tools to aerospace vehicles.

This book is a thorough, easy-to-read introduction to automatic control engineering, emphasizing the way physical properties of individual interconnected devices in a system influence the dynamic performance of the entire control system.

The author begins with the fundamentals of modeling mechanical, electrical, and electromechanical systems in the state-variable format. The emphasis is on classical feedback control theory and design, and their application to practical electromechanical and aerospace problems. Following a careful grounding in classical control theory, the author provides an introduction to modern control theory, including digital control and nonlinear system analysis. Problems at the end of each chapter (more than 230 altogether) help the reader apply principles discussed in the text to practical engineering situations.

Engineering students and practicing engineers will find what they need to know about control system analysis and design in this clear and comprehensive text.

T0318650

Control System Dynamics

Control System Dynamics

Robert N. Clark

University of Washington

CAMBRIDGE UNIVERSITY PRESS
Cambridge, New York, Melbourne, Madrid, Cape Town, Singapore, São Paulo

Cambridge University Press
The Edinburgh Building, Cambridge CB2 2RU, UK

Published in the United States of America by Cambridge University Press, New York

www.cambridge.org
Information on this title: www.cambridge.org/9780521472395

© Cambridge University Press 1996

This book is in copyright. Subject to statutory exception
and to the provisions of relevant collective licensing agreements,
no reproduction of any part may take place without
the written permission of Cambridge University Press.

First published 1996
This digitally printed first paperback version 2005

A catalogue record for this publication is available from the British Library

Library of Congress Cataloguing in Publication data

Clark, Robert N.
Control system dynamics / Robert N. Clark.
p. cm.
Includes bibliographical references and index.
ISBN 0-521-47239-3 (hc)
1. Feedback control systems. 2. Control theory. I. Title.
TJ216.C55   1995
629.8′312 – dc20   95-14945
                                                  CIP

ISBN-13 978-0-521-47239-5 hardback
ISBN-10 0-521-47239-3 hardback

ISBN-13 978-0-521-01793-0 paperback
ISBN-10 0-521-01793-9 paperback

Dedicated to my wife

Mary Quiatt Clark

and to our sons

Charles Winthrop Clark
John Robert Clark
Timothy James Clark
Franklin Thomas Clark

Contents

Preface   *page* xi

1 Introduction to Control System Engineering   1

   1.1 Background   1
   1.2 Modern Control Theory and the Digital Computer   1
   1.3 Organization of the Book   2
   1.4 Role of the Problems   3

2 Mathematical Models of Mechanical Systems   5

   2.1 Introduction   5
   2.2 Mass, Spring, and Damper System   6
   2.3 Simple Rate Gyroscope   11
   2.4 Pendulum with Moving Base   15
   2.5 Mechanical Drive Systems   21
   2.6 Single-Axis Rigid Spacecraft   25
   2.7 Longitudinal-Axis Dynamics of an Airplane   26
   2.8 Summary   34
   2.9 Problems   36

3 Mathematical Models of Electrical Systems   44

   3.1 Introduction   44
   3.2 Passive Circuits   45
   3.3 Active Circuits   51
   3.4 Transducers   58
   3.5 Noise in Electrical Signals   62
   3.6 Summary   63
   3.7 Problems   64

4 Mathematical Models of Electromechanical and Electrohydraulic Systems 70

  4.1 Introduction 70
  4.2 Fundamentals of the DC Machine 71
  4.3 Electromagnets 83
  4.4 Electrohydraulic Amplifier, Valve, and Ram System 88
  4.5 Summary 92
  4.6 Problems 93

5 Summary of Modeling Principles for Physical Systems 99

  5.1 Introduction 99
  5.2 Choice of State Variables 100
  5.3 The Output Equation 104
  5.4 Summary 106
  5.5 Problems 107

6 Solution to the Initial-Value Problem 110

  6.1 Introduction 110
  6.2 A First-Order Example 111
  6.3 The nth-Order Multivariable System 115
  6.4 A Single-Input–Single-Output System; the Transfer Function 117
  6.5 The nth-Order Single-Input–Single-Output System 122
  6.6 Composite Transfer Functions 126
  6.7 Summary 130
  6.8 Problems 131

7 Pole-Zero Methods of Analysis for Single-Input– Single-Output Systems 137

  7.1 Introduction 137
  7.2 Partial-Fraction Expansion of $Y(s)$ 141
  7.3 The Second-Order System 150
  7.4 A Third-Order System 162
  7.5 Higher-Order Systems 167
  7.6 Series Compensation 174
  7.7 Nonminimum Phase Systems 179
  7.8 Summary 180
  7.9 Problems 181

8 Automatic Feedback Control 189

  8.1 Automatic Control of Dynamic Systems 189
  8.2 Feedback Control of Dynamic Plants 191
  8.3 Block Diagrams for Control System Analysis 196

8.4 Control System Design   200
8.5 Historical Background and Full-State Feedback   202
8.6 Problems   205

9   Dynamic Analysis of Feedback Control Systems   210

9.1 The Stability Problem in Feedback Systems   210
9.2 Root-Locus Method for Factoring $1 + G(s)$ 214
9.3 Stability Analysis of an Electrohydraulic
Servomechanism   236
9.4 Control of Non-Integrating Processes   241
9.5 Non-Unity Feedback Systems   243
9.6 Negative-Gain Feedback Systems   246
9.7 Positive Feedback Systems   249
9.8 Summary   251
9.9 Problems   252

10   Design of Feedback Control Systems   260

10.1 Introduction   260
10.2 Series Compensation   261
10.3 The PID Controller   271
10.4 Pole-Zero Cancellation in Series Compensation   274
10.5 Parallel Compensation   276
10.6 Tracking Control   285
10.7 Design for Disturbance Rejection   291
10.8 Control of Unstable Plants   293
10.9 Control of Nonminimum Phase Plants   298
10.10 Summary   301
10.11 Problems   302

11   Frequency Response Analysis of Linear Systems   312

11.1 Definition of the Frequency Response for a Linear
System   312
11.2 Graphical Representation of Frequency Response Data   317
11.3 Frequency Response Obtained from Transfer Function   321
11.4 Asymptotic Approximation to Bode Plots   327
11.5 Frequency Response of Minimum Phase Systems   336
11.6 Frequency Response of Nonminimum Phase Systems   338
11.7 Experimental Measurements of the Frequency Response
of Unstable Systems   342
11.8 Summary   344
11.9 Problems   345

12   Stability Analysis by Nyquist's Criterion   352

12.1 Introduction   352
12.2 Cauchy's Fundamental Theorem   357

12.3 Nyquist's Stability Criterion   360
12.4 Summary   368
12.5 Problems   370

13 Dynamic Analysis of Feedback Systems by
        Frequency Response Methods   373

13.1 Introduction   373
13.2 Stability Margins Determined on Bode Diagrams   377
13.3 Closed-Loop M_p and ω_r Determined on the Nichols Chart 382
13.4 Control of Nonminimum Phase Plants   387
13.5 Time-Delay Elements in the Control Loop   393
13.6 Summary   400
13.7 Problems   401

14 Design of Feedback Systems by Frequency
        Response Methods   405

14.1 Introduction   405
14.2 Series Compensation   408
14.3 Parallel Compensation   419
14.4 Robustness   422
14.5 Summary   423
14.6 Problems   425

15 Advanced Topics   430

15.1 Introduction   430
15.2 MIMO Linear System Analysis   430
15.3 Discrete Time Systems and Digital Control   435
15.4 Nonlinear Analysis   442
15.5 Control over a Finite Interval   448
15.6 Alternate Control Law Structures   450
15.7 Summary   450
15.8 Problems   451

Appendices

A Physical Constants, Units, and Conversion Factors   453
B Trigonometric Formulas   458
C The Laplace Transformation and Tables   459
D Routh's Stability Criterion   472
E Normalized Time Response Curves   478
F Normalized Frequency Response Curves and Nichols Chart 487

Answers to Problems   491
Bibliography   499
Index   503

Preface

This textbook is designed to help the student learn the basic techniques necessary to begin the practice of automatic control engineering. The subject matter of the book, *classical control theory,* is a study of the dynamics of feedback control systems. Particular emphasis is placed upon the way in which the physical properties of the individual interconnected devices in a system influence the dynamic performance of the entire control system. Although classical control theory is normally taught to engineering students during their third or fourth year of study, it may also be learned by interested persons having a comparable background in mathematics, physics, and engineering.

Classical control theory was developed from the feedback amplifier technology of the 1920s and 1930s. It was first applied to automatic control systems for machine tools, industrial processes, and military equipment of many types. Although these applications bore little outward resemblance to their electronic amplifier antecedents, they all relied on the same principle of feedback for their remarkable performance. Increased demand for control systems in the 1940s gave rise to the branch of engineering known as *automatic control.* Research groups were formed in industry to develop and apply classical control theory. Engineering departments in universities offered seminars and courses in automatic control, and textbooks on the subject began to appear, including my earlier book which was published in 1962 (*Introduction to Automatic Control Systems,* Wiley & Sons, New York).

Classical control theory continues to be valid and useful, but since 1960 two developments in automatic control have necessitated a revised approach to the introductory study of the subject. One of these developments is *modern control theory,* based on the modeling of dynamic systems using differential equations in the state-variable form. This form of modeling is convenient for dealing with so-called multivariable devices, which are controlled by two or more driving inputs; classical control theory is convenient only for single-input devices. A second development that has profoundly influenced both the teaching and practice of control engineering is the digital computer, a powerful calculating tool for analysis and design which can also serve as an integral part of the physical system, processing dynamic signals on line.

This textbook also provides an introduction to modern control theory. It models dynamic systems in state-variable form and explores the initial-value problem in the time domain before moving on to frequency-domain methods. This approach to the subject clarifies the connection between classical and modern control theory. However, classical theory and design, which is the principal focus of the book, remains an excellent introduction to control engineering, both because it is applicable to many important engineering problems and because it is essential to the understanding of modern control theory.

Control System Dynamics is organized into four sections:

1. mathematical modeling and dynamic analysis of physical systems (Chapters 1 through 7);
2. analysis and design of feedback systems using pole-zero methods (Chapters 8, 9, and 10);
3. analysis and design of feedback systems using frequency response techniques (Chapters 11 through 14); and
4. advanced topics and appendices (Chapter 15 and Appendices A through F).

Chapters 2 through 7 model simple mechanical, electrical, electromechanical, electrohydraulic, and aircraft flight-control systems using the state-variable form of differential equations. These models are used throughout the book in examples and in the exercise problems. The transfer function concept – showing the relationship between the time-domain and frequency-domain representations of the models – is introduced, and pole-zero techniques are applied to the dynamic analysis of the models. Some of the material in Chapters 2, 3, and 4 can be covered quickly by students with previous experience in dynamic modeling.

Chapters 8, 9, and 10 introduce the concept of feedback control and the attendant problem of stability. The root-locus method, combined with pole-zero analysis, is used to analyze the stability and performance of single-loop control systems. Control of unstable plants and nonminimum phase systems are illustrated using practical examples drawn from earlier chapters. Positive feedback, series and parallel compensation, disturbance rejection, robustness, and other basic design ideas are developed.

Chapters 11 through 14 treat analysis and design using real frequency response methods. Bode techniques, the Nyquist stability criterion, and frequency-based performance measures are all used to treat many of the problems solved earlier by pole-zero methods.

Chapter 15 illustrates full-state feedback using the basic tools of modern control theory. It also discusses introductory ideas in digital control and in nonlinear system analysis. The appendices provide useful information on physical units and conversion factors between the English system of units and the SI system, key algebraic and trigonometric formulas, a brief introduction to the Laplace transformation (accompanied by a table of commonly used transform pairs), the complete form of Routh's stability criterion, and normalized time response and frequency response curves for first-, second-, and third-order systems.

The problems at the end of each chapter offer an opportunity for comprehensive application of the principles covered in the text. In some cases the student is asked to derive new formulas to be used in subsequent chapters or in professional practice. The student must have access to a computer equipped with one of the commercially available programs designed for, or adaptable to, control system calculations. More than twenty such programs suitable for personal computers are on the market. Although the 386-MATLAB® program was used to prepare and solve the problems, the problem statements are not keyed to this particular program. Answers to many of the problems appear in the back of the book.

Most of the contents of the book are covered in senior-level courses of 38 lectures in the Electrical Engineering Department and in the Department of Aeronautics and Astronautics at the University of Washington. I am grateful to my colleagues Professors E. Noges, R. B. Pinter, M. J. Damborg, and J. Vagners for constructive advice on the preliminary material for this book. Many students have also added to my perspective in teaching this subject. Professor D. B. DeBra and the late Professor H. H. Skilling, both of Stanford University, and Professor P. M. Frank of the Gerhard Mercator University of Duisburg have also given me valuable advice in developing this material. Dr. C. W. Clark of the National Institute of Standards and Technology and Mr. T. J. Clark each gave me helpful suggestions. I am also indebted to the University of Washington for granting me leave time for writing.

My wife, Mary Quiatt Clark, has made the book more readable by suggesting concise ways to express many of the key ideas in my early drafts. Her contributions are those of a co-author although she declines to be identified in that way.

Seattle, Washington
September 1995

Robert N. Clark

1 Introduction to Control System Engineering

1.1 Background

Dynamics plays the central role in automatic control engineering. The analytical techniques and design principles examined in this book are simply methods of dealing with dynamics problems from the specialized point of view of the automatic feedback control system. This collection of methods and procedures – known as *servomechanism theory, basic control theory,* and, in recent years, as *classical control theory* – constitutes the basic subject to be mastered by a beginning control system engineer.

A typical automatic control system consists of several interconnected devices designed to perform a prescribed task. For example, the task may be to move a massive object such as the table of a machine tool in response to a command. The interconnected devices of the system are typically electromechanical actuators, sensors that measure the position and velocity of the controlled object and the currents or voltages at the actuator, and a control computer that processes the sensor signals along with the command. These interconnected dynamic elements work simultaneously, and they also embody a feedback connection. Frequently the engineer must determine the dynamic response of the entire system to a given command when only the physical properties of the individual component elements of the system are known. This formidable task requires quantitative dynamic analysis even in relatively simple systems.

1.2 Modern Control Theory and the Digital Computer

Classical control theory is directly applicable to systems which have only single input variables and single response variables. These are called *single-input–single-output* (SISO) systems. A complex automatic control system, such as that for an airplane or for a chemical process plant, is normally driven by

1

several input commands. Many important variables within the plant respond to these multiple inputs. These *multiple-input–multiple-output* (MIMO) systems have given rise to the subject now called *modern control theory*. Engineers have been able to design increasingly complex MIMO automatic control systems using this advanced theory. This later theory encompasses classical control theory, and it gives the engineer a deeper understanding of control system dynamics. The concurrent development of the digital computer has given the engineer the tool that makes possible the practical application of modern control theory.

Modern control theory begins with writing the differential equations which represent the dynamics of a system in a way that conveniently accommodates multiple inputs and multiple outputs. This representation, called the *vector-matrix differential equation* or the *state-variable* form of the equations, is the format required for computer-aided calculations.

The digital computer has had a twofold impact on automatic control engineering. First, it has provided students and practitioners with powerful computing capacity. As a result, the computer has nearly eliminated the need to depend on the approximation methods that were formerly so useful. Nevertheless, approximation methods remain important in the initial stages of a design and to verify the results of computer-based analysis. Second, the high speed of operation of the digital signal processor makes it extremely useful as an integral operating element in the control system. It can perform not only the functions of electronic controllers but also additional functions that require data storage and logical operations.

1.3 Organization of the Book

The topics of the first few chapters will be generally familiar to many readers; however, the format, point of view, and terminology used in the analysis is that of the control engineer. Much of the material in these early chapters is essential to the later chapters.

The study of automatic control revolves around the concept of feedback and the methods of incorporating feedback into the dynamic analysis of physical systems. Mastering this subject must be approached with a firm understanding of the basic dynamics of physical systems. For this reason Chapters 2 through 7 are devoted to the modeling and analysis of specific physical systems not involving feedback. The mathematical models in these chapters are differential equations whose time domain solutions are obtained by elementary techniques, supplemented by computer calculations and by use of the Laplace transformation. These early chapters constitute an introduction to modern control theory because the models are derived in state-variable form.

Feedback is introduced in Chapter 8, and the principles of analysis developed in the earlier chapters using the time-domain approach are applied to feedback systems in Chapter 9. Chapter 10 deals with basic design principles applied to satisfy specified dynamic performance requirements of feedback systems. Feedback system stability emerges here as a pivotal problem.

An important point of view of dynamics, the *frequency response,* is introduced in Chapter 11. The specialized terminology of the frequency response method and several useful graphical techniques for displaying frequency response information are defined and applied to practical examples not involving feedback.

Chapters 12, 13, and 14 deal with the analysis and design of feedback systems using the frequency response method. Nyquist's stability criterion, the sole topic of Chapter 12, relates the stability of the closed-loop system to the frequency response of the elements in the loop. Chapters 13 and 14 apply the analytical techniques of Bode, Nyquist, and Nichols to the design problems treated in Chapter 10 from the time-domain point of view. Therefore, Chapters 12–14 may be regarded as the frequency-domain counterparts to Chapters 8–10, in which analysis and design were approached from the time domain.

Advanced topics that are essential to preliminary design, product development, research, or teaching of automatic control systems are introduced in Chapter 15. The relationship between *classical control theory,* the main subject of this book, and *modern control theory* is examined. A common dynamic phenomenon peculiar to nonlinear systems is demonstrated, and a basic model for a digital feedback control system is derived.

The appendices provide quantitative data on physical units, some elementary algebraic and trigonometric formulas, a review of the Laplace transformation including a useful table of transforms, a brief treatment of Routh's stability criterion, and some normalized time response and frequency response curves that are useful in practical analysis.

We concentrate our attention on classical control theory in the main body of the text for the following reasons.

1. Classical control theory provides a convenient way to understand stability, which is the most important problem in feedback control systems.
2. Classical control theory can be used successfully to design some multi-output feedback control systems.
3. Classical control theory, as a subset of modern control theory, is universally regarded as a prerequisite to modern control theory. Examples from classical control theory are frequently used to explain the advanced techniques of modern control theory.
4. The inclusion of a reasonably comprehensive treatment of modern control theory here would have doubled the size of this book.

1.4 Role of the Problems

The problems at the end of each chapter provide the reader with the opportunity to apply the principles covered in the text, to extend those principles, and to derive new relationships needed to solve problems in subsequent chapters. Students who believe they are already skilled at modeling simple dynamic systems should work some of the problems in Chapters 2, 3, and 4 to confirm their skill. Answers to many problems are found in the back of the book.

In order to solve many of the problems, the student needs access to a personal computer or work station and an analysis program suitable for control system work. A number of reliable and easy-to-use analysis programs are available. These require problem data to be in matrix form, the natural form of state-variable models. The problems and examples in this book were all worked on a personal computer using the 386-MATLAB® (version 3.5k) program. The answers given have been rounded down to four or five significant figures, a precision that is adequate for confirming answers to exercise problems and more than adequate for representing the physical data normally available in control engineering problems.

2 Mathematical Models of Mechanical Systems

2.1 Introduction

We first study a very simple class of mechanical systems – those consisting of a single rigid body or of two rigid bodies simply connected. The bodies will be restricted to planar motion in most cases, but they may rotate as well as translate in the plane. We recognize that most practical mechanical systems feature parts that move in three dimensions, but our restrictions are necessary here because the dynamics of massive bodies in three-dimensional motion is, in most cases, beyond the scope of this introductory book on automatic control. However, our use of simple systems has some advantages. We can use the free-body diagram to formulate directly from Newton's and Euler's laws the differential equations that describe the dynamics of the motion without resorting to the more abstract approach of using variational principles, which is usually necessary in the three-dimensional case. It is also possible in this simple setting to illustrate an important principle of mechanics that one must observe when expressing Euler's law for a body undergoing angular acceleration with respect to inertial space. Furthermore, since the motions of bodies in many practical applications *are* approximately planar, our simple approach in such cases will yield useful results.

The mathematical model we seek for each of the systems studied here is a set of differential equations that describe the physical system and its environment, plus certain auxiliary information that permits us to use these equations to determine the dynamic behavior of our system. This auxiliary information consists of the boundary conditions existing between the system and its environment, the identification of the exciting agents (inputs) to the system, and the interval of time during which the differential equations provide a valid description of the system dynamics. The model also includes any approximations we make to the physical laws governing the system in order to make the model mathematically tractable. In our mechanical systems, the differential equations will be analytical expressions of Newton's laws of dynamics and Euler's law pertaining to the rotary motion of the bodies. In most cases we will ignore the earth's rotation and assume that the earth is an inertial reference frame. This severe restriction on the general validity of our analytical approach simplifies our models and yields solutions to the equations that are accurate enough for many practical purposes.

Our simple mechanical systems and the circumstances in which we study their dynamic behavior will yield ordinary differential equations that usually have constant coefficients. Once we determine the time interval during which the equations are valid and the appropriate physical conditions of the system at the initial instant of that time interval, we will have a model for the system that matches precisely the simplest statement of the initial-value problem found in textbooks on differential equations. Therefore, the modeling process is the bridge between the physics and the mathematics of the problem.

We define *state variables* for our equations, and through a simple algebraic process we express the equations in a useful standard form called the *state-variable form*. This form provides insight into the dynamics of the system that is not apparent from the original form derived from the physics of the system, and it makes very convenient the numerical solution of the initial-value problem by computer. This state-variable form is important because it is the basis of modern control theory, a subject essential to control engineering practice but beyond the scope of this book.

In this chapter we concentrate on obtaining models of mechanical and aerospace systems; in Chapters 3 and 4 we add electrical systems, electromechanical systems, and electrohydraulic systems to our repertoire of important types of mathematical models. These models are derived both in the examples examined in the text and in the problems following each chapter. Please note that the problems are therefore an integral part of our study. Beginning with Chapter 6, we will use these models in the analysis and design of control systems.

2.2 Mass, Spring, and Damper System

A simple mass, spring, and damper system is shown in Fig. 2.1. The displacement of the cart from a reference point fixed to the level plane is $z(t)$. We take $z = 0$ at the position in which the spring is relaxed. The level plane is assumed to be an inertial reference frame. We ignore the mass of the wheels, and we assume that the friction in the wheel bearings is negligible. A force $f(t)$ is applied in the positive $z(t)$ direction, as shown. The cart has constant mass M, and we assume the motion of the cart to be in the horizontal direction only; it does not rotate, nor do the wheels lose contact with the plane. We wish to establish a mathematical model for this system that describes the relationship of the displacement $z(t)$ to the driving force $f(t)$. Newton's second law is the starting point in this analysis. We employ the free-body force–mass diagram of Fig. 2.2 to illustrate the elements of this approach.

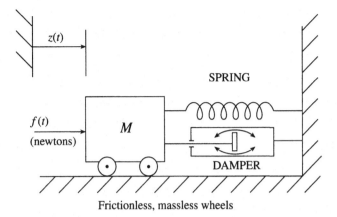

Fig. 2.1   The mass, spring, and damper system.

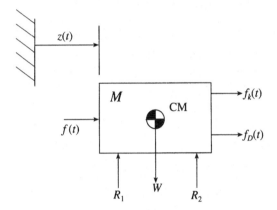

Fig. 2.2   Free-body force–mass diagram.

Each of five agents applies a force to the mass. These agents are gravity, the vertical contact forces supporting the cart, the external driving force $f(t)$, the contact force due to the spring $f_k(t)$, and the contact force due to the damper plunger $f_D(t)$. (Because we have neglected the mass of the wheels and the friction in the wheel bearings, there are no horizontal components of contact force accompanying the two vertical components R_1 and R_2.) The sum of the gravity force W and the other two vertical forces R_1 and R_2 will be zero under our assumption that the cart does not accelerate in the vertical direction. If we are not interested in calculating R_1 and R_2 then we need not write an equation for the vertical component of the acceleration of the cart.

The two forces R_1 and R_2 also have moments about the center of mass, as do the three horizontal forces. These moments depend on the points on the cart at which the forces are applied, as well as on the forces themselves; however, the sum of the moments will be zero because we have assumed that the cart will not rotate about its center of mass. The terms R_1 and R_2 will fluctuate as the cart moves to and fro, in order that the sum of these moments remains zero at all times. However, if we are not interested in calculating these two dynamic vertical forces we need not write the moment equation for the angular acceleration of the cart.

Under these assumptions, it is only the motion in the z direction that is of interest. Newton's second law for this motion, that the rate of change of the linear momentum of the mass in the z direction must equal the sum of all the forces applied to the mass in the z direction, is expressed analytically as

$$M\ddot{z} = f(t) + f_k(t) + f_D(t). \tag{2.1}$$

In this equation the unit for mass, M, is kilograms [kg], the unit for displacement is meters [m], the unit of force is newtons [N], and the unit of time is seconds [s]. Consequently, $\ddot{z}(t)$ has units of meters per second per second [(m/s)/s].

In this case both the spring force and the damper force are related to the motion of the cart. We assume that the spring force depends only on the displacement $z(t)$ and that the damper force depends only on the relative velocity between the plunger and the cylinder, which in this case is $\dot{z}(t)$ [m/s]. These dependencies are determined by the physical properties of the spring and damper. The simplest type of spring model incorporates the ideal (lossless) Hooke's law, in which the force is proportional to the deflection. In our coordinate system this may be written as

$$f_k(t) = -kz(t), \tag{2.2}$$

where k is the *spring constant* having units of N/m. The minus sign is necessary because the arrow denoting $f_k(t)$ in Fig. 2.2 points in the $+z$ direction. In the damper, a viscous fluid working around the plunger as it moves back and forth in the cylinder transmits the force $f_D(t)$ to the cart. A laboratory test of the damper would reveal a force–velocity characteristic resembling that shown in Fig. 2.3. A model of this characteristic could be

$$f_D(t) = -[\alpha\dot{z}(t) + \beta\dot{z}^3(t)], \tag{2.3}$$

where α N/(m/s) and β N/(m/s)^3 are coefficients chosen so that Eqn. 2.3 will fit the laboratory data of Fig. 2.3. It is implicit in Eqn. 2.3 that if mechanical contact exists between plunger and cylinder it will not create a friction force independent of $\dot{z}(t)$.

We now substitute the expressions for $f_k(t)$ and $f_D(t)$ in Eqns. 2.2 and 2.3 into Eqn. 2.1 to obtain a differential equation with $z(t)$ as the dependent variable, the physical parameters M, k, α, and β as coefficients, and the external applied force $f(t)$ on the right-hand side:

$$M\ddot{z}(t) + \alpha\dot{z}(t) + \beta\dot{z}^3(t) + kz(t) = f(t). \tag{2.4}$$

Equation 2.4 governs the motion of the cart in response to the applied force, provided that $z(t)$ remains within the mechanical limits of its excursion. If the spring should become compressed to a solid column of metal, for example, or if it should be stretched to the breaking point, then Eqn. 2.4 would no longer give an accurate description of the dynamics of the system.

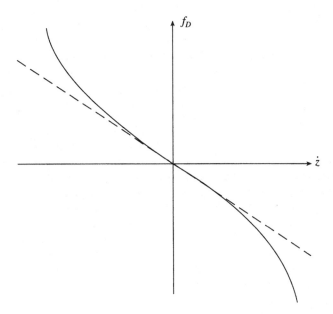

Fig. 2.3 Force–velocity characteristic of damper.

Divide Eqn. 2.4 on both sides by the nonzero constant M; this yields

$$\ddot{z}(t) + \left[\frac{\alpha}{M}\right]\dot{z}(t) + \left[\frac{\beta}{M}\right]\dot{z}^3(t) + \left[\frac{k}{M}\right]z(t) = \left[\frac{1}{M}\right]f(t). \qquad (2.5)$$

Equation (2.5) is an ordinary nonlinear differential equation of the second order having constant coefficients. The physical meaning of the coefficients $[\alpha/M]$, $[\beta/M]$, $[k/M]$, and $[1/M]$, which have positive real values, need not concern an analyst who is interested in determining only the mathematical implications of the equation and who might, for example, wish to investigate the solution of the equation when the bracketed terms are assumed to be negative real numbers. This could lead to interesting mathematics, but one physical implication of this mathematical assumption – that the cart has a negative mass – would invalidate the model for engineering purposes! This shows the distinction between mathematics and engineering, and it also shows why the engineer must use proper algebraic signs for terms in the equations that express physical laws.

The derivation of the state-variable form of Eqn. 2.5 is simply a convenient re-arrangement of the equation that involves no new physical principles and requires no abstract mathematical procedures. We simply introduce the notation: $z(t) = x_1(t)$ and $\dot{z}(t) = x_2(t)$. We therefore have $\ddot{z}(t) = \dot{x}_2(t)$ and $\dot{x}_1(t) = x_2(t)$. Now Eqn. 2.5, a second-order equation, can be represented equivalently by the following pair of first-order differential equations:

$$\dot{x}_1(t) = x_2(t),$$

$$\dot{x}_2(t) = -\left[\frac{k}{M}\right]x_1(t) - \left[\frac{\alpha}{M}\right]x_2(t) - \left[\frac{\beta}{M}\right]x_2^3(t) + \left[\frac{1}{M}\right]f(t), \qquad (2.6)$$

which we call the *state-variable* form of Eqn. 2.5; $x_1(t)$ and $x_2(t)$ are the state variables.

To complete the mathematical model of the mass–spring–damper system, we must decide what we want to know about the dynamics of the system and append those decisions to Eqn. 2.6 in suitable form. If, as in many cases, we are interested in the *initial-value problem,* then we apply a known force $f(t)$ beginning at a known instant t_0 and terminating at a known instant t_F. This establishes the time interval during which our equations are to be valid, $t_0 \leq t \leq t_F$. It is also necessary to know the *initial state* of our system, which is the set of values for the state variables at the initial instant t_0, $x_1(t_0)$, and $x_2(t_0)$. A restriction on the excursion of $z(t)$ is expressed as: $z_{min} \leq z(t) \leq z_{max}$. We now have a complete model for the mass–spring–damper system, which is required for the solution of the initial-value problem. The model is

$$\dot{x}_1(t) = x_2(t),$$

$$\dot{x}_2(t) = -\left[\frac{k}{M}\right]x_1(t) - \left[\frac{\alpha}{M}\right]x_2(t) - \left[\frac{\beta}{M}\right]x_2^3(t) + \left[\frac{1}{M}\right]f(t), \quad \text{where:}$$

$$t_0 \leq t \leq t_F \text{ and } f(t) \text{ is known,} \tag{2.7}$$

$$x_1(t_0) \text{ and } x_2(t_0) \text{ are known,}$$

$$z_{min} \leq x_1(t) \leq z_{max}.$$

We will return to this model in subsequent chapters and study the solution $z(t)$ in some detail. For the present we can use Eqn. 2.7 to make two calculations by inspection. First, let the initial time be $t_0 = 0$, and let the driving force be $f(t) = F\sigma(t)$ N, where $\sigma(t)$ is the unit step function occurring at $t = 0$. (The unit step function is devoid of physical units and is defined in Appendix C.) Let the initial conditions be $x_1(0) = z(0)$ m and $x_2(0) = \dot{z}(0)$ m/s. Inspecting Eqn. 2.7, we see that the initial acceleration of the cart $\ddot{z}(0^+) = \dot{x}_2(0^+)$ is

$$\ddot{z}(0^+) = \frac{1}{M}[F - kz(0) - \alpha\dot{z}(0) - \beta[\dot{z}(0)]^3] \text{ (m/s)/s.} \tag{2.8}$$

(The notation $\ddot{z}(0^+)$ is also explained in Appendix C.) Our second calculation is for the final value of the position of the cart, $z(t_F)$. We assume that t_F is great enough so that the cart will have settled to a stationary position at that final time. This implies that both $\dot{x}_1(t_F)$ and $\dot{x}_2(t_F)$ are zero, which in turn implies that $x_2(t_F)$ is also zero. Hence by inspection we have $z(t_F) = F/k$ m.

In later chapters we will want to deal with linear differential equations. Equation 2.7 is nonlinear due solely to the $\beta\dot{z}^3(t)$ term. If we place another restriction on the motion of $z(t)$, namely that the speed of the cart will remain small enough so that $f_D(t) \cong -\alpha\dot{z}(t)$ (see Fig. 2.3, in which the dashed line has a slope $-\alpha$), then $\beta\dot{z}^3(t)$ can be dropped from Eqn. 2.7 to yield a linear model of the mass–spring–damper system that will be approximately valid under the following restrictions:

$$\dot{x}_1(t) = x_2(t),$$

$$\dot{x}_2(t) = -\left[\frac{k}{M}\right]x_1(t) - \left[\frac{\alpha}{M}\right]x_2(t) + \left[\frac{1}{M}\right]f(t), \quad \text{where:}$$

$$\quad (2.9)$$

$t_0 \leq t \leq t_F$ and $f(t)$ is known,

$x_1(t_0)$ and $x_2(t_0)$ are known,

$z_{\min} \leq x_1(t) \leq z_{\max}$ and $\beta\dot{z}^3(t) \approx 0$.

2.3 Simple Rate Gyroscope

Gyroscopes of several types are used as sensors on aircraft, spacecraft, ships, and other vehicles. They are also found in fixed-base systems such as steerable antennas and astronomical observatories. Gyroscopes provide dynamic measurements of position, velocity, or acceleration that are essential for control of the system upon which they are mounted.

We now study a simple type of mechanical gyroscope, called a rate gyroscope, which is intended to provide a signal proportional to the angular rate of its moving base. A sketch of such a device is shown in Fig. 2.4. A rapidly spinning rotor is mounted in a frame, called a *gimbal,* which in turn is mounted on the moving base. The gimbal rotates around an axis (called the output axis) which is normal to the spin axis. The gimbal is restrained in its angular motion about the output axis by a linear spring and damper. In fact, the angular displacement of the gimbal from its rest position, $\theta(t)$, will not exceed a small

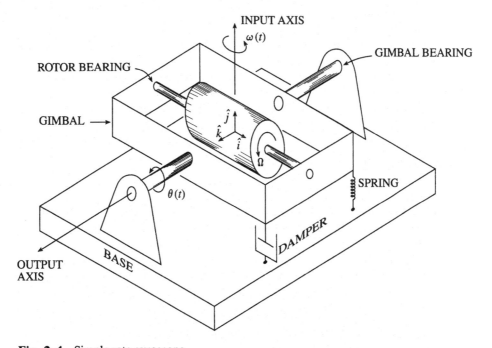

Fig. 2.4 Simple rate gyroscope.

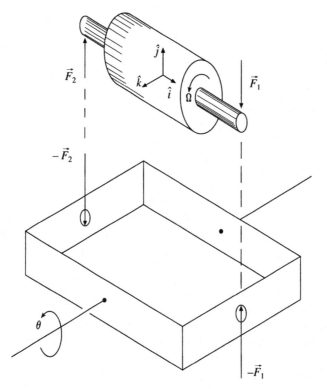

Fig. 2.5   Free-body diagram of rotor.

angle, say 0.1 rad (radian). The input axis is normal to both the output axis and the spin axis. The term $\omega(t)$ rad/s is the angular rate of the vehicle to which the base is rigidly attached, and it is $\omega(t)$ that the instrument is intended to measure.

Figure 2.5 is a free-body diagram for the rotor. The reference frame $\{\hat{i}, \hat{j}, \hat{k}\}$ is centered in the center of mass of the rotor but is fixed with respect to the gimbal, with the $\hat{k}$ axis remaining aligned with the output axis of the gimbal. Because the gimbal moves with respect to inertial space, the unit vectors $\hat{i}(t)$, $\hat{j}(t)$, and $\hat{k}(t)$ are time-dependent. The angular momentum of the rotor about its center of mass with respect to inertial space is essentially

$$\vec{H}(t) = I_R \Omega \hat{i}(t) + I_0 \dot{\theta}(t) \hat{k}(t) \text{ kg-m}^2/\text{s}, \tag{2.10}$$

where I_R is the moment of inertia of the rotor about the $\hat{i}$ axis (in kg-m^2), I_0 is the moment of inertia about the $\hat{k}$ axis, and Ω is the speed of the rotor in rad/s. Ω is kept constant by an electric motor or by an air motor. The gyroscope is designed so that the product $I_R \Omega$ is as large as the materials of the device will permit. Because $\omega(t)$ is very much smaller than Ω, $\omega(t)$ has been ignored in calculating $\vec{H}(t)$. Euler's law, that $\vec{H}(t)$ remains fixed with respect to inertial space if no forces act upon the rotor, reveals the fundamental property of gyroscopes that makes them so useful in practical applications. Euler's law gives the relationship between any forces acting on the rotor and the resultant motion of the spinning rotor:

$$\vec{M}(t) = \frac{d\vec{H}(t)}{dt} \text{ N-m}, \tag{2.11}$$

where $\vec{M}(t)$ is the sum of the moments due to all the forces acting on the rotor about the center of mass of the rotor, and $d\vec{H}(t)/dt$ is the rate of change of $\vec{H}(t)$ with respect to inertial space. Since Ω is constant, we have

$$\vec{M}(t) = \frac{d\vec{H}(t)}{dt} = I_R \Omega \frac{d\hat{i}(t)}{dt} + I_0 \ddot{\theta}(t) \hat{k}(t) + I_0 \dot{\theta}(t) \frac{d\hat{k}(t)}{dt}. \tag{2.12}$$

Assume that the vehicle undergoes a maneuver such that its angular velocity with respect to the earth (inertial space, actually) is

$$\vec{\omega}(t) = \omega(t) \hat{j}(t) \text{ rad/s}. \tag{2.13}$$

This motion reveals that $d\hat{i}(t)/dt = -\omega(t)\hat{k}(t)$ and that $d\hat{k}(t)/dt = \omega(t)\hat{i}(t)$, so the total moment on the rotor must be

$$\vec{M}(t) = [I_0 \ddot{\theta}(t) - I_R \Omega \omega(t)] \hat{k}(t) \text{ N-m}, \tag{2.14}$$

where the term $\dot{\theta}(t)\omega(t)$ has been taken to be zero. This $\hat{k}(t)$ moment comes from the side forces on the rotor bearings, shown as $\vec{F}_1$ and $\vec{F}_2$, which lie in the $\{\hat{i}, \hat{j}\}$ plane. The moment on the gimbal due to the side forces on the bearings is, by Newton's third law, $-\vec{M}(t)$, as indicated by the forces $-F_1$ and $-F_2$ in Fig. 2.5. In addition, there is a moment $-k\theta(t)\hat{k}(t)$ on the gimbal due to the spring restraint, and also a moment $-d\dot{\theta}(t)\hat{k}(t)$ due to the damper restraint. According to Euler's law, the sum of all these moments is equal to the moment of inertia of the gimbal multiplied by its angular acceleration around the $\hat{k}$ axis, $I_G \ddot{\theta}(t) = I_R \Omega \omega(t) - I_0 \ddot{\theta}(t) - k\theta(t) - d\dot{\theta}(t)$, which may be expressed as

$$[I_G + I_0]\ddot{\theta}(t) + d\dot{\theta}(t) + k\theta(t) = [I_R \Omega]\omega(t). \tag{2.15}$$

Here k is the equivalent elastic coefficient of the spring restraint, referred to the output axis, and its dimensions are N-m/rad; d is the equivalent rotary viscous coefficient of the damper, referred to the output axis, and its dimensions are N-m/(rad/s).

A sensor such as a shaft encoder, a potentiometer, a rotary variable differential transformer, or a mechanical indicating mechanism is mounted on the gimbal output axis to provide a measurement of $\theta(t)$. The sensor is not shown in Fig. 2.4. If the sensor is electromechanical, the measurement obtained could be a voltage

$$e_{\text{out}}(t) = K_S \theta(t) \text{ V}, \tag{2.16}$$

where K_S is the sensor scale factor, V/rad. The scale factor of the gyroscope-sensor combination is defined in terms of the steady-state response of $e_{\text{out}}(t)$ to a constant vehicle turning rate $\omega(t) = \omega_{SS}$. For this constant excitation, the steady-state solution to Eqn. 2.15, obtained by setting $\ddot{\theta}(t) = \dot{\theta}(t) = 0$, is $\theta_{SS} = [I_R \Omega/k]\omega_{SS}$. When this solution is combined with Eqn. 2.16, the overall instrument scale factor is found to be

$$\frac{e_{\text{outSS}}}{\omega_{SS}} = \frac{K_S I_R \Omega}{k} \text{ V/(rad/s)}. \tag{2.17}$$

We can establish a state-variable model of this system using the technique we employed for the mass–spring–damper system. Assume that we know the interval defined by the initial and final times t_0 and t_F and the initial conditions $\theta(t_0)$ and $\dot{\theta}(t_0)$. Define our two state variables $x_1(t) = \theta(t)$ and $x_2(t) = \dot{\theta}(t)$. Define $I_T = I_G + I_0$, and make the appropriate substitutions and algebraic re-arrangements of Eqn. 2.15 to obtain the state-variable model:

$$\dot{x}_1(t) = x_2(t),$$

DYNAMICS:

$$\dot{x}_2(t) = -\left[\frac{k}{I_T}\right]x_1(t) - \left[\frac{d}{I_T}\right]x_2(t) + \left[\frac{I_R\Omega}{I_T}\right]\omega(t);$$

SENSOR: $e_{\text{out}}(t) = K_S x_1(t);$ (2.18)

CONDITIONS: $t_0 \le t \le t_F$; $x_1(t_0)$, $x_2(t_0)$, and $\omega(t)$ are known.

Note the similarity between this model for the rate gyroscope and that for the mass–spring–damper system in Eqn. 2.9. Here we have added a sensor equation to our model. The sensor equation is an integral part of a state-variable model for a system that is to be controlled automatically.

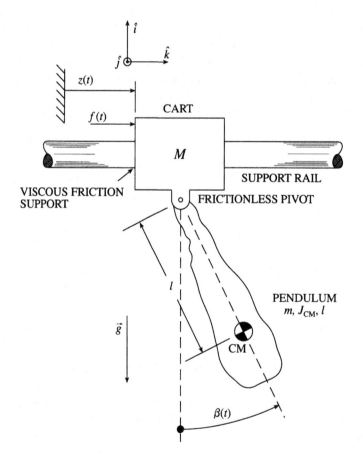

Fig. 2.6 Cart–pendulum system.

2.4 Pendulum with Moving Base

A cart of mass M supported on a horizontal rail is propelled along that rail by an applied force $f(t)$, as shown in Fig. 2.6. Viscous friction exists between the cart and the rail. The position of the cart with respect to a given reference point on the rail is $z(t)$. A pendulum of mass m hangs below the cart on a frictionless pivot. The center of mass of the pendulum lies a distance l from the pivot, and the moment of inertia of the pendulum about an axis normal to the plane of rotation passing through the center of mass is J_{CM}. $\beta(t)$ is the angular displacement of the pendulum from the vertical, with counterclockwise being the sense of positive $\beta(t)$ as indicated in Fig. 2.6. We wish to model this system so that we can calculate $z(t)$ and $\beta(t)$ in response to the driving force $f(t)$ and in response to nonzero initial conditions at the beginning of our response interval. We will use the free-body-diagram approach to obtain the correct expressions for Newton's and Euler's laws.

The triad of unit vectors $\{\hat{i}, \hat{j}, \hat{k}\}$ is fixed with respect to the rail, and the rail is fixed in inertial space; g is the gravitational constant, 9.81 (m/s)/s. Figure 2.7 shows a free-body diagram for the cart. The vertical forces acting on the cart are:

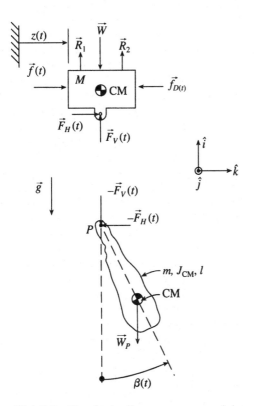

Fig. 2.7 Free-body diagrams, cart–pendulum system.

$\vec{W} = W(-\hat{i}) = -Mg\hat{i}$, the constant force of gravity;
$\vec{R}_1(t)$ and $\vec{R}_2(t)$ = contact forces of the rail supporting the cart; and
$\vec{F}_V(t) = F_V(t)\hat{i}$, the vertical component of the pivot contact force.

The horizontal forces acting on the cart are:

$\vec{f}(t) = f(t)\hat{k}$, the external applied or driving force;
$\vec{f}_D(t) = f_D(t)(-\hat{k}) = -D\dot{z}(t)\hat{k}$, the viscous friction force between the
    rail and the cart; and
$\vec{F}_H(t) = F_H(t)\hat{k}$, the horizontal component of the pivot contact force.

We assume that the rail supports the cart so as to constrain its motion to translation only along the horizontal axis. The cart does not rotate as it slides along the rail. Therefore, although the horizontal and vertical forces acting on the cart will produce moments about the center of mass, these moments – each of which may fluctuate while the cart is in motion – will sum to zero at all times. Because of this constraint, and because we are not interested in calculating $R_1(t)$, $R_2(t)$, $F_H(t)$, or $F_V(t)$, we need not write the moment equation for the angular acceleration of the cart, nor the force equation for the vertical acceleration of the cart. These circumstances are similar to those prevailing in the analysis of the spring–mass–damper system in Section 2.2.

It is now a simple matter to apply Newton's second law to the mass M. Because we need calculate only the horizontal motion, we may drop the vector notation and write

$$M\ddot{z}(t) = f(t) - D\dot{z}(t) + F_H(t). \qquad (2.19)$$

The pendulum motion consists of both translation and rotation with respect to inertial space. We must therefore write both force and moment equations for the pendulum free body. Newton's laws give us the proper expressions for the translatory motion. If we neglect air drag on the pendulum, only two agents act on the pendulum: gravity and the contact force of the pivot. Figure 2.7 shows this latter force resolved for convenience into its horizontal and vertical components. We note that Newton's third law reveals the contact force of the pivot on the cart to be the negative of the contact force of the pivot on the pendulum, as shown in Fig. 2.7. It is helpful at this point to review Newton's second law in a comprehensive verbal form, which, for a constant mass, is

$$\begin{bmatrix} \text{sum of all forces acting} \\ \text{on a body of mass } m \end{bmatrix} = (m) \times \begin{bmatrix} \text{acceleration of the center of mass} \\ \text{of the body with respect to inertial space} \end{bmatrix}.$$
$$(2.20)$$

The sum of all the forces acting on the pendulum of mass m is

$$\vec{W}_P - \vec{F}_V(t) - \vec{F}_H(t) = mg(-\hat{i}) - F_V(t)\hat{i} - F_H(t)\hat{k}, \qquad (2.21)$$

which provides us with the left-hand side of Eqn. 2.20. To obtain the right-hand side we must consider the kinematics of the pendulum motion, using the diagram in Fig. 2.8. Here $\dot{z}(t)\hat{k}$ is the velocity of point P with respect to inertial

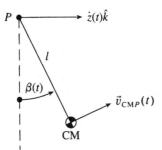

Fig. 2.8 Kinematics of pendulum motion.

space, and $\vec{v}_{\mathrm{CM}P}(t)$ is the velocity of the center of mass with respect to the point P. Therefore, the velocity of the center of mass of the pendulum with respect to inertial space is

$$\vec{v}_{\mathrm{CM}I}(t) = \dot{z}(t)\hat{k} + \vec{v}_{\mathrm{CM}P}(t). \tag{2.22}$$

The term $\vec{v}_{\mathrm{CM}I}(t)$ may be resolved into its horizontal and vertical components as follows:

$$\vec{v}_{\mathrm{CM}I}(t) = [l\dot{\beta}(t)\sin\beta(t)]\hat{i} + [\dot{z}(t) + l\dot{\beta}(t)\cos\beta(t)]\hat{k}. \tag{2.23}$$

The acceleration of the center of mass is obtained from Eqn. 2.23 by differentiating

$$\vec{a}_{\mathrm{CM}I}(t) = \frac{d\vec{v}_{\mathrm{CM}I}(t)}{dt}. \tag{2.24}$$

Because $\hat{i}$ and $\hat{k}$ do not depend on time, the differentiation of $\vec{v}_{\mathrm{CM}I}(t)$ is

$$\begin{aligned} \vec{a}_{\mathrm{CM}I}(t) &= [l\ddot{\beta}(t)\sin\beta(t) + l\dot{\beta}^2(t)\cos\beta(t)]\hat{i} \\ &\quad + [\ddot{z}(t) + l\ddot{\beta}(t)\cos\beta(t) - l\dot{\beta}^2(t)\sin\beta(t)]\hat{k}. \end{aligned} \tag{2.25}$$

We next address the rotary motion of the pendulum by applying Euler's law, whose comprehensive verbal expression, for constant mass bodies in planar motion, is

$$\begin{bmatrix} \text{sum of all moments of all forces} \\ \text{acting on the body about an axis} \\ \text{through the center of mass and} \\ \text{normal to the plane of motion} \end{bmatrix} = J_{\mathrm{CM}} \times \begin{bmatrix} \text{angular acceleration} \\ \text{of the body with} \\ \text{respect to inertial space} \end{bmatrix}. \tag{2.26}$$

We note that J_{CM} is the moment of inertia of the pendulum about the center of mass (not about the point P), and also that we are to calculate the moments of the forces with respect to the center of mass. The moment of $\vec{W}_P$ is zero, and the sum of the moments of the pivot contact forces is

$$[lF_H(t)\cos\beta(t) + lF_V(t)\sin\beta(t)]\hat{j}. \tag{2.27}$$

Equation 2.27 gives us the left-hand side of Eqn. 2.26, and the right-hand side is simply

$$J_{\text{CM}}\ddot{\beta}(t)\hat{j}. \tag{2.28}$$

We now have all the basic equations for our model. We proceed by equating the $\hat{i}$ component of Eqn. 2.21 with the $\hat{i}$ component of Eqn. 2.25, multiplied by m; we do the same with the $\hat{k}$ components of both equations. This gives us the following two scalar (nonvector) equations:

$$ml[\ddot{\beta}(t)\sin\beta(t) + \dot{\beta}^2(t)\cos\beta(t)] = -F_V(t) - mg, \tag{2.29}$$

$$m\ddot{z}(t) + ml\ddot{\beta}(t)\cos\beta(t) - ml\dot{\beta}^2(t)\sin\beta(t) = -F_H(t). \tag{2.30}$$

Equations 2.27 and 2.28 give another scalar equation:

$$lF_H(t)\cos\beta(t) + lF_V(t)\sin\beta(t) = J_{\text{CM}}\ddot{\beta}(t). \tag{2.31}$$

We now have four scalar equations (Eqns. 2.19, 2.29, 2.30, and 2.31) which describe the dynamics of the cart–pendulum system.

Euler's law also applies to the cart, since each of the seven forces on the cart shown in Fig. 2.7 has a moment about the center of mass. If we assume that the cart does not rotate but only translates, these moments will sum to zero at all times. Further, since we are not interested in calculating $\vec{R}_1$, $\vec{R}_2$, $\vec{F}_H$, $\vec{F}_V$, or $\vec{f}_D$, we need not write the equation for Euler's law for the cart.

Next we solve Eqn. 2.29 for $F_V(t)$, and substitute $F_V(t)$ and $F_H(t)$ (which are available directly from Eqn. 2.30) for these expressions in Eqns. 2.19 and 2.31. We have now reduced our set to two nonlinear differential equations that do not involve the reaction forces between either the rail and the cart or the cart and the pendulum:

$$[M+m]\ddot{z}(t) + [ml\cos\beta(t)]\ddot{\beta}(t) = f(t) - D\dot{z}(t) + ml\dot{\beta}^2(t)\sin\beta(t), \tag{2.32}$$

$$[ml\cos\beta(t)]\ddot{z}(t) + [J_{\text{CM}} + ml^2]\ddot{\beta}(t) = -mgl\sin\beta(t). \tag{2.33}$$

These can be re-arranged into two equivalent equations, one with $\ddot{z}(t)$ and lower-order derivatives of $\beta(t)$ and $z(t)$ and the other with $\ddot{\beta}(t)$ and the lower-order derivatives of $\beta(t)$ and $z(t)$:

$$\ddot{z}(t) = \frac{1}{N}\{(J_{\text{CM}} + ml^2)[f(t) - D\dot{z}(t) + ml\dot{\beta}^2(t)\sin\beta(t)]$$
$$+ (ml)^2 g\cos\beta(t)\sin\beta(t)\}, \tag{2.34}$$

$$\ddot{\beta}(t) = -\frac{ml}{N}\{(M+m)g\sin\beta(t) + ml\dot{\beta}^2(t)\cos\beta(t)\sin\beta(t)$$
$$+ \cos\beta(t)[f(t) - D\dot{z}(t)]\}, \tag{2.35}$$

where

$$N = (M + m) J_{CM} + ml^2(M + m \sin^2 \beta(t)). \tag{2.36}$$

This is a useful form for obtaining a state-variable model for the cart–pendulum system.

In setting up our state-variable model, we identify the force $f(t)$ as our input quantity; $z(t)$ and $\beta(t)$ constitute the dynamic response of the cart–pendulum system to this input force. Because our basic equations 2.34 and 2.35 are second-order differential equations, we make the following selection for the state variables:

$$x_1(t) = z(t), \quad x_2(t) = \dot{z}(t), \quad x_3(t) = \beta(t), \quad \text{and} \quad x_4(t) = \dot{\beta}(t). \tag{2.37}$$

Using this notation, we write our basic physical equations as

$$\dot{x}_1(t) = x_2(t),$$
$$\dot{x}_2(t) = a_{22} x_2(t) + a_{23} \sin x_3(t) + a_{2NL}(t) + b_2 f(t),$$
$$\dot{x}_3(t) = x_4(t),$$
$$\dot{x}_4(t) = a_{42} x_2(t) + a_{43} \sin x_3(t) + a_{4NL}(t) + b_4 f(t). \tag{2.38}$$

The eight coefficients are:

$$a_{22} = -\left[\frac{D(J_{CM} + ml^2)}{N}\right], \quad a_{23} = \left[\frac{(ml)^2 g \cos x_3(t)}{N}\right], \quad b_2 = \left[\frac{J_{CM} + ml^2}{N}\right],$$

$$a_{42} = \left[\frac{mlD \cos x_3(t)}{N}\right], \quad a_{43} = -\left[\frac{mlg(M + m)}{N}\right], \quad b_4 = -\left[\frac{ml \cos x_3(t)}{N}\right],$$

$$a_{2NL} = \left[\frac{(J_{CM} + ml^2) mlx_4^2(t) \sin x_3(t)}{N}\right],$$

$$a_{4NL} = -\left[\frac{(ml)^2 x_4^2(t) \cos x_3(t) \sin x_3(t)}{N}\right]. \tag{2.39}$$

The reason for the notation a_{2NL} and a_{4NL} in Eqns. 2.38 and 2.39 is that these terms include nonlinear functions of the state variables in their numerators.

The cart–pendulum system represents a class of practical devices. An overhead crane used in construction of buildings is one example. In many of these applications the angle $\beta(t)$ is permitted only a small excursion from its equilibrium value of zero. This is especially true in control applications where a load must be positioned accurately. If $|\beta(t)|$ never exceeds 0.1 radian, for example, then both $\cos \beta(t)$ and $\sin \beta(t)$ may be approximated very closely by $\cos \beta(t) \approx 1$ and $\sin \beta(t) \approx \beta(t)$. Furthermore, if $|\beta(t)|$ is small, the term $\dot{\beta}^2(t) \sin \beta(t)$ will be very small, and the two nonlinear coefficients a_{2NL} and a_{4NL} may be dropped from Eqn. 2.38. N can also be approximated as $[(M + m) J_{CM} + mMl^2]$. By making these approximations in Eqn. 2.38 we obtain the following set of linear state-variable equations:

$$\dot{x}_1(t) = x_2(t),$$
$$\dot{x}_2(t) = a_{22}x_2(t) + a_{23}x_3(t) + b_2 f(t),$$
$$\dot{x}_3(t) = x_4(t),$$
$$\dot{x}_4(t) = a_{42}x_2(t) + a_{43}x_3(t) + b_4 f(t). \tag{2.40}$$

Here the six coefficients are now constants:

$$a_{22} = -\left[\frac{D(J_{CM} + ml^2)}{(M+m)J_{CM} + mMl^2}\right], \quad a_{23} = \left[\frac{(ml)^2 g}{(M+m)J_{CM} + mMl^2}\right],$$

$$b_2 = \left[\frac{J_{CM} + ml^2}{(M+m)J_{CM} + mMl^2}\right], \quad a_{42} = \left[\frac{mlD}{(M+m)J_{CM} + mMl^2}\right], \tag{2.41}$$

$$a_{43} = -\left[\frac{mgl(M+m)}{(M+m)J_{CM} + mMl^2}\right], \quad b_4 = -\left[\frac{ml}{(M+m)J_{CM} + mMl^2}\right].$$

Here is a particularly useful way to write the set of four linear first-order differential equations given in Eqn. 2.40, using matrix notation:

$$\underbrace{\begin{bmatrix} \dot{x}_1(t) \\ \dot{x}_2(t) \\ \dot{x}_3(t) \\ \dot{x}_4(t) \end{bmatrix}}_{4\times 1} = \underbrace{\begin{bmatrix} 0 & 1 & 0 & 0 \\ 0 & a_{22} & a_{23} & 0 \\ 0 & 0 & 0 & 1 \\ 0 & a_{42} & a_{43} & 0 \end{bmatrix}}_{4\times 4} \underbrace{\begin{bmatrix} x_1(t) \\ x_2(t) \\ x_3(t) \\ x_4(t) \end{bmatrix}}_{4\times 1} + \underbrace{\begin{bmatrix} 0 \\ b_2 \\ 0 \\ b_4 \end{bmatrix}}_{4\times 1} \underbrace{[f(t)]}_{1\times 1}. \tag{2.42}$$

Compare Eqn. 2.42 to Eqn. 2.40 to see how two matrices multiply. A further efficiency in writing Eqn. 2.42 can be realized by using the notation

$$\dot{x}(t) = Ax(t) + Bf(t), \tag{2.43}$$

where we understand that the matrices $\dot{x}(t)$, $x(t)$, A, and B have these meanings:

$$\dot{x}(t) = \begin{bmatrix} \dot{x}_1(t) \\ \dot{x}_2(t) \\ \dot{x}_3(t) \\ \dot{x}_4(t) \end{bmatrix}, \quad x(t) = \begin{bmatrix} x_1(t) \\ x_2(t) \\ x_3(t) \\ x_4(t) \end{bmatrix}, \quad A = \{a_{ij}\} = \begin{bmatrix} 0 & 1 & 0 & 0 \\ 0 & a_{22} & a_{23} & 0 \\ 0 & 0 & 0 & 1 \\ 0 & a_{42} & a_{43} & 0 \end{bmatrix}, \quad B = \begin{bmatrix} 0 \\ b_2 \\ 0 \\ b_4 \end{bmatrix}. \tag{2.44}$$

The notation $A = \{a_{ij}\}$ is read as "A is the matrix having the element a_{ij} in its ith row and jth column." In terms of linear algebra, the 4×1 matrices $\dot{x}(t)$ and $x(t)$ represent elements (called *vectors*) in a *four-dimensional vector space*. The 4×4 matrix A represents a *linear transformation* from the four-dimensional vector space containing $x(t)$ to the four-dimensional vector space containing $\dot{x}(t)$. The 4×1 matrix B represents a linear transformation from the one-dimensional vector space containing $f(t)$ to the four-dimensional vector space containing $\dot{x}(t)$. It is useful for an analyst to understand the distinction between a matrix and the underlying mathematical concept which that matrix represents because, as we have just seen, matrices are used to represent a variety of different mathematical concepts. These mathematical considerations are important in modern control theory, but for our present purpose we will use matrix notation simply as a short method for representing our dynamic models. The matrix representation is also required for entering data into computer files in most linear system analysis programs.

2.5 Mechanical Drive Systems

In most mechanical control systems, power must be transmitted from a prime mover, such as an electric or hydraulic motor, to a driven mechanism that requires a torque level and/or a speed level different from that of the prime mover. Typically, the motor is capable of less torque than that required by the load, but it can turn at a rate faster than is necessary at the load. In this case a speed reduction mechanism between motor and load is used to match the speed and torque capabilities of the motor to the speed and torque requirements of the load. We now study two such drive mechanisms, the gear train and the belt-drive system.

Figure 2.9 is a schematic sketch of a single-mesh spur gear train coupled to an inertia-spring load. We assume there is an ideal mesh: the teeth of the small gear engage precisely those of the large gear so that no lost motion or backlash exists between the teeth. The ideal mesh also assumes that no friction losses occur. Neither idealization can be obtained in practice; if manufacturing measures are taken to reduce the backlash, the friction will increase, and vice versa. In a good-quality gear set in which the backlash has been made negligible for the application at hand, the efficiency of a single mesh can approach 95 percent. We temporarily neglect this loss and assume that the contact force between the gear teeth is directed tangentially to the gear circumference. Newton's third law asserts that the contact forces are equal and opposite. For shaft 1 we have

$$T(t) - r_1 f_C(t) = J_1 \ddot{\theta}_1(t), \tag{2.45}$$

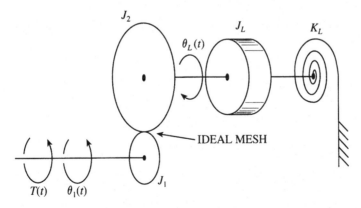

Fig. 2.9 Schematic of gear train.

where $T(t)$ is the applied torque, $f_C(t)$ is the contact force between touching teeth, r_1 is the radius of the small gear, and J_1 is the moment of inertia of the small gear. For the load shaft the corresponding relationship is

$$r_2 f_C(t) - K_L \theta_L(t) = (J_2 + J_L) \ddot{\theta}_L(t), \tag{2.46}$$

where r_2 is the radius of the large gear, $(J_2 + J_L)$ is the total moment of inertia of the load shaft, and K_L is the spring constant of the elastic load, with units of N-m/rad. Note that the rotational sense of $\theta_1(t)$ is opposite to that of $\theta_L(t)$. We define the *gear ratio* to be n, where

$$n = \frac{r_2}{r_1} = \frac{[\text{number of teeth on gear 2}]}{[\text{number of teeth on gear 1}]}. \tag{2.47}$$

By neglecting the backlash, we have a simple kinematic relationship between $\theta_1(t)$ and $\theta_L(t)$,

$$n\theta_L(t) = \theta_1(t), \tag{2.48}$$

from which it follows directly that

$$n\dot{\theta}_L(t) = \dot{\theta}_1(t) \quad \text{and} \quad n\ddot{\theta}_L(t) = \ddot{\theta}_1(t). \tag{2.49}$$

We eliminate $f_C(t)$ between Eqns. 2.45 and 2.46, and represent the resultant equation in terms of $\theta_L(t)$ and $\ddot{\theta}_L(t)$ to obtain

$$nT(t) = [n^2 J_1 + (J_2 + J_L)] \ddot{\theta}_L(t) + K_L \theta_L(t). \tag{2.50}$$

Equation 2.50 is the dynamic description of the equivalent mechanical system shown in Fig. 2.10. The equivalent moment of inertia of the whole system, referred to $\theta_L(t)$, is $[n^2 J_1 + (J_2 + J_L)]$. The influence of J_1 on this total moment of inertia is magnified by the square of the gear ratio as compared to the influence of $(J_2 + J_L)$. In many cases J_1 is small compared to $(J_2 + J_L)$, whereas $n^2 J_1$ is not. In these cases J_1 cannot be ignored in the analysis of the whole system. We also note that the torque on the output shaft is n times the input torque. The speed of the load shaft is also reduced from that of the input shaft by the factor n. Another point of interest is that in the equivalent system the sense of the input torque $nT(t)$ agrees with the sense of the load-shaft displacement $\theta_L(t)$, but it is opposite to that of the input torque $T(t)$ of the actual system shown in Fig. 2.9.

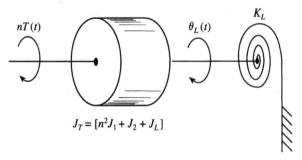

Fig. 2.10   Equivalent mechanical system.

The consequence of assuming an ideal mesh of the gears is twofold. First, without backlash we enjoyed the simple kinematic relationship between $\theta_1(t)$ and $\theta_L(t)$ given in Eqn. 2.48, which led to the relationship in Eqn. 2.50. If backlash occurs, the analysis will be nonlinear and therefore more involved. Because backlash can cause undesired oscillations in a feedback control system, gear meshes are manufactured for minimal backlash, but the resulting increase in the friction between gear teeth reduces power transmission efficiency, the second consequence of our assumption. If this efficiency is included in the analysis, but backlash is not, Eqn. 2.50 is modified to

$$n\eta T(t) = [n_2\eta J_1 + (J_2 + J_L)]\ddot{\theta}_L(t) + K_L\theta_L(t); \tag{2.51}$$

here η is the efficiency, which can be in the neighborhood of 0.95 for a single mesh. If there are m meshes in a gear train then the overall efficiency is

$$\eta_1 \times \eta_2 \times \cdots \times \eta_m. \tag{2.52}$$

The power supplied at the input shaft is $T(t) \times \dot{\theta}_1(t)$ watts [W] if torque is in N-m and speed is in rad/s. But the power delivered to the output shaft, for unaccelerated motion, is

$$n\eta T(t)\dot{\theta}_L(t) = \eta T(t)\dot{\theta}_1(t), \tag{2.53}$$

and the efficiency is defined as

$$\text{EFFICIENCY:} \qquad \eta = \left\{ \frac{\text{power delivered to load shaft}}{\text{power supplied at the input shaft}} \right\}_{\ddot{\theta}_L=0}. \tag{2.54}$$

If we define our state variables for the geared system to be $x_1(t) = \theta_L(t)$ and $x_2(t) = \dot{\theta}_L(t)$, then Eqn. 2.51 can be written in state-variable form:

$$\dot{x}(t) = Ax(t) + BT(t), \quad \text{where}$$

$$A = \begin{bmatrix} 0 & 1 \\ -K_L/J_T & 0 \end{bmatrix}, \quad B = \begin{bmatrix} 0 \\ n\eta/J_T \end{bmatrix}, \tag{2.55}$$

$$x(t) = \begin{bmatrix} x_1(t) \\ x_2(t) \end{bmatrix}, \quad \text{and} \quad J_T = n^2\eta J_1 + J_2 + J_L.$$

The belt-drive power transmission system is illustrated in Fig. 2.11. In many industrial applications the load shaft turns in only one direction; an exhaust fan drive is one such example. But in many control applications the direction of rotation of the load shaft must be reversed in accordance with a control command signal; therefore the shaft will turn clockwise as often as it does counterclockwise. The belt–pulley design for control applications must minimize backlash for the same reasons that backlash is minimized in gear design, whereas this may not be important in unidirectional drives. Most belts in control applications are toothed so as to grip the pulleys without slipping. Where this is the case, we have a simple kinematic relationship between $\theta_1(t)$ and $\theta_L(t)$:

$$\theta_1(t) = n\theta_L(t), \quad \dot{\theta}_1(t) = n\dot{\theta}_L(t), \quad \text{and} \quad \ddot{\theta}_1(t) = n\ddot{\theta}_L(t); \tag{2.56}$$

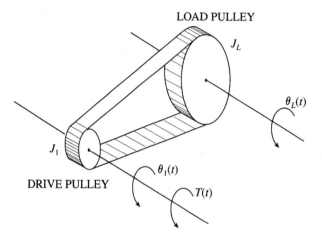

Fig. 2.11   Belt-drive system.

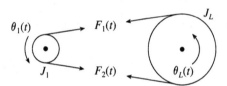

Fig. 2.12   Free-body diagram of belt-drive system.

n is the belt drive ratio: [radius of load pulley]/[radius of drive pulley] $= [r_2]/[r_1]$.

The dynamic equations for the belt-drive system can be derived with the aid of the free-body diagram in Fig. 2.12. If the mass of the belt is negligible, the tension in the upper half of the belt $F_1(t)$ is the same at all points of the belt between the two pulleys. The same is true in the lower half, but there the tension is $F_2(t)$. Euler's law gives us

DRIVE PULLEY: $[F_1(t) - F_2(t)]r_1 + T(t) = J_1\ddot{\theta}_1(t),$

LOAD PULLEY: $[F_2(t) - F_1(t)]r_2 = J_L\ddot{\theta}_L(t).$ (2.57)

Eliminating $[F_1(t) - F_2(t)]$ between these two equations gives the model for the belt-drive system under the assumption that there is no backlash, no slippage, and no lost power in the transmission:

$$nT(t) = [n^2 J_1 + J_L]\ddot{\theta}_L(t).$$ (2.58)

Because the efficiency of power transmission in the belt-drive system is approximately the same as that in the system using spur gears (≈ 0.95), Eqn. 2.58 can be modified to

$$n\eta T(t) = [n^2\eta J_1 + J_L]\ddot{\theta}_L(t).$$ (2.59)

In addition to the rotary drives analyzed here, various other types of mechanical drives are used in control systems – the rotary-to-linear power converter (ball–screw mechanism), for example. Space does not permit a description of these devices in this book.

2.6 Single-Axis Rigid Spacecraft

Figure 2.13 represents a spacecraft flying a near-earth or an interplanetary orbital trajectory. The craft is assumed to be a rigid body of constant mass. The body does not rotate, except for slow angular movement about a single axis. The angular attitude of the craft about this axis, $\theta(t)$, is to be controlled by a command signal $\theta_{com}(t)$ in order to fulfill a communications or surveillance requirement. The control is effected by an angular momentum exchange device or a set of gas expulsion thrusters. In addition to the desired control moments which position the spacecraft in the required direction, unwanted moments are also present. These unwanted moments are very small compared to the available control moment, but over a prolonged period they can cause a significant pointing error. These error-causing moments come from micrometeorite impact, unbalanced solar radiation pressure, or slight misalignments of the propulsion engine. When the spacecraft is in near-earth orbit, disturbance moments due to aerodynamic drag, gravity gradients, or interaction between the earth's magnetic field and currents in on-board electrical equipment may occur.

The basic model for the dynamics of the spacecraft is quite simple under the restrictive conditions assumed here. If $M_C(t)$ is the control moment on the rigid body, and if $M_D(t)$ is the sum of all the disturbance moments, we have

$$M_C(t) + M_D(t) = J_{CM}\ddot{\theta}(t) \tag{2.60}$$

where J_{CM} is the moment of inertia of the body about the $\theta(t)$ axis. A state-variable model for this system can be established using the notation

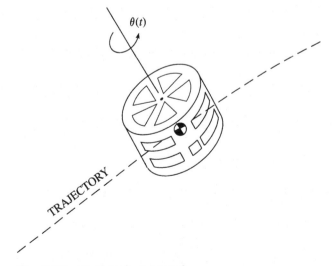

$\theta(t)$

TRAJECTORY

Fig. 2.13 Rigid-body spacecraft.

$$x(t) = \begin{bmatrix} x_1(t) \\ x_2(t) \end{bmatrix} = \begin{bmatrix} \theta(t) \\ \dot{\theta}(t) \end{bmatrix}, \tag{2.61}$$

in which case the matrix differential–equation version of Eqn. 2.60 is

$$\underbrace{\begin{bmatrix} \dot{x}_1(t) \\ \dot{x}_2(t) \end{bmatrix}}_{\substack{\dot{x}(t) \\ 2 \times 1}} = \underbrace{\begin{bmatrix} 0 & 1 \\ 0 & 0 \end{bmatrix}}_{\substack{A \\ 2 \times 2}} \underbrace{\begin{bmatrix} x_1(t) \\ x_2(t) \end{bmatrix}}_{\substack{x(t) \\ 2 \times 1}} + \underbrace{\begin{bmatrix} 0 & 0 \\ 1/J_{CM} & 1/J_{CM} \end{bmatrix}}_{\substack{B \\ 2 \times 2}} \underbrace{\begin{bmatrix} M_C(t) \\ M_D(t) \end{bmatrix}}_{\substack{u(t) \\ 2 \times 1}}. \tag{2.62}$$

Here the 2×1 matrix of the two driving (input) functions $M_C(t)$ and $M_D(t)$ has been denoted as $u(t)$. We now separate the total applied moment into the component that we can engineer, $M_C(t)$, and the component that comes from random outside disturbances, $M_D(t)$.

To derive a complete model for an initial-value solution to Eqn. 2.62, we must specify an interval of time over which our differential equation will adequately represent the physical situation we are modeling, the values of the two state variables at the initial time, and the two components of the input matrix $u(t)$ during the specified time interval:

$$\dot{x}(t) = Ax(t) + Bu(t), \quad \text{where}$$
$$t_0 \leq t \leq t_F, \quad x_1(t_0), x_2(t_0) \text{ are known, and} \tag{2.63}$$
$$M_C(t), M_D(t) \text{ are specified so } u(t) \text{ is known.}$$

The control moment input to the spacecraft body $M_C(t)$ will be produced by a momentum exchange device driven by a control computer designed using the engineering principles developed in Chapters 8 through 14. The random disturbance moment $M_D(t)$ must be somehow specified from knowledge of the environment in which the spacecraft travels during the time interval of interest.

2.7 Longitudinal-Axis Dynamics of an Airplane

A side view of an airplane in flight is shown in Fig. 2.14. We wish to establish a dynamic model that will permit us to calculate the response of the aircraft

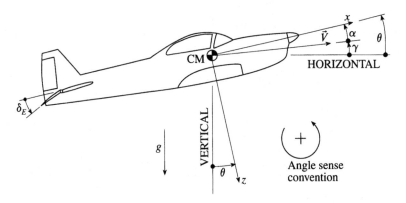

Fig. 2.14 Definition of θ, α, γ, and δ_E.

to elevator and throttle control inputs. This is a difficult task unless we place some severe restrictions on the scope of our analysis. We assume that the aircraft is a rigid body which is symmetrical about a plane passing through its center of mass, designated as the (x, z) plane. The (x, y, z) coordinate frame, having unit vectors $(\hat{i}, \hat{j}, \hat{k})$, is fixed in the body and centered in the center of mass, with the x axis forward, as shown. The z axis is taken as pointing "downward," to be consistent with common practice in flight-control system engineering. The line of thrust is not necessarily collinear with the x axis, nor is it necessarily parallel to it, although in many aircraft it may be nearly so. We will consider the effects of only two control inputs, the elevator deflection, $\delta_E(t)$, and the throttle deflection (or thrust), $\delta_T(t)$ in degrees [deg]. The elevator is restricted in its motion to approximately ±0.4 rad about its center position. The throttle angle deflection goes from zero to full throttle, which we will assume is 100 deg. Note that the sense of the elevator deflection is that positive δ_E means the trailing edge is down, as indicated in Fig. 2.14. We assume that the aircraft will undergo only planar motion; that is, the center of mass may translate in a plane vertical to the earth's surface, and the body may rotate about the y axis, but the (x, z) plane will remain coincident with the vertical plane in which the center of mass moves. The velocity of the center of mass with respect to the earth (taken to be inertial space) is $\vec{V}(t)$ ft/s. We also assume that the air mass in which the airplane flies is stationary with respect to the earth. The angle from the horizontal to the vector $\vec{V}(t)$, called the *flight-path angle,* is designated as $\gamma(t)$ rad, as shown. The angle from the horizontal to the x axis, called the *pitch angle,* is designated as $\theta(t)$ rad in Fig. 2.14. The *angle of attack,* $\alpha(t)$ rad, is the angle from the vector $\vec{V}(t)$ to the x axis. As the aircraft maneuvers, $\alpha(t)$, $\gamma(t)$, and $\theta(t)$ may fluctuate between positive and negative values, so it is important to have a clear definition of each of these angles, as established here. In Fig. 2.14 the angles are all shown as positive. The altitude of the center of mass above mean sea level is $h(t)$ ft.

We use the English system of units – pounds [lb] for force, feet [ft] for length, slugs [slug] for mass, and seconds [s] for time – in this example because much of the technical information on aircraft dynamics is given in these units. (The discussion of units and conversion factors in Appendix A should be consulted by those unfamiliar with the SI, CGS, and English systems. Numerical data often come in a mixture of units that the engineer must convert to a consistent set of units before use in dynamic analysis.)

We begin our modeling by identifying the kinematic relationships needed to apply Newton's and Euler's laws to this rigid body system in planar motion. The acceleration of the center of mass with respect to the earth is $d\vec{V}(t)/dt$. The x-axis and z-axis components of $\vec{V}(t)$ are denoted as the *airspeed,* $u_A(t)$, and the *vertical speed,* $w(t)$:

$$\vec{V}(t) = u_A(t)\hat{i} + w(t)\hat{k}, \tag{2.64}$$

so the acceleration of the center of mass with respect to inertial space is

$$\frac{d\vec{V}(t)}{dt} = \dot{u}_A(t)\hat{i} + u_A(t)\frac{d\hat{i}}{dt} + \dot{w}(t)\hat{k} + w(t)\frac{d\hat{k}}{dt} \text{ (ft/s)/s}, \tag{2.65}$$

where $d\hat{i}/dt$ and $d\hat{k}/dt$ are not zero, since the (x, y, z) reference frame rotates with respect to the earth. These are given by

$$\frac{d\hat{i}}{dt} = -\dot{\theta}(t)\hat{k} \quad \text{and} \quad \frac{d\hat{k}}{dt} = \dot{\theta}(t)\hat{i}, \tag{2.66}$$

so the acceleration is

$$\frac{d\vec{V}(t)}{dt} = [\dot{u}_A(t) + \dot{\theta}(t)w(t)]\hat{i} + [\dot{w}(t) - \dot{\theta}(t)u_A(t)]\hat{k}. \tag{2.67}$$

The angular momentum of the vehicle with respect to inertial space is

$$\vec{H}(t) = I_{yy}\dot{\theta}(t)\hat{j} \quad \text{slug(ft)}^2/\text{s} \ (\text{or lb-ft-s}). \tag{2.68}$$

Since $d\hat{j}/dt = 0$, the rate of change of $\vec{H}(t)$ is

$$\frac{d\vec{H}(t)}{dt} = I_{yy}\ddot{\theta}(t)\hat{j} \quad \text{ft-lb}, \tag{2.69}$$

where I_{yy} is the moment of inertia of the body around the y axis in slug(ft)^2.

The forces acting on the airframe are due to gravity and thrust, and to aerodynamic pressure and friction. The gravity force is

$$-mg \sin \theta(t)\hat{i} + mg \cos \theta(t)\hat{k} \quad \text{lb}, \tag{2.70}$$

where m is the mass of the body in slugs and g is the gravitational constant 32.174 (ft/s)/s; m and g are assumed to be constant. The thrust force and the aerodynamic forces are much more complex than the gravity force. They depend on the dynamic air pressure, the angle of attack, the pitch rate $\dot{\theta}(t)$, the elevator deflection $\delta_E(t)$, the throttle deflection $\delta_T(t)$, and other dynamic variables. We will treat them, and their moments about the y axis, in a special simplified manner in this section.

Because all the vectors representing forces on the airframe lie in the (x, z) plane, we can consider the projections of the force vectors onto the x and z axes, obtaining two scalar differential equations for the translational motion of the center of mass. Incorporating the acceleration components of Eqn. 2.67 with the force components of Eqn. 2.70, we have, according to Newton's second law,

$$\begin{aligned} x \text{ AXIS:} &\quad -mg \sin \theta(t) + X(u_A, w, \delta_E, \delta_T) = m[\dot{u}_A(t) + \dot{\theta}(t)w(t)], \\ z \text{ AXIS:} &\quad mg \cos \theta(t) + Z(u_A, w, \dot{\theta}, \delta_E, \delta_T) = m[\dot{w}(t) - \dot{\theta}(t)u_A(t)]. \end{aligned} \tag{2.71}$$

Here $X(u_A, w, \delta_E, \delta_T)$ and $Z(u_A, w, \dot{\theta}, \delta_E, \delta_T)$ represent the sum of all the thrust and aerodynamic forces acting on the body in the x direction and in the z direction, respectively. These forces depend in very complex ways on the dynamic variables u_A, w, $\dot{\theta}$, δ_E, and δ_T (and possibly others to a negligible degree). In a similar manner, we apply Euler's law to obtain the y-axis dynamic equation:

$$y \text{ AXIS:} \quad M(u_A, w, \dot{w}, \dot{\theta}, \delta_E, \delta_T) = I_{yy}\ddot{\theta}(t), \tag{2.72}$$

where $M(u_A, w, \dot{w}, \dot{\theta}, \delta_E, \delta_T)$ represents the sum of the moments about the y axis of all the thrust and aerodynamic forces acting on the body.

We account for the complexity of $X(u_A, w, \delta_E, \delta_T)$, $Z(u_A, w, \dot{\theta}, \delta_E, \delta_T)$, and $M(u_A, w, \dot{w}, \dot{\theta}, \delta_E, \delta_T)$ by restricting our dynamic analysis to an approximation technique based on *perturbation theory*. In this approach we study a limited but important problem, namely that of small motions of the aircraft about an equilibrium flight condition. This equilibrium condition is taken here to be straight and level flight at a given altitude, airspeed, mass, and center-of-mass location, with given positions of flaps and landing gear. The controls $\delta_E(t)$ and $\delta_T(t)$ are "trimmed" to the specific values, denoted as δ_{E0} and δ_{T0}, which are required to maintain the equilibrium flight condition. We then allow only small deviations of the control inputs from these trim values, so that we have

$$\text{ELEVATOR:} \quad \delta_E(t) = \delta_{E0} + \Delta\delta_E(t),$$
$$\text{THROTTLE:} \quad \delta_T(t) = \delta_{T0} + \Delta\delta_T(t). \tag{2.73}$$

The small deviations in the control inputs are $\Delta\delta_E(t)$ and $\Delta\delta_T(t)$; $\Delta\delta_E(t)$ will be limited to, for example, ± 0.04 rad, and $\Delta\delta_T(t)$ to ± 2 deg. In a similar manner, the response variables will be:

$$\theta(t) = \theta_0 + \Delta\theta(t), \quad \dot{\theta}(t) = \Delta\dot{\theta}(t), \quad \ddot{\theta}(t) = \Delta\ddot{\theta}(t),$$
$$\gamma(t) = \Delta\gamma(t), \quad \dot{\gamma}(t) = \Delta\dot{\gamma}(t);$$
$$h(t) = h_0 + \Delta h(t), \quad \dot{h}(t) = \Delta\dot{h}(t), \quad \ddot{h}(t) = \Delta\ddot{h}(t);$$
$$\alpha(t) = \alpha_0 + \Delta\alpha(t), \quad \dot{\alpha}(t) = \Delta\dot{\alpha}(t), \tag{2.74}$$
$$u_A(t) = u_{A0} + \Delta u_A(t), \quad \dot{u}_A(t) = \Delta\dot{u}_A(t),$$
$$w(t) = w_0 + \Delta w(t), \quad \dot{w}(t) = \Delta\dot{w}(t).$$

We note that the assumed equilibrium flight condition dictates that $\gamma_0 = 0$. The small changes $\Delta\theta(t)$, $\Delta h(t)$, $\Delta u_A(t)$, etc. are called the *perturbations* in $\theta(t)$, $h(t)$, $u_A(t)$, etc. Now the aerodynamic and thrust force terms $X(u_A, w, \delta_E, \delta_T)$ and $Z(u_A, w, \dot{\theta}, \delta_E, \delta_T)$ and the moment term $M(u_A, w, \dot{w}, \dot{\theta}, \delta_E, \delta_T)$ can be expressed in similar fashion:

$$X(u_A, w, \delta_E, \delta_T) = X_0 + \Delta X(u_A, w, \delta_E, \delta_T),$$
$$Z(u_A, w, \dot{\theta}, \delta_E, \delta_T) = Z_0 + \Delta Z(u_A, w, \dot{\theta}, \delta_E, \delta_T), \tag{2.75}$$
$$M(u_A, w, \dot{w}, \dot{\theta}, \delta_E, \delta_T) = M_0 + \Delta M(u_A, w, \dot{w}, \dot{\theta}, \delta_E, \delta_T).$$

Substitute $X(u_A, w, \delta_E, \delta_T)$, $\theta(t)$, $\dot{u}_A(t)$, $\dot{\theta}(t)$, and $w(t)$ into the x-axis force Eqn. 2.71:

$$X_0 + \Delta X(u_A, w, \delta_E, \delta_T) = mg \sin(\theta_0 + \Delta\theta(t)) + m[\Delta\dot{u}_A(t) + \Delta\dot{\theta}(t)(w_0 + \Delta w(t))]. \tag{2.76}$$

We expand this equation and make the approximations

$$\cos\Delta\theta(t) \cong 1, \quad \sin\Delta\theta(t) \cong \Delta\theta(t), \quad \text{and} \quad \Delta\dot{\theta}(t)\Delta w(t) \cong 0, \tag{2.77}$$

so that Eqn. 2.76 may be written approximately as

$$[X_0 - mg \sin \theta_0] + \Delta X(u_A, w, \delta_E, \delta_T) \cong m[g \cos \theta_0 \Delta \theta(t) + \Delta \dot{u}_A(t) + w_0 \Delta \dot{\theta}(t)].$$
$$(2.78)$$

Treating the z-axis force equation in the same way, we have

$$[Z_0 + mg \cos \theta_0] + \Delta Z(u_A, w, \dot{\theta}, \delta_E, \delta_T) \cong m[g \sin \theta_0 \Delta \theta(t) + \Delta \dot{w}(t) - u_{A0} \Delta \dot{\theta}(t)],$$
$$(2.79)$$

where the approximation $\Delta \dot{\theta}(t) \Delta u_A(t) \cong 0$ has been made. The y-axis moment equation is also treated similarly:

$$M_0 + \Delta M(u_A, w, \dot{w}, \dot{\theta}, \delta_E, \delta_T) = I_{yy} \Delta \ddot{\theta}(t). \qquad (2.80)$$

To achieve straight and level equilibrium flight, the trimmed values of the elevator and throttle controls, δ_{E0} and δ_{T0}, must be such that the y-axis moment, M_0, is zero. These control settings also produce the equilibrium airspeed and airframe attitude required for the equilibrium aerodynamic forces of lift, drag, and thrust on the airframe to exactly balance the airframe weight. This equilibrium condition means that the bracketed terms on the left sides of Eqns. 2.78 and 2.79 are each zero. Therefore, our dynamic equations reduce to the set

$$\Delta X(u_A, w, \delta_E, \delta_T) \cong m[g \cos \theta_0 \Delta \theta(t) + \Delta \dot{u}_A(t) + w_0 \Delta \dot{\theta}(t)],$$

$$\Delta Z(u_A, w, \dot{\theta}, \delta_E, \delta_T) \cong m[g \sin \theta_0 \Delta \theta(t) + \Delta \dot{w}(t) - u_{A0} \Delta \dot{\theta}(t)], \qquad (2.81)$$

$$\Delta M(u_A, w, \dot{w}, \dot{\theta}, \delta_E, \delta_T) = I_{yy} \Delta \ddot{\theta}(t).$$

The terms $\Delta X(u_A, w, \delta_E, \delta_T)$, $\Delta Z(u_A, w, \dot{\theta}, \delta_E, \delta_T)$, and $\Delta M(u_A, w, \dot{w}, \dot{\theta}, \delta_E, \delta_T)$ are the very small forces and moments that produce the very small accelerations of the airframe. These in turn lead to very small fluctuations of the airframe attitude and airspeed about their equilibrium values; it is these fluctuations in which we are interested.

The functions $\Delta X(u_A, w, \delta_E, \delta_T)$, $\Delta Z(u_A, w, \dot{\theta}, \delta_E, \delta_T)$, and $\Delta M(u_A, w, \dot{w}, \dot{\theta}, \delta_E, \delta_T)$, while small, are nevertheless complicated; they must be approximated to be tractable in engineering analysis. We therefore expand these functions in Taylor series about their equilibrium conditions. As an example, consider the expansion of $\Delta X(u_A, w, \delta_E, \delta_T)$:

$$\Delta X(u_A, w, \delta_E, \delta_T)$$

$$= \left[\frac{\partial X}{\partial u_A} \right]_0 \Delta u_A(t) + \left[\frac{\partial X}{\partial w} \right]_0 \Delta w(t) + \left[\frac{\partial X}{\partial \delta_E} \right]_0 \Delta \delta_E(t) + \left[\frac{\partial X}{\partial \delta_T} \right]_0 \Delta \delta_T(t)$$

$$+ \{\text{terms involving higher-order}$$
$$\text{partial derivatives of } \Delta X(u_A, w, \delta_E, \delta_T)\}, \qquad (2.82)$$

where $[\partial X/\partial u_A]_0$ is the constant obtained by evaluating the partial derivative of $\Delta X(u_A, w, \delta_E, \delta_T)$ with respect to u_A at the equilibrium flight condition. If the control inputs $\Delta \delta_E(t)$ and $\Delta \delta_T(t)$ remain small, and if the dynamic response variables $\Delta u_A(t)$ and $\Delta w(t)$ are also small, then usually that part of Eqn. 2.82 labeled {terms involving higher-order partial derivatives of $\Delta X(u_A, w, \delta_E, \delta_T)$}

will contribute only a negligible amount to $\Delta X(u_A, w, \delta_E, \delta_T)$ as compared to the amount due to the terms involving the first-order partial derivatives. Assuming this to be the case, we make the approximation

$$\Delta X(u_A, w, \delta_E, \delta_T) \cong \left[\frac{\partial X}{\partial u_A}\right]_0 \Delta u_A(t) + \left[\frac{\partial X}{\partial w}\right]_0 \Delta w(t)$$

$$+ \left[\frac{\partial X}{\partial \delta_E}\right]_0 \Delta \delta_E(t) + \left[\frac{\partial X}{\partial \delta_T}\right]_0 \Delta \delta_T(t). \qquad (2.83)$$

Following the same procedure for $\Delta Z(u_A, w, \dot{\theta}, \delta_E, \delta_T)$ and $\Delta M(u_A, w, \dot{w}, \dot{\theta}, \delta_E, \delta_T)$, we have

$$\Delta Z(u_A, w, \dot{\theta}, \delta_E, \delta_T) \cong \left[\frac{\partial Z}{\partial u_A}\right]_0 \Delta u_A(t) + \left[\frac{\partial Z}{\partial w}\right]_0 \Delta w(t) + \left[\frac{\partial Z}{\partial \dot{\theta}}\right]_0 \Delta \dot{\theta}(t)$$

$$+ \left[\frac{\partial Z}{\partial \delta_E}\right]_0 \Delta \delta_E(t) + \left[\frac{\partial Z}{\partial \delta_T}\right]_0 \Delta \delta_T(t),$$

$$\qquad (2.84)$$

$$M(u_A, w, \dot{w}, \dot{\theta}, \delta_E, \delta_T) \cong \left[\frac{\partial M}{\partial u_A}\right]_0 \Delta u_A(t) + \left[\frac{\partial M}{\partial w}\right]_0 \Delta w(t)$$

$$+ \left[\frac{\partial M}{\partial \dot{w}}\right]_0 \Delta \dot{w}(t) + \left[\frac{\partial M}{\partial \dot{\theta}}\right]_0 \Delta \dot{\theta}(t)$$

$$+ \left[\frac{\partial M}{\partial \delta_E}\right]_0 \Delta \delta_E(t) + \left[\frac{\partial M}{\partial \delta_T}\right]_0 \Delta \delta_T(t).$$

The force and moment coefficients,

$$\left[\frac{\partial X}{\partial u_A}\right]_0, \left[\frac{\partial X}{\partial w}\right]_0 \cdots \left[\frac{\partial Z}{\partial u_A}\right]_0, \left[\frac{\partial Z}{\partial w}\right]_0 \cdots \left[\frac{\partial Z}{\partial \delta_T}\right]_0, \left[\frac{\partial M}{\partial u_A}\right]_0 \cdots \left[\frac{\partial M}{\partial \delta_T}\right]_0,$$

are determined by calculations based on the configuration of the aircraft using the appropriate aerodynamic theory, on experimental measurements in a wind tunnel, and on data obtained in flight tests.

We now introduce the notation

$$X_{uA} = \frac{1}{m}\left[\frac{\partial X}{\partial u_A}\right]_0, X_w = \frac{1}{m}\left[\frac{\partial X}{\partial w}\right]_0, \ldots, X_{\delta E} = \frac{1}{m}\left[\frac{\partial X}{\partial \delta_E}\right], X_{\delta T} = \frac{1}{m}\left[\frac{\partial X}{\partial \delta_T}\right]_0;$$

$$Z_{uA} = \frac{1}{m}\left[\frac{\partial Z}{\partial u_A}\right]_0, Z_w = \frac{1}{m}\left[\frac{\partial Z}{\partial w}\right]_0, \ldots, Z_{\delta T} = \frac{1}{m}\left[\frac{\partial Z}{\partial \delta_T}\right]_0; \qquad (2.85)$$

$$M_{uA} = \frac{1}{I_{yy}}\left[\frac{\partial M}{\partial u_A}\right]_0, M_w = \frac{1}{I_{yy}}\left[\frac{\partial M}{\partial w}\right]_0, \ldots, M_{\delta T} = \frac{1}{I_{yy}}\left[\frac{\partial M}{\partial \delta_T}\right]_0.$$

These coefficients are called *stability derivatives* because they determine the stability of the aircraft in the absence of control inputs.

We now substitute these abbreviated terms into Eqns. 2.83 and 2.84 and the resulting expressions into the three dynamic equations given in Eqn. 2.81. This gives us a set of linear differential equations with constant coefficients:

$$\Delta \dot{u}_A(t) = X_{uA} \Delta u_A(t) + X_w \Delta w(t) - g \cos \theta_0 \Delta \theta(t)$$
$$- w_0 \Delta \dot{\theta}(t) + X_{\delta E} \Delta \delta_E(t) + X_{\delta T} \Delta \delta_T(t),$$

$$\Delta \dot{w}(t) = Z_{uA} \Delta u_A(t) + Z_w \Delta w(t) - g \sin \theta_0 \Delta \theta(t)$$
$$+ (Z_{\dot{\theta}} + u_{A0}) \Delta \dot{\theta}(t) + Z_{\delta E} \Delta \delta_E(t) + Z_{\delta T} \Delta \delta_T(t),$$

$$\Delta \ddot{\theta}(t) = M_{uA} \Delta u_A(t) + M_w \Delta w(t) + M_{\dot{w}} \Delta \dot{w}(t)$$
$$+ M_{\dot{\theta}} \Delta \dot{\theta}(t) + M_{\delta E} \Delta \delta_E(t) + M_{\delta T} \Delta \delta_T(t).$$

$$(2.86)$$

Note that the $\cong$ signs have been replaced by $=$ signs in this analysis. Nevertheless, we must understand that Eqns. 2.86 are approximations to the physics of aircraft dynamics and that they are valid only for small excursions in the dynamic variables.

We now obtain a state-variable model from Eqns. 2.86, using the definitions

$$x(t) = \begin{bmatrix} x_1(t) \\ x_2(t) \\ x_3(t) \\ x_4(t) \end{bmatrix} = \begin{bmatrix} \Delta u_A(t) \\ \Delta w(t) \\ \Delta \theta(t) \\ \Delta \dot{\theta}(t) \end{bmatrix} \quad \text{so that} \quad \dot{x}(t) = \begin{bmatrix} \Delta \dot{u}_A(t) \\ \Delta \dot{w}(t) \\ \Delta \dot{\theta}(t) \\ \Delta \ddot{\theta}(t) \end{bmatrix}. \qquad (2.87)$$

Thus we can write Eqns. 2.86 in the standard matrix form, $\dot{x}(t) = Ax(t) + Bu(t)$,

$$\underset{\substack{\downarrow \\ 4 \times 1}}{\dot{x}(t)} = \underbrace{\begin{bmatrix} a_{11} & a_{12} & a_{13} & a_{14} \\ a_{21} & a_{22} & a_{23} & a_{24} \\ 0 & 0 & 0 & 1 \\ a_{41} & a_{42} & a_{43} & a_{44} \end{bmatrix}}_{\substack{A \\ 4 \times 4}} \underset{\substack{\downarrow \\ 4 \times 1}}{x(t)} + \underbrace{\begin{bmatrix} b_{11} & b_{12} \\ b_{21} & b_{22} \\ 0 & 0 \\ b_{41} & b_{42} \end{bmatrix}}_{\substack{B \\ 4 \times 2}} \underbrace{\begin{bmatrix} \Delta \delta_E(t) \\ \Delta \delta_T(t) \end{bmatrix}}_{\substack{u(t) \\ 2 \times 1}} \qquad (2.88)$$

where the elements a_{ij} and b_{ij} of the A and B matrices are:

$$a_{11} = X_{uA}, \ a_{12} = X_w, \ a_{13} = -g \cos \theta_0, \ a_{14} = -w_0, \quad b_{11} = X_{\delta E}, \ b_{12} = X_{\delta T},$$

$$a_{21} = Z_{uA}, \ a_{22} = Z_w, \ a_{23} = -g \sin \theta_0, \ a_{24} = (Z_{\dot{\theta}} + u_{A0}), \quad b_{21} = Z_{\delta E}, \ b_{22} = Z_{\delta T},$$

$$a_{41} = (M_{uA} + M_{\dot{w}} Z_{uA}), \ a_{42} = (M_w + M_{\dot{w}} Z_w), \ a_{43} = -M_{\dot{w}} g \sin \theta_0,$$

$$a_{44} = [M_{\dot{w}}(Z_{\dot{\theta}} + u_{A0}) + M_{\dot{\theta}}], \quad b_{41} = (M_{\dot{w}} Z_{\delta E} + M_{\delta E}), \ b_{42} = (M_{\dot{w}} Z_{\delta T} + M_{\delta T}).$$

Note that the 2×1 matrix having the control inputs $\Delta \delta_E(t)$ and $\Delta \delta_T(t)$ as elements is designated as $u(t)$. As standard notation for the matrix of input variables in the state-variable form of the model, $u(t)$ is sometimes called the *input vector*.

The numerical values of the stability derivatives depend on the aircraft configuration and the flight condition. We can judge the relative magnitudes and signs of some of these coefficients from the physics of the aircraft. For example, the aerodynamic and thrust forces in the z-axis direction $Z(u_A, w, \dot{\theta}, \delta_E, \delta_T)$ indicate that these forces depend on $w(t)$ but not on $\dot{w}(t)$. The dependence on $w(t)$ is due to aerodynamic drag, which is a function of airspeed in that direction; however, because the aerodynamic force due to acceleration through the air in that direction is negligible in most airplanes, the dependence on $\dot{w}(t)$ is not included. In our airplane we would expect the z-axis forces to depend significantly on the elevator deflection, and we would expect the sign of $[\partial Z/\partial \delta_E]_0$ to be negative because positive δ_E generates a force in the negative z direction. We might also expect the magnitude of $[\partial Z/\partial \delta_T]_0$ to be very small if the thrust line is nearly normal to the z axis. Similarly, if the thrust line passes through the center of mass, we expect the y-axis moment due to thrust forces, as reflected in the coefficient $[\partial M/\partial \delta_T]_0$, to be zero. If the thrust line passed below the center of mass we would expect $[\partial M/\partial \delta_T]_0$ to be positive. The y-axis moment due to elevator deflection is very significant, and the sign of the coefficient $[\partial M/\partial \delta_E]_0$ is negative because the elevator is in the rear of the aircraft.

Dynamic systems subject to automatic control require sensors mounted to measure the state variables or variables dependent on the state variables. Assume that we mount an airspeed sensor oriented to respond only to $u_A(t)$ and having a sensitivity or scale factor of k_{uA} V/(ft/s). Assume that we also mount a pitch-attitude gyro with a scale factor of k_θ V/rad. These two instruments would provide the following output voltages:

$$\text{AIRSPEED:} \quad e_{uA}(t) = k_{uA}[u_{A0} + \Delta u_A(t)] \text{ V.}$$
$$\text{PITCH ATTITUDE:} \quad e_\theta(t) = k_\theta[\theta_0 + \Delta\theta(t)] \text{ V.} \tag{2.89}$$

If we define the perturbations in these signals to be $\Delta e_{uA}(t)$ and $\Delta e_\theta(t)$, we can define a matrix of output signals $y(t)$ that is compatible with our state-variable dynamic equation, Eqn. 2.88, as follows:

$$y(t) = \begin{bmatrix} y_1(t) \\ y_2(t) \end{bmatrix} = \begin{bmatrix} \Delta e_{uA}(t) \\ \Delta e_\theta(t) \end{bmatrix} = \begin{bmatrix} k_{uA} & 0 & 0 & 0 \\ 0 & 0 & k_\theta & 0 \end{bmatrix} \begin{bmatrix} x_1(t) \\ x_2(t) \\ x_3(t) \\ x_4(t) \end{bmatrix}. \tag{2.90}$$

This is represented in matrix notation as

$$y(t) = Cx(t), \quad \text{where } C = \{c_{ij}\} = \begin{bmatrix} c_{11} & 0 & 0 & 0 \\ 0 & 0 & c_{23} & 0 \end{bmatrix} = \begin{bmatrix} k_{uA} & 0 & 0 & 0 \\ 0 & 0 & k_\theta & 0 \end{bmatrix}. \tag{2.91}$$

We now have the complete perturbation model in matrix differential–equation form, including aircraft longitudinal dynamics and the instrumentation that we have installed:

$$\dot{x}(t) = Ax(t) + Bu(t),$$
$$y(t) = Cx(t),$$
$$t_0 \le t \le t_F, \quad u(t) \text{ is known,} \quad \text{and} \tag{2.92}$$
$$x(t_0) = x_0 \text{ is known.}$$

All the terms are defined in Eqns. 2.87–2.91. Note that the model is set up as the initial-value problem in which we must know the initial values of the state variables, and also the two control inputs during the time interval of interest. We address the solution to this problem in Chapter 6.

2.8 Summary

We have applied Newton's second and third laws and Euler's law to single rigid bodies and to two simply connected rigid bodies in planar motion. We summarize the analysis of this chapter by considering the free-body diagram in Fig. 2.15. Here four forces are applied to the body at known points. These forces are time-variable vector quantities, some of which may be due to contact with other bodies connected to the one shown. The body has constant mass M, with the center of mass CM. A second point B is also fixed in the body. Under the influence of the four applied forces $\vec{F}_1(t)$, $\vec{F}_2(t)$, $\vec{F}_3(t)$, and $\vec{F}_4(t)$, the body will experience both translation and rotation in the plane of motion. The acceleration of the CM with respect to inertial space is denoted by the vector $\vec{a}_{CM}(t)$. The velocity of point B with respect to inertial space is denoted as $\dot{\vec{z}}(t)$ in the diagram, and the acceleration of point B with respect to inertial space is denoted as $\ddot{\vec{z}}(t)$. Note that the position of the CM with respect to point B is given by the vector $\vec{L}$, which is fixed in the body but is not fixed in inertial space. The translational motion is governed by Newton's second law:

$$\begin{bmatrix} \text{sum of all the} \\ \text{forces acting on} \\ \text{the body} \end{bmatrix} = M \times \begin{bmatrix} \text{acceleration of CM} \\ \text{with respect to} \\ \text{inertial space} \end{bmatrix}.$$

We represent this law in equation form as

$$\vec{F}_1(t) + \vec{F}_2(t) + \vec{F}_3(t) + \vec{F}_4(t) = M \times [\vec{a}_{CM}(t)]_{\text{inertial space}}. \tag{2.93}$$

The rotational motion of the body is governed by Euler's law, which requires the calculation of the moments of the applied forces with respect to an

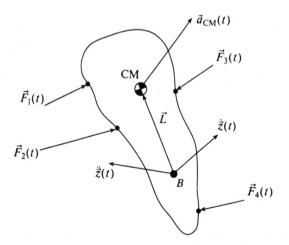

Fig. 2.15 Free-body diagram, rigid body in planar motion at time *t*.

axis normal to the plane of motion and passing through a designated point in the body. If that point is B, then Euler's law is expressed as

$$\vec{T}_B(t) = J_B \vec{\alpha}(t) + \{\text{additional terms}\}; \tag{2.94}$$

$\vec{T}_B(t)$ is the sum of the moments of the four forces about the axis through point B, and $\vec{\alpha}(t)$ is the angular acceleration of the body with respect to inertial space. J_B is the moment of inertia of the body around the axis through point B:

$$J_B = J_{\text{CM}} + M|\vec{L}|^2, \tag{2.95}$$

where J_{CM} is the moment of inertia of the body about an axis perpendicular to the plane of motion and passing through the center of mass.

The {additional terms} in Eqn. 2.94 depend on $\vec{L}$, $\vec{\omega}(t)$, $\dot{\vec{z}}(t)$, $\ddot{\vec{z}}(t)$, and M ($\vec{\omega}(t)$ is the angular velocity of the body with respect to inertial space). The {additional terms} quantity is zero if point B is fixed in inertial space ($\dot{\vec{z}}(t) = \ddot{\vec{z}}(t) = 0$) or if point B coincides with the center of mass, in which case $J_B = J_{\text{CM}}$. In expressing Euler's law, analysts normally select the axis through the center of mass about which to calculate the sum of the moments of the applied forces in order to ensure that the {additional terms} need not be calculated. In cases where point B is not the center of mass but is known to remain stationary with respect to inertial space, B can be used as the reference point for the axis about which the sum of the moments of the applied forces are calculated. The formula in Eqn. 2.94 can then be used with the assurance that the {additional terms} will be zero.

There are other conditions under which the quantity {additional terms} will be zero even though $\vec{L}$, $\dot{\vec{z}}(t)$, and $\ddot{\vec{z}}(t)$ are nonzero. However, since these conditions usually occur only momentarily during the course of the motion, it would unduly complicate the analysis if a point B other than the center of mass or a stationary point were chosen as a reference for calculating moments of the forces.

We have made distinct efforts to obtain linear differential equations as models of the various systems analyzed in this chapter, and we will continue to do so throughout the book. Because all physical systems are nonlinear, this emphasis on linear models for physical systems must be justified. Three reasons can be offered.

1. Many physical systems are approximately linear, provided the dynamic excursions of the state variables remain within prescribed (usually small) limits. The character of the motion within these small limits frequently gives a qualitative indication of what will happen for excursions beyond the linear range. Here, the stability of the motion is a prime consideration. Linear models are very useful for predicting the stability of nonlinear systems in many but not in all instances. The results obtained from linear models, in the initial-value problem for example, are only approximate because of the linearization of physical properties of the system. However, in most engineering situations there are uncertainties in the numerical values of physical parameters which cause even an exact, nonlinear analysis to yield results that only approximate the performance of the real system. In many control engineering problems approximation is acceptable, provided adequate margins of performance exist.

2. Some forms of nonlinear analysis are extensions of linear analysis. The piecewise linear method of calculating the response of nonlinear systems is an example, and the *describing function* method of analysis in nonlinear control systems is another. To understand those techniques one must know linear analysis, which, in this sense, is the first step toward nonlinear analysis.

3. The existence and uniqueness of solutions to the initial-value problem for linear differential equations are well understood, and they do not complicate the analysis of linear systems, whereas those properties for some nonlinear equations and two-point boundary-value problems can add complications to the analysis. It is sometimes better to have a thorough understanding of the theory behind an approach that leads quickly to an approximate answer than it is to be uncertain of an approach that may require more time to achieve an answer. However, considerable insight into a problem is required to invoke this justification for a linear analysis, because it is equally true in some cases that a quick but inaccurate answer is worse than no answer. Moreover, computer-based methods for the analysis of linear models can perform calculations of spectacular scope. The data-entry techniques for these methods are based on the state-variable format which we have begun to develop in this chapter, and which is developed further in subsequent chapters.

The value of the exercise problems at the end of the chapter cannot be overemphasized. These draw on all the principles illustrated in the chapter, and many of them require some extension of those principles to situations not explicitly treated in the chapter. The models developed in the problems are used in later chapters for the main controlled elements in automatic control systems. The problems also require the proper use of units and dimensions. (See Appendix A for a review of units and dimensions, and also for a table of conversion factors and numerical values for the basic physical units used in control engineering.) Answers to some of the problems are provided for guidance to the solutions.

2.9 Problems

Problem 2.1
A thin disk of uniform material weighs 25 lb and is 50 cm in diameter. How much torque (measured in N-m) is required to accelerate the disk 10 (deg/s)/min about its polar axis?

Problem 2.2
The rotor of a large machine is rotating at a speed of ω_0 rad/s. The polar moment of inertia of the rotor is J kg-m^2. A constant retarding torque of T N-m is applied to the rotor. How long will it take to bring the rotor to a stop?

Problem 2.3
In Fig. P2.3, $z(t)$ is the displacement of the mass from its rest position. A viscous friction characterized by the parameter α N/(m/s) exists between the mass and its otherwise

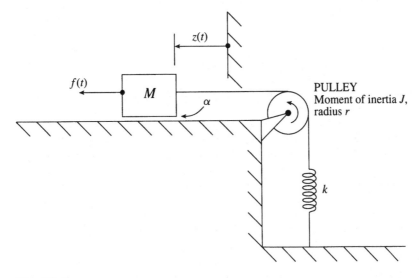

Fig. P2.3

frictionless and level support surface. The massless cable moves the pulley without slipping. Establish a state-variable model for this system in the pattern of Eqn. 2.92, where the applied force $f(t)$ is the input and $z(t)$ is the output. You must identify the state variables of your model and specify all elements of the A, B, and C matrices as algebraic combinations of the physical parameters of the system. The cable remains in tension during the motion.

Problem 2.4

Refer to Fig. 2.1 in the text. The state-variable model for the linearized M, k, D system with the state vector $x(t) = [z(t) \; \dot{z}(t)]^{\mathrm{T}}$, the input variable $f(t)$ N, and the output variable $z(t)$ m, is:

$$\dot{x}(t) = \begin{bmatrix} a_{11} & a_{12} \\ -15 & -2 \end{bmatrix} x(t) + \begin{bmatrix} b_{11} \\ b_{21} \end{bmatrix} f(t),$$

$$z(t) = [c_{11} \; c_{12}] x(t).$$

If k is known to be 30 N/m, find the numerical values of M and α, stating the physical units of each. Also, find the numerical values for a_{11}, a_{12}, b_{11}, b_{21}, c_{11}, and c_{12}. What are the physical units for b_{21}?

Problem 2.5

A piston of mass M is fitted loosely in a cylinder and is driven by input force $f(t)$, as shown in Fig. P2.5. The viscous drag coefficient between piston and cylinder is α N/(m/s). The cylinder is restrained by the spring having elastic coefficient k. The displacement of the piston with respect to the stationary reference is $z_m(t)$, and that of the cylinder is $z_c(t)$. Derive a state-variable model for this system that is valid until the time the piston strikes an end of the cylinder. Use the standard form

$$\dot{x} = Ax + Bu,$$

$$y = Cx + Du,$$

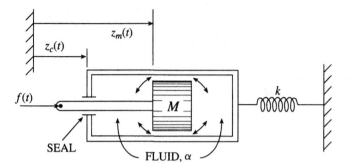

Fig. P2.5

where the input u is taken as $f(t)$ and $y(t)$ is the output vector, which we want to be

$$y(t) = [z_m(t) \; z_c(t)]^T.$$

You must define your state vector $x(t)$ in terms of the physical variables of the system, and you must give values for all the elements in the four system matrices A, B, C, and D. We neglect the mass of the cylinder and the friction of the seal.

Problem 2.6
A metallic sphere has a diameter of 4 cm and a moment of inertia about any axis through its center of 420 g-cm^2. Choose the best answer from the following list.

- **(a)** The sphere is solid aluminum.
- **(b)** The sphere is a steel shell.
- **(c)** The sphere may be either steel or aluminum.
- **(d)** The sphere is solid steel.
- **(e)** The sphere is an aluminum shell.

Problem 2.7
A cart is pushed along a horizontal surface by an applied force $f(t)$, as indicated in Fig. P2.7. The wheels do not skid with respect to the surface. The total mass of the cart, M, includes both the front and rear wheel and axle assemblies. The rear-wheel axle assembly is characterized by its mass M_R, the radius of the wheel r_R, and the moment of inertia

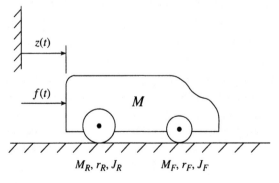

Fig. P2.7

of the assembly about the axle J_R. The front-wheel axle assembly is similarly characterized by r_F, M_F, and J_F. Derive a state-variable model for this system with input $f(t)$ and
output $z(t)$, the position of the cart with respect to the surface. Identify your state variables and give the values of all elements in your four system matrices.

Problem 2.8

The cart in Problem P2.7 is used as the carriage for the inverted-pendulum balancing
system depicted in Fig. P2.8. For simplicity, assume that the two wheel–axle assemblies
are identical. The pendulum angle from the vertical is $\beta(t)$, and the position of the cart
with respect to a reference point on the ground is $z(t)$. The pendulum is characterized by
the three parameters m, l, and J_{CM}, as in Section 2.4 of the text. The inverted pendulum
is unstable, but it can be balanced to remain near the upright position by moving the
cart back and forth with a suitable applied force $f(t)$. Write the differential equations
of motion, valid for $-\pi/2 < \beta(t) < \pi/2$, necessary to describe the dynamic behavior
of this pendulum–cart system.

Now assume that the balancing act is so successful that $\beta(t)$ will actually deviate
from the vertical by no more than 0.1 rad. Simplify your differential equations so that
they are sufficiently comprehensive to describe the motion of the system under this
assumption. Derive a linear state-variable model for this system in the standard form,
using the following notation:

$$u(t) = \text{input} = f(t),$$

$$y(t) = \text{output vector} = [z(t) \;\; \beta(t)]^{\text{T}},$$

$$x(t) = \text{state vector} = [z(t) \;\; \dot{z}(t) \;\; \beta(t) \;\; \dot{\beta}(t)]^{\text{T}}.$$

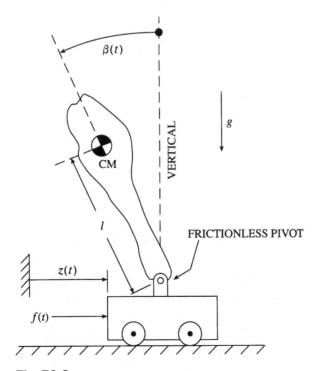

Fig. P2.8

Be sure to identify each element of your four system matrices with an algebraic combination of the physical parameters M, m, r_R, J_R, J_{CM}, g, and l.

Problem 2.9

Refer to the diagram of the belt-drive system which appears as Fig. 2.11 of the text. Assume that the belt has a total length of L and a uniform mass density. Let the efficiency of power transmission be η. The ratio of the diameter of the load pulley to that of the drive pulley is n. Let $J_1 = 400$ g-in^2 and the load pulley moment of inertia be $J_L = 7$ kg-cm^2. The drive ratio n is 5, the efficiency η is 0.9, L is 50 in, and the mass density of the belt is given as 0.5 kg/in. The load pulley diameter is 10 in. How much torque T in N-m must be applied at shaft 1 in order to accelerate the load shaft 35 rpm/s? Refer to Appendix A for the conversion factors required to express your parameters in a consistent set of units.

Problem 2.10

Refer to Fig. 2.14 in the text, which depicts a Navion aircraft flying in a vertical plane. We employ the assumptions and notation of Section 2.7 for this problem. The aircraft is cruising at a low altitude with the flaps and landing gear retracted. It has a mass of 85.4 slug with the CM at 29.5 percent of the mean aerodynamic chord. The engine is working at 60 percent of its rated power ($\delta_{T0} = 60°$) and the aircraft is trimmed for straight and level flight, which gives $\theta_0 = 0.6$ deg, $u_{A0} = 176$ ft/s, and $w_0 = 1.84$ ft/s. The moment of inertia of the rigid craft about the y axis is $I_{yy} = 3000$ slug-ft^2. The gravitational constant is 32.174 ft/s^2. The flight control inputs are limited to fluctuations in $\Delta\delta_E$ and $\Delta\delta_T$ of a few degrees. The stability derivatives for this aircraft are:

$$X_{uA} = -0.0451, \qquad Z_{uA} = -0.3697, \qquad M_{uA} = 0,$$

$$X_w = 0.03607, \qquad Z_w = -2.0244, \qquad M_w = -0.05,$$

$$X_{\delta E} = 0, \qquad Z_{\dot\theta} = 0, \qquad M_{\dot w} = -0.005165,$$

$$X_{\delta T} = 0.1, \qquad Z_{\delta E} = -28.17, \qquad M_{\dot\theta} = -2.0767,$$

$$Z_{\delta T} = 0, \qquad M_{\delta E} = -11.1892,$$

$$M_{\delta T} = 0.$$

Derive a state-variable model for the longitudinal axis dynamics for the Navion, taking $u(t) = [\Delta\delta_E(t) \ \Delta\delta_T(t)]^T$ as the input vector and $y(t) = [\Delta\theta(t) \ \Delta\dot\theta(t)]^T$ as the output vector. For the state vector use $\Delta x(t) = [\Delta u_A(t) \ \Delta w(t) \ \Delta\theta(t) \ \Delta\dot\theta(t)]^T$.

Problem 2.11

Use the state-variable model derived in Problem 2.10 to determine what the ultimate state of the flight will be if a one-degree "up" elevator angle is applied, holding the throttle constant at 60 percent of engine power. Stated analytically, this question is: Find $\lim_{t\to\infty}[x(t)]$ when $\Delta\delta_E(t) = -(1/57.3)$ rad and $\Delta\delta_T = 0$. Note that you are not asked to find $x(t)$ for all t from 0 to ∞; you are asked only for the final value of $x(t)$.

You can find $\Delta x(\infty)$ almost by inspection of the model; you need not solve the differential equations. Remember, $x(\infty) = x_0 + \Delta x(\infty)$, and our trim conditions give us $x_0 = [176 \ 1.84 \ 0.6/57.3 \ 0]^T$. Draw a diagram showing the beginning aircraft velocity vector $\vec{V}(0)$ and the final velocity vector $\vec{V}(\infty)$. Judging from the physics of the problem, does your model give a reasonable result?

Problem 2.12

Let the altitude above sea level of the CM of the Navion aircraft be denoted as $h(t)$. Then $h(t) = h_0 + \Delta h(t)$, where h_0 is the trim altitude (1,000 ft), and $\Delta h(t)$ is the small deviation in altitude from h_0 resulting from the maneuvers of the aircraft. Note that $\dot{h}(t) = \Delta \dot{h}(t)$.

Show that $\dot{h}(t)$ is given approximately by $u_{A0}\Delta\theta - w(t)$, provided that $\Delta\theta$ remains small and therefore $\Delta\dot{h}(t) = u_{A0}\Delta\theta - \Delta w(t)$.

Now establish a fifth-order state-variable model for the aircraft dynamics by augmenting the state vector with $\Delta h(t)$: $x(t) = [\Delta u_A(t) \ \Delta w(t) \ \Delta\theta(t) \ \Delta\dot{\theta}(t) \ \Delta h(t)]^T$. Assume that five sensors are mounted in the aircraft to measure its dynamic response during flight tests. These sensors are: an airspeed indicator, a pitch-angle attitude sensor, a pitch-rate sensor, an altimeter, and a vertical speed sensor. Assume unity scale factors for all five sensors, so that the output vector is: $y(t) = [\Delta u_A(t) \ \Delta\theta(t) \ \Delta\dot{\theta}(t) \ \Delta h(t) \ \Delta\dot{h}(t)]^T$. Write the complete state-variable model as follows:

DYNAMICS: $\qquad \dot{x}(t) = Ax(t) + Bu(t),$

SENSORS: $\qquad y(t) = Cx(t) + Du(t).$

Be sure to specify the numerical values for all the elements in A, B, C, and D.

Problem 2.13

Refer to Fig. 2.11 in the text. Neglect the mass of the belt. $J_1 = 1$ kg-m^2, $J_2 = 40$ kg-m^2, and $n = 5$. A state-variable model for this belt-drive system, where the input torque is measured in N-m, is:

$$\underbrace{\begin{bmatrix} \dot{\theta}_L(t) \\ \ddot{\theta}_L(t) \end{bmatrix}}_{\dot{x}(t)} = \underbrace{\begin{bmatrix} a_{11} & a_{12} \\ a_{21} & a_{22} \end{bmatrix}}_{A} \underbrace{\begin{bmatrix} \theta_L(t) \\ \dot{\theta}_L(t) \end{bmatrix}}_{x(t)} + \underbrace{\begin{bmatrix} b_{11} \\ 0.07 \end{bmatrix}}_{B} T(t),$$

$$\theta_L(t) = Cx(t).$$

Identify the C matrix, find the numerical values for the elements of the A matrix, find b_{11}, and calculate the efficiency of the belt transmission.

Problem 2.14

A rate gyroscope similar to that described in Section 2.3 is mounted in an airplane, with its input axis aligned with the yaw axis of the craft (the z axis in Fig. 2.14). The moment of inertia of the rotor about its spin axis is 1,400 g-cm^2, the spring restraint coefficient on the gimbal is 3×10^5 dyne-cm/deg, and the voltage gradient of the output shaft sensor is 50 mV/deg. If the aircraft is turning at a steady rate of 10 deg/s and the output voltage is 100 mV, calculate the angular speed of the rotor and the deflection of the gimbal from its center position.

Problem 2.15

A ball (body B) is free to roll without slipping on a straight rack, as shown in Fig. P2.15. The rack is a pendulum with its pivot point P fixed with respect to inertial space. An actuator applies torque $T(t)$ to the rack pendulum (body A) about the pivot axis, as indicated in the sketch. $z(t)$ is the displacement of the CM of the ball to the right of the center line of the pendulum. Here d is the distance from P to the CM of body A, and h is the distance from P to the CM of the ball when the ball is centered on the rack. Note

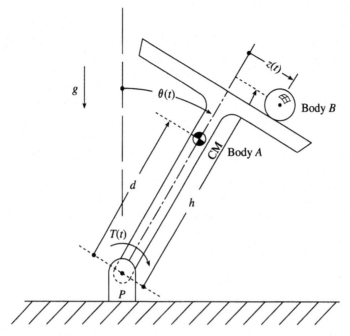

Fig. P2.15

that this system could be constructed so that both h and d could be adjusted to have either positive or negative values.

The term $\theta(t)$ denotes the clockwise displacement of the pendulum from its vertical position; g, the gravitational constant, is 981 cm/s². Body A has a mass M_A g and a moment of inertia about an axis through its CM parallel to the pivot axis of J_A g-cm². Body B, whose radius is r cm, is similarly characterized by M_B and J_B. Show that the equations of motion for this system are:

$$[I(t)]\ddot{\theta}(t)+\left[M_Bh+\frac{J_B}{r}\right]\ddot{z}(t)+2M_Bz(t)\dot{z}(t)\dot{\theta}(t)-[M_Agd\sin\theta(t)]-M_Bg[L(t)]=T(t),$$

and

$$\left[M_Bh+\frac{J_B}{r}\right]\ddot{\theta}(t)+\left[M_B+\frac{J_B}{r^2}\right]\ddot{z}(t)=M_B[z(t)\dot{\theta}^2(t)+g\sin\theta(t)],\quad\text{where}$$

$$I(t)=J_A+M_Ad^2+M_B(h^2+z^2(t))+J_B\quad\text{and}\quad L(t)=h\sin\theta(t)+z(t)\cos\theta(t).$$

Problem 2.16

(a) Define the state vector for the ball–rack–pendulum system as $x(t)=[x_1(t)\ x_2(t)\ x_3(t)\ x_4(t)]^T=[\theta(t)\ \dot{\theta}(t)\ z(t)\ \dot{z}(t)]^T$. Derive a set of four first-order nonlinear differential equations in the form $\dot{x}_i(t)=f_i(x_1(t),x_2(t),x_3(t),x_4(t),T(t))$, $i=1,2,3,4$, where the four $f_i(\cdot)$ are nonlinear functions of the four state variables and where the four first-order equations, taken together as a set, are equivalent to the two second-order equations given in Problem 2.15.

(b) Check your result in (a) by considering the following practical questions:

1. Does each term in $f_i(\cdot)$, $i = 1, 2, 3, 4$, have the same physical dimension?
2. Do all of the parameters M_A, M_B, J_A, J_B, g, r, h, and d appear in your equations?
3. Assume that $h > d > 0$. Let $x(0^+) = [0\ 0\ e\ 0]^T$, where $e > 0$ but is small. Let $T(t) = 0$. Calculate the values for each $\dot{x}_i(0^+)$. Determine whether the values are correct from a consideration of the physical situation represented by these initial conditions.
4. See if the equilibrium state $x(t) = [0\ 0\ 0\ 0]^T$ dictates that all four $\dot{x}_i(t)$ must be zero.
5. Make $h < d < 0$ in your equations and repeat step 3. Do you get results which are physically sensible?

Problem 2.17

Obtain a linear state-variable model for the ball–rack–pendulum system that is valid for small departures from the equilibrium state $[\theta = 0, \dot{\theta} = 0, z = 0, \dot{z} = 0]$. Let the input to your model be $T(t)$, and let the output be $z(t)$. Express each element of the A, B, C, and D matrices in terms of the eight physical parameters of the system.

Problem 2.18

The axes of the two rotors in Fig. P2.18 are collinear, and the whole system is driven by the applied torque $T(t)$. The two linear springs are characterized by their elastic coefficients k_1 and k_2. Both rotors are subject to viscous damping torques, assumed to be linear and characterized by the damping constants d_1 and d_2. Establish a linear state-variable model for the dynamics of this system, where the applied torque is considered to be the input variable.

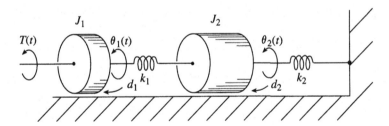

Fig. P2.18

3 Mathematical Models of Electrical Systems

3.1 Introduction

Electric circuits composed of discrete elements in which the voltages and currents are described in continuous time are sometimes called *analog circuits,* to distinguish them from circuits used in digital signal processors. We now model some analog circuits by writing ordinary differential equations in the state-variable form just as we modeled mechanical systems in Chapter 2. We apply the basic principles of circuit analysis, which are Kirchhoff's voltage and current laws, Ohm's law, Faraday's law, and the law relating the voltage across a capacitor to the current through it. We consider only circuits composed of the following elements: voltage sources, current sources, resistors, capacitors, inductors, and operational amplifiers. Our treatment here is limited to the type of circuits commonly encountered in the signal-processing parts of control systems, as opposed to those in power supplies or in other high-powered devices.

In Section 3.2 we confine our attention to passive circuits. These circuits contain no energy sources (such as operational amplifiers or batteries) except for the voltage or current sources that drive them. We can apply nearly all the basic circuit laws in this simple setting, and we can demonstrate an important principle of control engineering pertaining to interconnected elements, called the *loading effect.*

In Section 3.3 we model active circuits having operational amplifiers. This extends significantly our capability to design dynamic circuits for control system purposes. We also study simple instrumentation circuits for measuring dynamic variables. In Section 3.4, we discuss briefly the problem of noise in electrical measuring devices and develop a state-variable model showing how noise enters a linear dynamic system.

The problems at the end of the chapter require derivations beyond those appearing in the text. The results of these problem solutions are used in later chapters.

3.2 Passive Circuits

A linear resistor of R ohms [Ω] is connected in series with an inductor of L henries [H], and the combination is driven by a voltage source with terminal voltage $e(t)$ volts [V], as shown in Fig. 3.1. The current, $i(t)$ amperes [A], is delivered by the voltage source and flows through both the resistor and the inductor. The purpose of our analysis is to calculate the current $i(t)$ that results from the applied voltage $e(t)$ and from the initial value of the current. Knowing this resultant $i(t)$ permits us to calculate any other voltage or power of interest in the system. We proceed by applying the basic principles of electric circuit theory to this system.

Ohm's law states that the voltage across the resistor, in the sense indicated by the $+$ and $-$ signs, is

$$e_R(t) = Ri(t) \text{ V.} \tag{3.1}$$

The voltage across the inductor, in the sense indicated, is $e_L(t)$. If we assume that the magnetic field associated with the inductor is only that caused by the current $i(t)$, then Faraday's law gives the volt–ampere relationship for the inductor:

$$e_L(t) = L\frac{di(t)}{dt} \text{ V.} \tag{3.2}$$

We frequently use this special case of Faraday's law. We will use Faraday's law again in Chapter 4, where it will be necessary to state it in its basic form.

We now apply Kirchhoff's voltage law to obtain the relationship between $e(t)$, $e_R(t)$, and $e_L(t)$. Starting at point A and proceeding clockwise, we simply tabulate the voltage across each element of the circuit as we encounter it in traversing the entire circuit: $e(t) - e_R(t) - e_L(t)$. This gives us the voltage from point A around the circuit back to point A, which is zero. Kirchhoff declared that

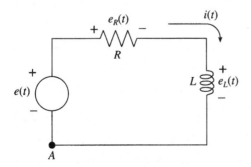

Fig. 3.1 Series RL circuit.

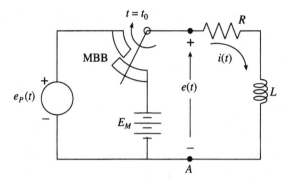

Fig. 3.2   RL circuit with initial current.

the sum of the voltages around a closed path in a circuit must be zero. Applying this statement to our present example, we arrive at the essential result

$$e(t) - e_R(t) - e_L(t) = 0. \tag{3.3}$$

To obtain the required relationship between the driving voltage $e(t)$ and the current $i(t)$, we substitute the constitutive relationships of Eqns. 3.1 and 3.2 into Eqn. 3.3:

$$e(t) = Ri(t) + L\frac{di(t)}{dt}. \tag{3.4}$$

Assume that the voltage source that produces $e(t)$ at its terminals is composed internally of a battery of voltage E_M and a time variable source with terminal voltage $e_P(t)$, along with a switch having a *make-before-break* (MBB) characteristic, as shown in Fig. 3.2. The switch is thrown instantly at time $t = t_0$. We assume that prior to t_0 the switch has been in place for a long time so that at the time the switch is thrown the current in the circuit is E_M/R A. Because the switch is thrown instantly, and because it is an MBB switch, the current in the circuit is uninterrupted at the time of switching, an important consideration in a circuit with an inductor. Combining this information with the dynamic relationship in Eqn. 3.4, we have an engineering model for our circuit:

$$\frac{di(t)}{dt} = -\frac{R}{L}i(t) + \frac{1}{L}e_P(t), \quad \text{where}$$

$$t_0 \le t \le t_F, \tag{3.5}$$

$$i(t_0) = \frac{E_M}{R}, \quad \text{and} \quad e_P(t) \text{ is known.}$$

The time interval of interest begins at the instant the switch is thrown, and the terminal time t_F is determined by the context of the problem at hand. If we adopt state-variable notation for the model, taking $x(t) = [x_1(t)] = [i(t)]$, we have

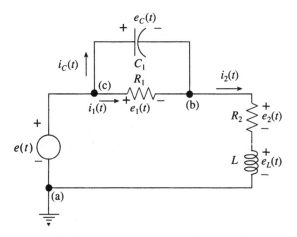

Fig. 3.3 Series–parallel RLC circuit.

$\dot{x}(t) = Ax(t) + Bu(t)$, where

$t_0 \le t \le t_F$, $x(t_0) = E_M/R$, and $u(t) = e_P(t)$, with (3.6)

$A = [-R/L]$ and $B = [1/L]$.

As a second example, we model the circuit shown in Fig. 3.3. A capacitor of value C_1 farads [F] shunts the resistor R_1 in this case. The voltage–current relationships for the resistors and the inductor are the same as in the first example;

$$e_1(t) = R_1 i_1(t), \quad e_2(t) = R_2 i_2(t), \quad e_L(t) = L\frac{di_2(t)}{dt}. \quad (3.7)$$

However, the capacitor voltage is given by

$$e_C(t) = \frac{1}{C_1} q(t) \text{ V}, \quad (3.8)$$

where $q(t)$ is the accumulated charge on the capacitor in coulombs [C]. The current into the capacitor $i_C(t)$ supplies the charge and is simply

$$i_C(t) = \frac{dq(t)}{dt} \text{ C/s or A}. \quad (3.9)$$

This permits us to rewrite Eqn. 3.8 as

$$C_1 \frac{de_C(t)}{dt} = i_C(t). \quad (3.10)$$

We now write Kirchhoff's voltage law, proceeding from point (a) to point (c) to point (b), and then back to point (a), as follows:

$$e(t) - e_1(t) - e_2(t) - e_L(t) = 0. \quad (3.11)$$

Note that we have marked point (a) as the reference point or the "ground point" in the circuit. Writing Kirchhoff's voltage law starting at point (b) to point (c) and back to point (b), we have $e_1(t) - e_C(t) = 0$, which may be written as

$$R_1 i_1(t) = e_C(t). \tag{3.12}$$

We next substitute the relations of Eqn. 3.7 into Eqn. 3.11:

$$e(t) = R_1 i_1(t) + R_2 i_2(t) + L \frac{di_2(t)}{dt}. \tag{3.13}$$

Now we invoke Kirchhoff's current law at point (b). Kirchhoff pointed out that since no electric charge can accumulate at a junction of conductors, the sum of all the currents flowing into the junction at any instant t must equal the sum of all the currents flowing out of the junction at t. In this case,

$$i_C(t) + i_1(t) = i_2(t). \tag{3.14}$$

Next we substitute the relations established in Eqns. 3.7, 3.10, and 3.12 into Eqn. 3.14:

$$C_1 \frac{de_C(t)}{dt} = -\frac{1}{R_1} e_C(t) + i_2(t). \tag{3.15}$$

We then substitute Eqn. 3.12 into Eqn. 3.13 to obtain

$$L \frac{di_2(t)}{dt} = -e_C(t) - R_2 i_2(t) + e(t). \tag{3.16}$$

At this point we should recognize that Eqns. 3.15 and 3.16 together constitute a state-variable representation of the electrodynamics of the circuit, as given by the fundamental laws of electric circuits. The state variables are $e_C(t)$ and $i_2(t)$. We note that if at time t the input voltage $e(t)$ is known, then every voltage and every current in the circuit at that instant t can be determined, provided the two variables $e_C(t)$ and $i_2(t)$ are also known. For example, we can obtain $i_1(t)$, $i_C(t)$, $e_2(t)$, and $e_L(t)$ from Kirchhoff's laws:

$$i_1(t) = \frac{e_C(t)}{R_1}, \qquad e_2(t) = R_2 i_2(t),$$

$$i_C(t) = i_2(t) - i_1(t), \qquad e_L(t) = e(t) - e_C(t) - e_2(t). \tag{3.17}$$

This example illustrates the central role that the state variables play in the dynamics of this circuit. This idea is pursued further in Chapter 5. We now define our state variables in the standard notation: $x_1(t) = e_C(t)$ and $x_2(t) = i_2(t)$. Re-arranging Eqns. 3.15 and 3.16, we have:

$$\dot{x}(t) = Ax(t) + Bu(t), \quad \text{where}$$

$$t_0 \le t \le t_F, \quad u(t) = e(t) \text{ is known,}$$

$$x(t_0) = \begin{bmatrix} e_C(t_0) \\ i_2(t_0) \end{bmatrix} \text{ is known, and} \tag{3.18}$$

$$A = \begin{bmatrix} -1/R_1 C_1 & 1/C_1 \\ -1/L & -R_2/L \end{bmatrix}, \quad B = \begin{bmatrix} 0 \\ 1/L \end{bmatrix}.$$

This is the complete model for the solution of the initial-value problem for the circuit in Fig. 3.3. The solution to this initial-value problem may be obtained by elementary means; we will consider it in Chapter 6.

If we connect a voltmeter between reference point (a) and point (b), and consider this voltage reading to be the "output" of our circuit, then we must supplement the model of Eqn. 3.18 with an output equation. Assume for simplicity that the voltmeter has infinite input resistance, so that it will not load the circuit; that is, it will draw no current from the circuit for its operation. (Most voltmeters approach this ideal performance level.) If the voltmeter reading is $y(t)$, we have

$$
\begin{aligned}
y(t) &= e_2(t) + e_L(t) \\
&= R_2 x_2(t) + L\dot{x}_2(t),
\end{aligned} \tag{3.19}
$$

which shows behavior typical of some sensors used in control systems – namely, that the sensor output is a linear combination of a state variable and the derivative of a state variable. As a consequence, the input $u(t)$ may appear directly in the output if it appears in the derivative term, which it does in this case; $L\dot{x}_2(t) = -x_1(t) - R_2 x_2(t) + u(t)$. Therefore $y(t)$ has this form:

$$
\begin{aligned}
y(t) &= Cx(t) + Du(t), \quad \text{where} \\
C &= [-1 \ 0] \quad \text{and} \quad D = [1].
\end{aligned} \tag{3.20}
$$

One often sees the pair of state-variable equations for linear systems written in the general form

$$
\begin{aligned}
\text{DYNAMICS:} \quad & \dot{x}(t) = Ax(t) + Bu(t), \\
\text{OUTPUT:} \quad & y(t) = Cx(t) + Du(t),
\end{aligned} \tag{3.21}
$$

accompanied perhaps by a "block diagram" of the system, as shown in Fig. 3.4.

A circuit driven by a current source is shown in Fig. 3.5. The current source delivers a prescribed current $I(t)$ A into node (b) and takes a similar current out of node (a). The current $I(t)$ may fluctuate between positive and negative values, so that the current "into" node (b) is not necessarily unidirectional. Kirchhoff's current law at node (b) gives one of the relationships needed to form a mathematical model for this circuit:

$$
I(t) = i_1(t) + i_2(t). \tag{3.22}
$$

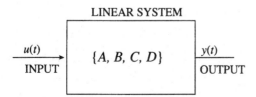

Fig. 3.4 Block diagram representing a linear system.

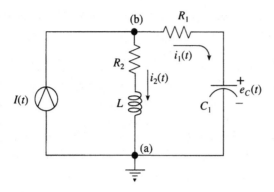

Fig. 3.5 Circuit driven by current source.

Kirchhoff's voltage law, written for the path proceeding from node (a) to node (b) and through the R_1-C_1 branch back to node (a), gives another relationship:

$$L\frac{di_2(t)}{dt} + R_2 i_2(t) = R_1 i_1(t) + e_C(t). \tag{3.23}$$

The third relationship is the constitutive equation for the capacitor:

$$C_1 \dot{e}_C(t) = i_1(t). \tag{3.24}$$

A good choice for the state-variable set for our model is $x_1(t) = e_C(t)$ and $x_2(t) = i_2(t)$. We seek two first-order differential equations having these two state variables, along with the driving current $I(t)$ on the right-hand sides. This indicates that we should eliminate $i_1(t)$ between the three defining equations. When this is done, the resultant two state-variable equations are

$$C_1 \dot{e}_C(t) = -i_2(t) - I(t) \quad \text{and}$$
$$L\frac{di_2(t)}{dt} = e_C(t) - (R_1 + R_2) i_2(t) + R_1 I(t), \tag{3.25}$$

which are expressed in standard form as

$$\dot{x}(t) = Ax(t) + Bu(t), \quad \text{where}$$

$$A = \begin{bmatrix} 0 & -1/C_1 \\ 1/L & -(R_1 + R_2)/L \end{bmatrix}, \quad B = \begin{bmatrix} 1/C_1 \\ R_1/L \end{bmatrix}, \quad \text{and} \tag{3.26}$$

$$u(t) = I(t), \quad x(t) = \begin{bmatrix} x_1(t) \\ x_2(t) \end{bmatrix} = \begin{bmatrix} e_C(t) \\ i_2(t) \end{bmatrix}.$$

Now let us add an output circuit to the network of Fig. 3.5 in order to facilitate measuring $e_C(t)$ as an output signal from our system. The voltage measuring device produces the signal $y(t)$ V and puts a load on the original circuit of the series combination of R_3 and R_4, as shown in Fig. 3.6. The load consists of the current $i_3(t)$, which is not present in the original circuit of Fig. 3.5. The output voltage is

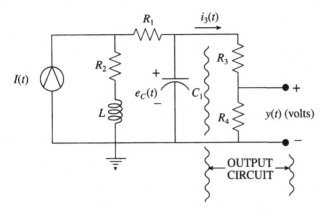

Fig. 3.6 Output circuit added to circuit of Fig. 3.5.

$$y(t) = \frac{R_4}{R_3 + R_4} e_C(t) = [c_{11} \; 0] x(t) = Cx(t), \quad \text{where}$$

$$c_{11} = \frac{R_4}{R_3 + R_4}.$$

(3.27)

But now the dynamics of the circuit are different from that described in Eqn. 3.26, owing to the load of the output circuit. An analysis of the loaded circuit shows that for the same state-variable choice, $x_1(t) = e_C(t)$ and $x_2(t) = i_2(t)$, the A and B matrices will be

$$A = \begin{bmatrix} -1/C_1(R_3 + R_4) & -1/C_1 \\ 1/L & -(R_1 + R_2)/L \end{bmatrix}, \qquad B = \begin{bmatrix} 1/C_1 \\ R_1/L \end{bmatrix}.$$

(3.28)

The presence of the load resistors is reflected in the nonzero a_{11} term in the A matrix. Because the dynamic model of the loaded circuit is different from that of the unloaded circuit, the response $e_C(t)$ to a given input $I(t)$ will be different in the two cases. It is for this reason that the modeling of a dynamic system must be done with the output circuit in place, as it will be used in service, to ensure that an accurate model will result. Of course, if the resistance $(R_3 + R_4)$ becomes very large, then the loading effect becomes small and a_{11} approaches 0. Many output circuits are intentionally designed to have a high input resistance to minimize the effect of loading, but this is not always possible.

3.3 Active Circuits

An active circuit is one that contains one or more sources of energy. Typically these devices are *controlled sources* in which the terminal voltage (or current) is a function of a voltage (or a current) in a different branch of the circuit. Thus we have voltage-controlled voltage sources (VCVS), current-controlled voltage sources (CCVS), voltage-controlled current sources (VCCS), or current-controlled current sources (CCCS).

An example of an active circuit is shown in Fig. 3.7. Here the external driving source is the voltage source $E(t)$. A voltage-controlled voltage source in

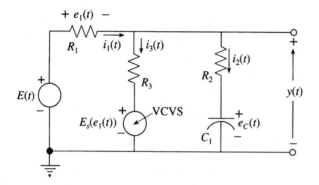

Fig. 3.7 Active circuit with a voltage-controlled voltage source (VCVS).

the center branch has terminal voltage $E_s(e_1)$ in the sense indicated by the $+$ and $-$ signs. The control voltage for the VCVS is $e_1(t)$, the voltage across resistor R_1 in the sense indicated. This VCVS might be a voltage amplifier, using $e_1(t)$ as its input. Kirchhoff's laws applied to this circuit yield the following equations governing the circuit dynamics:

$$E(t) = R_1 i_1(t) + R_3 i_3(t) + E_s(e_1(t)),$$
$$R_3 i_3(t) + E_s(e_1(t)) = R_2 i_2(t) + e_C(t), \quad \text{and}$$
$$i_1(t) = i_2(t) + i_3(t), \quad \text{where}$$
$$e_1(t) = R_1 i_1(t) \quad \text{and} \quad C_1 \dot{e}_C(t) = i_2(t).$$

$$(3.29)$$

The VCVS function $E_s(e_1(t))$ depends on the characteristics of the devices used as the VCVS. The simplest such device is a voltage amplifier for which

$$E_s(e_1(t)) = K_0 e_1(t) = K_0 R_1 i_1(t) \tag{3.30}$$

where K_0 is the *voltage gain* of the amplifier, V/V. Because in this example $e_1(t) = R_1 i_1(t)$, we could also consider the controlled source to be a CCVS with the property

$$E_s(i_1(t)) = K i_1(t), \quad \text{where}$$
$$K = K_0 R_1 \text{ V/A}.$$

$$(3.31)$$

It is worthwhile to note that the coefficient K may be designed to be either positive or negative.

Using the definition in Eqn. 3.30, we may reduce the equation set in Eqn. 3.29 to

$$C_1 \beta_R \dot{e}_C(t) = -(R_1 + R_0) e_C(t) + R_0 E(t), \quad \text{where}$$
$$R_0 = R_3 + K \quad \text{and} \quad \beta_R = [R_1 R_2 + R_1 R_3 + R_0 R_2].$$

$$(3.32)$$

This is a first-order dynamic system having a state-variable representation

$$\dot{x}(t) = A x(t) + B u(t), \quad \text{where}$$
$$x(t) = e_C(t), \quad u(t) = E(t), \quad \text{and}$$
$$A = \left[\frac{-(R_1 + R_0)}{C_1 \beta_R} \right], \quad B = \left[\frac{R_0}{C_1 \beta_R} \right].$$

$$(3.33)$$

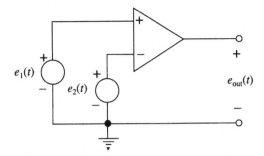

Fig. 3.8 Schematic representation of an operational amplifier.

If we consider the voltage $y(t)$ to be the "output" of this system, then the state-variable form for $y(t)$ is

$$y(t) = Cx(t) + Du(t), \quad \text{where}$$
$$C = [R_2 C_1 A + 1] \quad \text{and} \quad D = [R_2 C_1 B]. \tag{3.34}$$

It is interesting to note that the D matrix in this system is nonzero. This occurs because the output $y(t)$ is a linear combination of $x(t)$ and $\dot{x}(t)$, and $u(t)$ appears in $\dot{x}(t)$. The physical reason for $y(t)$ having a component directly proportional to $u(t)$ is obvious from the circuit diagram.

A very important active circuit element in automatic control systems is the operational amplifier (op-amp). Figure 3.8 is a schematic representation of a typical operational amplifier, showing neither the utility connections (such as the power supply) nor the internal circuitry. Two signal voltages, $e_1(t)$ and $e_2(t)$, are connected to terminals labeled $+$ and $-$ at the input of the amplifier. The output voltage of the amplifier is called $e_{\text{out}}(t)$. The amplifier is designed so that $e_{\text{out}}(t)$ is restricted to fluctuate between limits that are typically -10 V to $+10$ V. The relationship between the input voltages and $e_{\text{out}}(t)$ is

$$e_0(t) = K_A[e_1(t) - e_2(t)] \quad \text{where} \quad K_A \gg 1, \tag{3.35}$$

provided that:

    (a) -10 volts $< e_{\text{out}}(t) < +10$ volts;
    (b) $e_{\text{out}}(t)$ does not fluctuate too rapidly;
    (c) no current flows out of the amplifier; and
    (d) no current flows into the amplifier from either of the two input signal sources.

Since the amplifier gain K_A is actually as high as 10^6, we see from restriction (a) that the difference voltage $[e_1(t) - e_2(t)]$ must always be very close to zero when the amplifier is in use. Normally the amplifier is used in a manner such that $e_{\text{out}}(t)$ is impressed across a high resistance so that restriction (c) is essentially satisfied. The amplifier is also designed to have a high input resistance, so that restriction (d) is satisfied for all reasonable values of $e_1(t)$ and $e_2(t)$ that will occur in use. Because restriction (b), which pertains to the dynamic performance of the amplifier, has little practical impact in most control system applications, discussion of this restriction is deferred to later chapters.

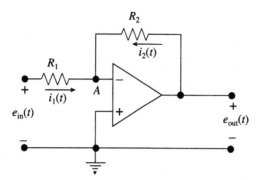

Fig. 3.9   Operational amplifier connected as a precision voltage amplifier.

The operational amplifier is used as the central element in hundreds of different circuit configurations. One of the simplest is shown in Fig. 3.9. Here we are interested in the relationship between the input voltage $e_{in}(t)$ and the output voltage $e_{out}(t)$. This relationship is determined very easily from the characteristics of the amplifier given in Eqn. 3.35. The voltage at the + terminal is zero in this case; therefore, since $e_{out}(t)$ must be less than 10 V in magnitude and because K_A is so large, the voltage at point A will be very nearly zero. The current into the amplifier from point A is virtually zero because of the high input resistance of the amplifier and because the voltage at A is nearly zero. By Kirchhoff's current law we have

$$i_1(t) + i_2(t) = 0, \tag{3.36}$$

and by Ohm's law we have

$$i_1(t) = \frac{e_{in}(t)}{R_1} \quad \text{and} \quad i_2(t) = \frac{e_{out}(t)}{R_2}. \tag{3.37}$$

Combining Eqns. 3.36 and 3.37, we obtain the desired relationship

$$e_{out}(t) = -\frac{R_2}{R_1} e_{in}(t). \tag{3.38}$$

It is possible to obtain a very precise and stable ratio, R_2/R_1, even in production-line quantities, so the circuit in Fig. 3.9 is used where a voltage amplifier having a precise gain is required. Equation 3.38 is valid only if the operational amplifier gain remains very high; however, the gain need not be precisely known nor remain fixed, so long as it remains very high.

We use the − terminal for point A rather than the + terminal to keep the circuit dynamically stable. This aspect of circuit design is explored in depth in Chapters 8 through 15. In fact, it is a central topic of automatic control theory. We need not pursue the dynamic stability question at this point.

A useful variation on the precision amplifier design is shown in Fig. 3.10. Here two input voltages and two input resistors are attached to point A. We have

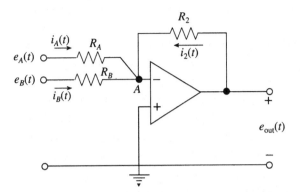

Fig. 3.10 Operational amplifier connected as a summing amplifier.

$$i_A(t) + i_B(t) + i_2(t) = 0, \quad \text{where}$$

$$i_A(t) = \frac{e_A(t)}{R_A}, \quad i_B(t) = \frac{e_B(t)}{R_B}, \quad i_2(t) = \frac{e_{\text{out}}(t)}{R_2}, \tag{3.39}$$

from which we obtain the result

$$e_{\text{out}}(t) = -\left[\frac{R_2}{R_A} e_A(t) + \frac{R_2}{R_B} e_B(t)\right]. \tag{3.40}$$

The ratios of resistances may be obtained precisely, and the circuit, called a *summing circuit,* can obviously be an amplifier as well as a summer. Furthermore, several input resistors could be added to the circuit so that the output could be the sum of several inputs, rather than two as shown in this example.

If the negative sign in Eqn. 3.40 is undesirable, a second precision amplifier can be cascaded with the first, as shown in Fig. 3.11. In this system we must be careful to prevent the second circuit from loading the first by keeping the current $i_3(t)$ below the specified capability of the operational amplifier. Provision (c) of Eqn. 3.35 can be violated up to a design limit of the operational amplifier

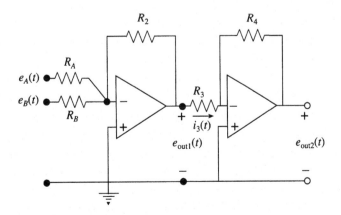

Fig. 3.11 Two-stage summing amplifier.

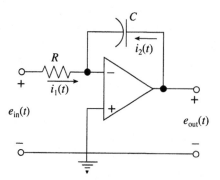

Fig. 3.12 Operational amplifier connected as an integrator.

(usually a fraction of an ampere). This means that R_3 must be greater than a few hundred ohms. Alternately, adjustments can be made in the value of R_3 or R_4 in order to obtain the desired overall gain even if $i_3(t)$ does load the first amplifier. The overall relationship in this circuit has no negative sign:

$$e_{\text{out}2}(t) = \frac{R_4}{R_3}\left[\frac{R_2}{R_A}e_A(t) + \frac{R_2}{R_B}e_B(t)\right]. \tag{3.41}$$

Another extremely useful operational amplifier circuit, shown in Fig. 3.12, is called an *integrating circuit*. The equations for this circuit are obtained as in the previous examples:

$$i_1(t) + i_2(t) = 0,$$

$$i_1(t) = \frac{e_{\text{in}}(t)}{R}, \tag{3.42}$$

$$C\dot{e}_{\text{out}}(t) = i_2(t).$$

Eliminating $i_1(t)$ and $i_2(t)$, we obtain the differential equation of motion for the circuit

$$\dot{e}_{\text{out}}(t) = -\frac{1}{RC}e_{\text{in}}(t). \tag{3.43}$$

An alternate way of writing Eqn. 3.43 is

$$e_{\text{out}}(t) = -\frac{1}{RC}\int e_{\text{in}}(t)\,dt, \tag{3.44}$$

which may be expressed as: The output voltage is a constant times the integral of the input voltage. The constant in this case is $-1/RC$. If we add a second circuit, as in Fig. 3.11, we can avoid the minus sign. If we set the multiplying constant to unity, we will have a circuit for which the output voltage is the integral of the input voltage: $e_{\text{out}}(t) = 1\int e_{\text{in}}(t)\,dt$. We note that the coefficient 1 has the units 1/s in this case. If we wish to use a definite integral starting at $t = 0$, then

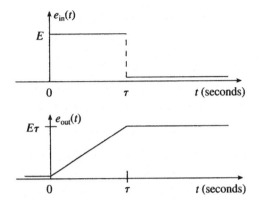

Fig. 3.13 Operation of an integrator.

$$e_{out}(t) = 1 \int_0^t e_{in}(t)\,dt + e_{out}(0), \qquad (3.45)$$

where $e_{out}(0)$ is the initial value of $e_{out}(t)$ because of a nonzero charge on the capacitor at the starting instant $t = 0$.

We now illustrate the main reason for the importance of the integrator as a circuit element. Consider the integrator represented in the block diagram in Fig. 3.13. Let us say the output voltage is zero prior to the start of our experiment at $t = 0$. At $t = 0$ a pulse of magnitude E and duration τ is applied to the input, as shown. The circuit integrates the input function so that, during the interval $0 < t < \tau$, the output of the integrator is the ramp function

$$e_{out}(t) = Et \quad \text{for } 0 < t < \tau. \qquad (3.46)$$

At $t = \tau$ the input drops to zero, so that the *slope* of the output also drops to zero, since for the integrator we have $e_{in}(t) = \dot{e}_{out}(t)$. The output itself remains at the level $E\tau$, as shown.

For the period $t > \tau$, the output is nonzero whereas the input is zero, so the ratio e_{out}/e_{in} during this period is infinite. Electronics engineers sometimes describe the integrator as an "infinite-gain amplifier" because it is possible for the integrator to maintain a nonzero output for a prolonged period during which the input is zero. (This is possible only when the output voltage is built up prior to the prolonged interval, as illustrated in Fig. 3.13.) This property is very useful in automatic control systems, where we find not only electronic integrating devices but also mechanical, electromechanical, and fluid integrators.

3.4 Transducers

Control systems require *sensors* (also called *transducers* or *instruments*) to measure the important physical variables of the system and to make those measurements available almost instantly. This usually means that the measurements must be electrical voltages or currents that can be easily transmitted and easily processed by computing equipment. Here we discuss four of these simple electromechanical transducers as typical examples. The first of these, a potentiometer circuit, is designed to provide a voltage proportional to the angular position of a mechanical shaft. The shaft has a zero reference position from which it can move both clockwise and counterclockwise. The voltage must also be bipolar – for example, positive for counterclockwise displacements and negative for clockwise displacements.

Figure 3.14 shows the rotary potentiometer and the measuring circuit producing the output signal $e_{out}(t)$, which must be proportional to the shaft position $\theta(t)$. The shaft can be positioned anywhere between the two limits $-\theta_{max}$ and $+\theta_{max}$. The potentiometer is a resistance element of R ohms equipped with a sliding contact which is mechanically fixed to the shaft, so that there are three electrical connections to the potentiometer, A, B, and W (for "wiper") as shown. The total voltage impressed across terminals A–B is $2E$ V (E V on each side of the zero reference point). The output signal $e_{out}(t)$ is the voltage at terminal W with respect to the zero reference point. The resistive element might be fine resistance wire wound tightly on a circular mandrel, or it might be a solid material such as carbon or conductive plastic. If we assume that the element

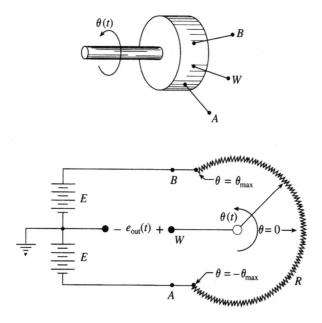

Fig. 3.14 A rotary position potentiometer and its exciting circuit.

is homogeneous and that it provides an infinite resolution to our transducer, then – provided that no current flows through the wiper – the output voltage will be

$$e_{\text{out}}(t) = \left[\frac{2E}{2\theta_{\max}}\right]\theta(t) = K_P\theta(t). \tag{3.47}$$

The constant K_P, called the *voltage gradient* or the *transducer gain,* has units of V/rad. The ideal conditions assumed here of infinite resolution, uniform resistive material, and no current loading can be realized closely enough by practical devices to make Eqn. 3.47 valid for many applications. However, if extreme precision is required in the measurement of $\theta(t)$, it will be necessary to assess the influence of the granularity and nonhomogeneity of the resistance element, the contact resistance of the sliding contact, the temperature sensitivity and noise properties of the resistance element, and the loading effect of a current in the wiper. When all these factors are assessed quantitatively it will be found that $e_{\text{out}}(t) = f(\theta(t))$, where $f(\theta(t))$ is a very complicated function that may not be sufficiently close to the relationship in Eqn. 3.47 to satisfy the accuracy requirement under all operating conditions. In this case a different type of position transducer must be considered. It is necessary in any case to calculate the power dissipated in the resistive element, which is $4E^2/R$ watts [W], in order to be sure that the element can withstand the thermal stress of that dissipation.

An important type of shaft-position transducer is the digital shaft encoder, which provides a digital word coded to minimize the ambiguities inherent in digital measurements. The number of digits in the word determines the resolution of the device, which can be as fine as a small fraction of an arc-second. The digital word can be utilized directly in a computer or it can be converted to a continuous time voltage by a digital-to-analog converting circuit.

Shaft encoders come in two varieties, absolute encoders and incremental encoders. The absolute encoder operates as just described, but the incremental encoder provides only pulses that are synchronized with the shaft position to produce a pulse for each small change in shaft angle. The pulses are counted by an external digital circuit to determine the angular displacement of the shaft from a reference point established at the outset of the operation.

Another resistance-based transducer, used to sense very small mechanical displacements, is the *strain gauge*. This consists of a length of very fine resistance wire laid up in loops, as illustrated in Fig. 3.15, and firmly attached to a substrate. The substrate is cemented firmly to a structural member which is subject to variable stress and consequently undergoes slight deformation (*strain*). The deformation changes the total length of the resistance wire, thereby also changing the resistance of the wire between terminals A and B:

$$R_{AB} = R_0 + \Delta R. \tag{3.48}$$

Here ΔR is the resistance change due to the small deformation Δl:

$$\Delta R = k\,\Delta l, \tag{3.49}$$

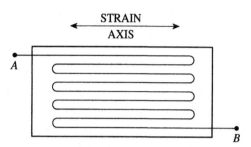

Fig. 3.15 Strain gauge.

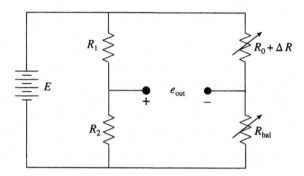

Fig. 3.16 Bridge circuit.

where k has units of Ω/m. To produce a voltage proportional to Δl we use the *bridge circuit* of Fig. 3.16. Here the strain-gauge resistance comprises one arm of the bridge, and an adjustable resistor R_{bal} comprises another arm. The bridge is excited with a constant and known voltage E. The voltage e_{out} will depend on ΔR as follows, provided no current flows into the voltmeter measuring e_{out}:

$$e_{out} = \frac{E}{R_1 + R_2} \left[\frac{R_0 R_2 - R_1 R_{bal} + R_2 \Delta R}{R_0 + R_{bal} + \Delta R} \right]. \tag{3.50}$$

The bridge is balanced at zero strain ($\Delta R = 0$) by adjusting R_{bal} so that e_{out} is zero when the strain is known to be zero. This means that $R_{bal} = R_0 R_2 / R_1$. Note that the two resistors R_1 and R_2 need not be precision resistors, but they should be invariable and of known value.

With the bridge balanced for zero strain, we have

$$e_{out} = \frac{E R_1 R_2}{R_1 + R_2} \left[\frac{\Delta R}{R_0(R_1 + R_2) + R_1 \Delta R} \right]. \tag{3.51}$$

This is a nonlinear function of ΔR, but if ΔR is very small (as it would be for small strains) then

$$e_{out} = \frac{E R_1 R_2}{R_0(R_1 + R_2)^2} [\Delta R] \quad \text{provided } \Delta R \ll \frac{R_0(R_1 + R_2)}{R_1}. \tag{3.52}$$

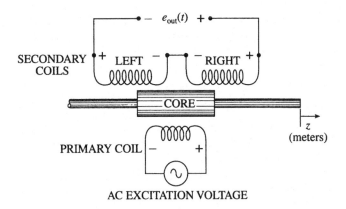

Fig. 3.17 Linear variable differential transformer (LVDT).

Substituting Eqn. 3.49 for ΔR, we obtain an expression for e_{out} in terms of the strain:

$$e_{out} = \frac{ER_1R_2k}{R_0(R_1+R_2)^2}[\Delta l] = K_s[\Delta l] \tag{3.53}$$

where K_s V/m is a calibration constant for the bridge–strain gauge circuit. The strain Δl depends on the stress in the structural member, and the stress in turn may depend on an applied load of interest in the experimental set-up. Further calibration would be required to make e_{out} directly proportional to the applied load of interest. If the strain is large, then it is necessary to use the nonlinear relationship (Eqn. 3.51) to obtain e_{out} as a nonlinear function of Δl. Because the strain gauge is sensitive to temperature, that effect must also be considered in the design of experiments.

A transducer intended for precise measurement of linear displacement is the linear variable differential transformer (LVDT). A schematic diagram of this device is shown in Fig. 3.17. A small iron core slides linearly in a cylindrical structure. The core displacement is $z(t)$ m from a known center position. The primary winding is energized by an alternating voltage, $E_P \sin(2\pi ft)$ V, where f is the frequency of the energizing voltage, typically 2500 hertz [Hz] (hertz was formerly known as cycles/second, [cps]). The term 2π has units of radians/cycle, which means $2\pi ft$ has units of radians [rad]. The core magnetically couples the primary winding to the two secondary windings with the polarities indicated by the + and − signs in Fig. 3.17. When the core is centered, $z(t) = 0$. The output voltage $e_{out}(t)$ will also be zero, because of the way the two secondary windings are connected. When the core is displaced from its null position, the output voltage is

$$e_{out}(t) = [Kz(t)]\sin(2\pi ft) \text{ V}. \tag{3.54}$$

The sensitivity factor of the device, K V/m, has a numerical value depending on the design of the instrument and on E_P and f. If $z(t)$ is negative – that is, if the core is displaced to the left of its null position – then $e_0(t)$ will also be "negative," which means that $e_{out}(t)$ will be 180° out of phase with the primary-coil energizing voltage. The detecting circuit driven by $e_{out}(t)$ must be capable of sensing the sign (phase) of $e_{out}(t)$ as well as its magnitude $|Kz(t)|$.

The LVDT is used as the position-to-voltage transducer in many different types of sensors, including gyroscopes, accelerometers, pressure gauges, load cells, and linear position sensors. The magnetic coupling principle is also used for angular position sensing, in which case the device is known as a rotary variable differential transformer (RVDT).

3.5 Noise in Electrical Signals

Electrical signals in control systems are usually contaminated by noise. A signal voltage $e(t)$ can be represented as

$$e(t) = e_{desired}(t) + e_{noise}(t), \tag{3.55}$$

where $e_{desired}(t)$ is what $e(t)$ would be if no contamination were present, and $e_{noise}(t)$ denotes the contamination. An example is shown in Fig. 3.18.

Electrical noise may originate from the environment and enter the system through capacitive or inductive coupling. Sophisticated electrical design of the grounding and shielding of the sensitive elements in the system can minimize this component of $e_{noise}(t)$.

Some noise is generated inherently in electronic circuits. This component of $e_{noise}(t)$ is usually a random fluctuation that may be reduced by lowering the operating temperature of the circuits, an approach which is seldom practical. When the statistical properties of $e_{noise}(t)$ are known, it is possible to minimize the undesirable effects of the noise by filtering $e(t)$. Those parts of electrical engineering known as *communication theory* and *signal processing* are devoted to the art of filtering $e(t)$ in a manner that reduces $e_{noise}(t)$ as much as possible while distorting $e_{desired}(t)$ as little as possible. An analytical treatment of the filtering problem is beyond the scope of this book.

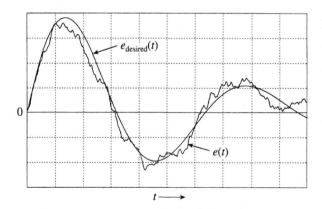

Fig. 3.18 $e(t) = $ signal + noise.

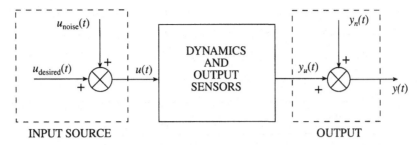

Fig. 3.19 Input and output noise sources.

The signal and noise components may be incorporated into a state-variable model of a dynamic system, as shown in Fig. 3.19. Two noise sources are shown here, $u_{\text{noise}}(t)$ and $y_n(t)$, because noise arising from two separate physical sources may be statistically uncorrelated. If the dynamic system is linear, we can write:

$$
\begin{aligned}
\text{DYNAMICS:} \quad & \dot{x}(t) = Ax(t) + B[u_{\text{desired}}(t) + u_{\text{noise}}(t)], \\
\text{OUTPUT:} \quad & y(t) = y_u(t) + y_n(t) \\
& \quad = Cx(t) + D[u_{\text{desired}}(t) + u_{\text{noise}}(t)] + y_n(t),
\end{aligned}
\tag{3.56}
$$

where $y_n(t)$ is the noise component of $y(t)$ due to the source associated with the sensors. But now $x(t)$ itself is contaminated by the input noise. It can be represented as

$$
x(t) = x_d(t) + x_n(t),
\tag{3.57}
$$

where, because the system is linear, $x_d(t)$ is the dynamic response of the state vector if $u_{\text{noise}}(t)$ were zero and $x_n(t)$ is the response if $u_{\text{desired}}(t)$ were zero. Consequently, we can write

$$
y(t) = \underbrace{Cx_d(t) + Du_{\text{desired}}(t)}_{y_{\text{desired}}(t)} + \underbrace{Cx_n(t) + Du_{\text{noise}}(t) + y_n(t)}_{y_{\text{noise}}(t)},
\tag{3.58}
$$

where $y_{\text{desired}}(t)$ is what the output $y(t)$ would be if both noise sources $u_{\text{noise}}(t)$ and $y_n(t)$ were zero.

The noise content of a signal such as $y(t)$ is frequently quite small as compared to the desired component, and for some practical purposes it may be ignored. However, where the desired component is comparable to the noise component, advanced techniques of analysis and design are required.

3.6 Summary

In this chapter we have applied some of the fundamentals of electrical circuit theory to derive dynamic models for passive and active circuits. We have also seen that the differential equation models have the same form as the models for the mechanical systems in Chapter 2. A few simple transducers, or sensors, have been introduced to show how some dynamic quantities are measured in automatic control systems. Finally, we have recognized that noise is an inherent property of electrical signals, and have briefly outlined the problem of dealing with noise in linear systems.

3.7 Problems

Problem 3.1

For the circuit in Fig. P3.1, choose the state vector to be $x(t) = [e_{C1}(t) \ i_2(t)]^T$ and establish a state-variable model in the standard form of Eqn. 3.21, where the current source $I(t)$ is the input variable $u(t)$. Express all elements of the A, B, C, and D matrices in terms of the circuit parameters R_1, R_2, R_3, C_1, and L.

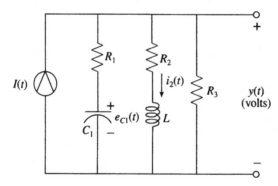

Fig. P3.1

Problem 3.2

For the circuit in Fig. P3.2, let the state variables be $x_1(t) = e_{C1}(t)$ and $x_2(t) = i_2(t)$. Obtain the state-variable model in the standard form of Eqn. 3.21 in which the applied voltage $e(t) = u(t)$. Express the elements of A, B, C, and D in terms of the circuit parameters.

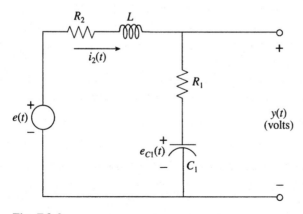

Fig. P3.2

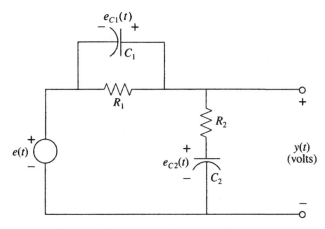

Fig. P3.3

Problem 3.3
Obtain the standard-form state-variable model for the circuit in Fig. P3.3, using the notation $x_1(t) = e_{C1}(t)$, $x_2(t) = e_{C2}(t)$, and $u(t) = e(t)$. Express the elements of A, B, C, and D in terms of R_1, R_2, C_1, and C_2.

Problem 3.4
The circuit in Fig. P3.4 incorporates a current-controlled current source (CCCS) that produces a current $I(t) = Ki_L(t)$ in the sense indicated. Derive a state-variable model for this circuit in which the voltage $u(t)$ is the input and the voltage $y(t)$ is the output. Identify your state variables in terms of the circuit variables, and identify each element of your A, B, C, and D matrices in terms of the circuit parameters R_1, R_2, C_1, L, and K.

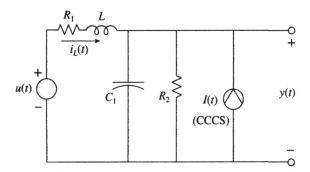

Fig. P3.4

Problem 3.5
For the passive circuit shown in Fig. P3.5, obtain a state-variable model in the standard form $\dot{x}(t) = Ax(t) + Bu(t)$ and $y(t) = Cx(t) + Du(t)$, where the state vector is taken as $x(t) = [e_1(t)\ e_2(t)\ i_3(t)]^T$. Specify each element in the four system matrices A, B, C, and D in terms of the circuit parameters R_1, R_2, R_3, C_1, C_2, and L.

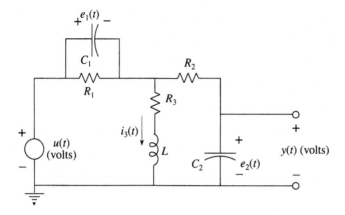

Fig. P3.5

Problem 3.6

Consider the circuit diagramed in Fig. P3.6.

(a) Derive a single differential equation with $e_C(t)$ as the dependent variable and $u(t)$ as the driving input function. Starting with this differential equation, identify a set of state variables and the state vector $x(t)$, and establish a vector-matrix differential equation in the standard format: $\dot{x}(t) = Ax(t) + Bu(t)$ and $y(t) = Cx(t) + Du(t)$, where $y(t) = e_C(t)$. Specify the element values in the four system matrices in terms of R_1, R_2, L, and C_1.

(b) Now derive a single differential equation with $i(t)$ as the dependent variable and $u(t)$ as the driving input function. Are you able to select a set of state variables so that this differential equation can be written in the vector-matrix form: $\dot{\hat{x}}(t) = \hat{A}\hat{x}(t) + \hat{B}u(t)$ and $\hat{y}(t) = \hat{C}\hat{x}(t) + \hat{D}u(t)$, where $\hat{y}(t) = i(t)$? If so, express the elements of your new system matrices $\hat{A}$, $\hat{B}$, $\hat{C}$, and $\hat{D}$ in terms of the parameters R_1, R_2, L, and C_1. If you are unable to find such a set of state variables, explain why.

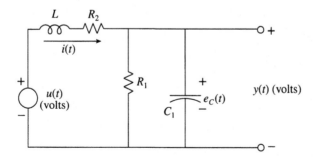

Fig. P3.6

Problem 3.7

A typical operational amplifier circuit is drawn schematically in Fig. P3.7. We assume that the amplifier has a very high gain. A state-variable model for this circuit is known to be:

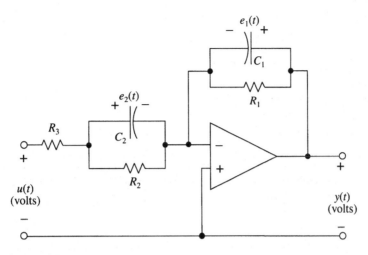

Fig. P3.7

$$\dot{x}(t) = \begin{bmatrix} -1 & a_{12} \\ a_{21} & -3 \end{bmatrix} x(t) + \begin{bmatrix} -2 \\ b_{21} \end{bmatrix} u(t) \quad \text{and}$$

$$y(t) = [c_{11} \ c_{12}] x(t) + [d_{11}] u(t), \quad \text{where}$$

$$x(t) = \begin{bmatrix} e_1(t) \\ e_2(t) \end{bmatrix}.$$

If $C_1 = C_2 = 1$ microfarad [μF], find the values for R_1, R_2, R_3, a_{12}, a_{21}, b_{21}, c_{11}, c_{12}, and d_{11}.

Problem 3.8

The operational amplifier circuit in Fig. P3.8 is useful in the design of automatic control systems. Show that the output voltage is related to the input voltage according to the integro-differential equation

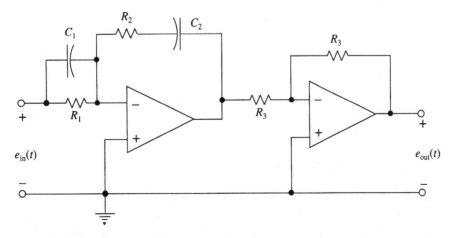

Fig. P3.8

$$e_{\text{out}}(t) = \underbrace{K_P e_{\text{in}}(t)}_{\substack{\text{proportional} \\ \text{term}}} + \underbrace{K_I \int e_{\text{in}}(t)\, dt}_{\substack{\text{integral} \\ \text{term}}} + \underbrace{K_D \frac{de_{\text{in}}(t)}{dt}}_{\substack{\text{derivative} \\ \text{term}}},$$

and evaluate K_P, K_I, and K_D in terms of R_1, R_2, C_1, and C_2. Why does R_3 not appear in the input–output relationship?

Problem 3.9

A schematic diagram of an operational amplifier circuit is shown in Fig. P3.9. The output resistance R_o of the amplifier is included, as is a load resistance R_L. The voltage gain of the amplifier itself is K. The two resistors R_1 and R_2 are external to the amplifier and are connected to obtain a circuit gain G:

CIRCUIT GAIN: $G = \dfrac{e_{\text{out}}(t)}{e_{\text{in}}(t)}.$

We neglect the dynamic properties of the amplifier, so G will be constant if all the circuit parameters remain constant. Calculate G as a function of the circuit parameters R_o, R_1, R_2, R_L, and K. You may assume that $|K| \ggg 1$ in your calculation, so that the voltage $\epsilon(t)$ will be virtually zero.

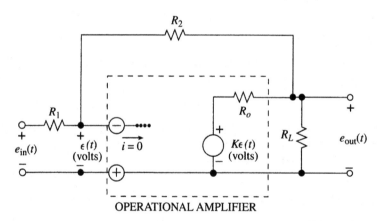

Fig. P3.9

Problem 3.10

Nominal values for the resistors in the operational amplifier circuit in Problem 3.9 are $R_1 = R_2 = R_L = 5000\ \Omega$ and $R_o = 100\ \Omega$. Use these nominal values in your formula for G, and calculate G for values of K ranging from -100 to -10^7. Plot $G(K)$ versus K, using a logarithmic axis for K.

Problem 3.11

The operational amplifier in Fig. P3.9 is very useful in mass-producing circuits having a precise gain G that remains virtually constant even if K changes due to aging of the circuit components or to fluctuations in the operating temperature, provided only that K remains very high in absolute value. Further, this precision can be obtained using nonprecision resistors for R_1 and R_L, provided that the resistor for R_2 has been precisely trimmed at the factory to yield the specified value for G. Let $K = -10^6$ and with the nominal values $R_o = 100\ \Omega$, $R_1 = 5000\ \Omega$, and $R_L = 5000\ \Omega$, find the range of values for R_2 for which the gain G will lie in the range of -9.9996 to -10.0004, so that we can guarantee the gain to be 10.000.

Problem 3.12

In the op-amp circuit of Fig. P3.12, $C_1 = 1\ \mu\text{F}$. Can you find positive values for R_1 and R_2 that will make the circuit behave according to $\dot{e}_{\text{out}}(t) + 50e_{\text{out}}(t) = Ke_{\text{in}}(t)$? If so, find those values and determine the coefficient K. If not, explain why not.

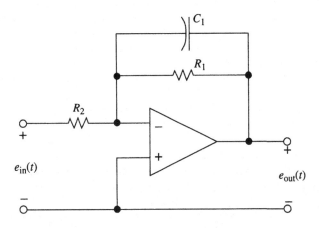

Fig. P3.12

Problem 3.13

Refer to the strain-gauge bridge circuit in Fig. 3.16. Let $E = 10$ V. If the strain-gauge element R_0 has a resistance of $350\ \Omega$, $R_1 = 492\ \Omega$, and $R_2 = 511\ \Omega$, then what must be the value of the balancing resistance R_{bal} in order that $|e_{\text{out}}|$ be less than $100\ \mu\text{V}$ when $\Delta R = 0$?

4 Mathematical Models of Electromechanical and Electrohydraulic Systems

4.1 Introduction

We now derive state-variable models for several types of actuators that are used in automatic control systems to drive massive loads that must be precisely positioned. These models are used in subsequent chapters, where they appear as vital components in automatic systems.

One of the most commonly used actuators is the DC motor, an ideal subject for our study because of its immediate practical application in many engineering design problems. It is the simplest device that requires the simultaneous application of Newton's laws of mechanics, Kirchhoff's laws of electric circuits (subjects fundamental to Chapters 2 and 3), and the first principles of electromechanical energy conversion manifested in Faraday's law and Ampere's rule (which are reviewed in this chapter).

We then continue our study with the analysis of another commonly used system, the electromagnet driven by an electronic amplifier. This combination of two basically nonlinear devices is used to drive loads which require substantial forces but which are intended to travel only small distances from an equilibrium position. Because only limited displacements are required, our linear approximations to the nonlinear force–displacement–current relationships of the amplifier–magnet combination yield a useful linear state-variable model for this system.

The chapter concludes with the analysis of an electrohydraulic actuating system that combines the amplifier–magnet system with an hydraulic servo-valve–ram system. This type of system is used in applications where high power (greater than several horsepower) is required in a small space and where the actuator is commanded by an electrical driving signal, in many cases delivered from a remote site. Because significant power amplification occurs within the valve–ram combination, this system shows some interesting dynamic phenomena that do not occur in simpler actuating systems such as the DC motor.

70

4.2 Fundamentals of the DC Machine

Consider the elementary machine illustrated in Fig. 4.1. Two parallel conducting rails separated by L m and level with respect to gravity lie in the plane of the paper. A uniform magnetic field of intensity B webers/(meter)^2 [Wb/m^2], directed normal to the plane of the conducting rails and having polarity such that the B vector is directed upward from the plane, exists between the rails but not outside them, as indicated by the single shading in the diagram. A perfect conductor slides without friction on the rails and remains normal to both rails. The conductor carries current $i(t)$ A in the direction indicated. Flexible wires having no resistance lie in the plane of the rails and are connected to the ends of the sliding conductor. The terminals of the wires lie close together. Ampere's rule for this configuration gives us

$$f(t) = BLi(t) \text{ N,} \tag{4.1}$$

where $f(t)$, the total force on the conductor, is directed parallel to the rails and to the right if the current has the direction shown. If the magnetic field polarity is reversed, or if the direction of the current is reversed, the force will be directed to the left. Note that this force does not depend on the voltage $e(t)$ or on the velocity $v(t)$. We also note that Ampere's rule relates the *mechanical* force $f(t)$ to the *electrical* current $i(t)$. This is the first of our two electromechanical principles required for modeling the DC machine.

The second of our electromechanical principles is Faraday's law, which, for the configuration given in Fig. 4.1, is

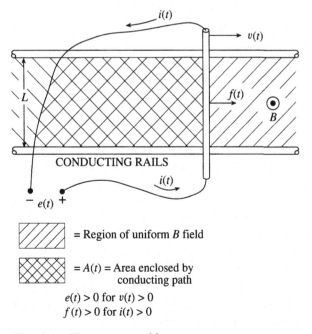

= Region of uniform B field

= $A(t)$ = Area enclosed by conducting path

$e(t) > 0$ for $v(t) > 0$
$f(t) > 0$ for $i(t) > 0$

Fig. 4.1 Elementary machine.

$$e(t) = \frac{d\Phi(t)}{dt} \text{ V.} \tag{4.2}$$

Here $\Phi(t)$ is the total magnetic flux encircled by the conducting path from one of the terminals through the flexible wires and the rigid conductor back to the other terminal. In this case, because the constant magnetic flux density B Wb/m^2 exists only between the rails, the total magnetic flux is

$$\Phi(t) = BA(t) \text{ Wb,} \tag{4.3}$$

where $A(t)$ is the area between the rails indicated by the double shading in Fig. 4.1. Combining Eqn. 4.3 with Eqn. 4.2, we have

$$e(t) = \frac{d[BA(t)]}{dt} = B\frac{dA(t)}{dt} \text{ V,} \tag{4.4}$$

since B is constant. But now $dA(t)/dt$ may be written as $Lv(t)$, so that Faraday's law for this configuration becomes

$$e(t) = BLv(t) \text{ V,} \tag{4.5}$$

where $v(t)$ is the velocity of the conductor along the rails in m/s, and $e(t)$ has the polarity indicated in Fig. 4.1.

We note three significant features of Faraday's law as it appears in Eqn. 4.5. First, the voltage appearing at the terminals $e(t)$ does not depend on the current $i(t)$, provided B is constant and independent of $i(t)$, and provided the current path has zero resistance. Second, Faraday's law, like Ampere's rule, relates a *mechanical* quantity $v(t)$ to an *electrical* quantity $e(t)$. Third, the constant $[BL]$ appears in both Faraday's law and Ampere's rule:

$$\begin{aligned} \text{AMPERE'S RULE:} \quad & f(t) = [BL]i(t) \text{ N,} \\ \text{FARADAY'S LAW:} \quad & e(t) = [BL]v(t) \text{ V.} \end{aligned} \tag{4.6}$$

We see that the units of this *machine constant* $[BL]$ can be designated either as N/A or as V/(m/s). The reader should verify that these two physical unit designations for the machine constant are equivalent. The fundamental physical basis for electromechanical energy conversion is expressed in Eqn. 4.6. The remainder of our modeling of the DC machine involves only the familiar principles of electric circuit analysis and mechanics, plus some details on construction of the machine.

All of the different design configurations for DC machines – each having its own advantages in performance, weight, size, or cost – depend upon the physical principles of Eqn. 4.6. Furthermore, most of these configurations lead the control engineer to the same mathematical model for the dynamic relationships among the terminal variables of the machine: applied voltage, input current, torque developed, acceleration, velocity, and position of the output shaft. Accordingly, we now treat a very common type of DC machine, the separately excited, armature-controlled, commutator machine.

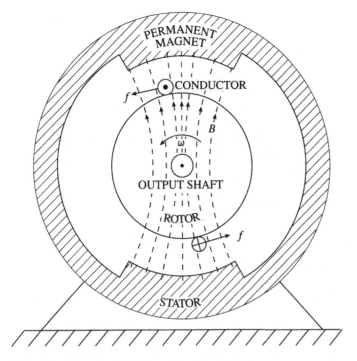

Fig. 4.2 Elements of DC machine (end view).

Figure 4.2 shows an elementary sketch of the principal features of such a machine. This end view shows a cylindrical rotor supported by the output shaft bearings and rotating with angular velocity $\omega(t)$ rad/s in a positive (counterclockwise) sense. The rotor has a depth (into the page) of L m. A conductor, carrying current $i(t)$ A and mounted lengthwise on the rotor, is wound around the rotor end so that two conducting paths on opposite sides of the rotor, each of length L, carry the same current $i(t)$. The conductor carries the current "out of the page" on its upper path and "into the page" on the lower path. The current is conducted out of the machine through sliding contacts (brushes) between the rotor and the stator. The stator, or stationary member of the machine, is a permanent magnet (or an electromagnet energized separately from the rotor circuit) supporting a magnetic field that we will assume has a uniform intensity B Wb/m^2 in the vicinity of the conductors.

Focusing attention on the upper conductor, we see that the current-magnetic field configuration matches exactly that shown in Fig. 4.1. Therefore, the force on the upper conductor has the direction indicated and a magnitude given by Ampere's rule, Eqn. 4.1. Similarly, on the lower conductor, we have a force of equal intensity in the direction shown. Note that both of these forces combine to produce a torque on the rotor in the counterclockwise sense. This torque has a magnitude of $2 \times f(t) \times r$ N-m, where r is the radius of the conductor coil. In most machines the conductor coil consists of many turns, rather than the single

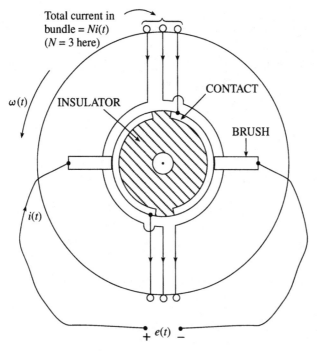

Fig. 4.3   Schematic of DC machine.

turn we have assumed here, with the current $i(t)$ in each turn. A sketch of the rotor with three turns is given in Fig. 4.3. Since the total force developed on the bundle of N turns is N times that for a single turn, the resulting torque on the rotor is

$$T(t) = [2NBLr]i(t) \text{ N-m.} \tag{4.7}$$

We consider torque to be positive in the counterclockwise sense.

The magnetic flux $\Phi(t)$ enclosed by the conducting circuit changes as the rotor turns counterclockwise. The term $\Phi(t)$ can be approximated from a vertical projection of the rectangular coil onto a horizontal plane, which gives the equivalent area containing the flux $A(t)$. In accordance with Faraday's law, the voltage at the terminals $e(t)$ changes in synchronism with dA/dt, not with $A(t)$. When the plane of the coil is aligned with the B vector, $A(t)$ is zero but dA/dt is at its largest value, hence $e(t)$ is at its maximum. When the + voltage terminal is connected to the lower conductor bundle and the B vector is directed upward, as shown in Fig. 4.3, $e(t)$ will be at its maximum positive value, $2NBL \times v(t)$, where $v(t)$ is the velocity of the conductor bundle. The kinematic relationship between $v(t)$ and the angular velocity of the output shaft is

$$\omega(t) = \frac{v(t)}{r} \text{ rad/s.} \tag{4.8}$$

Therefore we have

$$e(t) = [2NBLr]\omega(t) \text{ V.} \tag{4.9}$$

We note that the constant $2NBLr$ appears in the torque equation as well as in the voltage equation, which we might expect from our analysis of the elementary machine as summarized in Eqn. 4.6. We designate this as the *machine constant K*. For our DC machine we thus have the following fundamental relationships, coming from Ampere's rule and Faraday's law:

$$T(t) = Ki(t) \text{ N-m} \quad \text{and}$$
$$e(t) = K\omega(t) \text{ V}, \quad \text{where} \tag{4.10}$$
$$K = 2NBLr \text{ N-m/A or V/(rad/s).}$$

The machine constant K is called the *torque constant* or the *back-voltage constant*; however, if units other than the SI system are used, it is difficult to recognize that these two terms refer to the same constant. Laboratory measurements of torque, voltage, and current on a small DC machine might yield a torque constant of 14 oz-in/A and a back-voltage constant of 10.35 V/(thousand rpm). The two numbers 14 and 10.35, each bearing seemingly different units, suggest that the two constants are unrelated physically. However, if the measurements were expressed in the SI system then both constants would have the same numerical value, 0.09886 N-m/A (or V/(rad/s)).

In Fig. 4.2 we see that the magnetic field is very much stronger at the top and at the bottom of the stator where the air gap between stator and rotor is narrowest. Because the field is weak at the right and left parts of the stator, the current will produce negligible torque on the rotor when the conducting bundles are in this position. This is one reason why the simple configuration in Fig. 4.2 is not a practical one for an efficient motor design, even though it clearly illustrates the principles of operation of real machines.

Real machines are designed so that current-carrying bundles are immersed in strong magnetic fields at all positions of the rotor. This can be accomplished with the stator configuration of Fig. 4.2 if we include several additional coils spaced around the periphery of the rotor, each coil having its own isolated set of sliding contacts. Only the coil in the intense part of the field will be conducting the current from the external drive source, all the other coils being temporarily disconnected. The process of switching coils, called *commutation,* is done either mechanically by the brush–commutator arrangement in conventional machines, or electronically in brushless machines. Each coil in its turn conducts the current for a brief period of the cycle as the rotor revolves. Owing to this feature of the machine's construction, the voltage–speed relationship in Eqn. 4.9 is valid for all positions of the rotor. This scheme has many further refinements in electric machine design that are beyond the scope of our study here.

We now proceed to derive a very useful dynamic model for the DC machine, focusing on its terminal characteristics rather than on the electromechanical properties of its internal features. We use the *circuit-model* approach, which represents the internal electromechanical behavior of the machine as an electric circuit, rather than an approach involving a study of the electromagnetic fields within the machine. The circuit approach is much simpler than the field-theory

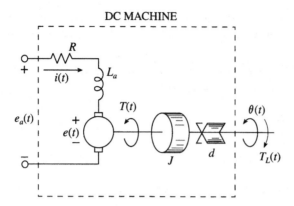

Fig. 4.4   Circuit model of DC machine.

approach. It gives remarkably accurate results for our purpose, which is to establish the dynamic relationships among the terminal, or external, variables of the machine. We can establish the numerical values for our circuit parameters by simple laboratory measurements of the terminal variables on the machine itself – voltage, current, torque, speed – without resorting to measurements or calculations on the internal variables, such as magnetic field distributions. Such internal relationships are important to machine designers because they affect the efficiency and other performance factors of the machine. As control system engineers we must be aware of those design problems, but we can normally assume that the machine will function properly if we restrict the terminal variables to the prescribed limits of operation established by the machine designers.

Figure 4.4 is the circuit model for a DC machine having a constant magnetic field provided either by a permanent magnet structure (typical in small machines) or by an electromagnet energized by a power source separate from that which supplies the armature voltage and current. The magnetic field is not shown in the figure, where R is the resistance of the circuit in which the armature current $i(t)$ circulates. This includes the resistance of the wires in the coil plus the contact resistance of the brushes wiping the sliding contacts. We assume R to be constant when the rotor is turning. The term L_a denotes the *self-inductance* of the armature circuit, included here because we are dealing with a coil of wire. The voltage $e(t)$ is that expected from Faraday's law. For the electrical side of the machine we can write

$$e_a(t) - e(t) = L_a \frac{di(t)}{dt} + Ri(t) \text{ V,} \tag{4.11}$$

where $e_a(t)$ is the control voltage applied at the terminals of the machine. From Eqn. 4.10, $e(t)$ is given as

$$e(t) = K\omega(t) = K \frac{d\theta(t)}{dt}, \tag{4.12}$$

where K is the machine constant and $\theta(t)$ rad is the angular position of the output shaft with respect to a given reference. Equation 4.11 can now be written as

$$e_a(t) = L_a \frac{di(t)}{dt} + Ri(t) + K \frac{d\theta(t)}{dt} \text{ V.} \tag{4.13}$$

The torque developed on the shaft by the current–magnetic field interaction (Ampere's rule) is $T(t) = Ki(t)$ (Eqn. 4.10). Euler's law applied to the output shaft gives us

$$\left[\begin{array}{c} \text{total torque on the shaft} \\ \text{in the } +\theta \text{ direction} \end{array}\right] = J \frac{d^2\theta(t)}{dt^2} \text{ N-m,}$$

$$[Ki(t) - T_L(t) - d(\omega(t))] = J \frac{d^2\theta(t)}{dt^2}. \tag{4.14}$$

Here $T_L(t)$, the *load torque,* is the reaction torque on the shaft (in the $-\theta$ sense) due to a load driven by the shaft. $T_L(t)$ can be a complicated function of the shaft motion, the nature of the load, and environmental factors. $d(\omega(t))$, the velocity-dependent torque on the shaft, is in general a nonlinear function of $\omega(t)$. J kg-m^2 is the moment of inertia of the machine shaft and all massive parts that are rigidly attached to it, excluding those of the load. We introduce simplified notation and re-arrange Eqns. 4.13 and 4.14 to yield

$$L_a \frac{di(t)}{dt} + Ri(t) + K\omega(t) = e_a(t),$$

$$Ki(t) - d(\omega(t)) - J \frac{d\omega(t)}{dt} = T_L(t). \tag{4.15}$$

Let us now regard $e_a(t)$ and $T_L(t)$ as control inputs to the machine that we can vary at will in a laboratory setting. $i(t)$ is a physical quantity that "goes into" the machine in the electrical sense, but here we regard it as a response rather than as an input, because if we manipulate both inputs $e_a(t)$ and $T_L(t)$ arbitrarily then $i(t)$ and $\omega(t)$ will respond as governed by Eqn. 4.15. We can now see that the dynamic behavior of the machine in response to the inputs $e_a(t)$ and $T_L(t)$ will be determined by five characteristics: the four constant coefficients K, R, L_a, and J, plus the torque function $d(\omega(t))$. We determine the numerical values for these parameters by performing measurements using a dynamometer, a laboratory fixture upon which we mount the machine so that we can vary our control inputs $e_a(t)$ and $T_L(t)$ in order to measure and record them. At the same time, we will measure and record both $i(t)$ and $\omega(t)$.

We first lock the rotor tight, so that $\omega(t) = 0$ during the test we perform to determine the electrical parameters L_a and R. Equation 4.15 reduces to

$$L_a \frac{di(t)}{dt} + Ri(t) = e_a(t),$$

LOCKED-ROTOR TEST: $\tag{4.16}$

$$Ki(t) = T_L(t).$$

We then apply a known test voltage – for example, a step input of E V – and carefully measure the dynamic response of the current $i(t)$. This response is the solution to the first-order differential equation in Eqn. 4.16, which for a constant input voltage E and for an initial current of zero is

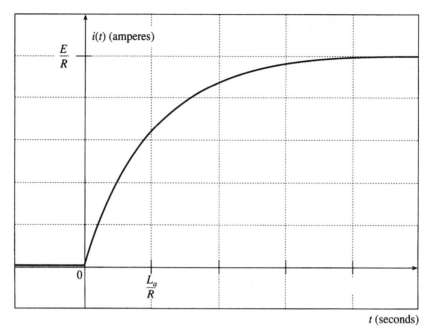

Fig. 4.5 $i(t)$ in locked-rotor test.

$$i(t) = \frac{E}{R}[1 - \epsilon^{-(R/L_a)t}] \text{ A},\tag{4.17}$$

a record of which would appear as in Fig. 4.5. Since we know the value of E, we determine R simply by reading the steady-state value of the current:

$$R = \frac{E}{i_{SS}} \ \Omega.\tag{4.18}$$

Since the rotor is locked, we need not measure $T_L(t)$ during this test. Once R is known, the inductance L_a is determined by measuring the *time constant* L_a/R of the dynamic response:

$$L_a = R \times [\text{time constant}] \text{ H}.\tag{4.19}$$

The locked-rotor test should be repeated several times, using a range of values for E and locking the rotor in several different positions to observe the effect of the brush contact resistance. The power dissipated in the machine during the locked-rotor test is $E \times i(t)$ watts [W]. Both E and $i(t)$ must be monitored to avoid overheating the machine during these tests.

The machine constant K may be determined by another simple test in which the rotor turns under a load provided by $T_L(t)$. In this test we make only steady-state measurements. Because both the voltage and the load torque are held constant, the current and the speed will also remain constant. From Eqn. 4.15 we have, for $di/dt = 0$,

$$i = \frac{e_a}{R} - \frac{K}{R}\omega, \quad \text{where } i, e_a, \text{ and } \omega \text{ are constant.}\tag{4.20}$$

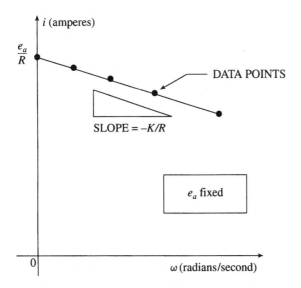

Fig. 4.6 Determining K from data for i versus ω.

We repeat the test several times, each time with a different load torque but with the same voltage applied. A plot of the current versus speed obtained in this series of tests would appear as in Fig. 4.6. K is determined from the slope of this plot once R has been established from the locked-rotor tests. We note that it is unnecessary to actually measure the load torque in these tests; it is required only to hold the torque constant at a convenient level while the current and speed are measured.

The nature of the torque term $d(\omega(t))$ is also easily determined using steady-state measurements of current and shaft speed. In these tests the shaft is released completely ($T_L = 0$), and several measurements of i and ω are made by varying e_a. Equation 4.15, for $d\omega/dt = 0$ and $T_L = 0$, reveals that

$$Ki = d(\omega). \tag{4.21}$$

A plot of the $i - \omega$ data from these measurements will typically look like that in Fig. 4.7. The nature of this data near the origin of the graph is significant. The static friction on the shaft Ki_0 N-m is due to the bearing friction and to the contact force of the brushes against the commutator, and therefore the current is not zero at $\omega = 0$. At very low speeds, ω_1 for example, the running friction is less than the static friction, and at higher speeds the windage losses and viscous friction losses in the bearings add to the torque required to maintain the higher speeds. The torque-reversal characteristic of $d(\omega)$ near the origin is not encountered in machines used in *speed control* systems, where the shaft speed is regulated around an operating value greater than zero so that the shaft always turns in the same direction. However, in a *position control* system, where the position of the shaft $\theta(t)$ (and not the speed $\omega(t)$) is regulated around an operating value, the speed will be operating in the vicinity of zero most of the time. Since the shaft will be rotating clockwise as often as it rotates counterclockwise, the speed will frequently pass through zero.

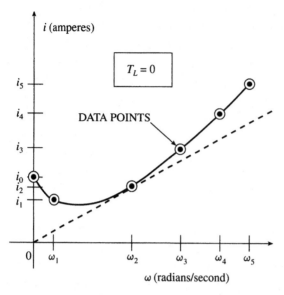

Fig. 4.7   Measurements to determine $d(\omega)$.

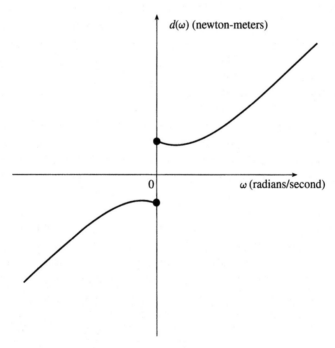

Fig. 4.8 $d(\omega)$ versus ω.

Figure 4.8 shows the $d(\omega)$ function for both directions of shaft rotation. The static friction on the shaft can cause undesirable behavior in the control performance. For this reason DC machines intended for servomechanism applications (shaft *position* control) are designed with special low-friction bearings and with special attention to the brush friction characteristics. Brushless motors, where the commutation is done electronically, have been developed to relieve several problems of mechanical commutation, including the static friction problem.

The moment of inertia of the rotor J can be determined in several ways.

1. The manufacturer provides a value, which is determined at the factory but which obviously cannot account for any attachments the user might wish to make. Also, many manufacturers prefer to express J not in units of (mass) $\times$ (length)^2 but in units of (torque)/(angular acceleration). An example which requires care in converting to a consistent set of units is: $J = 2.97 \times 10^{-3}$ in-oz/ (rpm/s), which is approximately 2×10^{-4} kg-m^2.

2. If the dimensions and material mass density of the rotor are known, J can be calculated. This usually requires approximations to the geometry of the rotor in order to simplify the calculations, but the result can be used as an approximation to the value that is more accurately determined by dynamic measurements.

3. A torsional spring of known compliance might be attached to the shaft in order to make the rotor into a torsional pendulum. Relieving the brush contact force and measuring the oscillations of the pendulum, assuming they are very lightly damped, permits the moment of inertia to be calculated as

$$J = \frac{k}{(\omega_{\mathrm{osc}})^2} = \frac{k\tau^2}{4\pi^2} \text{ kg-m}^2, \tag{4.22}$$

where k is the spring compliance [N-m/rad], ω_{osc} is the frequency of the oscillations [rad/s], and τ is the period of the oscillations [s].

4. Other dynamic measurements suggested by Eqn. 4.15, can yield a numerical value for J.

The control of the shaft position or of the velocity of a DC machine in response to a command signal is effected by varying the terminal voltage $e_a(t)$. The normal procedure is to drive the motor with an electronic amplifier having the capability to deliver the required $e_a(t)$ in response to a low-level voltage at the input of the amplifier. In a position control application, the amplifier must be capable of supplying both positive and negative $e_a(t)$. This requirement demands careful electronic design to achieve a smooth transition between the positive and negative regions of operation without a *dead zone* in the static $e_a \Leftrightarrow e$ relation in the transition region. This type of servo amplifier is sometimes called a *four-quadrant* amplifier. (In unidirectional speed control applications, the amplifier need not reverse the sign of $e_a(t)$; it merely modulates that voltage in accordance with changes in the load torque and in the set-point speed.) The amplifier must also be capable of supplying the instantaneous power $[e_a(t) \times i(t)]$ W required to move the load.

Figure 4.9 is a circuit diagram for the amplifier–motor combination, with the amplifier represented by its "equivalent circuit" characterized by the two parameters K_A and R_o. The possibility of a "loading" problem is evident from this diagram. If the amplifier is tested and qualified for its $e(t)$-to-$e_a(t)$ performance

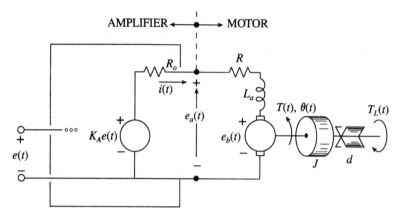

Fig. 4.9   Circuit diagram for amplifier–motor.

with a load different from that of the motor (which can happen if the amplifier and motor are manufactured in different factories), the amplifier may not qualify when it is loaded by the motor. This is a common problem in control system engineering, when two components – manufactured by two different companies, either or both of whom may have misinterpreted the interface specifications – are brought together to perform a system function.

The model for the amplifier–motor combination is easily obtained by substituting $K_A e(t)$ for $e_a(t)$ and $(R + R_o)$ for R in Eqn. 4.15:

$$L_a \frac{di(t)}{dt} + (R + R_o)i(t) + K\omega(t) = K_A e(t),$$
$$J \frac{d\omega(t)}{dt} + d(\omega(t)) - Ki(t) = -T_L(t). \tag{4.23}$$

We now establish a state-variable model for the amplifer–motor combination by defining the state vector $x(t)$ and the input vector $u(t)$ as follows:

$$x(t) = \begin{bmatrix} x_1(t) \\ x_2(t) \\ x_3(t) \end{bmatrix} = \begin{bmatrix} i(t) \text{ A} \\ \theta(t) \text{ rad} \\ \omega(t) \text{ rad/s} \end{bmatrix}, \quad u(t) = \begin{bmatrix} u_1(t) \\ u_2(t) \end{bmatrix} = \begin{bmatrix} e(t) \text{ V} \\ T_L(t) \text{ N-m} \end{bmatrix}. \tag{4.24}$$

Equation 4.23, when re-arranged and augmented by the relationship $d\theta(t)/dt = \omega(t)$, gives the state-variable form

$$\begin{bmatrix} \dot{x}_1(t) \\ \dot{x}_2(t) \\ \dot{x}_3(t) \end{bmatrix} = \begin{bmatrix} -((R+R_o)/L_a)x_1(t) - (K/L_a)x_3(t) \\ x_3(t) \\ (K/J)x_1(t) - (1/J)d(x_3(t)) \end{bmatrix}$$
$$+ \begin{bmatrix} (K_A/L_a) & 0 \\ 0 & 0 \\ 0 & -(1/J) \end{bmatrix} \begin{bmatrix} u_1(t) \\ u_2(t) \end{bmatrix}. \tag{4.25}$$

The interaction between the amplifier and motor components is evidenced in this state-variable model by the two matrix elements (K_A/L_a) and $-((R+R_o)/L_a)$, each of which is dependent on physical properties of *both* the amplifier and the motor.

Equation 4.25 is nonlinear owing to the $d(x_3(t))$ term in Fig. 4.8. The term $d(x_3(t))$ is often approximated by a linear function – a straight line through the origin – with the following justification. Servomotors, which are machines intended for position control applications, are designed with low friction bearings and with attention given to minimizing brush friction. In many operating environments the servomotor is subject to vibration, which also has the effect of reducing static friction. Therefore, the static friction level displayed in Fig. 4.8 will be small enough to be ignored for some purposes. Also, in most servomotors the portion of the curve away from the origin is nearly a straight line, at least in the moderate speed range. Hence the model we obtain when we approximate $d(x_3(t))$ by a linear function produces analytical results that predict experimental behavior to a satisfactory degree. Nevertheless, in some other important cases the presence of a small amount of static friction, not included in the linear model, yields experimental dynamic behavior that degrades the system performance. In these cases a nonlinear simulation of the friction characteristic may be used to predict experimental dynamic behavior, but the nonlinear simulator is less amenable to analysis than is the linear model.

The state-variable model of Eqn. 4.25 is made linear by the approximation

$$d(\omega(t)) \cong b\omega(t), \tag{4.26}$$

where b, the *viscous damping coefficient* [N-m/(rad/s)], is the slope of the straight line approximation to the torque–speed curve of Fig. 4.8. Given this approximation, the model of Eqn. 4.25 takes on the standard linear state-variable form:

$$\dot{x}(t) = Ax(t) + Bu(t), \quad \text{where}$$

$$A = \underbrace{\begin{bmatrix} -((R+R_o)/L_a) & 0 & -(K/L_a) \\ 0 & 0 & 1 \\ (K/J) & 0 & -(b/J) \end{bmatrix}}_{3 \times 3} \text{ and } B = \underbrace{\begin{bmatrix} (K_A/L_a) & 0 \\ 0 & 0 \\ 0 & -(1/J) \end{bmatrix}}_{3 \times 2}. \tag{4.27}$$

We will make further use of this linear model of the DC machine in the chapters to follow.

4.3 Electromagnets

An electromagnet is frequently employed where it is necessary to provide a mechanical force depending on an electric current. Figure 4.10 shows the elements of such a device. A magnetic structure supports a circuit of magnetic flux driven by the coil of N turns carrying current i A. Part of the magnetic circuit is a movable armature that slides smoothly on the support member. An air gap (or possibly a vacuum gap) of length g m is also in the magnetic circuit, as shown. We assume that g is small enough for the magnetic flux density to be

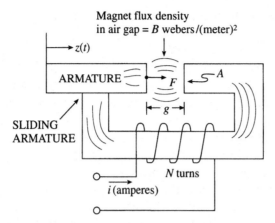

Magnet flux density
in air gap = B webers/(meter)^2

ARMATURE

F

A

g

SLIDING
ARMATURE

N turns

i (amperes)

$z(t)$

Fig. 4.10 Elements of an electromagnet.

essentially constant at B Wb/m^2 across the face of the armature. The cross-sectional area of the armature face is A m^2. A tractive force F N developed on the face of the armature is related to the area and field strength as

$$F = \frac{AB^2}{2\mu_0},\tag{4.28}$$

where μ_0 is the magnetic permeability of the air gap having a value of $4\pi \times 10^{-7}$ henries/meter [H/m]. The magnetic field intensity is proportional to the magnetomotive force Ni ampere turns. For an efficient magnetic structure we have

$$B = \frac{\mu_0}{g} Ni,\tag{4.29}$$

which combined with Eqn. 4.28 gives the useful relationship

$$F = \left[\frac{A\mu_0 N^2}{2}\right]\left(\frac{i^2}{g^2}\right) \text{N}.\tag{4.30}$$

We note that F is positive when i is either positive or negative. Equation 4.30 is valid only for combinations of N, i, and g for which B is given in Eqn. 4.29. The maximum value for B, which can be realized using modern magnetic materials in normal environments, is approximately 1 Wb/m^2.

Tractive electromagnets patterned after the scheme depicted in Fig. 4.10 are used in small current-to-force transducers of many types. In most such transducers it is necessary to apply the force in both the positive and the negative directions. In this case, the device is constructed with two air gaps having the corresponding tractive forces opposed to one another, as shown in Fig. 4.11. The force on the left side of the sliding member is, from Eqn. 4.30,

$$F_L = K_M\left[\frac{i_L^2}{g_L^2}\right], \quad \text{where } K_M = \frac{A\mu_0 N^2}{2},\tag{4.31}$$

and the force on the right side is

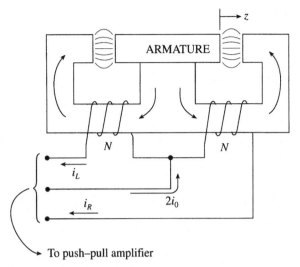

Fig. 4.11 Push–pull electromagnet.

$$F_R = K_M \left[\frac{i_R^2}{g_R^2} \right]. \tag{4.32}$$

Therefore, the net force on the armature in the positive z direction is

$$F = K_M \left[\frac{g_L^2 i_R^2 - g_R^2 i_L^2}{g_R^2 g_L^2} \right]. \tag{4.33}$$

Let $z(t)$ be the displacement of the armature to the right of the position in which both air gaps are equal. We then have

$$g_R = g_0 - z(t) \quad \text{and} \quad g_L = g_0 + z(t) \tag{4.34}$$

where g_0 is the air-gap lengths with the armature centered. The right and left coil currents are supplied by a *push–pull amplifier* having an equivalent circuit shown in Fig. 4.12. Neglecting the dynamic effects of coil inductance for the moment, we calculate $i_R(t)$ and $i_L(t)$ to be

$$i_L(t) = \frac{E - K_A e_{\text{in}}(t)}{R_o + R_c} = i_0 - \Delta i(t),$$

$$\tag{4.35}$$

$$i_R(t) = \frac{E + K_A e_{\text{in}}(t)}{R_o + R_c} = i_0 + \Delta i(t).$$

The purpose of the amplifier is to provide the bias current i_0 and the differential current $\Delta i(t)$, which is proportional to the time-varying voltage at the input to the amplifier $e_{\text{in}}(t)$. We note that the circuit shown in Fig. 4.12 is valid only for values of $\Delta i(t)$ smaller in magnitude than i_0, because the amplifier can supply only unidirectional currents to the magnet coils. We now express the net force, Eqn. 4.33, in terms of g_0, $z(t)$, i_0, and $\Delta i(t)$:

$$F = 4 K_M \left[\frac{(g_0 \Delta i(t) + i_0 z(t))(g_0 i_0 + z(t) \Delta i(t))}{(g_0^2 - z^2(t))^2} \right]. \tag{4.36}$$

PUSH–PULL AMPLIFIER

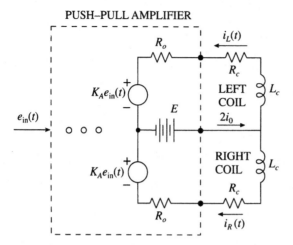

Fig. 4.12 Equivalent circuit for amplifier–electromagnet system.

It is interesting to determine the static characteristics of Eqn. 4.36, that is, to observe how F depends on z and Δi when these are constant. Let us choose the following numerical values for the physical parameters:

$$A = 2 \times 10^{-4} \text{ m}^2, \quad \mu_0 = 4\pi \times 10^{-7} \text{ H/m}, \quad N = 300 \text{ turns},$$

$$i_0 = 2 \text{ A}, \quad g_0 = 0.005 \text{ m}, \quad K_M = 1.131 \times 10^{-5} \text{ N-m}^2/\text{A}^2.$$

We now plot F versus z, with Δi as a parameter, in Fig. 4.13. Similarly, we plot F versus Δi, with z as a parameter, in Fig. 4.14.

First, we note that although $F(z, \Delta i)$ is a nonlinear function, it is approximately linear for values of Δi that are much smaller than i_0 and for values of z that are much smaller than g_0. Under these restrictions, Eqn. 4.36 becomes

$$F \cong \frac{4K_M i_0}{g_0^3}[g_0 \Delta i + i_0 z]. \tag{4.37}$$

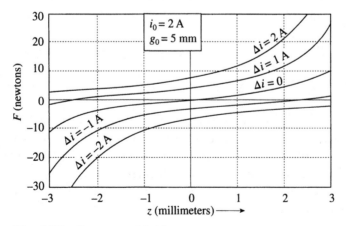

Fig. 4.13 F versus z with Δi as a parameter.

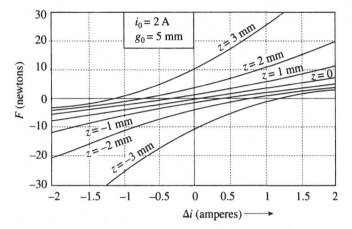

Fig. 4.14 F versus Δi with z as a parameter.

Since the character of $F(z, \Delta i)$ in the region near $(0, 0)$ is quite important in the stability studies to be undertaken in later chapters, the approximation given in Eqn. 4.37 will be very useful.

Second, we note that the F-versus-z curve indicates that, for z positive (for displacements of the armature toward the right), F is also positive. This characterizes a condition of *static instability,* which means that the force developed on the armature due to a small displacement from the equilibrium position is in the direction to *increase* that small displacement. This exceedingly important property of the electromagnet system requires careful engineering in applications where such a device is used as an actuator. The amplifier–magnet system can be rendered statically stable by the addition to the armature of a spring restraint of strength sufficient to reverse the composite force–displacement character. Alternately, feedback control can be used to stabilize this inherently unstable device.

The air gaps in this example range only from 5 millimeters [mm] (for $z = 0$) to 2 mm (for $z = 3$ mm). This is a representative range for many applications of electromagnetic actuators in small devices where the purpose of the device is to keep the controlled element (the armature) centered in the presence of disturbing forces on the element, or in devices where only small displacements of the element are required, as in high-performance hydraulic servovalves.

A linear state-variable model for the amplifier–magnet system can be obtained by selecting the input variable to be $e_{\text{in}}(t)$ and selecting the state vector as

$$x(t) = \begin{bmatrix} x_1(t) \\ x_2(t) \end{bmatrix} = \begin{bmatrix} z(t) \\ \dot{z}(t) \end{bmatrix}. \tag{4.38}$$

We neglect the inductance of the coils so that $\Delta i(t)$ is proportional to $e_{\text{in}}(t)$. Let the mass of the armature be m kg, and let the viscous friction on the sliding surface have a coefficient d N/(m/s). The linear vector-matrix differential equation is

$$\dot{x}(t) = \begin{bmatrix} a_{11} & a_{12} \\ a_{21} & a_{22} \end{bmatrix} x(t) + \begin{bmatrix} b_{11} \\ b_{21} \end{bmatrix} e_{\text{in}}(t), \quad \text{where:}$$

$$a_{11} = 0, \quad a_{12} = 1, \quad b_{11} = 0, \tag{4.39}$$

$$a_{21} = \left[\frac{4K_M i_0^2}{mg_0^3} \right], \quad a_{22} = -\left[\frac{d}{m} \right], \quad b_{21} = \left[\frac{4K_M K_A i_0}{m(R_o + R_c)g_0^2} \right].$$

This model is used in the electrohydraulic valve-and-ram system described in the next section.

4.4 Electrohydraulic Amplifier, Valve, and Ram System

Hydraulic actuators in the multihorsepower class can be made quite small compared to electric motors of the same power. It is usually necessary to drive these actuators from electrical signals, so there must be an electrohydraulic device to convert the low-powered signals to high-powered hydraulic fluid flow. Figure 4.15 is a schematic sketch of such a device, the electrohydraulic servovalve. The valve depicted here is a single-stage spool valve. It is typical of early developments in the electrohydraulic servomechanism field, and it has been largely superseded by other types of valves such as the two-stage valve using a nozzle and flapper as the first stage. Nevertheless, this simple design exhibits the basic principles, inherent in most types of servovalves, of conversion from electromagnetic to hydraulic energy.

The valve spool is actuated by an electromagnet operating on the principles developed in Section 4.3. A push–pull electric power amplifier delivers a

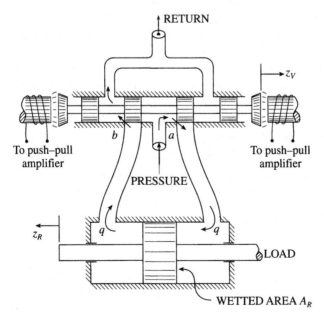

Fig. 4.15 Schematic of servovalve and ram.

differential current to the two coils of the magnet, which in turn provides a net force on the spool. If the current in the right-hand coil is higher than that in the left-hand coil, then the net force will be directed to the right, as shown in Eqn. 4.36 in which $z(t)$ is denoted as $z_V(t)$ (Fig. 4.15). As the spool is displaced to the right, the metering port a is uncovered, admitting high-pressure hydraulic fluid from the center chamber of the spool into the line leading to the main power element – in this case, a piston-driven ram. The spool and its corresponding cylinder and metering ports are precisely machined to minimize leakage through the ports when the valve is in the centered (closed) position, and to provide the desired porting area as a function of valve displacement $z_V(t)$. In most valves the displacement rarely exceeds 1 mm. The hydraulic fluid flow, denoted as $q(t)$ m³/s, displaces the ram piston toward the left, and the same flow exits the ram cylinder through metering port b, to the low-pressure side of the main pump. This assumes that the fluid is incompressible, which is nearly the case for hydraulic oils but which is not the case for gases. The ram piston and cylinder are also designed and machined to minimize friction and leakage of fluid from one side of the piston to the other. Thus the ram velocity is proportional to the fluid flow:

$$\dot{z}_R(t) = \frac{1}{A_R} q(t), \tag{4.40}$$

where A_R is the wetted area of the ram piston. When it is desired to move the ram toward the right, the left-hand magnet coil will have the greater current. The spool will be displaced to the left, and the flow will be metered across the opposite sides of the spool lands. The sign of $e_{in}(t)$ determines which coil gets the higher current, with e_{in} positive giving more current to the right coil and e_{in} negative giving more to the left coil.

We use Eqn. 4.39 to relate the control voltage $e_{in}(t)$ to the electromagnetic force developed on the spool. At the center position of the spool, $z_V(t) = 0$, we have

$$F(t) = \frac{4K_M K_A i_0}{(R_o + R_c)g_0^2} e_{in}(t), \tag{4.41}$$

which neglects the dynamic influence of the inductance of the magnet coils.

Several other forces on the valve spool must be considered in a dynamic analysis of the valve motion. First we consider those forces that depend on the valve displacement $z_V(t)$. These include electromagnetic forces, forces provided by mechanical springs, and forces caused by fluid flow within the valve. We saw in Section 4.3 that the effect of a change in air gap (for small displacements of the spool from its center position) is an increase in the force proportional to the air-gap change. Mechanical centering springs (not shown in Fig. 4.15) are frequently attached to the spool; these provide a force which is proportional to the displacement but opposite to that due to the changing air gap. As the metering ports open, a region is created in the vicinity of the ports in which the fluid velocity is high. This causes a reaction force on the spool that varies with the valve displacement. In some cases this variation is almost proportional to the displacement, giving it the same effect as a spring restraint on the spool. However, in some spool-and-orifice designs the variation is nonlinear, and might even reverse sign for large spool displacements. In addition to this orifice reaction force, some spool designs are such that flow forces exist that depend on spool displacement due to the configuration of the valve chambers between the lands. For small displacements of the valve spool, the combined influence of the electromagnetic force, the mechanical spring restraint force, and the flow forces on the spools can usually be represented by the linear relation

$$f_{\text{elastic}}(t) = k_e z_V(t); \tag{4.42}$$

k_e can be positive or negative, depending upon the design of the electromagnet, the centering springs, and the valve spool. It is therefore possible that these elastic forces (forces proportional to displacement) can have the effect opposite to that of a passive mechanical spring. This possibility exists because the magnet and valve are energy-conversion, or *active,* devices. In our later study of dynamic stability we will see the considerable advantage of having the total elastic force act as a passive spring.

A second class of forces on the valve spool depends on the velocity of the spool. The origins of these forces lie both in the viscous friction drag on the spool as it moves in the hydraulic fluid and in a flow force that depends on the design of the center and end chambers between lands. This flow force can be in either the same or the opposite direction as that of the viscous force. Normally, the valve is designed for the total velocity-dependent force to appear as a passive viscous damping force, again for dynamic stability. For small velocities, this force can also be approximately represented by the linear relation

$$f_{\text{viscous}} = k_v \dot{z}_V(t). \tag{4.43}$$

Finally, the friction force on the spool is represented here as a constant, f_{fric}.

The dynamic equation for the motion of the spool, for small displacements from its center position, is formed by summing all of these forces in the usual way:

$$F - k_e z_V(t) - k_v \dot{z}_V(t) - f_{\text{fric}} = m \ddot{z}_V(t), \tag{4.44}$$

where m is the mass of the spool. The friction force, caused mainly by minute contaminating particles in the fluid, can lead to a serious loss of performance of the servovalve. The oil is filtered, and other design precautions are taken, to minimize this undesirable force. The friction may also be reduced somewhat if the valve operates in a vibration environment, but it is primarily because of this friction that the two-stage valve was developed. In this type of valve the magnet actuates a small flapper, which diverts high-pressure fluid to the end chambers of the spool and thus greatly increases the actuating force available at the spool.

If we assume that the pressure drop across the metering port is constant and that the area of the port is proportional to $z_V(t)$, then the fluid flow into the ram cylinder will be proportional to the valve displacement,

$$q(t) = K_q z_V(t) \tag{4.45}$$

where K_q has units of $(m^3/s)/m$.

To obtain a linear state-variable model for the amplifier–magnet–valve–ram system, we first make the assumption that the friction coefficient f_{fric} is negligible. Further, we assume that the push–pull amplifier shown in Fig. 4.12 is connected to the valve-actuating magnet shown in Fig. 4.15. We define the input variable of the entire system to be $u(t) = e_{in}(t)$ and the output variable to be $y(t) = k_0 z_R(t)$, where k_0 is the scale-factor coefficient of a translational position transducer attached to the ram. Assume that $y(t)$ has units of V, so that k_0 has units of V/m. Establish the state vector $x(t)$ as

$$x(t) = \begin{bmatrix} x_1(t) \\ x_2(t) \\ x_3(t) \end{bmatrix} = \begin{bmatrix} z_V(t) \\ \dot{z}_V(t) \\ z_R(t) \end{bmatrix}. \tag{4.46}$$

Equations 4.40 through 4.45, when combined and arranged in matrix form, produce the linear state-variable model:

$$\dot{x}(t) = \begin{bmatrix} 0 & 1 & 0 \\ a_{21} & a_{22} & 0 \\ a_{31} & 0 & 0 \end{bmatrix} x(t) + \begin{bmatrix} 0 \\ b_{21} \\ 0 \end{bmatrix} u(t),$$

$$y(t) = [0 \ 0 \ k_0] x(t), \quad \text{where} \tag{4.47}$$

$$a_{21} = -\frac{k_e}{m}, \quad a_{22} = -\frac{k_v}{m}, \quad a_{31} = \frac{K_q}{A_R}, \quad b_{21} = \frac{4 K_M K_A i_0}{m(R_o + R_c) g_0^2}.$$

We frequently use a block diagram to depict the connections among the various devices that make up a system whose analytical model is the standard A, B, C, D state-variable model of Eqn. 4.47. (The 1×1 matrix D happens to be 0 in this example.) A device-based block diagram for the amplifier–magnet–valve–ram–sensor system is shown as Fig. 4.16. Each block lists the physical parameters associated with the device identified. Several features of this device-based diagram must be noted.

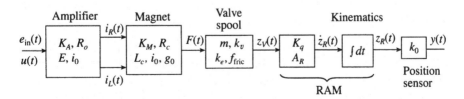

Fig. 4.16 Device-based block diagram.

1. It is not always possible to identify a physical parameter as being associated with a single block. The bias current i_0, for example, plays a key role in both the amplifier and the magnet. The flow gain K_q is a parameter that depends on the working pressure of the hydraulic supply, the design of the valve-metering orifices, and the characteristics of the conduits carrying the fluid to the ram, but it is listed as a parameter associated with the ram.

2. A block may list a parameter that does not appear in the linear state-variable model. This may happen if the manufacturer of the device specifies a parameter that is essential to the operation of the device, such as the voltage E in the amplifier, but that does not ultimately appear in the equations describing the linear behavior of the device. Those equations are based on an assumption that the device is working as it was intended to work. Often, experimental results on systems of this sort do not match the predictions based on the linear model, and the discrepancy between the two may be due to a "hidden" parameter such as E. A second example is the parameter L_c in the magnet block. L_c is the inductance of the magnet coil, which we assumed to be negligible to keep our analysis simple, but which might prove to be significant. Similarly, the friction coefficient f_{fric} in the valve has been taken to be zero.

3. A block representing a physical device might include kinematic relationships in addition to parameter values. The block labeled RAM shows the kinematic relationship between $\dot{z}_R(t)$ and $z_R(t)$, namely, $z_R(t) = \int \dot{z}_R(t)\,dt$.

4. The state-variable model has mathematical parameters that are algebraic composites of the physical parameters of the devices. An example is the coefficient b_{21} in Eqn. 4.47, $4K_M K_A i_0 / m(R_o + R_c) g_0^2$. We note that this parameter depends on the physical characteristics of three different physical devices – the amplifier, the magnet, and the valve! This dependence illustrates an important fact in control system engineering, that the dynamic performance of the overall system depends on the interaction of the devices. For example, in comparing the experimental performance of this system with the performance expected from the model, it might be determined that the numerical value of b_{21} is only a fraction of what was expected, and the source of this error must be identified. This means that the physical characteristics of all three devices contributing to b_{21} must be checked, since it is not apparent which of the three is at fault. Further, the checks must be made with the devices interconnected as they are in service; separate bench checks on the individual devices often fail to reveal the cause of the problem.

4.5 Summary

We have developed models for several devices commonly found in automatic control systems. These models are developed from the salient physical characteristics of the devices to the point where we can understand manufacturers' data sheets and specifications normally accompanying the devices from the factory. Furthermore, the principles of modeling employed here are applicable to many other types of devices that are important in control system engineering. The problems in Chapter 4 require extensions of the analysis covered in the text, and these extensions involve principles from Chapters 2 and 3. Subsequent work in the later chapters of the book will require the information and techniques resulting from these extensions.

The electromagnet is an example from a class of devices – the unstable actuator or the unstable plant – that is significant to automatic control engineering. This class of device has become increasingly common in several branches of control engineering during the past thirty years, and poses significant limits on the performance of feedback control systems in which such a device is employed. Furthermore, the existence of these devices requires the engineer to have a comprehensive understanding of the stability theory of linear systems introduced in Chapter 8.

We have illustrated several principles pertinent to control system engineering in the discussion of Fig. 4.16. The reader should assimilate these ideas before proceeding.

4.6 Problems

Problem 4.1

Assume that a dynamometer for testing electric machines permits simultaneous measurements of voltage, current, shaft speed, and torque in the shaft. You run a series of steady-state speed–torque tests on DC machine model UWEE. The data for these tests are plotted in Fig. P4.1.

Refer to Fig. 4.4 and Eqn. 4.15 in the text. Assume that $d(\omega(t)) = b = $ a constant, with units of N-m/(rad/s). Which of the five machine parameters K, R, L_a, J, and b can be determined from these speed–torque test data? Find their numerical values. Why is it not possible to determine the other parameters from these tests?

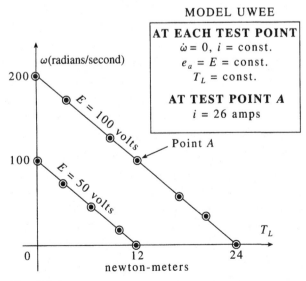

Fig. P4.1

Problem 4.2

With $E = 100$ V, find the speed–torque combination at which the model UWEE motor will deliver maximum power to the load. What is that power? What is the electrical power delivered to the machine at this maximum load power condition? Would you classify this machine as a fractional horsepower, a 1-horsepower, or a multihorsepower machine?

Problem 4.3

The model UWEE motor is connected to a driving amplifier characterized by its gain K_A and its output resistance R_o (see Fig. 4.9). The state-variable model for the amplifier–motor combination is

$$\dot{x}(t) = \begin{bmatrix} -1000 & 0 & -160 \\ 0 & 0 & 1 \\ 1200 & 0 & -12 \end{bmatrix} x(t) + \begin{bmatrix} 1300 \\ 0 \\ 0 \end{bmatrix} e(t), \quad \text{where } x(t) = \begin{bmatrix} i(t) \\ \theta(t) \\ \dot{\theta}(t) \end{bmatrix}.$$

If the armature inductance is known to be 3 mH, find the numerical values for J, K_A, and R_o.

Problem 4.4

A mobile cart is propelled by a DC motor using a belt-drive scheme, as indicated in Fig. P4.4. The motor is driven by an amplifier using the circuit arrangement shown in Fig. 4.9 in the text. The total mass of the amplifier, motor, drive mechanism, wheels, and cart is M kg. The drive-wheel assembly has mass M_R, moment of inertia J_R, and radius r. The corresponding parameters for the other wheel assembly are M_f, J_f, and r. The motor parameters are K, R, J, b, and L_a; those for the amplifier are K_A and R_o. The drive-belt ratio is n; that is, the motor shaft turns n times as fast as the drive wheels. The drive belt does not slip on the pulleys, nor do the wheels slip on the ground, which is level. Derive a state-variable model for this system using $e(t)$ as the input, $x(t) = [i(t) \; z(t) \; \dot{z}(t)]^{\mathrm{T}}$ as the state vector, and $y(t) = z(t)$ as the output quantity. The elements of your A, B, C, and D matrices must be shown as combinations of the physical parameters of this system.

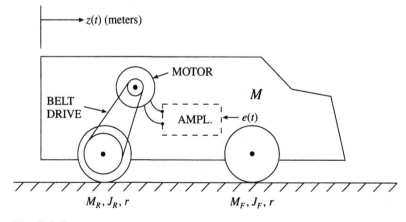

Fig. P4.4

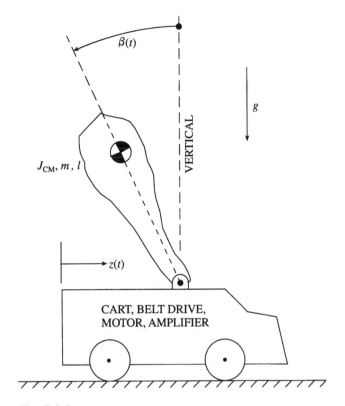

Fig. P4.5

Problem 4.5

We now use the cart-belt drive–motor–amplifier system in Problem 4.4 to mount an inverted pendulum having physical parameters J_{CM}, m, g, and l on the top of the cart (refer to Fig. P2.8 and to Section 2.4 of the text). The composite system, now characterized by eighteen parameters, is shown in Fig. P4.5. We have the means to balance the unstable pendulum by applying the appropriate voltage $e(t)$ to the input of the amplifier. We will approach this problem later in the book, but the first step is to obtain an accurate dynamic model that relates the driving voltage to the state vector and to its derivative. With the pendulum added we must augment our state vector as follows: $x(t) = [i(t) \; z(t) \; \dot{z}(t) \; \beta(t) \; \dot{\beta}(t)]^T$. You may assume that $\beta(t)$ and $\dot{\beta}(t)$ remain small enough for a linear model to represent the system dynamics adequately. Obtain a state-variable model in the normal $\{A, B, C, D\}$ format, but now let the output quantity be $y(t) = [z(t) \; \beta(t)]^T$.

Problem 4.6

Refer to the equivalent circuit diagram for the amplifier–electromagnet system shown in Fig. 4.12 of the text. Let the magnet be characterized by its mass m and the viscous drag coefficient d, as explained in Section 4.3. Obtain a state-variable model for this system in which the inductance of the magnet coils, L_c, is included as a physical parameter. It will be necessary to augment the state vector given in Eqn. 4.38 for this model.

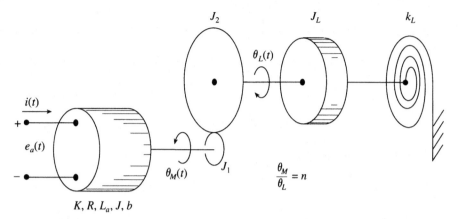

Fig. P4.7

Problem 4.7
A DC motor characterized by its five parameters K, R, L_a, J, and b is connected to
a gear-and-load combination characterized by J_1, J_2, J_L, k_L, and the gear ratio n, as
shown in Fig. P4.7. Write the differential equations relating $e_a(t)$ to $\theta_L(t)$.

Problem 4.8
A locked-rotor test is run on the model UWEE machine in which a 10-V source is
switched on to the armature and the resultant transient current is recorded. How long
will it take for the current to reach 4 A?

Problem 4.9
The ball–rack–pendulum system depicted in Fig. P2.15 is characterized by the parameters M_B, J_B, M_A, J_A, d, h, r, and g. The driving torque $T(t)$ is provided by a DC motor
coupled to body A by a belt drive, as shown in Fig. P4.9. The motor is characterized by
K, R, L_a, J, and b, and the belt-drive ratio is n. The belt does not slip, and its mass may
be neglected. Obtain a state-variable model for this system with $e_a(t)$ as the input and
the vector $[\theta(t)\ z(t)]^{\mathrm{T}}$ as the output. Identify your choice of state variables and specify
the elements of the A, B, C, and D matrices as algebraic combinations of the fourteen
system parameters. Make the linearizing assumptions that $\theta(t)$ and $z(t)$ each remain
close to their equilibrium values of zero.

Problem 4.10
Refer to Problem 4.6. Let the parameters of the amplifier–electromagnet system have
the following numerical values:

$$\frac{4K_M i_0}{g_0^2} = 7.238 \text{ N/A}, \quad \frac{4K_M i_0^2}{g_0^3} = 1{,}448 \text{ N/m}, \quad L_c = 0.002 \text{ H}, \quad m = 0.004 \text{ kg},$$

$$R_o = 2\ \Omega, \quad R_c = 3\ \Omega, \quad K_A = 1 \text{ V/V}, \quad d = 0.24 \text{ N/(m/s)}.$$

Evaluate the A and B matrices of your state-variable model using these values.

Problem 4.11
The amplifier–electromagnet system is used to drive the hydraulic valve–ram system as
described in Section 4.4. However, the coil inductance L_c is to be included in the model.

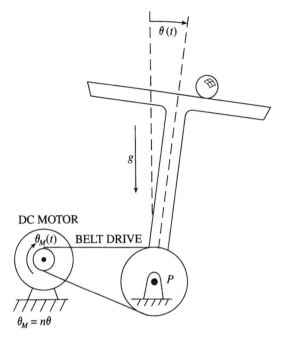

Fig. P4.9

Let *m* represent the combined mass of the magnet armature and the valve spool, and let $z_V(t)$ denote their displacement from the valve-centered position. Assume that the friction force f_{fric} is negligible. Consider $e_{in}(t)$ to be the input function for the model and $y(t)$ the output (see Fig. 4.16). Take the state vector to be $[z_V(t)\ \dot{z}_V(t)\ \Delta i(t)\ z_R(t)]^{\mathsf{T}}$ and derive the state-variable model. Be sure to include k_e, k_v, K_q, A_R, and k_0 in your model.

Problem 4.12

Evaluate your model derived in Problem 4.11 using the parameter values given in Problem 4.10 along with the following

$$k_e = 2{,}900 \text{ N/m}, \quad k_v = 0.4 \text{ N/(m/s)}, \quad k_0 = 2 \text{ V/m},$$

$$K_q = 3{,}000 \text{ (cm}^3/\text{s)/cm}, \quad A_R = 5 \text{ cm}^2.$$

If a constant 1 V is applied to the input, the system will settle into a steady state in which Δi, z_V, and $\dot{z}_R$ will all assume constant values. Use your model to calculate those values.

Problem 4.13

An electromagnet supports an armature that is free to slide in the vertical direction under the force of gravity, as shown in Fig. P4.13. The pole face has an area of 1 cm² and the coil has 300 turns.

(a) If the armature has a mass of 100 g, how much current is required to support the armature when the air gap is 2 mm? Assume $g = 9.81$ N/kg.

(b) Let your answer to (a) be termed i_0. Let $z(t)$ be the displacement of the armature above the point at which the air gap is 2 mm. Establish differ-

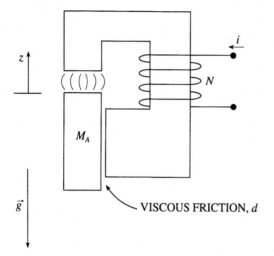

Fig. P4.13

ential equations of motion for the armature that are valid for small displacements about its equilibrium position. Let $i(t) = i_0 + \Delta i(t)$, and express your equations in terms of $\Delta i(t)$. For parameters, use the terms M_A,
d, μ_0, A, N, i_0, and g_0 as they are defined in Section 4.3.

5 Summary of Modeling Principles for Physical Systems

5.1 Introduction

In the type of dynamic system analysis that concerns us here, we normally identify the input quantity (or quantities) and the system state variables as our first step. We then write the differential equation (or equations) that describe the relationships between the input variables, the state variables, and their derivatives. The process of establishing these equations requires an understanding of the physical principles that govern the dynamics of our system. We have seen in the first four chapters that the principles of mechanics (the laws of Newton and Euler), those of electromechanics (the laws of Faraday, Ampere, Ohm, and others), and those of fluid mechanics, including aerodynamics, are all basic to the systems of interest here. The equations that result are generally nonlinear, ordinary, differential equations. In our work the differential equations are also restricted to those having constant physical parameters. In much of our work we also concentrate on the study of the dynamics of systems in a restricted regime of operation, usually for motions of the system near an equilibrium state (called a *bias point* in electronic circuits, or a *trim condition* in aircraft flight-control systems). With this further restriction on our analysis, the nonlinear differential equations may usually be approximated by linear differential equations having constant coefficients. We have seen several examples of this form of approximation in the first four chapters, and in the ensuing chapters our attention will be focused almost exclusively on linear systems of differential equations. The behavior of the linearized equations is usually, but not always, quite close to that of the nonlinear equations for motion in the immediate neighborhood of the equilibrium state, and is reasonably close for moderate excursions away from the equilibrium state.

The linearized differential equations can always be arranged in the convenient vector-matrix form $\dot{x}(t) = Ax(t) + Bu(t)$, where $x(t)$ is the $n \times 1$ matrix

99

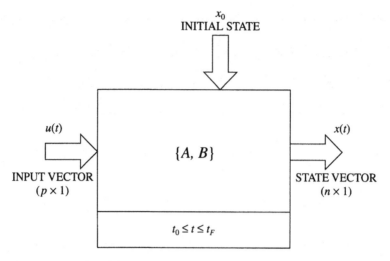

Fig. 5.1   Model of linear system.

whose elements are the state variables and $u(t)$ is the $p \times 1$ matrix whose elements are the input variables, with p separate driving input functions to the system. A is the $n \times n$ state distribution matrix, and B is the $n \times p$ input distribution matrix. We always have in mind that the equations are valid only for a certain interval of time, starting at an initial instant t_0 and terminating at an instant t_F. The state variables may have nonzero values at, or immediately prior to, the initial time. These are called initial conditions, and the state vector at the initial time is denoted as $x(t_0) = x_0$. We are usually free to choose $t_0 = 0$. The mathematical model of our dynamic system, for the initial-value problem, is given by the set of relationships

$$\dot{x}(t) = Ax(t) + Bu(t), \quad \text{where}$$

$u(t)$ is known in the interval $t_0 \le t \le t_F$ and   (5.1)

$x(t_0) = x_0$ is known.

This model is often represented by the block diagram shown in Fig. 5.1, which demarcates the system from its environment. In Section 5.3 we add features to this model to account for the mounting of sensors that provide output signals from the system. Later, we add further features to accommodate input disturbances and output disturbances, as was suggested in Section 3.4.

5.2   Choice of State Variables

In previous chapters we studied examples of the modeling process outlined in Section 5.1. In those examples we chose the state variables without formally asking what characteristics a physical quantity must possess to qualify as a state variable. We selected physical variables, such as the velocity of a massive object or the voltage across a capacitor, for two reasons. First, the selected variables were related directly to the dynamic properties of practical interest in

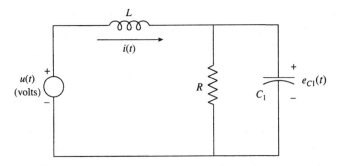

Fig. 5.2   RLC circuit.

the system – the velocity of an object that is driven by an input force, or the energy stored in a capacitor as a consequence of an electrical input. Second, the set of variables defined as state variables permit us to find a unique solution to the differential equations, although this latter point remains to be demonstrated. The nature of a set of state variables is such that at any given instant, say $t = t_1$, each state variable $x_1(t_1)$, $x_2(t_1)$, ..., $x_n(t_1)$ can be assigned an arbitrary numerical value without violating the underlying physical principles upon which the system is founded. As an example, in the RLC circuit in Fig. 5.2, the current $i(t_1)$ and the voltage $e_{C1}(t_1)$ can be assigned any numerical values, regardless of the value of the driving voltage $u(t_1)$, without violating Kirchhoff's voltage or current laws. These two variables, $i(t)$ and $e_{C1}(t)$, constitute a legitimate set of state variables. However, in the circuit shown in Fig. 5.3, the two voltages $e_{C1}(t)$ and $e_{C2}(t)$ do not constitute a legitimate set of state variables because Kirchhoff's voltage law does not allow arbitrary values to be assigned to these two voltages at an arbitrary instant, since the input $u(t)$ is specified at all instants.

The property of *valid arbitrary assignment* is a necessary but not a sufficient condition for a set of state variables. It is also necessary that the state-

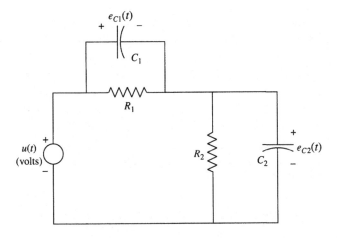

Fig. 5.3   RC circuit.

variable set be such that, when values are assigned to the variables at $t = t_0$, the given input function $u(t)$ will yield a *unique* solution to the differential equations. An example is the differential equation for the dynamics of the RLC circuit in Fig. 5.2:

$$\ddot{e}_{C1}(t) + \left(\frac{1}{RC_1}\right)\dot{e}_{C1}(t) + \left(\frac{1}{LC_1}\right)e_{C1}(t) = \left(\frac{1}{LC_1}\right)u(t). \qquad (5.2)$$

If we select only $e_{C1}(t)$ as the entire state-variable set, then the property of valid arbitrary assignment is satisfied but the equation will have a different solution for each different value of the initial condition $\dot{e}_{C1}(t_0)$. Thus we must also include $\dot{e}_{C1}(t)$ in our set. If we then make the notational changes $e_{C1}(t) = x_1(t)$ and $\dot{e}_{C1}(t) = x_2(t)$, we can write Eqn. 5.2 as

$$\dot{x}_2(t) = -\left(\frac{1}{LC_1}\right)x_1(t) - \left(\frac{1}{RC_1}\right)x_2(t) + \left(\frac{1}{LC_1}\right)u(t), \qquad (5.3)$$

which gives our state-variable vector-matrix differential equation in the standard form:

$$\begin{bmatrix} \dot{x}_1(t) \\ \dot{x}_2(t) \end{bmatrix} = \begin{bmatrix} 0 & 1 \\ -1/LC_1 & -1/RC_1 \end{bmatrix} \begin{bmatrix} x_1(t) \\ x_2(t) \end{bmatrix} + \begin{bmatrix} 0 \\ 1/LC_1 \end{bmatrix} u(t). \qquad (5.4)$$

Another important consideration in selecting a proper set of state variables arises in cases illustrated by the following example. Consider once again the circuit in Fig. 5.2. Assuming we are particularly interested in the current $i(t)$, we will write a differential equation with $i(t)$ as the dependent variable:

$$\frac{d^2i(t)}{dt^2} + \left(\frac{1}{RC_1}\right)\frac{di(t)}{dt} + \left(\frac{1}{LC_1}\right)i(t) = \left(\frac{1}{RLC_1}\right)u(t) + \left(\frac{1}{L}\right)\frac{du(t)}{dt}. \qquad (5.5)$$

Now, with the notation $i(t) = x_1(t)$ and $di(t)/dt = x_2(t)$, we have the state-variable form

$$\begin{bmatrix} \dot{x}_1(t) \\ \dot{x}_2(t) \end{bmatrix} = \begin{bmatrix} 0 & 1 \\ -1/LC_1 & -1/RC_1 \end{bmatrix} \begin{bmatrix} x_1(t) \\ x_2(t) \end{bmatrix} + \begin{bmatrix} 0 \\ 1/RLC_1 \end{bmatrix} u(t) + \begin{bmatrix} 0 \\ 1/L \end{bmatrix} \dot{u}(t). \qquad (5.6)$$

This is a legitimate vector-matrix differential–equation representation of Eqn. 5.5, and our selection of state variables satisfies both the "valid arbitrary assignment" principle and the "unique solution to a given $u(t)$" principle. But Eqn. 5.6 is not in the standard form $\dot{x} = Ax + Bu$ because it has the extra term $[0\ 1/L]^T\dot{u}(t)$ on the right-hand side! Thus we see that a set of variables satisfying both of our principles may not lead to a vector-matrix equation of the standard form $\dot{x} = Ax + Bu$.

However, it is possible to represent the basic differential Eqn. 5.5 in the standard form by modifying the selection for the state-variable set. This is done by re-arranging Eqn. 5.5 as

$$\left[\frac{d^2i(t)}{dt^2} - \frac{1}{L}\frac{du(t)}{dt}\right] + \left(\frac{1}{RC_1}\right)\frac{di(t)}{dt} + \left(\frac{1}{LC_1}\right)i(t) = \left(\frac{1}{RLC_1}\right)u(t) \qquad (5.7)$$

and defining the state-variable set to be

$$x_1(t) = i(t),$$

$$x_2(t) = \frac{di(t)}{dt} - \frac{1}{L}u(t),$$ (5.8)

so that Eqn. 5.7, written in state-variable notation, becomes

$$\dot{x}_2(t) = -\left(\frac{1}{LC_1}\right)x_1(t) - \left(\frac{1}{RC_1}\right)x_2(t).$$ (5.9)

The vector-matrix differential–equation representation of the basic differential Eqn. 5.5 now has the standard form

$$\begin{bmatrix} \dot{x}_1(t) \\ \dot{x}_2(t) \end{bmatrix} = \begin{bmatrix} 0 & 1 \\ -1/LC_1 & -1/RC_1 \end{bmatrix} \begin{bmatrix} x_1(t) \\ x_2(t) \end{bmatrix} + \begin{bmatrix} 1/L \\ 0 \end{bmatrix} u(t).$$ (5.10)

In order to obtain the standard form for the dynamic equation, it was necessary to incorporate the input into our state vector. This is a legitimate procedure, which (we have now seen) is necessary in cases where time derivatives of the input appear on the right-hand side of the fundamental physical equations. It is interesting to note that the state distribution matrices are identical in all three versions of our description of this circuit, Eqns. 5.4, 5.6, and 5.10.

Another property of a properly chosen set of state variables for a given system is that it is always possible to define a new set of state variables for the system which is analytically equivalent to the original set. The change in state variables, which is simply a change in the coordinate system of the state space, is accomplished as follows.

Assume that we have expressed the physical principles of our dynamic system in terms of n state variables $x_1(t), x_2(t), ..., x_n(t)$, each of which is a physical variable such as the current in an inductor, the voltage across a capacitor, the velocity of a massive object, and so forth. We could call this the *natural* set of state variables, and our standard form for the dynamic equation would be

$$\dot{x}(t) = Ax(t) + Bu(t).$$ (5.11)

Now define a *new* state vector, $q(t)$, using a matrix P:

$$q(t) = Px(t);$$ (5.12)

P must be an $n \times n$ matrix that has an inverse (is nonsingular). Let the inverse of P be Q; that is, $Q = P^{-1}$ and $P = Q^{-1}$, so that $x(t) = Qq(t)$. Eqn. 5.11 can then be written equivalently as

$$\dot{q}(t) = [Q^{-1}AQ]q(t) + [Q^{-1}B]u(t).$$ (5.13)

If we use the notation $\hat{A} = Q^{-1}AQ$ and $\hat{B} = Q^{-1}B$, we have

$$\dot{q}(t) = \hat{A}q(t) + \hat{B}u(t),$$ (5.14)

which is expressed in the standard form, having new A and B matrices ($\hat{A}$ and $\hat{B}$) to go along with the new state vector $q(t)$.

The purpose of making the change in state vectors from $x(t)$ to $q(t)$ is that the form of the new matrix $\hat{A}$ can offer analytical advantages over the form of the old matrix A through an appropriate choice of P. Several such forms, called *canonical forms,* are used in modern control theory. Most of these canonical forms are characterized by having many zero elements in the $\hat{A}$ matrix, the non-zero elements being distributed in patterns that are convenient for special purposes. One important canonical form is the *Jordan normal form* in which the $\hat{A}$ matrix is diagonal, or nearly so. The price paid for the analytical convenience offered by a canonical form of $\hat{A}$ is that the new state variables $q_1(t), q_2(t), \ldots,$ $q_n(t)$ are not identified in a one-to-one relationship with the physical variables of the system, but rather each $q_i(t)$ is a linear combination of the natural state variables. This becomes clear when Eqn. 5.12 is expanded:

$$\begin{bmatrix} q_1(t) \\ q_2(t) \\ \vdots \\ q_i(t) \\ \vdots \\ q_n(t) \end{bmatrix} = \begin{bmatrix} p_{11} & p_{12} & \cdots & p_{1n} \\ p_{21} & p_{22} & \cdots & p_{2n} \\ \vdots & \vdots & & \vdots \\ p_{i1} & p_{i2} & \cdots & p_{in} \\ \vdots & \vdots & & \vdots \\ p_{n1} & p_{n2} & \cdots & p_{nn} \end{bmatrix} \begin{bmatrix} x_1(t) \\ x_2(t) \\ \vdots \\ x_i(t) \\ \vdots \\ x_n(t) \end{bmatrix}. \tag{5.15}$$

We see that the new state variable $q_i(t)$ is the linear combination of the natural, or physical, state variables:

$$q_i(t) = p_{i1} x_1(t) + p_{i2} x_2(t) + \cdots + p_{in} x_n(t) = \sum_{j=1}^{n} p_{ij} x_j(t). \tag{5.16}$$

For this reason the new set $\{q_1(t), q_2(t), \ldots, q_n(t)\}$ is considered to be an *abstract* set. This terminology does not obstruct the usefulness of the change of coordinates in analysis, and it helps to explain the appearance of a complex-valued state variable $q_i(t)$ in those cases where some of the p_{ij} happen to be complex-valued; all of the physical state variables are, of course, real-valued.

5.3   The Output Equation

Equation 5.1 expresses the dynamic relationship between the input variables and the state variables. To complete the mathematical model of our system we must identify the output quantities. These quantities depend not only on the instrumentation used to measure the motions of the system, but also on the way in which those instruments are mounted. If the instruments themselves affect the motions of the system, their characteristics must be incorporated into the state variables. An instrument output typically produces a signal that is a linear combination of the state variables and, in some cases, of their derivatives.

Consider the spring–mass–damper system in Fig. 5.4. Here the input variable is the applied force $f(t)$, and the two state variables are the position and the velocity of the mass, $z(t)$ and $\dot{z}(t)$. The dynamic state variable equation is

$$\begin{bmatrix} \dot{x}_1(t) \\ \dot{x}_2(t) \end{bmatrix} = \begin{bmatrix} 0 & 1 \\ -K/M & -d/M \end{bmatrix} \begin{bmatrix} x_1(t) \\ x_2(t) \end{bmatrix} + \begin{bmatrix} 0 \\ 1/M \end{bmatrix} u(t). \tag{5.17}$$

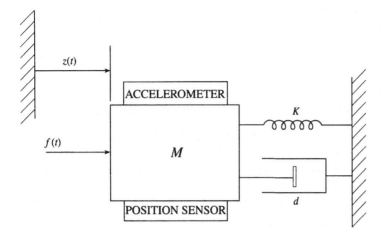

Fig. 5.4 Spring, mass, and damper system.

Now assume that we mount a sensor in our system to measure the position of the mass. The sensor provides a voltage $y_1(t)$ that is proportional to the position:

$$y_1(t) = k_1 x_1(t) \text{ V}, \tag{5.18}$$

where k_1 is the instrument scale factor, V/m. We now mount a second sensor on the mass, an accelerometer, which provides a voltage $y_2(t)$ proportional to the acceleration of the mass with respect to inertial space (taken to be the ground in this case):

$$y_2(t) = k_2 \ddot{z}(t) = k_2 \dot{x}_2(t) \text{ V}, \tag{5.19}$$

where k_2 is the accelerometer scale factor, V/(m/s^2). We see that the accelerometer output signal depends on the *derivative* of one of the state variables. Expanding $y_2(t)$, we find that

$$y_2(t) = k_2 \left[-\frac{K}{M} x_1(t) - \frac{d}{M} x_2(t) + \frac{1}{M} u(t) \right]. \tag{5.20}$$

The output vector $y(t)$ is composed of the two instrument signals:

$$\underbrace{\begin{bmatrix} y_1(t) \\ y_2(t) \end{bmatrix}}_{y(t)} = \underbrace{\begin{bmatrix} k_1 & 0 \\ -k_2 K/M & -k_2 d/M \end{bmatrix}}_{C} \underbrace{\begin{bmatrix} x_1(t) \\ x_2(t) \end{bmatrix}}_{x(t)} + \underbrace{\begin{bmatrix} 0 \\ k_2/M \end{bmatrix}}_{D} u(t). \tag{5.21}$$

This example illustrates why some systems have an output vector with an additive term $Du(t)$ whereas others do not, depending upon the type and the mounting of the sensors used.

In the general case where there are m sensors, we can say that since sensors respond to a linear combination of the states and the derivatives of the states, we have

$$y(t) = K_1 x(t) + K_2 \dot{x}(t); \tag{5.22}$$

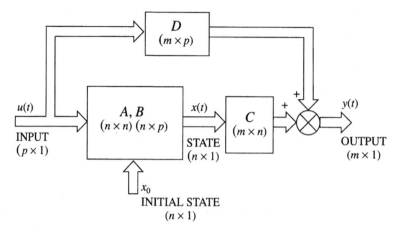

Fig. 5.5   Model of linear system including output sensors.

K_1 and K_2 are matrices of dimension $m \times n$ whose elements are the sensor scale factors. Substitute $\dot{x}(t)$ from the dynamic equation into Eqn. 5.22 to obtain

$$y(t) = K_1 x(t) + K_2 [Ax(t) + Bu(t)]$$
$$= \underbrace{[K_1 + K_2 A]}_{\substack{C \\ m \times n}} x(t) + \underbrace{[K_2 B]}_{\substack{D \\ m \times p}} u(t). \tag{5.23}$$

In some systems the matrix product $K_2 B$ will be zero when K_2 is not zero. In this event, a D matrix will not appear in the system model. The block diagram in Fig. 5.1 can now be augmented to show the role of the sensors in the overall system; see Fig. 5.5.

5.4 Summary

In this chapter we have explored several fine points pertinent to the process of modeling physical systems in the standard state-variable form. The result of a successful modeling process is a valid set of first-order differential equations in vector-matrix form, along with the initial conditions necessary to solve the initial-value problem uniquely:

DYNAMICS: $\dot{x}(t) = Ax(t) + Bu(t)$;

OUTPUT: $y(t) = Cx(t) + Du(t)$;

$t_0 \leq t \leq t_F$ is specified,

$x(t_0) = x_0$ and $u(t)$ are known.

$$\tag{5.24}$$

The solution of the initial-value problem is the response of all n state variables $x_1(t), x_2(t), ..., x_n(t)$ over the given time interval. The m output variables $y_1(t), y_2(t), ..., y_m(t)$ are also produced. These solutions are studied to determine whether they are satisfactory for the purpose at hand; if they are not, the solutions may help to decide what change should be made in the system to improve the responses.

Chapter 6 is devoted to the solution of the initial-value problem. Several potent computer programs designed to operate reliably on personal computers (as well as on larger machines) offer remarkable means for obtaining numerical solutions to the initial-value problem. These programs require the describing equations to be entered into the computer in the standard form given in Eqn. 5.24, and it is for this reason that our modeling process has emphasized this particular formulation of the dynamic equations. Further discussion of this approach appears in Chapter 6, where a second form of modeling, the *transfer function,* is introduced. This second approach to dynamic system modeling is also called *frequency-domain* analysis.

5.5 Problems

Problem 5.1
A linear system whose input is $u(t)$ and whose output is $y(t)$ behaves dynamically according to the following differential equation: $\dddot{y}(t) + 2\ddot{y}(t) + 3\dot{y}(t) + 4y(t) = u(t) + 5\ddot{u}(t)$. Define a set of state variables and find four matrices A, B, C, and D that represent this system in the standard form:

$$\dot{x}(t) = Ax(t) + Bu(t), \qquad y(t) = Cx(t) + Du(t).$$

Problem 5.2
The state-variable model for a linear system is

$$\dot{x}(t) = Ax(t) + B\begin{bmatrix} u_1(t) \\ u_2(t) \end{bmatrix}, \quad \begin{bmatrix} y_1(t) \\ y_2(t) \end{bmatrix} = Cx(t) + D\begin{bmatrix} u_1(t) \\ u_2(t) \end{bmatrix}.$$

The input and output variables are related by the following set of differential equations:

$$\ddot{y}_1(t) + 2\dot{y}_1(t) - 3(y_1(t) + y_2(t)) = 4\dot{u}_1(t),$$

$$\dot{y}_2(t) + 5y_1(t) - 3y_2(t) = u_1(t) + u_2(t).$$

Define a set of state variables for this system and evaluate A, B, C, and D.

Problem 5.3
The pair of differential equations

$$\dot{y}_1(t) + 2y_2(t) = \dot{u}(t) - 2u(t),$$

$$2\dot{y}_2(t) - 2y_1(t) - 4y_2(t) = 3u(t),$$

describe the dynamics of a SIMO system. Define a state vector and establish a state-variable model for this system.

Problem 5.4
Refer to Fig. 5.3 in the text. Let the output function be $y(t) = e_{C2}(t)$. Define a state vector for the SISO system and express the elements of A, B, C, and D as combinations of R_1, R_2, C_1, and C_2.

Problem 5.5

Work part (b) of Problem 3.6 (in Chapter 3).

Problem 5.6

Consider the electromagnet system in Problem 4.13(b). A proximity sensor is incorporated into this system to provide a voltage $y(t)$ that is proportional to the displacement $z(t)$, where $y(t) = k_0 z(t)$. The sensor does not load the armature or the magnet. Derive a state-variable model in the form

$$\dot{x}(t) = Ax(t) + B\Delta i(t),$$

$$y(t) = Cx(t) + D\Delta i(t),$$

from the equations of motion you found in Problem 4.13(b). Identify your state variables and express each element of the system matrices in terms of the physical parameters k_0, M_A, d, μ_0, A, N, i_0, and g_0.

Problem 5.7

For the RLC circuit in Fig. 5.2: Take the state variables to be $x_1(t) = i(t)$ and $x_2(t) = e_{C1}(t)$, and find the A and B matrices for the dynamic expression $\dot{x}(t) = Ax(t) + Bu(t)$. Is the A matrix the same as those in Eqns. 5.4, 5.6, and 5.10?

Problem 5.8

For the RLC circuit in Fig. 5.2, define the output $y(t)$ to be the current in the resistor and the state variables to be $x_1(t) = i(t)$ and $x_2(t) = e_{C1}(t)$, as in Problem 5.7. Find the C and D matrices for the output equation $y(t) = Cx(t) + Du(t)$.

Problem 5.9

(a) For the RLC circuit in Fig. 5.2, assume that $L = 1$ H, $R = 1$ Ω, and $C_1 = 0.2$ F. Let the state variables be $x_1(t) = i(t)$ and $x_2(t) = e_{C1}(t)$, and let the output be the current in the resistor, as in Problem 5.8. Evaluate the four system matrices numerically.

(b) Now define a new state vector $q(t) = [q_1(t)\ q_2(t)]^T$, where the new state variables are related to the original state variables as follows: $q_1(t) = 3x_1(t) - x_2(t)$ and $q_2(t) = -2x_1(t) + 2x_2(t)$. Find the representation of the system dynamics in terms of the new state vector

$$\dot{q}(t) = \hat{A}q(t) + \hat{B}u(t),$$

$$y(t) = \hat{C}q(t) + Du(t),$$

and evaluate $\hat{A}$, $\hat{B}$, and $\hat{C}$ numerically.

(c) Evaluate the P and Q matrices involved in the transformation between the $x(t)$ representation of the state vector and the $q(t)$ representation, and calculate the matrix product PQ.

Problem 5.10

Use the same numerical values for R, L, and C_1 as in Problem 5.9. Are you able to define a new set of state variables as $q_1(t) = 3x_1(t) - x_2(t)$ and $q_2(t) = -5x_1(t) + \frac{5}{3}x_2(t)$ such that there will be a new state-variable model

$$\dot{q}(t) = \hat{A}q(t) + \hat{B}u(t),$$

$$y(t) = \hat{C}q(t) + Du(t)?$$

If so, evaluate $\hat{A}$, $\hat{B}$, and $\hat{C}$ numerically. If not, explain why you are unable to find such a model.

Problem 5.11

We have used several different sets of state variables for the circuit in Fig. 5.2. For this problem identify three of them as follows:

$$x(t) = \begin{bmatrix} i(t) \\ e_{C1}(t) \end{bmatrix}, \quad z(t) = \begin{bmatrix} i(t) \\ di(t)/dt - (1/L)u(t) \end{bmatrix}, \quad w(t) = \begin{bmatrix} e_{C1}(t) \\ de_{C1}(t)/dt \end{bmatrix}.$$

(a) Find the transformation matrix between the $x(t)$ and $z(t)$ representations.
(b) Find the transformation matrix between the $x(t)$ and $w(t)$ representations.
(c) Find the transformation matrix between the $z(t)$ and $w(t)$ representations.

Problem 5.12

Refer to the information on the Navion aircraft in Chapter 2 and in Problem 2.10. We wish to add an accelerometer to the flight instrumentation, mounted at the CM, so that the output vector will be $y(t) = [\Delta\theta(t) \ \Delta\dot{\theta}(t) \ n_a(t)]^T$. Here the sensitive axis of the accelerometer is aligned with the z axis of the aircraft to measure the *normal acceleration*, $n_a(t) = -\Delta\dot{w}(t) + u_{A0}\Delta\dot{\theta}(t)$. Use the numerical values for the stability derivatives given in Problem 2.10 and obtain the A, B, C, and D matrices for the state-variable model using $u(t) = [\Delta\delta_E(t) \ \Delta\delta_T(t)]^T$ as the input vector.

Problem 5.13

Refer to the information on the ball–rack–pendulum system in Chapter 2 (see Fig. P2.15 and Problems 2.15, 2.16, and 2.17). The physical parameters of the system are:

$$M_A = 1,000 \text{ g}, \quad J_A = 130,000 \text{ g-cm}^2, \quad h = 9 \text{ cm}, \quad d = 5 \text{ cm},$$

$$M_B = 260 \text{ g}, \quad J_B = 416 \text{ g-cm}^2, \quad r = 2 \text{ cm}, \quad g = 981 \text{ cm/s}^2.$$

(a) Is the ball made of aluminum or steel?
(b) Is the ball a hollow shell or is it solid?
(c) Evaluate numerically the A and B matrices for the linear model.
(d) Evaluate the C and D matrices if the system is instrumented to provide three output voltages:

$$y(t) = \begin{bmatrix} k_1 z(t) \\ k_2 \dot{z}(t) \\ k_3 \theta(t) \end{bmatrix}.$$

State the units for the three instrument scale factors k_1, k_2, and k_3.

Problem 5.14

Refer to Problem 2.18. Two position sensors are incorporated into the system to provide two output voltages,

$$y(t) = \begin{bmatrix} \Psi_1\theta_1(t) \\ \Psi_2\theta_2(t) \end{bmatrix}.$$

Evaluate the C and D matrices for the state-variable model of this system.

6 Solution to the Initial-Value Problem

6.1 Introduction

A linear dynamic system may be modeled as a set of differential equations using the following standard vector-matrix form:

$$\dot{x}(t) = Ax(t) + Bu(t), \quad \text{where}$$
$$t_0 \leq t \leq t_F \quad \text{and} \quad u(t) \text{ is known.} \tag{6.1}$$

If the initial conditions of all n state variables are known at (or just prior to) the initial time t_0, we add to Eqn. 6.1 the further constraint that $x(t_0)$ is also known. We then have the initial-value problem identified in Section 5.4. If the matrices A and B are constant, this is a well-circumscribed problem, since a solution to this set of equations not only exists but is also unique. These existence and uniqueness properties form the theoretical foundation that makes the state-variable model for linear systems so useful in engineering.

If fewer than n of the state variables are specified at the initial time t_0, and some are specified at the final time t_F, the problem is called a *two-point boundary-value problem*. This important class of dynamics problem is not as simply circumscribed as is the initial-value problem. Whether a solution exists and, if so, whether it is unique, are questions that arise because of the mixed boundary conditions. Answers to these questions must be determined by further analysis of the coefficients of the equation. For this reason, we delay consideration of the two-point boundary problem until Chapter 15. Because similar calculation requirements can arise if the A matrix has elements that depend on the independent variable t, we will not address this class of model, even though it is also important in engineering.

In this chapter we first identify the solution to the initial-value problem as the sum of two terms, called the *zero-input response* and the *zero-state response,* where the word "response" implies that the calculated solution is the

110

consequence of the stimulus due to both $u(t)$ and $x(t_0)$. In most practical engineering problems the solution is obtained by numerical integration of the differential equation set. Commercial computer programs especially designed for control system work are the most reliable tools for this purpose, and today it is very difficult to pursue the study of control system dynamics without this computer aid. The analytical description of how the vector-matrix differential equations are integrated involves linear algebra mixed with calculus. This is discussed briefly in Section 6.3 for a multi-input–multi-output (MIMO) system, but a detailed consideration is postponed to Chapter 15.

In this chapter we introduce *transfer function analysis*, an analytical approach to solving differential equations which does not require matrix formulation of the equations. Transfer function analysis is especially useful for studying the dynamics of single-input–single-output (SISO) systems, and is also amenable to the same computer-based tools used in the matrix formulation of the MIMO problem. Because it depends on the Laplace transformation, a brief review of that topic is given in Appendix C.

6.2 A First-Order Example

An elementary example of a first-order dynamic system will illustrate nearly all the basic structural properties of the solution to the initial-value problem. Let the RC circuit shown in Fig. 6.1 serve as the physical model for this example. Assume that the capacitor has an initial charge just prior to the closing of the switch, so that the initial condition on the capacitor voltage is $e_{C1}(t_0) = E_0$ V. We identify the state vector $x(t)$ as the capacitor voltage $x(t) = e_{C1}(t)$.

We take $t_0 = 0$ for convenience, and assume the switch to be closed at $t = 0$. The input function is designated as $u(t)$, and is an AC voltage:

$$u(t) = E_{\max} \cos(\omega t) \text{ V.} \tag{6.2}$$

The initial-value problem takes the standard formulation

$$\dot{x}(t) = ax(t) + bu(t) \quad \text{for } 0 \le t \le \infty, \quad \text{where} \tag{6.3}$$

$$x_0 = E_0 \quad \text{and} \quad u(t) = E_{\max} \cos(\omega t).$$

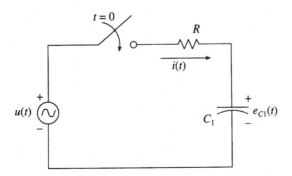

Fig. 6.1 RC circuit for first-order example.

The final time is not specified as a finite value, and the A and B matrices are designated by lower-case a and b to emphasize that we are dealing with a first-order system. In terms of the physical parameters of the circuit, we have

$$a = -\frac{1}{RC_1} \quad \text{and} \quad b = \frac{1}{RC_1}. \tag{6.4}$$

The differential equation in Eqn. 6.3 may be solved for $x(t)$ by the *integrating factor* method. First, multiply both sides of the equation by ϵ^{-at} and re-arrange the result:

$$\frac{dx(t)}{dt}\epsilon^{-at} - a\epsilon^{-at}x(t) = bE_{\max}\epsilon^{-at}\cos(\omega t). \tag{6.5}$$

Now recognize that the left-hand side of Eqn. 6.5 may be written as a total derivative:

$$\frac{d}{dt}[x(t)\epsilon^{-at}] = bE_{\max}\epsilon^{-at}\cos(\omega t). \tag{6.6}$$

Write Eqn. 6.6 as

$$d[x(t)\epsilon^{-at}] = bE_{\max}\epsilon^{-at}\cos(\omega t)\,dt, \tag{6.7}$$

and integrate both sides of Eqn. 6.7 over the interval 0 to t:

$$\int_{x(0)\epsilon^{-0}}^{x(t)\epsilon^{-at}} d[x(t)\epsilon^{-at}] = bE_{\max}\int_{t=0}^{t=t}\epsilon^{-at}\cos(\omega t)\,dt, \tag{6.8}$$

which yields

$$x(t)\epsilon^{-at} - E_0 = \left[\frac{bE_{\max}}{a^2+\omega^2}\right][\epsilon^{-at}[\omega\sin(\omega t) - a\cos(\omega t)]]_0^t. \tag{6.9}$$

Now evaluate the bracketed term according to the indicated limits, multiply both sides of Eqn. 6.9 by ϵ^{at}, and re-arrange the result to show the solution:

$$x(t) = E_0\epsilon^{at} + \frac{bE_{\max}}{a^2+\omega^2}[\omega\sin(\omega t) - a\cos(\omega t) + a\epsilon^{at}]. \tag{6.10}$$

Verify $x(t)$ as the unique solution by observing that it satisfies both the differential equation and the initial condition in the model given by Eqn. 6.3. When expressed in terms of the physical parameters of the system, the solution may be written as:

$$e_{C1}(t) = \underbrace{\{E_0\epsilon^{-t/RC_1}\}}_{\text{zero-input response}}$$

$$+ \underbrace{\left\{\frac{E_{\max}}{1+(\omega RC_1)^2}[(\omega RC_1)\sin(\omega t) + \cos(\omega t) - \epsilon^{-t/RC_1}]\right\}}_{\text{zero-state response}}. \tag{6.11}$$

We see that the solution $e_{C1}(t)$ is represented as the sum of two terms – the zero-input response (ZIR) and the zero-state response (ZSR). Because the ZIR represents what the total response would be if the input $u(t)$ were zero, only the system parameters R and C_1 and the initial capacitor voltage E_0 appear in the ZIR. Similarly, because the ZSR represents what the total response would be if the initial capacitor voltage E_0 were zero, it contains only the system parameters and the input parameters. This solution illustrates several significant properties of the initial-value problem, which are discussed in the following remarks.

Remark 1 The terms "zero-input response" and "zero-state response" are peculiar to control system analysis. The zero-input response is known in common mathematical terms as *the homogeneous solution* to the differential equation; the zero-state response is known as *a particular solution* to the differential equation.

Remark 2 The solution $e_{C1}(t)$ in Eqn. 6.11 may be written alternately as the sum of a *transient* and a *steady-state* response by expanding it into the form

$$e_{C1}(t) = \underbrace{\{E_0 \epsilon^{-t/RC_1}\}}_{\text{ZIR}} - \overbrace{\left\{\frac{E_{max}\epsilon^{-t/RC_1}}{1+(\omega RC_1)^2}\right\}}^{\text{transient response}} + \underbrace{\overbrace{\left\{\frac{E_{max}}{1+(\omega RC_1)^2}[(\omega RC_1)\sin(\omega t)+\cos(\omega t)]\right\}}^{\text{steady-state response}}}_{\text{ZSR}}.$$

In this example the transient response dies out as time increases, whereas the steady-state response continues to oscillate with a frequency ω rad/s. Note that the transient response depends both on the initial voltage of the capacitor and on the input amplitude and frequency. The terms "transient" and "steady-state" come from electrical engineering. Electric circuits driven by AC or DC voltage sources have a steady-state response that resembles the driving voltage. Momentary disturbances caused by switching die away quickly. Figure 6.2 shows the response $e_{C1}(t)$ as the RC circuit is switched onto an ordinary 60-Hz power line at the instant the line voltage is at its maximum value. The influences of both the transient and steady-state components of $e_{C1}(t)$ are apparent here.

In control and communication systems, the dynamic response of a system to input excitations and to initial conditions are better described by the ZIR and ZSR terminology for the following reasons. In systems that are unstable or neutrally stable, the ZIR will not die away with increasing time and therefore it is not properly characterized as a "transient." Moreover, in a stable system the input excitation may be a brief pulse, so here the ZSR *will* die away with increasing time and therefore lacks a "steady-state" nature.

Remark 3 Because we recognize that the right side of Eqn. 6.8 is a definite integral and that therefore its value is independent of the variable of integration, we can write the right side as

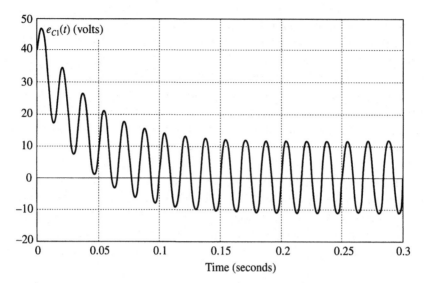

Fig. 6.2   Transient and steady-state response of RC circuit.

$$\int_0^t \epsilon^{-a\tau} b[E_{\max} \cos(\omega\tau)] \, d\tau, \tag{6.12}$$

where τ, the dummy variable of integration, drops out of the expression for the integral, which is the last term in Eqn. 6.10. If this expression (Eqn. 6.12) is substituted for the right-hand side of Eqn. 6.9, and the equation is then multiplied on both sides by ϵ^{at}, the solution may be written as

$$x(t) = E_0 \epsilon^{at} + \int_0^t \epsilon^{a(t-\tau)} b[E_{\max} \cos(\omega\tau)] \, d\tau. \tag{6.13}$$

Here ϵ^{at} has been taken inside the integral sign, since the integration is performed with respect to the variable τ and not with respect to t. The corresponding solution for the general input $u(t)$ is

$$x(t) = x_0 \epsilon^{at} + \int_0^t \epsilon^{a(t-\tau)} b u(\tau) \, d\tau, \tag{6.14}$$

where E_0 has been replaced by x_0. Equation 6.14 is the general solution to the initial-value problem for a first-order system having state variable $x(t)$, initial state x_0, and input $u(t)$. The integral in Eqn. 6.14 is called a *convolution integral;* the operation of convolving two functions of time is defined in Appendix C.

Remark 4   At the switching moment ($t = 0$) in this RC circuit, a discontinuity in $i(t)$ occurs that is not apparent in the solution for $e_{C1}(t)$. The current $i(t)$ is zero when the switch is open, so that $i(0^-) = 0$. But immediately following the closure of the switch we have $i(0^+) = (E_{\max} - E_0)/R$, which will not be zero when $E_{\max} \neq E_0$. In this book we call the value of variables just *prior to*

t_0 initial *conditions,* and the value of variables just *following* t_0 initial *values.* In many cases the state variables are continuous at t_0, so their initial conditions are the same as their initial values. But because the state variables are discontinuous at t_0 in some cases, it is necessary to distinguish the physical conditions of the system at $t = 0^-$ from those at $t = 0^+$ by use of this terminology. In the state-variable model, the initial condition x_0 is $x(t_0^-)$, also designated as $x(t_0)$; the initial value is denoted as $x(t_0^+)$.

Remark 5   A system is linear with respect to an input excitation provided that it possesses the properties of homogeneity and additivity. A system has *homogeneity* if its response to the input excitation $\Omega u(t)$ is $\Omega y(t)$, where $y(t)$ is the response to the excitation $u(t)$ and Ω is an arbitrary constant. A system has *additivity* if its response to the input excitation $[u_1(t) + u_2(t)]$ is $[y_1(t) + y_2(t)]$, where $y_1(t)$ and $y_2(t)$ are the individual responses to inputs $u_1(t)$ and $u_2(t)$ respectively. These two properties taken together constitute the *superposition* property of linear systems. It sometimes comes as a surprise to a student that the solution of this simple linear differential equation, $e_{C1}(t)$ in Eqn. 6.11, is not linear with respect to the input when the initial condition is nonzero; neither is it linear with respect to the initial condition if the input is nonzero.

Remark 6   The solution to the initial-value problem for a MIMO system governed by the matrix equation $\dot{x} = Ax + Bu$ is derived in precisely the same way that Eqn. 6.14 was derived for the scalar case, and the solution has exactly the same form. However, it is a matrix equation:

$$x(t) = [\epsilon^{At}]x_0 + \int_0^t [\epsilon^{A(t-\tau)}]Bu(\tau)\,d\tau, \tag{6.15}$$

where the dimensions of the matrices are

$\quad x(t)$: $n \times 1$,

$\quad u(t)$: $p \times 1$,

$\quad\;\; A$: $n \times n$,

$\quad\;\; B$: $n \times p$.

The *state transition matrix* $[\epsilon^{At}]$ has dimension $n \times n$. We discuss this matrix solution further in Section 6.3 and again in Chapter 15, where $[\epsilon^{At}]$ is defined by matrix analysis.

6.3   The *n*th-Order Multivariable System

Equation 6.1 is the mathematical model that describes the general linear dynamic system upon which we focus our attention in automatic control engineering. When the state-variable matrix $x(t)$ is of dimension $n \times 1$, we say that the *dynamic order* of the system is n. When the matrix of input variables $u(t)$ is of dimension $p \times 1$ with $p > 1$, the system is called a *multi-input* or a *multivariable* system. As we have previously seen, the dynamic Eqn. 6.1 in most situations is augmented by the output equation, which relates the output variables (typically sensor signals) to the state variables and, in some cases, to the input variables. The output variables may also be arranged in a matrix $y(t)$, which is related to $x(t)$ and $u(t)$ by the output equation

$$y(t) = Cx(t) + Du(t). \tag{6.16}$$

If there are m output variables then $y(t)$ will have dimension $m \times 1$; we call this system a *multi-output* system.

The analytical solution of the initial-value problem $x(t)$ is given in Eqn. 6.15. In most practical situations $x(t)$ is evaluated numerically by a computer programed to integrate the basic equation

$$\dot{x}(t) = Ax(t) + Bu(t) \quad \text{given } u(t) \text{ and } x_0. \tag{6.17}$$

As an example of this numerical approach for the amplifier–servomotor system modeled in Section 4.2, we have

$$x(t) = \begin{bmatrix} x_1(t) \\ x_2(t) \\ x_3(t) \end{bmatrix} = \begin{bmatrix} i(t) \text{ A} \\ \theta(t) \text{ rad} \\ \dot{\theta}(t) \text{ rad/s} \end{bmatrix} \quad \text{and} \quad u(t) = \begin{bmatrix} u_1(t) \\ u_2(t) \end{bmatrix} = \begin{bmatrix} e(t) \text{ V} \\ T_L(t) \text{ N-m} \end{bmatrix}. \tag{6.18}$$

The A and B matrices for this model are given in terms of the system physical parameters R, R_o, K_A, K, J, L_a, and b in Eqn. 4.27. Typical numerical values of these parameters for a small servomotor system are:

$$R = 1\,\Omega, \quad R_o = 2\,\Omega, \quad K = 0.1\,\text{N-m/A}, \quad L_a = 5\,\text{mH},$$

$$J = 7 \times 10^{-4}\,\text{kg-m}^2, \quad b = 0.004\,\text{N-m/(rad/s)}, \quad K_A = 5\,\text{V/V}.$$

The initial condition of the state is taken to be zero, $x_0 = [0\ 0\ 0]^T$; the input voltage is taken as a step of 5 V applied at $t = 0$; and the input load torque is taken as a short pulse of 1 N-m applied at $t = 0.05$ s. Figure 6.3 displays the results of the numerical integration for the interval $0 \le t \le 0.25$, showing the principal features of the response to these inputs.

Numerical solutions like this one are obtained quickly by the computer, and in most cases the results are both accurate and useful in practical engineering work. In some cases, however, numerical difficulties in the algorithms that integrate the state-variable equations cause serious errors in the results. For this reason the engineer must be able to assess the results presented by the computer. Inconsistencies between computer results and the physical principles of the problem often reveal computer errors. An alternate analytical approach to the problem might also resolve the inconsistency.

We now turn to such an alternate analytical approach, known variously as *classical control theory,* or *transform techniques,* or *frequency-domain techniques.* This approach is essentially restricted to linear systems and is primarily applied to single-input–single-output (SISO) systems, that is, to nth-order systems in which $p = 1$ and $m = 1$.

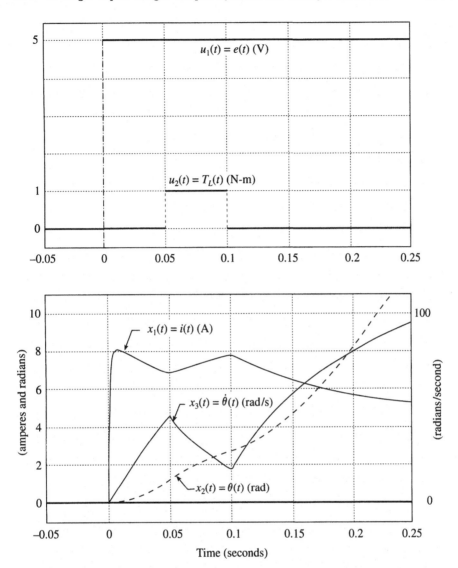

Fig. 6.3 State-variable responses to step voltage input and pulse torque input.

6.4 A Single-Input–Single-Output System; the Transfer Function

A massive cylinder with moment of inertia J kg-m^2 about its axis is restrained by a rotary spring having an elastic coefficient of K N-m/rad and by viscous friction with coefficient d N-m/(rad/s), as shown in Fig. 6.4. We define the output of this system to be $y(t)$ rad and the input to be the torque applied to the shaft, $u(t)$ N-m. With the input and output so defined, this system falls into the SISO class. We bypass the state-variable approach for modeling this dynamic system, using instead the classical differential equation approach. In

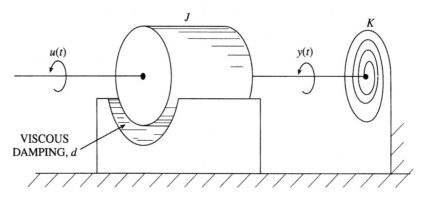

Fig. 6.4   *J*, *K*, and *d* system.

this classical approach we obtain a single differential equation with the output $y(t)$ as the dependent variable and the input $u(t)$ on the right-hand side of the equation. This yields a second-order differential equation:

$$\ddot{y}(t) + \left(\frac{d}{J}\right)\dot{y}(t) + \left(\frac{K}{J}\right)y(t) = \left(\frac{1}{J}\right)u(t), \quad \text{where}$$

$u(t)$ is to be specified for $t_0 \le t \le t_F$,   and

$y(t_0) = y_0$ and $\dot{y}(t_0) = \dot{y}_0$ are given.

(6.19)

Let us select $t_0 = 0$ for simplicity and employ the *Laplace transformation* technique to find the solution $y(t)$, which satisfies both the initial conditions and the differential equation given in Eqn. 6.19. A review of the Laplace transformation and its properties appears in Appendix C, along with a table of transforms. The Laplace transform of the *derivative* of a time function is essential to our present analysis; this operation is summarized as pair 33 in the table.

We begin our solution of Eqn. 6.19 with the assumption that $u(t)$, whatever we might choose it to be, has a Laplace transform. The properties of the transform imply that our solution $y(t)$, whatever it might turn out to be, also has a Laplace transform. We denote the transform of $u(t)$ as $U(s)$ and that of $y(t)$ as $Y(s)$. We take the Laplace transform of both sides of Eqn. 6.19, term-by-term. The transform operation produces an algebraic equation:

$$[s^2 Y(s) - sy_0 - \dot{y}_0] + \left(\frac{d}{J}\right)[sY(s) - y_0] + \left(\frac{K}{J}\right)Y(s) = \left(\frac{1}{J}\right)U(s). \qquad (6.20)$$

We re-arrange this equation so that

$$\left[s^2 + \left(\frac{d}{J}\right)s + \left(\frac{K}{J}\right)\right]Y(s) = \left(\frac{1}{J}\right)U(s) + \left[s + \frac{d}{J}\right]y_0 + \dot{y}_0. \qquad (6.21)$$

Note that all the terms on the right-hand side will be known when $u(t)$ and the initial conditions are specified. The bracketed term on the left-hand side is also known, since the two coefficients d/J and K/J are defined in terms of the three parameters describing the physics of this system. In this case, both d/J and K/J

have positive real values because the parameters are all real and positive. We factor this bracketed term as

$$\left[s^2 + \frac{d}{J}s + \frac{K}{J} \right] = (s + P_1)(s + P_2), \tag{6.22}$$

where the two roots $-P_1$ and $-P_2$ either are real-valued or, if complex, are complex conjugates of one another, depending on the relative values of the three parameters J, K, and d. They will be real-valued if $d \geq 2\sqrt{KJ}$, otherwise they will be complex. In the borderline case, $d = 2\sqrt{KJ}$, the two real roots will be equal.

We next divide both sides of Eqn. 6.21 by the quadratic term in Eqn. 6.22:

$$Y(s) = \underbrace{\left\{ \frac{1/J}{(s+P_1)(s+P_2)} \right\} U(s)}_{Y_{\text{ZSR}}(s)} + \underbrace{\left\{ \frac{(s+d/J)}{(s+P_1)(s+P_2)} \right\} y_0 + \left\{ \frac{1}{(s+P_1)(s+P_2)} \right\} \dot{y}_0}_{Y_{\text{ZIR}}(s)}. \tag{6.23}$$

Equation 6.23 gives us $Y(s)$, which is not the solution but is instead the Laplace transform of the solution. Note that $Y(s)$ is expressed here as the sum of $Y_{\text{ZSR}}(s)$ and $Y_{\text{ZIR}}(s)$, the Laplace transforms of the zero-state response and the zero-input response. We find the time-domain solution $y(t)$ for the given input $u(t)$ and the given initial conditions by substituting $U(s)$ and y_0 and $\dot{y}_0$ into Eqn. 6.23 and then taking the inverse Laplace transform of $Y(s)$. The inverse transform may be taken term-by-term, using a table of transforms such as that in Appendix C. The partial-fraction technique, which is described in Section 6.5, is also useful in finding the inverse transform.

As an example, take the input torque $u(t)$ to be a step function of T_0 N-m:

$$u(t) = T_0 \sigma(t), \quad \text{so that } U(s) = T_0\left(\frac{1}{s}\right). \tag{6.24}$$

The expression for $Y(s)$ then becomes

$$Y(s) = \underbrace{\left\{ \frac{(1/J)T_0}{s(s+P_1)(s+P_2)} \right\}}_{\text{pair 17 or 25}} + \underbrace{\left\{ \frac{(s+d/J)}{(s+P_1)(s+P_2)} \right\} y_0}_{\text{pair 15 or 24}} + \underbrace{\left\{ \frac{1}{(s+P_1)(s+P_2)} \right\} \dot{y}_0}_{\text{pair 19 or 26}}. \tag{6.25}$$

We now find $y(t)$, the solution to the initial-value problem, from Eqn. 6.25 by taking the inverse Laplace transform of $Y(s)$ term-by-term, using the designated transform pairs from the table in Appendix C. If P_1 and P_2 are real and unequal, pairs 25, 24, and 26 (with $a = 0$) are used; if they are complex conjugates, pairs 17, 15, and 19 (with $a = 0$) apply.

As an example, let $J = 0.5$ kg-m^2, $K = 50$ N-m/rad, and $T_0 = 100$ N-m. We calculate $y(t)$ for two cases, an overdamped case with $d = 12$ N-m/(rad/s) and an underdamped case with $d = 3$ N-m/(rad/s). Let the initial conditions be $y_0 = -1$ rad and $\dot{y}_0 = -30$ rad/s. For the underdamped case, the solution is

$$y(t) = 2 + 2.097\epsilon^{-3t}\sin(9.539t - 107.5°)$$
$$-\epsilon^{-3t}[1.048\sin(9.539t + 72.54°) + 3.145\sin(9.539t)], \qquad (6.26)$$

where the zero-state response is on the first line and the zero-input response on the second. The numerical values shown are rounded off to four significant digits. For the overdamped case, the solution is

$$y(t) = 2 + 0.809\epsilon^{-18.63t} - 2.809\epsilon^{-5.367t}$$
$$+ 2.666\epsilon^{-18.63t} - 3.666\epsilon^{-5.367t}. \qquad (6.27)$$

Both the underdamped and the overdamped responses are displayed in Fig. 6.5, which shows $y_{ZIR}(t)$ and $y_{ZSR}(t)$ separately in the lower graphs, and the sum $y(t)$ in the upper graph.

A major advantage of the Laplace transform approach to the dynamics problem is illustrated by Eqn. 6.23. This algebraic equation in the *Laplace variable s* comes from the transformation of the differential equation, Eqn. 6.19. Each of the bracketed terms in Eqn. 6.23 is a function involving only the

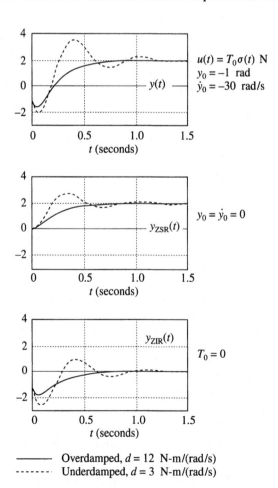

Fig. 6.5 $y(t)$, $y_{ZSR}(t)$, and $y_{ZIR}(t)$ for overdamped and underdamped cases.

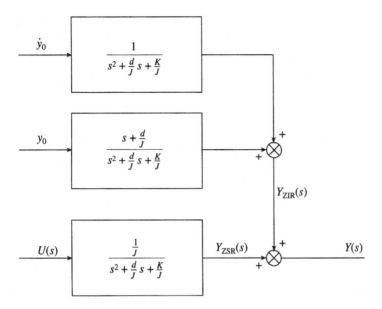

Fig. 6.6 Block diagram equivalent to Eqn. 6.23.

physical parameters of the system and the variable s. Each is multiplied by a term corresponding to the exciting agents of the system: the Laplace transform of the driving force $U(s)$ in the ZSR, and the numerical values of y_0 and $\dot{y}_0$ in the two terms in the ZIR. The solution to the initial-value problem defined in Eqn. 6.19 is therefore obtained by the process of multiplying two functions of s, which yields $Y(s)$. Of course, $Y(s)$ must be inverse-transformed to obtain the actual solution $y(t)$.

Each of the bracketed terms in Eqn. 6.23 is called a *transfer function*. The first bracketed term relates $Y_{ZSR}(s)$ to the input function $U(s)$. The second and third term each relate one component of $Y_{ZIR}(s)$ to its respective initial condition.

Figure 6.6 shows a block diagram for Eqn. 6.23. This diagram conveys exactly the same information as does Eqn. 6.23. Each block has an input at one end (indicated by an arrowhead) and an output at the other end. The output of a block is the product of the input and the transfer function written in the block. Each summing junction has two inputs and one output, with arrows and algebraic signs indicating how the diagram is to be interpreted.

Note that each of the three transfer functions for this system has the same denominator. The reason for this is clear from the steps taken in going from Eqn. 6.21 to Eqn. 6.23. This denominator, the second-order polynomial $s^2 + (d/J)s + K/J$, is called the *characteristic polynomial* for the system. The origin of this term will become clear as we progress, but we see here that the coefficients of the terms in this polynomial are algebraic combinations of the three characterizing parameters of the physical system – J, d, and K. It is the order of the characteristic polynomial, in this case 2, that defines the *dynamic order* of the system. This is the same as both the order of the original differential equation and the dimension of the state-variable matrix $x(t)$ in the state-variable model for the system. In Section 6.5 we generalize our analysis to the nth-order SISO system, where some structure not appearing in this example is added to our analysis.

6.5 The nth-Order Single-Input–Single-Output System

The differential equation for the linear nth-order single-input–single-output dynamic system with input $u(t)$ and output $y(t)$ is

$$\beta_n \frac{dy^n(t)}{dt^n} + \beta_{n-1} \frac{dy^{n-1}(t)}{dt^{n-1}} + \cdots + \beta_1 \frac{dy(t)}{dt} + \beta_0 y(t)$$

$$= \alpha_m \frac{du^m(t)}{dt^m} + \alpha_{m-1} \frac{du^{m-1}(t)}{dt^{m-1}} + \cdots + \alpha_1 \frac{du(t)}{dt} + \alpha_0 u(t), \qquad (6.28)$$

where $m \leq n$ in most physical systems and in many cases $m < n$. To derive the transfer functions for this system, take the Laplace transform of both sides of Eqn. 6.28:

$$\left\{ \begin{array}{l} \beta_n [s^n Y(s) - s^{n-1} y_0 - \cdots - s y_0^{(n-2)} - y_0^{(n-1)}] \\ + \beta_{n-1} [s^{n-1} Y(s) - s^{n-2} y_0 - \cdots - y_0^{(n-2)}] \\ \vdots \\ + \beta_0 [Y(s)] \end{array} \right\}$$

$$= \left\{ \begin{array}{l} \alpha_m [s^m U(s) - s^{m-1} u_0 - \cdots - u_0^{(m-1)}] \\ + \alpha_{m-1} [s^{m-1} U(s) - s^{m-2} u_0 - \cdots - u_0^{(m-2)}] \\ \vdots \\ + \alpha_0 [U(s)] \end{array} \right\}, \qquad (6.29)$$

where the notation $y_0^{(n-1)}$ means $(d^{n-1} y(t)/dt^{n-1})|_{t=0}$. Collect the terms of like powers of s and re-arrange Eqn. 6.29 to obtain:

$$[\beta_n s^n + \beta_{n-1} s^{n-1} + \cdots + \beta_1 s + \beta_0] Y(s)$$
$$= [\alpha_m s^m + \alpha_{m-1} s^{m-1} + \cdots + \alpha_1 s + \alpha_0] U(s)$$
$$+ [\text{terms due to the initial conditions}]. \qquad (6.30)$$

Divide both sides of Eqn. 6.30 by $[\beta_n s^n + \beta_{n-1} s^{n-1} + \cdots + \beta_1 s + \beta_0]$. Then

$$Y(s) = \underbrace{\left\{ \frac{K[s^m + a_{m-1} s^{m-1} + \cdots + a_1 s + a_0]}{s^n + b_{n-1} s^{n-1} + \cdots + b_1 s + b_0} \right\} U(s)}_{\text{ZSR}}$$

$$+ \underbrace{\left\{ \frac{Z_0^{(n-1)} s^{n-1} + \cdots + Z_0^{(1)} s + Z_0}{s^n + b_{n-1} s^{n-1} + \cdots + b_1 s + b_0} \right\}}_{\text{ZIR}}, \qquad (6.31)$$

where $K = \alpha_m/\beta_n$, $a_{m-1} = \alpha_{m-1}/\alpha_m$, ..., $a_0 = \alpha_0/\alpha_m$, and $b_{n-1} = \beta_{n-1}/\beta_n$, ..., $b_0 = \beta_0/\beta_n$. The constant terms Z_0, $Z_0^{(1)}$, ..., $Z_0^{(n-1)}$ depend on the initial conditions appearing in Eqn. 6.29.

A third-order example illustrates this general case. Let the differential equation be

$$2\dddot{y}(t) + 4\ddot{y}(t) + 6\dot{y}(t) + 8y(t) = 10u(t) + 20\dot{u}(t). \tag{6.32}$$

The corresponding $Y(s)$ is

$$Y(s) = \underbrace{\left\{ \frac{10(s+0.5)}{s^3 + 2s^2 + 3s + 4} \right\} U(s)}_{\text{ZSR}}$$

$$+ \underbrace{\left\{ \frac{s^2 + 2s + 3}{s^3 + 2s^2 + 3s + 4} \right\} y_0 + \left\{ \frac{s+2}{s^3 + 2s^2 + 3s + 4} \right\} \dot{y}_0 + \left\{ \frac{1}{s^3 + 2s^2 + 3s + 4} \right\} (\ddot{y}_0 - 10u_0).}_{\text{ZIR}}$$

$$\tag{6.33}$$

We note again that each of the four transfer functions in this system has the same characteristic polynomial, $[s^3 + 2s^2 + 3s + 4]$. This third-order polynomial signifies that the original differential equation is of third order, establishing the dynamic order of the system as 3. Furthermore, each transfer function is a *rational algebraic fraction,* that is, a ratio of polynomials in *s*. Also, each transfer function in this example is a *proper* rational algebraic fraction because the order of the numerator polynomial is less than that of the denominator polynomial. If the order of the numerator is equal to or greater than that of the denominator, the transfer function would be termed an *improper* rational algebraic fraction, which would occur only if $m \geq n$ in the original differential Eqn. 6.28.

The coefficients in the numerator and denominator polynomials (in this case, 10, 0.5, 1, 2, 3, and 4) are all algebraic combinations of the coefficients of the original differential equation,

$$10 = \frac{20}{2}, \quad 0.5 = \frac{10}{2 \times 10}, \quad 1 = \frac{2}{2}, \quad 2 = \frac{4}{2}, \quad 3 = \frac{6}{2}, \quad 4 = \frac{8}{2},$$

illustrating again that the transfer functions depend only on the physical properties of the system and *not* on the specific inputs to the transfer functions.

Transfer functions are often represented by a shortened notation. In this third-order example such a notation might be:

$$G(s) = \frac{10(s+0.5)}{s^3 + 2s^2 + 3s + 4},$$

$$\tag{6.34}$$

$$I_0(s) = \frac{s^2 + 2s + 3}{s^3 + 2s^2 + 3s + 4}, \quad I_1(s) = \frac{s+2}{s^3 + 2s^2 + 3s + 4}, \quad I_2(s) = \frac{1}{s^3 + 2s^2 + 3s + 4},$$

and Eqn. 6.33 could be diagramed as shown in Fig. 6.7.

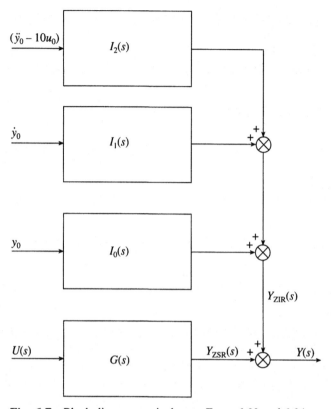

Fig. 6.7 Block diagram equivalent to Eqns. 6.33 and 6.34.

We use a standard form for transfer functions throughout the book. In this form, the denominator is written as a *monic polynomial,* which means that the coefficient of the highest power of s is 1. The numerator is written as a monic polynomial multiplied by a constant. Each of the transfer functions in Eqn. 6.34 appears in this standard form. Any rational algebraic fraction can be written in this form. As an example, the rational algebraic fraction

$$\frac{15s^2+375s+1500}{3s^4+27s^3+135s^2+261s+150}$$

can be written in standard form by dividing each term in the numerator and in the denominator by 3 to give:

$$\frac{5(s^2+25s+100)}{s^4+9s^3+45s^2+87s+50}.$$

In all the rational fractions encountered in this book, the multiplying constants and the coefficients in the polynomials are real-valued because these terms are algebraic combinations of the real-valued physical parameters of the system. Therefore, the polynomials can always be factored into real polynomials of order 1 or 2. For example,

$$\frac{5(s^2+25s+100)}{s^4+9s^3+45s^2+87s+50}=\frac{5(s+5)(s+20)}{(s+1)(s+2)[s^2+6s+25]}. \tag{6.35}$$

A second-order polynomial might be factorable into two first-order polynomials having real-valued constant terms, as in the numerator of Eqn. 6.35. However, it might also occur that the two first-order factors have complex-valued constant terms, as in the quadratic term $[s^2 + 6s + 25]$, which is factored as $[s^2 + 6s + 25] = (s + 3 + j4)(s + 3 - j4)$, where $j = \sqrt{-1}$. In this latter case the two constants are always complex conjugates of one another. The task of factoring a polynomial of order 3 or higher is tedious without the aid of a computer program designed for this task. Since much of the analytical work in control system engineering is critically dependent on factoring polynomials, an engineer must have such a computing facility readily available.

In many analysis problems the initial conditions are zero. In these cases the problem is simplified, involving a single transfer function. In the example of Eqn. 6.34 that transfer function is $G(s)$, the one relating $Y_{ZSR}(s)$ to $U(s)$:

$$G(s) = \frac{Y_{ZSR}(s)}{U(s)} = \frac{10(s+0.5)}{s^3 + 2s^2 + 3s + 4} = \frac{10(s+0.5)}{(s+1.651)[s^2 + 0.3494s + 2.423]}.$$
(6.36)

Frequently the subscript on $Y_{ZSR}(s)$ is dropped, and $G(s)$ is called "the" transfer function of the system, even though it is only one of several transfer functions necessary to completely characterize the initial-value problem.

Computer programs for control system analysis can convert a state-variable vector-matrix model of a linear system into the corresponding transfer function form when numerical entries for the A, B, C, and D matrices are provided. These programs, which work for multivariable systems as well as for SISO systems, are discussed in Chapter 15.

The Laplace transforms of all the input functions $u(t)$ used in this book are rational algebraic functions. Since $Y_{ZSR}(s)$ is the product of two rational algebraic functions, it is also a rational algebraic function having the same mathematical form as the transfer function. As an example, for the system in Eqn. 6.36 let $u(t) = 3[1 - \epsilon^{-2t}]$. Use the table of Laplace transforms to find that $U(s) = 6/[s(s+2)]$. Then $Y_{ZSR}(s)$ is

$$Y_{ZSR}(s) = \frac{60(s+0.5)}{s(s+2)(s+1.651)[s^2 + 0.3494s + 2.423]}.$$
(6.37)

The term $y_{ZSR}(t)$ is obtained by taking the inverse Laplace transform of $Y_{ZSR}(s)$. Consulting Appendix C, we see that $Y_{ZSR}(s)$ does not match any of the forms in the table of Laplace transforms given there. We may therefore choose either to consult a more comprehensive table, hoping to find an entry that matches our $Y_{ZSR}(s)$, or to expand $Y_{ZSR}(s)$ in *partial fractions,* as follows:

$$Y_{ZSR}(s) = \underbrace{\frac{3.75}{s}}_{\text{pair 2}} + \underbrace{\frac{26.19}{s+1.651}}_{\text{pair 5}} - \underbrace{\frac{22.5}{s+2}}_{\text{pair 5}} - \underbrace{\frac{7.439(s+0.5875)}{(s+0.1747)^2 + (1.547)^2}}_{\text{pair 15}}.$$
(6.38)

Now $Y_{ZSR}(s)$ is expressed as the sum of four simple terms, each of which corresponds to one of the factors of the denominator of $Y_{ZSR}(s)$, and each of which

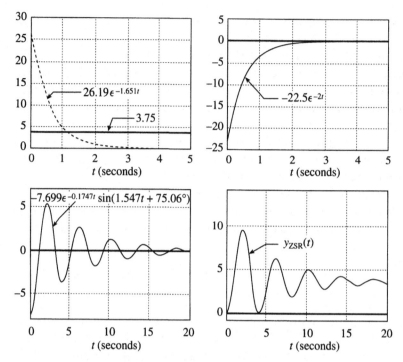

Fig. 6.8 $y_{ZSR}(t)$ (lower right) and its four components (Eqn. 6.39). Note difference in time and amplitude scales.

appears in the table of transforms in Appendix C. The time-domain response $y_{ZSR}(t)$ is easily obtained as the sum of the inverse transforms of the individual terms in Eqn. 6.38:

$$y_{ZSR}(t) = 3.75 + 26.19\epsilon^{-1.651t} - 22.5\epsilon^{-2t}$$
$$- 7.699\epsilon^{-0.1747t} \sin(1.547t + 75.06°). \tag{6.39}$$

Each of these four terms is plotted in Fig. 6.8, as is the composite $y_{ZSR}(t)$.

The partial-fraction expansion of rational algebraic fractions is fully developed in Chapter 7, where notation peculiar to control system engineering is introduced.

6.6 Composite Transfer Functions

An engineering system is usually made up of a collection of physical devices, which are interconnected so that the whole system will behave in a prescribed manner in response to the several inputs and initial conditions that excite it. One of the important features offered by the transfer function approach in analyzing the behavior of such a composite collection of devices is illustrated in Fig. 6.9. Here three devices are connected so that the output of device 1, labeled $Y_1(s)$, is the input to device 2. The output of device 2 is the input to device 3, and the output of the whole system is $Y_3(s)$. Each device is modeled

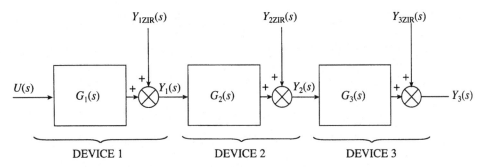

Fig. 6.9 Composite transfer function for three devices in series, including zero-input response for each device.

by the transfer functions relating its output to its input and to its initial conditions. These algebraic relationships are derived in Sections 6.4 and 6.5. The composite system model, relating the system input $U(s)$ and the various non-zero initial conditions of the three devices, also incorporates transfer functions. For this example we have

$$Y_3(s) = \underbrace{U(s) \times [G_1(s)G_2(s)G_3(s)]}_{Y_{3\mathrm{ZSR}}(s)}$$

$$+ \underbrace{Y_{3\mathrm{ZIR}}(s) + Y_{2\mathrm{ZIR}}(s) \times G_3(s) + Y_{1\mathrm{ZIR}}(s) \times G_2(s)G_3(s)}_{\text{terms due to initial conditions}}. \qquad (6.40)$$

If the initial conditions are zero then the composite relationship becomes very simple, and the overall system transfer function becomes the product of the three individual transfer functions, $G_1(s)G_2(s)G_3(s)$.

Figure 6.10 shows a second configuration of three interconnected devices, characterized by the three transfer functions $G(s)$, $P(s)$, and $Q(s)$. (The initial conditions are all assumed to be zero.) Here the overall system transfer function is

$$\frac{Y(s)}{U(s)} = [G(s) - P(s)Q(s)]. \qquad (6.41)$$

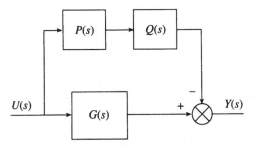

Fig. 6.10 Series–parallel connection of three zero-state transfer functions (Eqn. 6.41).

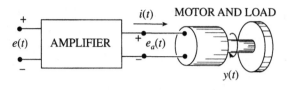

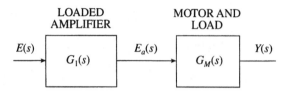

Fig. 6.11 Schematic diagram of amplifier–motor system and corresponding block diagram.

An important precaution against error when analyzing two interconnected devices is illustrated as follows, using the amplifier–servomotor combination shown in Fig. 4.9 and analyzed in Section 6.3. Figure 6.11 depicts this system in two ways. The upper diagram is a schematic picture which says little more than that an amplifier with output voltage $e_a(t)$ is connected to a motor by two wires. This is useful and perhaps even necessary information, but the diagram is not suitable for dynamic analysis. The lower diagram is devoid of some technical details concerning both amplifier and motor, but it has the necessary mathematical information (except for the initial-condition transfer functions, which are omitted) for dynamic analysis of the composite system. (The input load torque is assumed to be zero for this illustration and so it, too, is omitted.)

First we derive the transfer function for the motor. Under the assumptions we have made, the two differential equations defining the dynamic relationship between the voltage $e_a(t)$ and the shaft position $y(t)$ are

$$e_a(t) = L_a \frac{di(t)}{dt} + Ri(t) + K\frac{dy(t)}{dt},$$

$$Ki(t) = J\frac{d^2y(t)}{dt^2} + b\frac{dy(t)}{dt}.$$

(6.42)

We take the Laplace transform of each of these equations, assuming that the initial conditions i_0, y_0, and $\dot{y}_0$ are all zero. This yields two algebraic equations involving $E_a(s)$, $I(s)$, $Y(s)$, the Laplace variable s, and the motor parameters K, J, b, R, and L_a; $I(s)$ is eliminated between these equations because it is not a terminal variable in the transfer function $G_M(s)$. The single equation that results is

$$E_a(s) = \frac{JL_a}{K}\left[s^3 + \left(\frac{RJ+bL_a}{JL_a}\right)s^2 + \left(\frac{Rb+K^2}{JL_a}\right)s\right]Y(s).$$

(6.43)

Re-arranging Eqn. 6.43 yields the desired transfer function

$$
G_M(s) = \frac{Y(s)}{E_a(s)} = \frac{\dfrac{K}{JL_a}}{\left[s^3 + \left(\dfrac{RJ+bL_a}{JL_a} \right) s^2 + \left(\dfrac{Rb+K^2}{JL_a} \right) s \right]}. \tag{6.44}
$$

Next we derive the transfer function of the amplifier *with the motor connected*. The output circuit of the amplifier behaves according to

$$
K_A e(t) - R_o i(t) = e_a(t). \tag{6.45}
$$

Transformed, this becomes

$$
K_A E(s) - R_o I(s) = E_a(s). \tag{6.46}
$$

The $I(s)$ term is obtained from the analysis used to calculate $G_M(s)$, and is substituted into Eqn. 6.46 to give a single equation involving only $E(s)$ and $E_a(s)$, which in turn is re-arranged to give the transfer function of the amplifier as it is connected to the motor:

$$
\frac{E_a(s)}{E(s)} = G_1(s) = \frac{K_A \left[s^2 + \left(\dfrac{RJ+bL_a}{JL_a} \right) s + \left(\dfrac{Rb+K^2}{JL_a} \right) \right]}{\left[s^2 + \left(\dfrac{(R+R_o)J+bL_a}{JL_a} \right) s + \left(\dfrac{(R+R_o)b+K^2}{JL_a} \right) \right]}. \tag{6.47}
$$

We see that the transfer function of the amplifier, relating the input voltage $e(t)$ to the output voltage $e_a(t)$, involves not only the physical parameters associated with the amplifier unit, K_A and R_o, but all the parameters of the motor as well! Because the motor places a load on the amplifier, the amplifier will not perform as it would if it were unloaded. The loading effect between physical elements should always be evaluated in the design or analysis process. The manufacturer of an amplifier, typically not knowing how the device is to be used, may quote the "transfer function" to be the gain K_A, a quote that may mislead an unwary design engineer.

In some cases the loading problem is relatively minor. In our present example, if the output resistance R_o is very much smaller than the motor resistance R, the quadratic term in the numerator of $G_1(s)$ will be almost the same as the quadratic term in the denominator, and the transfer function will be approximately K_A.

This example has also shown that there may not be a one-to-one correspondence between the interconnected physical devices of a system and the block diagrams that describe the dynamic performance of the devices. This is actually an advantage in analysis because it permits one to study critical features of system behavior, either by splitting the transfer function of a single device into its mathematical components to reveal an important dynamic property of the system, or by adding a transfer function of a hypothetical device in a chain of blocks to discover possible improvements in design.

6.7   Summary

The solution to the initial-value problem for an nth-order multivariable system modeled as $\dot{x}(t) = Ax(t) + Bu(t)$ was shown to be similar in form to that of a first-order single-variable system $\dot{x}(t) = ax(t) + bu(t)$. The difference in the two forms lies in the matrix notation (A and B) used in the multivariable case and the scalar notation (a and b) used in the single-variable case. These solutions revealed both the distinction between the zero-state response (ZSR) and the zero-input response (ZIR), and also the difference between the transient response and the steady-state response in systems where that traditional terminology applies. The formal definition of a linear system was given in Remark 5 following Eqn. 6.14.

Analytical solutions to initial-value problems can be easily obtained only for relatively simple systems. For systems of dynamic order higher than 3, or for those whose input functions are multiple-term expressions, a solution is usually obtained by numerical methods using a standardized computer program. The solution for the transient response of the amplifier–motor system in Section 6.3 demonstrated both the practical value of the computer solution and the value of understanding the principles behind the analytical approach.

The transfer function method, based on the Laplace transformation of differential equations, is especially useful in single-input–single-output systems. The Laplace transform of the solution to the initial-value problem, $Y(s)$, was derived in Sections 6.4 and 6.5; $Y(s)$ is the sum of a zero-state response $Y_{ZSR}(s)$ and a zero-input response $Y_{ZIR}(s)$. The transfer function relating $U(s)$, the Laplace transform of the input $u(t)$, to $Y_{ZSR}(s)$ was defined and shown to be a rational algebraic function of s whose denominator, the characteristic polynomial, contains all the system parameters. The transfer functions relating $Y_{ZIR}(s)$ to the initial conditions appearing in the original differential equation were similarly derived. The block diagrams in Figs. 6.6 and 6.7 portray these relationships. Laplace transform tables may be used to evaluate solutions to specific problems. The technique of the partial-fraction expansion of $Y(s)$ to aid this process was also shown. The partial-fraction expansion, which reveals significant features of the solutions, is developed in greater detail in Chapter 7.

The transfer function of a system composed of the interconnection of several SISO devices having unidirectional transmission characteristics is an algebraic combination of the transfer functions of the individual devices. The individual transfer functions for each device must be derived when it is loaded by the device that it drives.

In Chapter 7 we shall use the s-domain approach to continue our analysis of SISO linear systems. The central purpose of that analysis is to relate the physical parameters of a system to the dynamic response of the system to given inputs and initial conditions. In engineering, a calculated system response frequently turns out to be unstable, or to be too slow, or to oscillate excessively, or to be otherwise unsatisfactory. Usually any one of several changes could be made in the system parameters to improve the response, and the engineer must select the best of these possible changes. The talent for distinguishing good approaches from those unfitted to the problem is a hallmark of a professional engineer. This talent has its analytical foundation in the techniques of Chapter 7.

6.8 Problems

Problem 6.1

(a) If $f(t) = 3 + 5t - 2\sin(\omega_0 t)$, find $F(s) = \mathcal{L}\{f(t)\}$. Refer to Appendix C for a review of the Laplace transform and for a table of transform pairs.

(b) From the differential equation $\dddot{y}(t) - 2\ddot{y}(t) + 5\dot{y}(t) = 0$ with initial conditions $\ddot{y}(0) = 1$, $\dot{y}(0) = 0$, and $y(0) = 0$, find the solution $y(t)$. Is $y(t)$ a stable function?

(c) From the differential equation $\ddot{z}(t) + 2\dot{z}(t) + 6z(t) = 3\epsilon^{-3t}$ and initial conditions $\dot{z}(0) = 0$ and $z(0) = 1$, find the solution $z(t)$. Is $z(t)$ a stable function?

(d) Given the integral equation $f(t) = \int_0^t f(t)\,dt + \epsilon^{-4t}$, find $f(t)$.

Problem 6.2

(a) If $H(s) = 9/(s+3)^2$, find $h(t) = \mathcal{L}^{-1}\{H(s)\}$.

(b) If $B(s) = (3s+6)(s-1)/s(s+1)(s^2+4s+4)$, find $b(t) = \mathcal{L}^{-1}\{B(s)\}$.

(c) Given $D(s) = K(s+2)/s(s+3)^2(s+4)$, find the initial value of $d(t)$. Determine the value K should have if the final value of $d(t)$ is to be 1.

(d) Given $F(s) = 2(s+1)/(s+2)$, find $f(t) = \mathcal{L}^{-1}\{F(s)\}$.

Problem 6.3

In the circuit of Fig. 6.1 the switch is closed at $t = 0$, the initial charge on the capacitor is Q_0 C, and the applied voltage is $u(t) = E\epsilon^{-\alpha t}$ V. Find the current $i(t)$ if $\alpha \neq 1/RC_1$. Find $i(t)$ if $\alpha = 1/RC_1$.

Problem 6.4

(a) In Fig. P6.4, $e_{in}(t)$ is the input to the circuit. Assume that the initial condition $i(0)$ is zero. Obtain the transfer function $E_L(s)/E_{in}(s)$.

(b) If $i(t)$ is considered to be the input to the circuit, find the transfer function $E_L(s)/I(s)$.

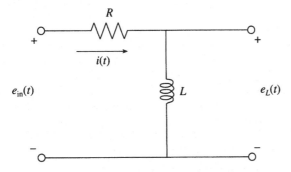

Fig. P6.4

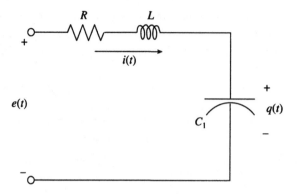

Fig. P6.5

Problem 6.5

Derive the transfer function for the zero-state response of the RLC circuit in Fig. P6.5. The charge on the capacitor $q(t)$ C is the output, and the applied voltage $e(t)$ V is the input. Compare your answer with $Y_{ZSR}(s)/U(s)$ in Fig. 6.6. Note the duality between the mechanical quantities (torque, moment of inertia, viscous damping, elasticity, etc.) and the electrical quantities (voltage, inductance, resistance, inverse capacitance, etc.). What is the mechanical dual of the current through the resistor in this comparison?

Problem 6.6

Refer to Problem 2.5. Let the input be $f(t)$ and the output be $z_c(t)$. The initial conditions are $\dot{z}_c(0) = \dot{z}_0$ and $z_c(0) = z_0$.

(a) Find $Z_c(s)$.
(b) Let $M = 1$ kg, $\alpha = 25/3$ N/(m/s), $k = 100$ N/m, and $f(t) = 10\epsilon^{-2t}$ N. Calculate $z_c(t)$ if $\dot{z}_0 = 0$ and $z_0 = 0.1$ m.

Problem 6.7

The input of a linear system is $u(t)$, and the output is $y(t)$. The transfer function relating $u(t)$ to the zero-state response (ZSR) of $y(t)$ is $G(s)$. Show that $g(t) = \mathcal{L}^{-1}\{G(s)\}$ is the ZSR of the system to a unit impulse input. Also, show that the ZSR to a unit ramp input is equal to the time integral of the ZSR to a unit step input.

Problem 6.8

How is the ZSR of a linear system to a unit ramp input ($u(t) = 1t$) related to its response to a unit parabolic input ($u(t) = 1t^2$)?

Problem 6.9

The switch in Fig. P6.9 is closed at time $t = 0$. The resultant output is one of the two curves shown. Find the value of R_2 and determine the time t_1 if $R_1 = 6\ \Omega$ and $L = 2$ H.

Problem 6.10

Obtain the zero-input transfer functions $Y(s)/i_L(0)$ and $Y(s)/q_C(0)$ for the circuit in Fig. P6.10, where $q_C(t)$ is the charge on capacitor C_1.

Problem 6.11

Obtain the zero-state transfer function $Y(s)/U(s)$ for the circuit in Fig. P6.10.

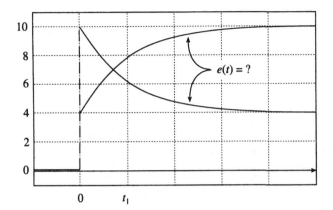

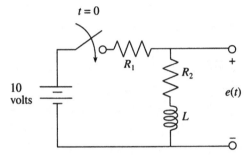

Fig. P6.9

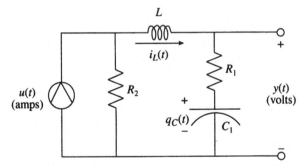

Fig. P6.10

Problem 6.12

Refer to the circuit in Fig. 3.3. Obtain the zero-state transfer function $I_2(s)/E(s)$ of this circuit. Also, obtain the zero-state transfer function $E_L(s)/E(s)$.

Problem 6.13

Refer to the circuit in Fig. 3.6. Obtain the zero-state transfer function $Y(s)/U(s)$, where we use the notation $u(t) = I(t)$ for the input to the system.

Problem 6.14

Refer to the circuit in Fig. 3.7. Let the VCVS be characterized by $E_s(t) = K_0 e_1(t)$. Use the notation $u(t) = E(t)$ for the input, and obtain the zero-state transfer function $Y(s)/U(s)$. Also, obtain the zero-input transfer function $Y(s)/e_C(0)$.

Problem 6.15

Refer to Problem 2.3. Derive the zero-state transfer function $Z(s)/F(s)$. If $M = 5$ kg, $r = 0.2$ m, $J = 0.2$ kg, and $k = 360$ N/m, what is the numerical value of α for which the system will be "critically damped," that is, on the borderline between overdamped and underdamped?

Problem 6.16

In the system of Problem 6.15, $f(t) = 100\sigma(t)$ N. Use the final-value theorem (see Appendix C) to find the steady-state displacement of the mass, that is, $\lim_{t \to \infty} z(t)$. Does your answer agree with your physical interpretation of this problem?

Problem 6.17

Put the following rational functions in the standard factored form illustrated in Section 6.5.

(a) $A(s) = \dfrac{12s+2}{2s^2+10s+12}$.

(b) $B(s) = \dfrac{2400s^2+7200s+4800}{2s^4+74s^3+844s^2+3520s+4800}$.

(c) $C(s) = \dfrac{K[s^3+3s^2+22s+20]}{2s^4+6s^3+72s^2+320s}$.

(d) $D(s) = \dfrac{20s^2+60s+40}{2s^3+6s^2+24s+20}$.

(e) $E(s) = \dfrac{90-72s-18s^2}{s^4+11s^3+19s^2+99s+90}$.

(f) $F(s) = \dfrac{4s+8}{(0.2s+1)[0.02s^2+0.04s+2]}$.

Problem 6.18

Let the Laplace transform of $c(t)$ be the function in Problem 6.17(c).

(a) Use the initial-value theorem to find the value of K for which $c(0^+) = 1$.
(b) Can you find a value of K for which the final value of $c(t)$ will be 1? If so, find that value; if not, explain why not.

Problem 6.19

Refer to Problem 4.5. Adopt the notation $u(t) = e(t)$, $y_1(t) = z(t)$, and $y_2(t) = \beta(t)$. Assume that the belt-drive efficiency is 1, and that the angular displacement of the pendulum remains small enough so that the linear model is valid. Derive the two zero-state transfer functions $Y_1(s)/U(s)$ and $Y_2(s)/U(s)$ in terms of the eighteen physical parameters of the amplifier, motor, belt drive, cart, and pendulum system.

Problem 6.20

Refer to Problems 4.5 and 6.19. The numerical values of the physical parameters of the amplifier–motor–belt drive–cart system are: $K_A = 3.9$, $R_o = 1\ \Omega$, $R = 2\ \Omega$, $L_a = 3$ mH, $K = 0.48$ N/A, $b = 0.0048$ N-m/(rad/s), $J = 4 \times 10^{-4}$ kg-m^2, $M = 12.5$ kg, $n = 5$, $J_r = 25 \times 10^{-4}$ kg-m^2, $J_f = 18 \times 10^{-4}$ kg-m^2, $r = 0.06$ m, and $g = 9.81$ m/s^2. The pendulum is a slim uniform rod 2.5 m in length with a mass of 2.5 kg. Evaluate the two zero-state transfer functions $Y_1(s)/U(s)$ and $Y_2(s)/U(s)$.

Problem 6.21
Refer to Problem 4.7. Use the notation $y(t) = \theta_L(t)$, and obtain the zero-state transfer function $Y(s)/E_a(s)$ in terms of the ten physical parameters of the system given in Problem 4.7. Also, obtain the zero-input transfer function $Y(s)/y(0)$.

Problem 6.22
Evaluate the two transfer functions in Problem 6.21 for the following numerical values of the physical parameters: $K = 0.48$ N-m/A, $R = 2\,\Omega$, $b = 0.0048$ N-m/(rad/s), $L_a = 3$ mH; $J_L = 150 \times 10^{-4}$, $J_2 = 16 \times 10^{-4}$, $J_1 = 0.4 \times 10^{-4}$, $J = 4 \times 10^{-4}$ kg-m²; $n = 6$, and $k_L = 40$ N-m/rad. If $e_a(t) = 50\sigma(t)$ V, use the final-value theorem (see Appendix C) to find the steady-state displacement of the output shaft resulting from this input.

Problem 6.23
Refer to Problems 4.11 and 4.12. Evaluate the zero-state transfer function $Z_R(s)/E_{in}(s)$.

Problem 6.24
Refer to Problems 4.13 and 5.6. Let the physical parameters be: $K_M = 5.65 \times 10^{-6}$ N-m²/ A², $i_0 = 0.833$ A, $g_0 = 0.002$ m, $M_A = 0.1$ kg, $d = 2$ N/(m/s), $A = 10^{-4}$ m², and $k_0 = 10$ V/m. Obtain the zero-state transfer function $Y(s)/\Delta I(s)$.

Problem 6.25
Let the current $\Delta i(t)$ in the electromagnet be produced by an amplifier such as that in Fig. 4.12, characterized by $K_A = 2$ and $R_o = 1\,\Omega$. The magnet coil is characterized by $R_c = 4\,\Omega$ and $L_c = 5$ mH. Combine your analysis in Problem 6.24 with these data to obtain the zero-state transfer function $Y(s)/E_{in}(s)$.

Problem 6.26
Refer to Problems 2.10 and 2.12. Use the numerical values for the stability derivatives of the Navion aircraft given in Problem 2.10. Adopt the notation $y_2(t) = \Delta\theta(t)$ and $y_4(t) = \Delta h(t)$, as indicated in Problem 2.12. Assume that the throttle remains fixed so that the input may be taken to be $u(t) = \Delta\delta_E(t)$. Obtain the two zero-state transfer functions $Y_2(s)/U(s)$ and $Y_4(s)/U(s)$.

Problem 6.27
Refer to Problem 2.18. Adopt the notation $x_1(t) = \theta_1(t)$, $x_2(t) = \theta_2(t)$, and $u(t) = T(t)$. Derive the two zero-state transfer functions $X_1(s)/U(s)$ and $X_2(s)/U(s)$.

Problem 6.28
In the system of Problem 6.27, derive the zero-input transfer functions $X_2(s)/\theta_1(0)$, $X_2(s)/\dot{\theta}_1(0)$, and $X_2(s)/\theta_2(0)$.

Problem 6.29
Refer to the linear state-variable model for the ball, rack, and pendulum obtained in Problem 5.13. Change the time-domain notations from $T(t)$ to $m(t)$ and from $\theta(t)$ to $x_1(t)$, so that the Laplace transform of these variables will be denoted $M(s)$ and $X_1(s)$. Let the eight physical parameters have the numerical values given in Problem 5.13. Derive the transfer functions for the zero-state response, $Z(s)/M(s)$ cm/dyne-cm and $X_1(s)/M(s)$ rad/dyne-cm. Express both of these transfer functions in the standard factored form shown in the right-hand side of Eqn. 6.36.

Problem 6.30

Refer to Problem 4.9. The DC motor parameters are $L_a = 0$, $K = 0.05$ N-m/A, $R = 3\,\Omega$, $b = (1/600)$ N-m/(rad/s), and $J = 1 \times 10^{-4}$ kg-m^2. The belt-drive ratio is $n = 5$. The motor is driven by an amplifier characterized as in Problem 4.4, with $K_A = 4$ and $R_o = 1\,\Omega$. The input voltage to the amplifier is $u(t)$. The physical parameter values for the ball, rack, and pendulum system are those given in Problem 6.29. Evaluate the zero-state transfer functions $X_1(s)/U(s)$ and $Z(s)/U(s)$.

7 Pole-Zero Methods of Analysis for Single-Input–Single-Output Systems

7.1 Introduction

We now develop methods for calculating the dynamic response of a linear system that is modeled by its transfer functions. The response depends upon the physical parameters of the system, the input function, and any nonzero initial conditions. We are particularly interested in relating the dynamic features of the response to the physical properties of the system. If the input is a step function, the response may fluctuate temporarily and eventually reach a constant value. The nature of the fluctuations – the length of time during which they persist, and whether they cause the response to overshoot its final value excessively or to oscillate with both positive and negative values – are dynamic features of vital importance. The initial values of the response and its derivatives and the final value of the response are also important characteristics that depend on the physical parameters. In simple systems with simple inputs – for example, a first- or second-order system with a step input – the dynamic features of the response are directly related to simple combinations of the parameter values. But in higher-order systems these important relationships are less obvious because the significant dynamic features of the response depend on complicated combinations of the parameter values. We must now employ a mixture of analysis tools, computer calculations, and approximation techniques to determine which of our system parameters have the most influence on the significant dynamic features of the response. In this chapter we develop some of the tools and techniques upon which the professional talent called "engineering judgment" is founded.

In our analysis, we will make extensive use of the zero-state response of our linear systems to simple input functions such as the step, ramp, exponential pulse, or sinusoid. However, the techniques we employ are also applicable to the zero-input response because the basic forms for these two components of the whole response are the same.

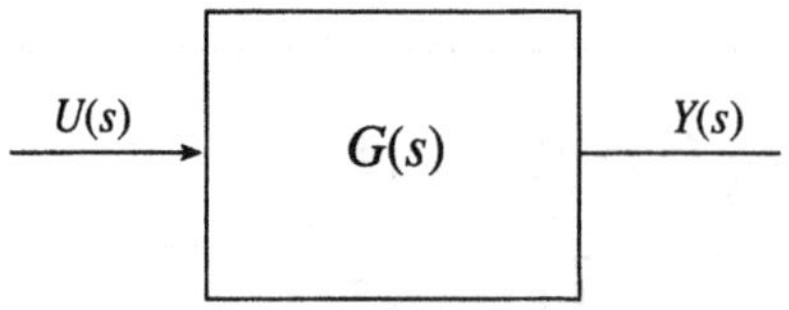

u(t) = input function
y(t) = zero-state response
$Y(s) = \mathcal{L}\{y(t)\}$
$U(s) = \mathcal{L}\{u(t)\}$
$$G(s) = \frac{Y(s)}{U(s)} = \frac{\mathcal{L}\{y(t)\}}{\mathcal{L}\{u(t)\}}$$

Fig. 7.1   Block diagram and notation for zero-state response transfer function $G(s)$.

We begin our analysis by considering the zero-state response of a linear system having a transfer function $G(s)$ that relates the Laplace transform of the input $U(s)$ to the Laplace transform of the zero-state response, denoted $Y(s)$. Figure 7.1 is a block diagram depicting the relationship $Y(s) = G(s) \times U(s)$. For the systems in this book, both $U(s)$ and $G(s)$ are rational functions of s; that is, they are ratios of polynomials in s. Therefore, since $Y(s)$ is also a ratio of polynomials in s, it can always be written in the standard form defined in Section 6.5:

$$Y(s) = \frac{K[s^m + a_{m-1}s^{m-1} + \cdots + a_1 s + a_0]}{s^n + b_{n-1}s^{n-1} + \cdots + b_1 s + b_0}. \tag{7.1}$$

The coefficients $a_0, a_1, a_2, ..., a_{m-1}$ and $b_0, b_1, b_2, ..., b_{n-1}$ and the constant K all depend on the physical parameters of the system and on the input function. Because here we are dealing with physical systems, these coefficients are all real-valued, as opposed to complex-valued. In many cases the coefficients are positive real numbers. Also, in most cases the order m of the numerator of $Y(s)$ is less than or equal to n, the order of the denominator. Gauss's fundamental theorem of algebra permits the numerator and denominator polynomials in Eqn. 7.1 to be expressed in factored form as follows:

$$\begin{aligned} Y(s) &= \frac{K[s^m + a_{m-1}s^{m-1} + \cdots + a_1 s + a_0]}{s^n + b_{n-1}s^{n-1} + \cdots + b_1 s + b_0} \\ &= \frac{K(s + Z_1)(s + Z_2) \cdots (s + Z_m)}{(s + P_1)(s + P_2) \cdots (s + P_n)}, \end{aligned} \tag{7.2}$$

where the numerator polynomial has m roots, $-Z_1, -Z_2, ..., -Z_m$, and the denominator polynomial has n roots, $-P_1, -P_2, ..., -P_n$. Because the m coefficients $a_0, a_1, a_2, ..., a_{m-1}$ are real-valued, the m roots $-Z_1, -Z_2, ..., -Z_m$ are either real-valued or, if complex-valued, they occur in complex conjugate pairs. Similarly, the n roots of the denominator $-P_1, -P_2, ..., -P_n$ either are real-valued or occur in complex conjugate pairs. An example is

$$Y(s) = \frac{50(s^2 + 3s + 2)}{s^3 + 12s^2 + 57s + 100} = \frac{50(s+1)(s+2)}{(s+4)(s+4+j3)(s+4-j3)}. \tag{7.3}$$

Several remarks concerning the algebraic relationships illustrated in Eqns. 7.2 and 7.3 are in order. These remarks are conditioned on the premise that the n roots in the denominator of $Y(s)$ are distinct from the m roots in the numerator.

Remark 1 The order of $Y(s)$ is the order n of the denominator polynomial. But n is also used to denote the dynamic order of the system, that is, the dimension of the A matrix in the state-variable model and consequently the order of the denominator of $G(s)$. So when we say elsewhere that our *system* is of nth order we will mean that the denominator of $G(s)$, not of $Y(s)$, is of order n.

Remark 2 The relationship between the coefficients $b_0, b_1, b_2, ..., b_{n-1}$ of the polynomial in the denominator of $Y(s)$ and the roots $-P_1, -P_2, ..., -P_n$ of that polynomial has concerned analysts for two hundred years. If $n = 1$, this relationship is trivial. If $n = 2$, the relationship can be expressed analytically by the well-known quadratic equation formula

$$-P_1, -P_2 = \frac{-b_1 \pm \sqrt{b_1^2 - 4b_0}}{2}, \tag{7.4}$$

so that the roots will be complex if $4b_0 > b_1^2$. If $n = 3$, the roots may also be expressed as algebraic combinations of the coefficients by the cubic equation formula found in mathematical handbooks. This formula is usually too cumbersome to use in practical situations where numerical approximations to the roots serve the purpose at hand. Similarly, for $n = 4$, the roots can be expressed as algebraic functions of the coefficients by the quartic equation formula also found in mathematical handbooks. However, if $n > 4$ then it is not possible to express the roots $-P_1, -P_2, ..., -P_n$ as algebraic functions of the coefficients $b_0, b_1, b_2, ..., b_{n-1}$. In this case the roots can be found only by assigning numerical values to the coefficients and proceeding numerically. Because much of automatic control system analysis depends on factoring polynomials, it is fortunate that standard computer application programs are readily available to perform this task quickly and nearly always accurately.

Remark 3 Two simple algebraic relationships between the coefficients $b_0, b_1, b_2, ..., b_{n-1}$ and the roots $-P_1, -P_2, ..., -P_n$, which exist for any value for n, are useful in practical situations:

$$b_{n-1} = P_1 + P_2 + \cdots + P_n = \sum_{i=1}^{n} P_i, \quad \text{and}$$

$$b_0 = P_1 P_2 \cdots P_n = \prod_{i=1}^{n} P_i. \tag{7.5}$$

For example, in Eqn. 7.3 $n = 3$, so we have

$$b_2 = 12 = 4 + (4 + j3) + (4 - j3) \quad \text{and}$$
$$b_0 = 100 = 4(4 + j3)(4 - j3). \tag{7.6}$$

The same relationships apply to the polynomial in the numerator:

$$a_{m-1} = Z_1 + Z_2 + \cdots + Z_m = \sum_{i=1}^{m} Z_i, \quad \text{and}$$

$$a_0 = Z_1 Z_2 \cdots Z_m = \prod_{i=1}^{m} Z_i. \tag{7.7}$$

Remark 4   The initial value of the response $y(t)$ may be determined by inspection of $Y(s)$, using the initial-value theorem:

$$y(0^+) = \lim_{s \to \infty} [sY(s)] = \lim_{s \to \infty} \left[\frac{K}{s^{(n-m-1)}} \right]. \tag{7.8}$$

From this we have

$$y(0^+) = \begin{cases} 0 & \text{if } n > m+1, \\ K & \text{if } n = m+1, \\ \infty & \text{if } n < m+1. \end{cases} \tag{7.9}$$

If the final value of $y(t)$ exists then it can also be determined by inspection of $Y(s)$, using the final-value theorem. But *whether* the final value exists cannot be determined simply by inspection of $Y(s)$. The formula for the final-value theorem is

$$y(\infty) = \lim_{s \to 0} [sY(s)], \quad provided\ y(\infty)\ \text{exists}. \tag{7.10}$$

If $y(\infty)$ exists, the response $y(t)$ is said to be a *stable* response. A necessary and sufficient condition for $y(t)$ to be stable is either that all of the roots of the denominator polynomial of $Y(s)$ have negative real parts, or that all but one of these roots have negative real parts with the remaining root being negative or zero. The roots may be determined by factoring the polynomial before the final-value theorem is applied. Alternatively, Routh's method can be applied to the denominator polynomial to determine whether $y(t)$ is stable (see Appendix D).

Remark 5   The response $y(t)$ can be calculated by taking the inverse Laplace transform of $Y(s)$. If this is done, the initial value $y(0^+)$ and the final value $y(\infty)$, if there is one, can be determined simply by inspecting $y(t)$. The initial-value theorem and the final-value theorem may be used to avoid the inverse transform process in cases where further knowledge of $y(t)$ is unnecessary. The partial-fraction expansion technique developed in Section 7.2 is often a convenient method for finding the inverse Laplace transform.

Remark 6   A *zero of* $Y(s)$ is defined as a value of s for which $Y(s) = 0$. We can identify such values by inspection of the factored form of $Y(s)$. For the $Y(s)$ in Eqn. 7.3, for example, it is clear that $s = -1$ and $s = -2$ are two zeros of $Y(s)$. The notation $-Z_1, -Z_2, \ldots$ is chosen for these zeros. A *pole of* $Y(s)$ is defined as a value of s for which $Y(s) = \infty$. For the $Y(s)$ in Eqn. 7.3 we

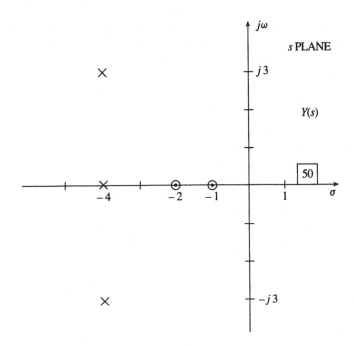

Fig. 7.2 Pole-zero plot of $Y(s)$ (Eqn. 7.3).

have poles for $s = -4$, $s = -4 + j3$, and $s = -4 - j3$. The notation $-P_1, -P_2, \ldots$ is chosen for the poles. It is also clear from the definition of a zero that since $s = \infty$ is also a zero for the $Y(s)$ in Eqn. 7.3, this particular $Y(s)$ has three finite poles and three zeros, one of which is infinite. In fact, any rational function of s has the same number of poles as it has zeros when those at infinity are included. This fact is important in the stability analysis of feedback control systems. The generic term for the poles and zeros of $Y(s)$ is *singularities of $Y(s)$*. We also make use of *pole-zero plots* of rational functions in which the finite zeros are indicated on the complex s plane by small circles and the finite poles by small crosses. Figure 7.2 shows a pole-zero plot of the $Y(s)$ given in Eqn. 7.3, with the multiplying constant 50 included on the small flag on the plot. The pole-zero plot of $Y(s)$, containing exactly the same information as the algebraic expression for $Y(s)$ in Eqn. 7.3, is a visual aid to engineering analysis. We use the notation $s = \sigma + j\omega$ for the complex variable s, where $j = \sqrt{-1}$, as noted in Fig. 7.2.

7.2 Partial-Fraction Expansion of $Y(s)$

Frequently the $Y(s)$ obtained while calculating the dynamic response of a linear system is not included in a short table of Laplace transforms. When this happens $Y(s)$ may be expanded in partial fractions into a sum of simple terms, each of which does appear in the short table. The inverse Laplace transform of $Y(s)$ may then be found almost by inspection. In the following examples we assume that $Y(s)$ has no common poles and zeros, and we first consider the case where $Y(s)$ is a *proper rational function* of s, that is, where $n > m$.

When $Y(s)$ is a proper rational function, we identify two cases: one in which the poles of $Y(s)$ are distinct, and one in which repeated poles of $Y(s)$ occur. We denote these as Cases I and II, and reserve Case III for the *improper rational functions* (where $m \geq n$).

Case I   Here $n > m$ and the poles of $Y(s)$ are distinct. In this case, the partial fraction expansion of $Y(s)$ has the form

$$Y(s) = \frac{K(s+Z_1)(s+Z_2)\cdots(s+Z_m)}{(s+P_1)(s+P_2)\cdots(s+P_n)} \equiv \frac{K_1}{s+P_1} + \frac{K_2}{s+P_2} + \cdots + \frac{K_n}{s+P_n}. \tag{7.11}$$

The symbol $\equiv$ indicates that the second equation, an identity, is valid for all values of the variable s. The n constant terms $K_1, K_2, ..., K_n$ are called the *residues of $Y(s)$ at the poles of $Y(s)$*, a term from the theory of complex variables. The residues may be evaluated by inspection, making use of the fact that Eqn. 7.11 is an identity. To evaluate K_1, for example, we first multiply both sides of the identity by $(s+P_1)$:

$$\frac{(s+P_1)K(s+Z_1)(s+Z_2)\cdots(s+Z_m)}{(s+P_1)(s+P_2)\cdots(s+P_n)} \equiv \frac{K_1(s+P_1)}{s+P_1}$$

$$+ \frac{K_2(s+P_1)}{s+P_2} + \cdots + \frac{K_n(s+P_1)}{s+P_n}. \tag{7.12}$$

We then cancel the common term $(s+P_1)$ on both sides of the identity to obtain

$$\frac{K(s+Z_1)(s+Z_2)\cdots(s+Z_m)}{(s+P_2)\cdots(s+P_n)} \equiv K_1 + \frac{K_2(s+P_1)}{s+P_2} + \cdots + \frac{K_n(s+P_1)}{s+P_n}. \tag{7.13}$$

Now, since Eqn. 7.13 is an identity, it must be valid for any value of s. If we substitute $s = -P_1$ on both sides of the identity, all the terms on the right side will vanish, except for K_1. We then have

$$K_1 = \frac{K(Z_1-P_1)(Z_2-P_1)\cdots(Z_m-P_1)}{(P_2-P_1)(P_3-P_1)\cdots(P_n-P_1)} = [(s+P_1)\times Y(s)]_{s=-P_1}. \tag{7.14}$$

Since $Y(s)$ is known in factored form, we can determine the value of K_1 by inspection.

The same procedure is used for $K_2, K_3, ..., K_n$, the formula for their evaluation being

$$K_i = [(s+P_i)\times Y(s)]_{s=-P_i}, \quad i = 1, 2, ..., n. \tag{7.15}$$

Pole-zero plots can be used to visualize the calculation of residues. As a simple example, consider a fourth-order $Y(s)$ expanded into partial fractions:

$$Y(s) = \frac{9(s+4)}{s(s+1)(s+3)(s+6)} = \frac{K_1}{s} + \frac{K_2}{s+1} + \frac{K_3}{s+3} + \frac{K_4}{s+6}. \tag{7.16}$$

Figure 7.3 is a pole-zero plot of $Y(s)$. Consider the calculation of K_2, the residue of $Y(s)$ at the pole $s = -1$. Using the formula in Eqn. 7.15, we have

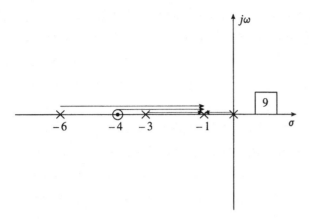

Fig. 7.3 Calculation of residue K_2 using the "arrow" method.

$$K_2 = \frac{9(4-1)}{(-1)(3-1)(6-1)} = \frac{9(+3)}{(-1)(+2)(+5)} = \frac{27}{-10} = -2.7. \tag{7.17}$$

If we draw a directed line segment (or arrow) from each of the poles and zeros of $Y(s)$ to the point at which K_2 is being calculated, $s = -1$, each of those line segments represents one of the terms in parentheses in Eqn. 7.17, with those segments directed toward the right being counted positive and the one directed to the left being negative, as shown in Fig. 7.3. Thus the graphical interpretation of the formula in Eqn. 7.15, applied to K_2, is

$$K_2 = \left. \frac{9 \prod_1 \vec{Z}_j}{\prod_3 \vec{P}_j} \right|_{s=-1} \tag{7.18}$$

where $\vec{Z}_j$ means the arrow drawn from the jth zero to the point $s = -1$, with the numerator having one such arrow in its product. (Note that j is used as an index in Eqn. 7.18, not as $\sqrt{-1}$.) Similarly, the denominator has the product of three arrows drawn from the poles to $s = -1$. The other three residues in this example can be calculated in this manner:

$$K_1 = \frac{9(+4)}{(+1)(+3)(+6)} = 2, \quad K_3 = \frac{9(+1)}{(-3)(-2)(+3)} = 0.5, \tag{7.19}$$

$$K_4 = \frac{9(-2)}{(-3)(-5)(-6)} = 0.2.$$

This example makes clear that the algebraic sign of the residue is the *same* as that of the flag constant (here the flag constant is 9), provided the pole at which the residue is calculated lies to the left of an *even* number of singularities of $Y(s)$. The sign of the residue is *opposite* to that of the flag constant if the pole at which the residue is calculated lies to the left of an *odd* number of singularities. The generalization of this graphical calculation for the case represented in Eqn. 7.11 is

$$K_i = \left. \frac{K \prod_m \vec{Z}_j}{\prod_{n-1} \vec{P}_j} \right|_{s=-P_i}, \quad i = 1, 2, \ldots, n. \tag{7.20}$$

The inverse Laplace transform of each term in the partial fraction expansion of $Y(s)$ in Eqn. 7.11 has the form of pair 5 in the table of transforms in Appendix C:

$$\mathcal{L}^{-1}\left\{\frac{K_i}{s+P_i}\right\} = K_i\epsilon^{-P_i t}. \tag{7.21}$$

Thus we have the solution $y(t) = \mathcal{L}^{-1}\{Y(s)\}$ as the sum of the n terms,

$$y(t) = K_1\epsilon^{-P_1 t} + K_2\epsilon^{-P_2 t} + \cdots + K_i\epsilon^{-P_i t} + \cdots + K_n\epsilon^{-P_n t}$$

$$= \sum_{i=1}^{n} K_i\epsilon^{-P_i t}. \tag{7.22}$$

Each of the terms in Eqn. 7.22 is called a *mode* of the dynamic response $y(t)$. Since each mode is associated with one of the poles of $Y(s)$, there are n modes, some of which have their origin in the input function and some of which come from the system itself. We also note that the general form of $y(t)$, that is, the number of modes and the exponential coefficient of each mode, is determined by the poles of $Y(s)$. The poles are the roots of the denominator polynomial of $Y(s)$, called the *characteristic polynomial of $Y(s)$*. However, the dynamic features of the complete response $y(t)$ depend on the residues as well as on the poles. The residues depend on the flag constant, on the zeros of $Y(s)$, and on the poles of $Y(s)$, as we see in Eqn. 7.20. The left side of Fig. 7.4 shows the four components of $y(t)$ for the example defined in Eqns. 7.16, 7.17, and 7.19. The sum of the four, $y(t)$ itself, is plotted on the right.

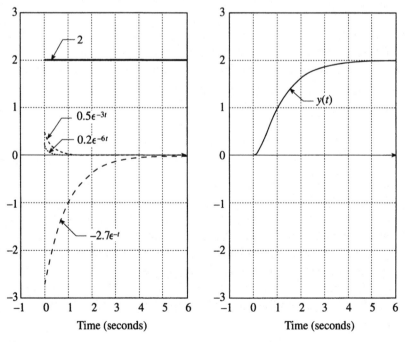

Fig. 7.4 The four components of $y(t)$ (left) and $y(t)$ (right).

It is important to recognize that two functions $Y(s)$ and $Q(s)$ can have the same denominator while the corresponding time functions $y(t)$ and $q(t)$ exhibit significantly different dynamic features due to different numerators of $Y(s)$ and $Q(s)$. As an illustration of this point, let $Y(s)$ be given by Eqn. 7.16, and let $Q(s)$ be a function having the same denominator but a different numerator:

$$Y(s) = \frac{9(s+4)}{s(s+1)(s+3)(s+6)} = \frac{2}{s} - \frac{2.7}{s+1} + \frac{0.5}{s+3} + \frac{0.2}{s+6}$$

$$Q(s) = \frac{72(s+0.5)}{s(s+1)(s+3)(s+6)} = \frac{2}{s} + \frac{3.6}{s+1} - \frac{10}{s+3} + \frac{4.4}{s+6}.$$

(7.23)

The flag constant of $Q(s)$ has been adjusted so that the leading residue, 2, will be the same as that in $Y(s)$ in order that the final values $y(\infty)$ and $q(\infty)$ will each be 2. Note the significant differences in the other three residues between $Y(s)$ and $Q(s)$. Figure 7.5 shows the four components of $q(t)$ plotted on the left and the sum of these four, $q(t)$ itself, plotted on the right; $y(t)$ is also plotted on the right to show the striking differences in the dynamic features of the two responses.

Consider an example in which $R(s)$, our response function of s, has a pair of complex-valued poles:

$$R(s) = \frac{2720(s+2)}{s(s+4)(s+20)[s^2+4s+68]}$$

$$= \frac{2720(s+2)}{s(s+4)(s+20)(s+2+j8)(s+2-j8)}.$$

(7.24)

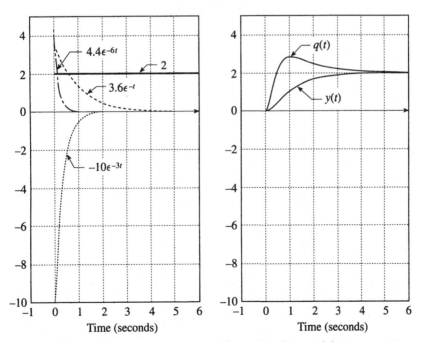

Fig. 7.5 The four components of $q(t)$ (left); $q(t)$ and $y(t)$ (right).

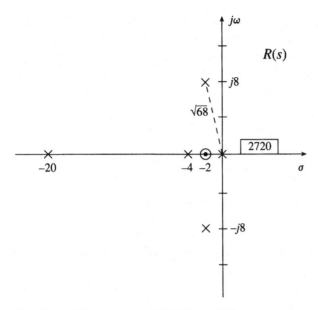

Fig. 7.6 Pole-zero plot of $R(s)$ (Eqn. 7.24).

A pole-zero plot of $R(s)$ is shown in Fig. 7.6. Note that 4, the coefficient of s in the quadratic term, is twice the negative of the real part of the complex pole, and that 68, the constant term, is the square of the distance from the pole to the origin. $R(s)$ may be expanded in partial fractions, using the formula given in Eqn. 7.15. The expansion has the regular form, but now we expect to see complex-valued residues at the two complex poles:

$$R(s) = \frac{1}{s} + \frac{1.25}{s+4} - \frac{0.3943}{s+20} + \frac{-0.9278 - j0.4124}{s+2+j8} + \frac{-0.9278 + j0.4124}{s+2-j8}. \quad (7.25)$$

The pole-zero plot in Fig. 7.7 shows the graphical calculation of the residue at the lower complex pole in the style of Eqn. 7.20:

$$\begin{bmatrix} \text{residue of } R(s) \\ \text{at } s = -2-j8 \end{bmatrix} = \frac{2720 \, \Pi_1 \, \vec{Z}_j}{\Pi_4 \, \vec{P}_j} \Bigg|_{s=-2-j8}$$

$$= \frac{2720(-j8)}{(-2-j8)(-j16)(2-j8)(18-j8)}$$

$$= -0.9278 - j0.4124. \quad (7.26)$$

The "arrows" $\vec{Z}_j$ and $\vec{P}_j$ in Fig. 7.7 are graphical representations of the complex numbers appearing in Eqn. 7.26. The residue at the upper pole is the complex conjugate of the residue at the lower pole because the upper pole is the mirror image of the lower pole about the real axis. The inverse Laplace transform of $R(s)$ can also be taken term-by-term, and because all the poles are distinct, each term has the form of pair 5 in the tables of Appendix C:

$$r(t) = 1 + 1.25\epsilon^{-4t} - 0.3943\epsilon^{-20t}$$

$$- (0.9278 + j0.4124)\epsilon^{-(2+j8)t} - (0.9278 - j0.4124)\epsilon^{-(2-j8)t}. \quad (7.27)$$

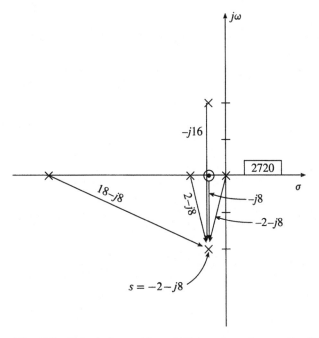

Fig. 7.7 Calculating residue of $R(s)$ at the pole $s = -2 - j8$ by the "arrow" method.

The last two complex-valued terms can be combined using Euler's formula (see Appendix B), or by using pair 16 from the table in Appendix C, with the result that these two terms reduce to a single real-valued term (as expected, since $r(t)$ is real-valued):

$$r(t) = 1 + 1.25\epsilon^{-4t} - 0.3943\epsilon^{-20t} + 2.0307\epsilon^{-2t} \sin(8t + 246.04°). \tag{7.28}$$

Figure 7.8 (left) shows each of the four real-valued components of $r(t)$ plotted versus time. Note that the time constant of the damping envelope, 1/2 second, is the reciprocal of the negative of the real part of the complex pole and that the frequency of oscillation, 8 radians/second, is the imaginary part of the complex pole. These two characteristics of the damped sinusoidal term are thus directly related to the positions of the complex pole pair in the s plane. The sum of these four, $r(t)$ itself, is plotted in Fig. 7.8 (at the right).

Case II Here $Y(s)$ has more poles than finite zeros, and one of the poles, $-P_1$, is repeated r times. $Y(s)$ has the form

$$Y(s) = \frac{K(s + Z_1)(s + Z_2)\cdots(s + Z_m)}{(s + P_1)^r(s + P_2)(s + P_3)\cdots(s + P_n)}, \tag{7.29}$$

where $(n - 1 + r) > m$. Now the partial-fraction expansion of $Y(s)$ is

$$Y(s) = \frac{K_1}{(s + P_1)} + \frac{C_2}{(s + P_1)^2} + \frac{C_3}{(s + P_1)^3} + \cdots + \frac{C_r}{(s + P_1)^r}$$

$$+ \frac{K_2}{(s + P_2)} + \frac{K_3}{(s + P_3)} + \cdots + \frac{K_n}{(s + P_n)}. \tag{7.30}$$

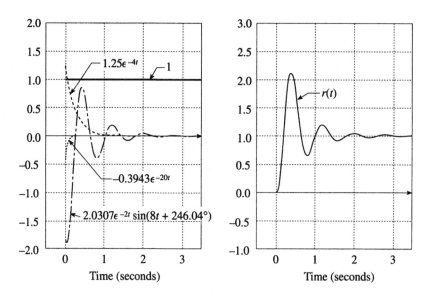

Fig. 7.8   The four real-valued components of $r(t)$ (left) and $r(t)$ (right).

The constants $K_1, K_2, ..., K_n$ are the residues of $Y(s)$ at the respective poles, and the constants $C_2, C_3, ..., C_r$ bear no special names. The residues $K_2, K_3, ..., K_n$ are computed as indicated in Eqn. 7.20 (for $i = 2, 3, ..., n$), but the residue K_1 and the constants $C_2, C_3, ..., C_r$ are computed as follows. Let $Q(s) = [(s + P_1)^r Y(s)]$, and perform the following calculations:

$$C_r = \{Q(s)\}_{s=-P_1},$$

$$C_{r-1} = \frac{1}{1!}\left\{\frac{dQ(s)}{ds}\right\}_{s=-P_1},$$

$$C_{r-2} = \frac{1}{2!}\left\{\frac{d^2Q(s)}{ds^2}\right\}_{s=-P_1},$$

$$\vdots$$

$$C_2 = \frac{1}{(r-2)!}\left\{\frac{d^{(r-2)}Q(s)}{ds^{(r-2)}}\right\}_{s=-P_1},$$

$$K_1 = \frac{1}{(r-1)!}\left\{\frac{d^{(r-1)}Q(s)}{ds^{(r-1)}}\right\}_{s=-P_1}.$$

(7.31)

After the residues and the constants are computed, obtain the inverse Laplace transform of $Y(s)$ by inspection of the partial-fraction expansion (Eqn. 7.30), where the terms involving $C_2, C_3, ..., C_r$ may be transformed, using pair 7 from the table:

$$\mathcal{L}^{-1}\left\{\frac{k!}{(s+P)^{k+1}}\right\} = t^k \epsilon^{-Pt}.$$

(7.32)

Here is an example that illustrates this case:

$$\Omega(s) = \frac{30(s+1)}{(s+2)^3(s+3)(s+5)}$$

$$= \frac{K_1}{s+2} + \frac{C_2}{(s+2)^2} + \frac{C_3}{(s+2)^3} + \frac{K_2}{s+3} + \frac{K_3}{s+5}. \tag{7.33}$$

Following the formulas in Eqn. 7.31, we obtain $Q(s) = 30(s+1)/(s+3)(s+5)$. The coefficients are:

$$K_3 = \{(s+5)\Omega(s)\}_{s=-5} = \frac{30(1-5)}{(2-5)^3(3-5)} = -\frac{20}{9},$$

$$K_2 = \{(s+3)\Omega(s)\}_{s=-3} = \frac{30(1-3)}{(2-3)^3(5-3)} = 30,$$

$$C_3 = \{(s+2)^3\Omega(s)\}_{s=-2} = \frac{30(1-2)}{(3-2)(5-2)} = -10,$$

$$C_2 = \frac{1}{1!}\left\{\frac{dQ(s)}{ds}\right\}_{s=-2}$$

$$= 30\left\{\frac{(s+3)(s+5)-(s+1)[(s+3)+(s+5)]}{(s+3)^2(s+5)^2}\right\}_{s=-2} = \frac{210}{9},$$

$$K_1 = \frac{1}{2!}\left\{\frac{d^2Q(s)}{ds^2}\right\}_{s=-2}$$

$$= \frac{30}{2}\left\{\frac{2[(s^2+2s-7)(s+4)-(s+1)(s+3)(s+5)]}{(s+3)^3(s+5)^3}\right\}_{s=-2} = -\frac{250}{9}.$$

Therefore, the time response is

$$\mathcal{L}^{-1}\{\Omega(s)\} = \omega(t) = \tfrac{1}{9}[270\epsilon^{-3t} - 20\epsilon^{-5t} - (250 - 210t + 45t^2)\epsilon^{-2t}]. \tag{7.34}$$

If the function $\Omega(s)$ is of higher order than that shown in this example, or if the root is repeated more than three times, then the labor involved in calculating the derivative terms may become excessive. This labor may be reduced considerably by approximating the repeated roots by separate roots having nearly the same value as the repeated root. If the $(s+2)^3$ term in Eqn. 7.33 is replaced in this manner we have

$$\Omega(s) \cong \frac{30(s+1)}{(s+1.9)(s+2)(s+2.1)(s+3)(s+5)}. \tag{7.35}$$

Since this expression has no repeated roots, the calculation of residues may be accomplished without taking derivatives of $Q(s)$. Although the form of this approximate version of $\Omega(s)$ will produce a form for $\omega(t)$ which differs from that in Eqn. 7.34, the instantaneous values of the approximate $\omega(t)$ will be close to the exact values in Eqn. 7.34. If a better approximation is required, the triple root can be replaced by $(s+1.99)(s+2)(s+2.01)$.

A useful graphical interpretation exists for the calculation of the residue of $Y(s)$ at a repeated pole of order 2. In this special case $Y(s)$ has the form

$$Y(s) = \frac{K(s+Z_1)\cdots(s+Z_m)}{(s+P_1)^2(s+P_2)\cdots(s+P_n)}$$

$$= \frac{C}{(s+P_1)^2} + \frac{K_1}{(s+P_1)} + \frac{K_2}{(s+P_2)} + \cdots + \frac{K_n}{(s+P_n)}. \qquad (7.36)$$

The residues $K_2, K_3, \ldots, K_n$ at the distinct poles may be determined by inspection, as explained previously. The constant C may also be determined by inspection:

$$C = \left. \frac{K \, \Pi_m \vec{Z}_j}{\Pi_{n-1} \vec{P}_j} \right|_{s=-P_1}. \qquad (7.37)$$

The residue K_1 may be calculated by the differentiation formula in Eqn. 7.31, but it is also given geometrically as

$$K_1 = C \left(\sum_{j=1}^m \frac{1}{\vec{Z}_j} - \sum_{j=2}^n \frac{1}{\vec{P}_j} \right)_{s=-P_1}. \qquad (7.38)$$

Case III Here $Y(s)$ is an improper rational function, $m \geq n$. In this case we simply divide the denominator of $Y(s)$ into its numerator until the remainder is a proper rational function. For $Y(s)$ in the standard form (Eqn. 7.11), this division yields

$$Y(s) = K \left[\alpha_1 + \alpha_2 s + \cdots + s^{m-n} + \frac{\beta(s^{n-1} + \gamma_{n-2}s^{n-2} + \cdots + \gamma_1 s + \gamma_0)}{s^n + b_{n-1}s^{n-1} + \cdots + b_1 s + b_0} \right],$$

where β and all α, γ are constants. The terms $K[\alpha_1 + \alpha_2 s + \cdots + s^{m-n}]$ correspond to what are known as *singularity functions*. Only rarely does $m - n$ exceed 1 in the response functions of physical systems, and for this reason the Laplace transform table in Appendix C lists only one singularity function, the impulse $\delta(t)$ whose Laplace transform is a constant. The remaining term in $Y(s)$,

$$K \left[\frac{\beta(s^{n-1} + \gamma_{n-2}s^{n-2} + \cdots + \gamma_1 s + \gamma_0)}{s^n + b_{n-1}s^{n-1} + \cdots + b_1 s + b_0} \right],$$

is a rational function that can be expanded in partial fractions under Case I or Case II.

The values of the residues, the poles, and (in Case III) the remainder term in $Y(s)$ can all be calculated by computer programs. These computer tools are very useful in most cases, but for some high-order functions, or for those having repeated poles, the algorithms may produce incorrect results. If this occurs the engineer can use the analytical techniques developed here to recognize the faulty calculations.

7.3 The Second-Order System

A second-order system has a defining differential equation of second order. This leads to a 2×2 A matrix in the state-variable model for the system and to a second-order denominator in the transfer functions of the system. Many practical devices are second-order systems, and because the dynamic performance of such systems sometimes resembles that of a higher-order system, it is useful to be familiar with the second-order system. Let us now study two examples of such systems in detail.

Consider the mechanical system described and illustrated in Section 6.4. The transfer functions relating the shaft position of this system to the applied torque and to the initial conditions are given in Eqn. 6.23. Because we are concerned here only with the zero-state response, we start with an abbreviated version of Eqn. 6.23:

$$Y(s) = \left[\frac{1/J}{s^2 + (d/J)s + K/J} \right] U(s). \tag{7.39}$$

It is useful to study the zero-state response $y(t)$ that results when a specific test input $u(t)$ is applied. The step function is a common test input in dynamic system analysis because it reveals most features of the response that are of interest in control engineering. Let $u(t) = T_0\sigma(t)$, where T_0 is the magnitude of the step input in N-m. We then have $U(s) = T_0/s$. The output is given by

$$Y(s) = \left[\frac{1/J}{s^2 + (d/J)s + K/J} \right] \frac{T_0}{s} = \frac{T_0[1/J]}{s(s + P_1)(s + P_2)}. \tag{7.40}$$

Here $Y(s)$ is characterized by having no finite zeros, two poles at $s = -P_1$ and $s = -P_2$ due to the transfer function, and a pole at $s = 0$ due to the step input. The poles at $s = -P_1$ and $s = -P_2$ depend on the three physical parameters of the system:

$$-P_1, -P_2 = \frac{-d/J \pm \sqrt{(d/J)^2 - 4(K/J)}}{2}. \tag{7.41}$$

We see that P_1 and P_2 are complex conjugates if the damping coefficient d is small enough, that is, if $d < 2\sqrt{KJ}$. When d is larger, the two poles are real-valued. The borderline, or *critical damping* case $d = 2\sqrt{KJ}$, renders $P_1 = P_2 = d/(2J)$. We now study the response for these three cases separately.

First We have $d > 2\sqrt{KJ}$, the overdamped case. The two poles are real-valued and unequal:

$$P_1 = \frac{d - \sqrt{d^2 - 4KJ}}{2J} \quad \text{and} \quad P_2 = \frac{d + \sqrt{d^2 - 4KJ}}{2J}. \tag{7.42}$$

Figure 7.9 shows a pole-zero plot of $Y(s)$. The partial-fraction expansion of $Y(s)$ is

$$Y(s) = \frac{K_0}{s} + \frac{K_1}{s + P_1} + \frac{K_2}{s + P_2}. \tag{7.43}$$

Calculate the residues by inspection to obtain

$$K_0 = \frac{T_0/J}{P_1 P_2}, \quad K_1 = \frac{T_0/J}{-P_1(P_2 - P_1)}, \quad K_2 = \frac{T_0/J}{-P_2[-(P_2 - P_1)]}. \tag{7.44}$$

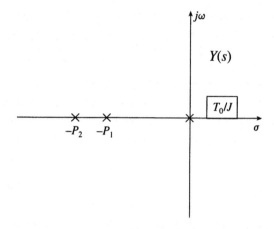

Fig. 7.9 Pole-zero plot of $Y(s)$ for the overdamped case.

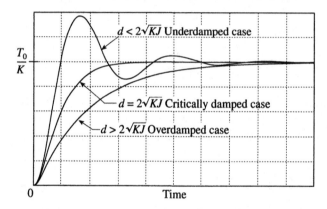

Fig. 7.10 $y(t)$ for the overdamped, critically damped, and underdamped cases.

It is clear from Eqn. 7.40 that two simple relationships exist between the physical parameters J, K, and d and P_1 and P_2. They are

$$P_1 P_2 = \frac{K}{J} \quad \text{and} \quad P_1 + P_2 = \frac{d}{J}. \tag{7.45}$$

K_0 is then T_0/K, so by inspection the time response is

$$y(t) = \frac{T_0}{K} + K_1 \epsilon^{-P_1 t} + K_2 \epsilon^{-P_2 t}. \tag{7.46}$$

This response has the form shown in Fig. 7.10 for the overdamped case.

Second We have $d = 2\sqrt{KJ}$, the critically damped case. If the damping coefficient happens to be exactly at its so-called critical value, the two pole values will be equal, $P_1 = P_2 = d/2J = \sqrt{K/J}$, and $Y(s)$ will be

$$Y(s) = \frac{T_0/J}{s(s + d/2J)^2} = \frac{T_0/J}{s(s + \sqrt{K/J})^2}, \tag{7.47}$$

so the partial-fraction expansion involves a repeated root:

$$Y(s) = \frac{K_0}{s} + \frac{K_1}{s+\sqrt{K/J}} + \frac{C_2}{(s+\sqrt{K/J})^2}. \tag{7.48}$$

Here K_0 has the same value as K_0 in the overdamped case, which is apparent from physical considerations as well as from the analysis. K_1 and C_2 are calculated from the formulas in Eqn. 7.31, with the following result:

$$Y(s) = \frac{T_0/K}{s} - \frac{T_0/K}{s+\sqrt{K/J}} - \frac{T_0/\sqrt{KJ}}{(s+\sqrt{K/J})^2}. \tag{7.49}$$

The inverse Laplace transform,

$$y(t) = \frac{T_0}{K}[1 - \epsilon^{-\sqrt{K/J}\,t} - \sqrt{K/J}\,t\epsilon^{-\sqrt{K/J}\,t}], \tag{7.50}$$

is also plotted in Fig. 7.10, as the middle curve.

Third   We have $d < 2\sqrt{KJ}$, the underdamped case. If the damping coefficient is less than the critical value, the poles will be complex conjugates:

$$P_1 = \frac{d}{2J} + j\frac{\sqrt{4KJ-d^2}}{2J} \quad \text{and} \quad P_2 = \frac{d}{2J} - j\frac{\sqrt{4KJ-d^2}}{2J}. \tag{7.51}$$

We introduce the notation $b = d/2J$ and $\omega_o = \sqrt{4KJ-d^2}/2J$, so that $Y(s)$ may be written as

$$Y(s) = \frac{T_0/J}{s(s+b+j\omega_o)(s+b-j\omega_o)} = \frac{K_0}{s} + \frac{K_1}{s+b+j\omega_o} + \frac{K_2}{s+b-j\omega_o}. \tag{7.52}$$

A pole-zero plot $Y(s)$ appears in Fig. 7.11, showing the locations of the complex poles. Using the graphical method for calculating K_0, K_1, and K_2, we have:

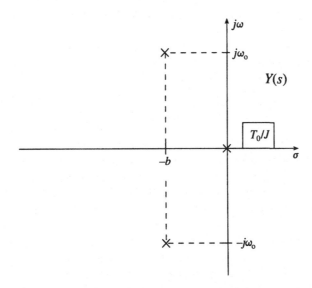

Fig. 7.11   Pole-zero plot of $Y(s)$ for the underdamped case.

$$K_0 = \frac{T_0}{J}\left[\frac{1}{(b-j\omega_o)(b+j\omega_o)}\right] = \frac{T_0}{J}\left[\frac{1}{b^2+\omega_o^2}\right] = \frac{T_0}{K},$$

$$K_1 = \frac{T_0}{J}\left[\frac{1}{(-b-j\omega_o)(-j2\omega_o)}\right] = \frac{T_0}{J}\left[\frac{-\omega_o-jb}{2\omega_o(b^2+\omega_o^2)}\right] = \alpha + j\beta, \quad \text{and} \qquad (7.53)$$

$$K_2 = \alpha - j\beta, \quad \text{where} \quad \alpha = -\frac{1}{2}\left[\frac{T_0}{K}\right] \text{ and } \beta = -\frac{b}{2\omega_o}\left[\frac{T_0}{K}\right].$$

We take the inverse Laplace transform of $Y(s)$, using pair 16 in the table of transforms:

$$y(t) = \frac{T_0}{K}\left[1 + \frac{\sqrt{\omega_o^2+b^2}}{\omega_o}\epsilon^{-bt}\sin(\omega_o t+\psi)\right], \quad \text{where} \quad \psi = \tan^{-1}\left(\frac{-\omega_o}{-b}\right). \quad (7.54)$$

Note that the angle ψ lies in the third quadrant, since both b and ω_o are positive numbers. This response is also plotted in Fig. 7.10 as the underdamped response curve.

Several interesting things about this second-order response can be observed from the analysis just completed. First, we can verify that the initial values of the position and the velocity of the system are zero by using the initial-value theorem on $Y(s)$ (Eqn. 7.40):

$$\lim_{t \to 0^+} y(t) = y(0^+) = \lim_{s \to \infty} [sY(s)] = 0,$$

$$\lim_{t \to 0^+} \dot{y}(t) = \dot{y}(0^+) = \lim_{s \to \infty} [s(sY(s)-y(0^+))] = 0. \qquad (7.55)$$

We can calculate the initial value of the acceleration as follows:

$$\lim_{t \to 0^+} \ddot{y}(t) = \ddot{y}(0^+) = \lim_{s \to \infty} [s(s^2Y(s)-sy(0^+)-\dot{y}(0^+))] = \frac{T_0}{J} \text{ (rad/s)/s.} \qquad (7.56)$$

We can also apply the final-value theorem to this example, because the response is a stable one:

$$y(\infty) = \lim_{s \to 0} [sY(s)] = T_0\left[\frac{1/J}{K/J}\right] = \frac{T_0}{K}. \qquad (7.57)$$

Next, if we set the damping coefficient d to zero, the completely undamped response will oscillate indefinitely, as can be observed from Eqn. 7.54:

$$y(t) = \frac{T_0}{K}\left[1 + \sin\left(\sqrt{\frac{K}{J}}t+270°\right)\right] = \frac{T_0}{K}\left[1 - \cos\left(\sqrt{\frac{K}{J}}t\right)\right]. \qquad (7.58)$$

This response is plotted in Fig. 7.12. From this response we define the *undamped natural frequency*, denoted ω_n, of the system.

$$\left[\begin{array}{c}\text{undamped natural}\\\text{frequency}\end{array}\right] \triangleq \omega_o|_{d=0} = \sqrt{\frac{K}{J}} = \omega_n \text{ rad/s.} \qquad (7.59)$$

Another parameter that is useful in describing the second-order response is the *damping ratio*, designated as ζ (zeta), and defined as the ratio of the damping coefficient d to that value of d which gives critical damping:

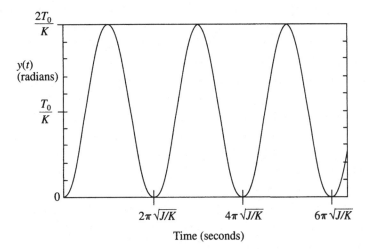

Fig. 7.12 $y(t)$ for the undamped case, $d = 0$.

$$\begin{bmatrix} \text{damping} \\ \text{ratio} \end{bmatrix} = \frac{d}{d_{\text{critical}}} = \frac{d}{2\sqrt{KJ}} = \zeta. \qquad (7.60)$$

It is convenient to write $Y(s)$ and $y(t)$ in terms of ζ and ω_n:

$$Y(s) = \frac{T_0}{K} \left[\frac{\omega_n^2}{s(s^2 + 2\zeta\omega_n s + \omega_n^2)} \right] \qquad (7.61)$$

$$y(t) = \frac{T_0}{K} \left[1 + \frac{\epsilon^{-\zeta\omega_n t}}{\sqrt{1-\zeta^2}} \sin(\omega_n \sqrt{1-\zeta^2}\, t + \psi) \right], \quad \text{where } \psi = \tan^{-1}\left(\frac{-\sqrt{1-\zeta^2}}{-\zeta} \right).$$

From this we see that the *frequency of oscillation* of the damped sinusoidal term in $y(t)$, which is designated as ω_0 in Eqn. 7.54, is related to the undamped natural frequency and the damping ratio:

$$\begin{bmatrix} \text{frequency of the} \\ \text{damped oscillations} \end{bmatrix} = \omega_0 = \omega_n \sqrt{1-\zeta^2} \text{ rad/s.} \qquad (7.62)$$

In control systems it is common to have values of ζ in the range 0.5 to 1, so that ω_0 is significantly lower than ω_n. However, in systems where the damping ratio is near zero (in electronics or in structures, for example), the distinction between ω_0 and ω_n is hardly noticeable. Figure 7.13 shows the relationship between ω_n, ω_0, and ζ.

For the underdamped case we can calculate the instant at which $y(t)$ reaches its peak value by setting $\dot{y}(t) = 0$ and finding the least value for t which satisfies this relationship. It turns out to be

$$\begin{bmatrix} \text{instant of peak} \\ \text{value of } y(t) \end{bmatrix} = t_p = \frac{\pi}{\omega_n \sqrt{1-\zeta^2}} \text{ s.} \qquad (7.63)$$

The maximum overshoot of $y(t)$ is determined to be

$$\begin{bmatrix} \text{percent of maximum} \\ \text{overshoot of } y(t) \end{bmatrix} = \frac{y(t_p) - y(\infty)}{y(\infty)} \times 100 = 100\epsilon^{-\pi\zeta/\sqrt{1-\zeta^2}}. \qquad (7.64)$$

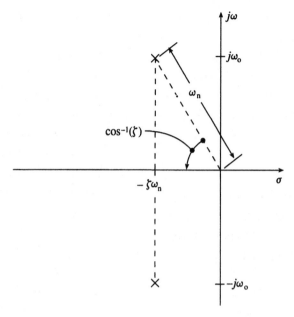

$$s^2 + 2\zeta\omega_n s + \omega_n^2 = (s + \zeta\omega_n + j\omega_o)(s + \zeta\omega_n - j\omega_o)$$
$$\omega_o = \omega_n \sqrt{1 - \zeta^2}$$

Fig. 7.13   Geometrical relationships between ζ, ω_n, and ω_o for the underdamped case.

(A note of caution is in order concerning Eqns. 7.63 and 7.64. These formulas pertain *only* to the overshoot properties of those second-order systems whose response functions have *exactly* the form given in Eqn. 7.61.)

Next, we normalize $y(t)$ with respect to its final value by dividing $y(t)$ by that final value:

$$\begin{bmatrix} y(t) \text{ normalized with} \\ \text{respect to } y(\infty) \end{bmatrix} = \frac{y(t)}{T_0/K} = \left[1 + \frac{\epsilon^{-\zeta(\omega_n t)}}{\sqrt{1-\zeta^2}} \sin(\sqrt{1-\zeta^2}(\omega_n t) + \psi) \right]. \quad (7.65)$$

Since the normalized final value is 1, this expression can be used for any second-order system whose response function has the form of Eqn. 7.61, regardless of the value of the constant multiplying the bracketed term.

Further normalization of $y(t)$ is possible. Note that t occurs in $y(t)$ only as a product with ω_n, so that by introducing the notation $\tau = \omega_n t$ we have $y(t)$ normalized with respect to both amplitude and time:

$$\begin{bmatrix} \text{normalized} \\ y(t) \end{bmatrix} = \hat{y}(\tau) = 1 + \frac{\epsilon^{-\zeta\tau}}{\sqrt{1-\zeta^2}} \sin(\sqrt{1-\zeta^2}\tau + \psi), \quad \text{where}$$
$$\psi = \tan^{-1}\left(\frac{-\sqrt{1-\zeta^2}}{-\zeta} \right). \quad (7.66)$$

Figure 7.14 shows $\hat{y}(\tau)$ plotted versus τ for five values of ζ. Appendix E contains a set of normalized curves covering a broader range of ζ.

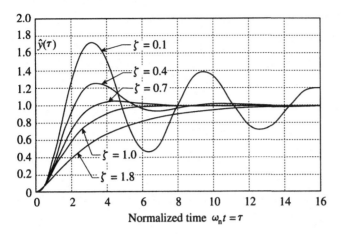

Fig. 7.14 Normalized response $y(\tau)$ versus τ for five values of ζ.

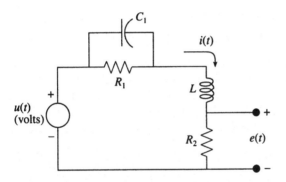

Fig. 7.15 RLC circuit.

We turn now to the second example of a second-order system, an RLC circuit excited by a voltage source $u(t)$ and having an output voltage $e(t)$, as diagramed in Fig. 7.15. The transfer function relating the input to the zero-state response is

$$\frac{E(s)}{U(s)} = \frac{\dfrac{R_2}{L}\left(s + \dfrac{1}{R_1 C_1}\right)}{s^2 + \left(\dfrac{R_1 R_2 C_1 + L}{L R_1 C_1}\right)s + \left(\dfrac{R_1 + R_2}{L R_1 C_1}\right)} = \frac{K(s+Z)}{s^2 + 2\zeta\omega_n s + \omega_n^2}. \tag{7.67}$$

The relationships between the physical parameters R_1, R_2, C_1, and L and the standard-form parameters K, Z, ζ, and ω_n are stated in Eqn. 7.67. We will apply a step input of E V to this circuit to compare the response to that of the previous example. Because $u(t) = E\sigma(t)$, $U(s) = E/s$ and the response function in the s domain is

$$E(s) = \frac{EK(s+Z)}{s(s^2 + 2\zeta\omega_n s + \omega_n^2)}. \tag{7.68}$$

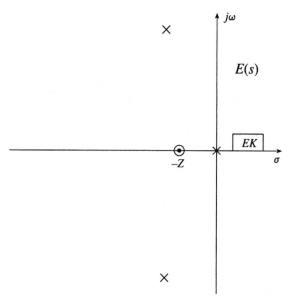

Fig. 7.16   Pole-zero plot of $E(s)$ for $U(s) = E/s$.

Figure 7.16 is a pole-zero plot of $Y(s)$, where the damping ratio ζ is in the neighborhood of 0.4. Note that the denominator of $E(s)$ in Eqn. 7.68 is the same as that of $Y(s)$ in Eqn. 7.61. In Eqn. 7.61 the parameters of the second-order mechanical system determine ζ and ω_n, and in Eqn. 7.67 the electrical parameters determine ζ and ω_n. $Y(s)$ in Eqn. 7.61 has no finite zeros, but $E(s)$ in Eqn. 7.68 has one finite zero. The zero in $E(s)$ can cause a significant difference in the corresponding time responses $e(t)$ in the RLC circuit and $y(t)$ in the K, d, J system. The first difference can be observed by calculating the initial value of $\dot{e}(t)$, using the initial-value theorem:

$$\dot{e}(0^+) = \lim_{t \to 0^+} \dot{e}(t) = \lim_{s \to \infty} s[sE(s)] = EK = \frac{ER_2}{L}. \tag{7.69}$$

The physical nature of the electrical circuit makes it clear why $\dot{e}(0^+)$ is nonzero when the input voltage is a step function; this fact is reflected in the finite zero in the transfer function. The physical nature of the mechanical system is such that a step input of torque does not cause a nonzero $\dot{y}(0^+)$ in that system, so the transfer function of that system has no finite zero. This difference in transfer functions is due to the definitions of input and output used in these two examples, and not to any fundamental difference in the dynamics between electrical and mechanical systems.

It is important to note that the nonzero initial *value* for $\dot{e}(t)$ in this example is inherent in the zero-state response, and is not caused by an initial *condition* $\dot{e}_0$. If we include a nonzero $\dot{e}_0$ in the analysis, which in this case would be the rate of change of the output voltage just prior to the application of the step input $u(t)$, then $E(s)$ would have an additional term, corresponding to the zero-input response, and the initial-value theorem would show $\dot{e}(0^+)$ to be $(ER_2/L) + \dot{e}_0$.

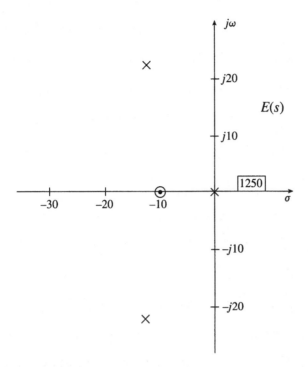

Fig. 7.17   Pole-zero plot of $E(s)$ for $E = 100$ V, $R_1 = 50 \ \Omega$, $R_2 = 12.5 \ \Omega$, $C_1 = 0.002$ F, and $L = 1$ H.

The inverse transformation of $E(s)$ may be found as pair 19 in the table if the damping ratio is less than 1:

$$e(t) = EK\left[\frac{Z}{\omega_n^2} + \frac{[Z^2 - 2Z\zeta\omega_n + \omega_n^2]^{1/2}}{\omega_n^2\sqrt{1-\zeta^2}}\epsilon^{-\zeta\omega_n t}\sin(\omega_o t + \psi)\right] \text{V}, \quad \text{where}$$

$$\psi = \tan^{-1}\left(\frac{\omega_o}{Z - \zeta\omega_n}\right) - \tan^{-1}\left(\frac{\sqrt{1-\zeta^2}}{-\zeta}\right). \tag{7.70}$$

Note that this is the same form as $y(t)$ in Eqn. 7.61, but here the coefficients depend on Z as well as on ζ and ω_n. A pole-zero plot of $E(s)$ is drawn in Fig. 7.17 for parameter values $R_1 = 50 \ \Omega$, $R_2 = 12.5 \ \Omega$, $C_1 = 0.002$ F, $L = 1$ H, and $E = 100$ V. The corresponding graph of $e(t)$ is shown in Fig. 7.18. The damping ratio of this system is 0.45. If there were no finite zero in the transfer function then the response would show an overshoot of 20.5 percent, but with the zero located at $s = -10$ the overshoot is more than 100 percent.

The effect of the finite zero on the dynamic response may be considered from an alternative point of view. First, define a new constant $K_0 = KZ$ and write $E(s)$ in the following form:

$$E(s) = \frac{EK_0}{Z}\left[\frac{s + Z}{s[s^2 + 2\zeta\omega_n s + \omega_n^2]}\right]$$

$$= \frac{EK_0}{s[s^2 + 2\zeta\omega_n s + \omega_n^2]} + \left(\frac{1}{Z}\right)\frac{sEK_0}{s[s^2 + 2\zeta\omega_n s + \omega_n^2]}. \tag{7.71}$$

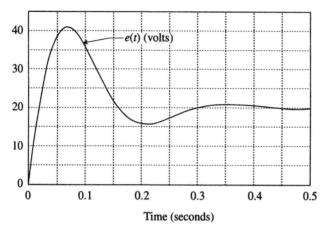

Fig. 7.18 $e(t)$ (Eqn. 7.70) for the parameter values given.

This is not a partial-fraction expansion of $E(s)$ but rather a simple algebraic separation of $Y(s)$ into two parts. Now $E(s)$ has the form

$$E(s) = E_2(s) + (1/Z)[sE_2(s)]. \tag{7.72}$$

Since $E_2(s)$ has the form already studied (Eqn. 7.61), we know that its inverse transform $e_2(t)$ is of the class illustrated in Fig. 7.10 and also in the normalized set of curves in Appendix E. Further, the inverse transform of $[sE_2(s)]$ is $\dot{e}_2(t)$, so the inverse transform $e(t)$ of $E(s)$ is the sum

$$e(t) = e_2(t) + (1/Z)\dot{e}_2(t). \tag{7.73}$$

The individual parts of $e(t)$ are plotted in Fig. 7.19, and the sum $e(t)$ was shown in Fig. 7.18. The effect of the zero in the transfer function is to add a derivative term to the basic response obtained from the transfer function not having the zero. The added derivative term is inversely proportional to Z. If the zero is

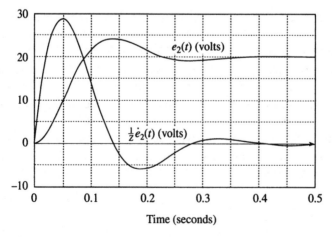

Fig. 7.19   The two parts of $e(t)$ shown in Fig. 7.18.

located close to the origin, the derivative term is large, but if the zero is located far from the poles, the derivative term will be small. For reference purposes, a set of normalized response curves for systems in this class of second-order systems (having a finite zero in the transfer function) is provided in Appendix E.

Because Z in this example is a positive number, the derivative term has a positive sign. The effect is that $e(t)$ has considerably more overshoot than does $e_2(t)$. If Z were negative (with the zero located in the right half of the s plane) then the derivative term would have a negative sign, and the resultant shape of the response $e(t)$ would be remarkably different from that in Fig. 7.18. This case, which lies in the category of *nonminimum phase* systems, is of major importance in control engineering. It appears again in Section 7.7.

Equation 7.67 shows how the zero and poles of the transfer function depend on the four physical parameters of this circuit. The zero depends only on R_1 and C_1, while the poles depend upon all four. Because the values of the physical parameters in a system determine the dynamic response of that system (as reflected in the positions of the poles and zeros on the s plane), we are very fortunate to have computer-based methods that can quickly factor the polynomials whose coefficients are functions of the parameters in order to determine the poles and zeros. It is convenient to display the results of such calculations graphically.

As an example of this technique, assume that R_1, C_1, and L have the fixed values given previously but that R_2 may be varied. The pole values and the constant K will change as R_2 changes. Figure 7.20 is a pole-zero plot of $E(s)$

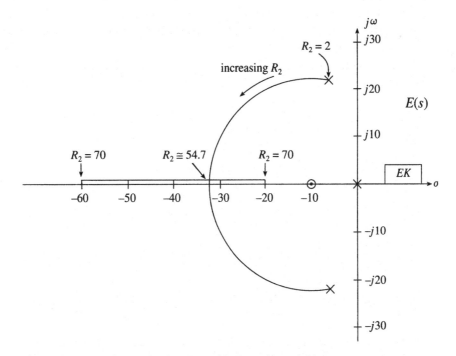

Fig. 7.20 Root locus plot for $E(s)$ with R_2 as the variable parameter.

showing how the pole locations depend on R_2 for values of R_2 in the range 2 Ω to 70 Ω. The pole locations migrate from positions of low damping for small values of R_2 along the locus to values of higher damping as R_2 approaches the value for which $\zeta = 1$; this occurs when $R_2 \cong 54.7$ Ω. For higher values of R_2, both roots are real-valued. This graph, called a *root-locus plot,* allows one to select the value of the adjustable parameter to achieve suitable pole locations. In this case, for example, if a damping ratio of about 0.7 is desired then R_2 would be set at 30 Ω. If the root-locus plot shows that none of the possible pole locations is suitable, it is then necessary to change the values of the "fixed" parameters and to generate a new pole-zero plot, or to perform an analysis equivalent to this graphical approach.

7.4 A Third-Order System

A good example of a third-order system is the electrohydraulic amplifier, valve, and ram system modeled in Sections 4.3 and 4.4. A transfer function model of this system is shown in Fig. 7.21. These transfer functions are based on linear models of the amplifier, magnet, valve, and ram. Friction is ignored, and it is assumed that only small displacements of the valve occur, so that the nonlinear relationships between valve displacement and the flow forces on the valve spool may also be neglected. The inductance of the electromagnet coils L_c is included in the transfer function relating the amplifier input voltage to the differential current. The *magnet force constant, K_F,* is a composite of the magnet parameters which relates the differential current to the net magnetic force applied to the valve spool. The output of the system is the velocity of the ram piston $\dot{z}_R(t)$, whose Laplace transform we denote as $V_R(s)$. The remaining notation in Fig. 7.21 is defined in Chapter 4. The overall transfer function relating $E_{in}(s)$ to $V_R(s)$ is the product of the five transfer functions shown in the block diagram:

$$\frac{V_R(s)}{E_{in}(s)} = \frac{\dfrac{2K_A K_F K_q}{L_c m A_R}}{\left(s + \dfrac{R_o + R_c}{L_c}\right)\left[s^2 + \dfrac{k_v}{m}s + \dfrac{k_e}{m}\right]} = \frac{K}{(s+P)[s^2 + 2\zeta\omega_n s + \omega_n^2]}. \qquad (7.74)$$

For convenience that will become apparent later, we write the gain K as

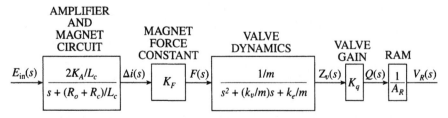

Fig. 7.21 Block diagram for amplifier, magnet, hydraulic valve, and ram system.

$$K = \underbrace{\left[\frac{2K_A K_F K_q}{(R_o + R_c) k_e A_R} \right]}_{K_c} P\omega_n^2 = K_c P\omega_n^2. \tag{7.75}$$

Note that the newly defined constant K_c is the steady-state calibration constant for the overall system, relating the steady-state ram velocity to the steady-state amplifier input voltage. K_c has units $(m/s)/V$. The transfer function is now written as

$$\frac{V_R(s)}{E_{in}(s)} = \frac{K_c [P\omega_n^2]}{(s+P)[s^2 + 2\zeta\omega_n s + \omega_n^2]}. \tag{7.76}$$

To study the dynamics of this system, we apply a step function of E volts at the input to the amplifier and calculate the resulting response of ram velocity. Again, we consider only the zero-state response:

$$V_R(s) = \frac{EK_c [P\omega_n^2]}{s(s+P)[s^2 + 2\zeta\omega_n s + \omega_n^2]}. \tag{7.77}$$

A pole-zero plot of $V_R(s)$ in Fig. 7.22 assumes the valve spool to be lightly damped. The inverse transform of $V_R(s)$ can be found either by expanding it in partial fractions or by using pair 27 in the table of transforms. The result is

$$v_R(t) = EK_c \left[1 - \frac{\omega_n^2 \epsilon^{-Pt}}{P^2 - 2P\zeta\omega_n + \omega_n^2} + \frac{P\epsilon^{-\zeta\omega_n t} \sin(\omega_o t - \psi)}{[(1-\zeta^2)(P^2 - 2P\zeta\omega_n + \omega_n^2)]^{1/2}} \right], \quad \text{where}$$

$$\psi = \tan^{-1}\left(\frac{\sqrt{1-\zeta^2}}{-\zeta} \right) + \tan^{-1}\left(\frac{\omega_n \sqrt{1-\zeta^2}}{P - \zeta\omega_n} \right) \quad \text{and} \quad \omega_o = \omega_n \sqrt{1-\zeta^2}. \tag{7.78}$$

It is interesting to study the dynamic behavior of this system as it responds to our choice of one of the physical parameters. For this illustration we will assign the following numerical values to the parameters, which are representative of practical electrohydraulic actuator systems:

$$m = 5 \text{ g} \qquad\qquad A_R = 8 \text{ cm}^2,$$
$$k_v = 600 \text{ dyne}/(cm/s), \qquad k_e = 45 \times 10^4 \text{ dyne/cm},$$
$$K_F = 10^6 \text{ dyne/A}, \qquad K_q = 8,000 \text{ (cm}^3/s)/cm,$$
$$L_c = 1 \text{ H}, \qquad\qquad R_c = 10 \text{ } \Omega,$$
$$R_o = \text{variable}.$$

Finally, K_A varies with R_o such that the ratio $K_A/(R_o + R_c)$ is always constant at 0.01.

This would be the situation where the magnet, valve, and ram are already designed but a suitable amplifier has not yet been found. With these numerical values and assumptions, the transfer function is

$$\frac{V_R(s)}{E_{in}(s)} = \frac{44.44[P(300)^2]}{(s+P)[s^2 + 2(0.2)(300)s + (300)^2]}, \quad \text{where } P = \frac{R_o + 10}{1}. \tag{7.79}$$

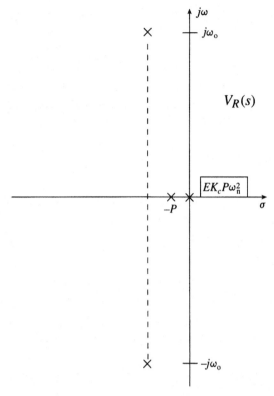

Fig. 7.22 Pole-zero plot of $V_R(s)$.

As a first trial we take $R_o = 10\,\Omega$, which gives a pole in the transfer function at $s = -20$. Notice that this places the real pole well to the right of the line connecting the complex poles (see Fig. 7.22). Let the input be a step function $e_{in}(t) = 4.5\sigma(t)$ V. This yields a response function of

$$V_R(s) = \frac{200[20(300)^2]}{s(s+20)[s^2+120s+(300)^2]}. \tag{7.80}$$

Expanding $V_R(s)$ in partial fractions, we have

$$V_R(s) = 200\left[\frac{1}{s} - \frac{1.0227}{s+20} + \frac{0.011364 - j0.032474}{s+60+j293.9} + \frac{0.011364 + j0.032474}{s+60-j293.9}\right]. \tag{7.81}$$

Note that the residues of $V_R(s)$ at $s = 0$ and at $s = -20$ are much larger than the residues at the complex poles. The geometry of the s-plane pole-zero plot shows why this is so. The inverse transform is taken by inspection, utilizing pair 16 in the inspection process:

$$v_R(t) = 200[1 - 1.0227\epsilon^{-20t} + 0.06881\epsilon^{-60t}\sin(293.9t + 160.7°)] \text{ cm/s.} \tag{7.82}$$

The time constant of the damped sinusoidal component of $v_R(t)$, 1/60 second, is only one third that of the exponential term, so the sinusoid dies out

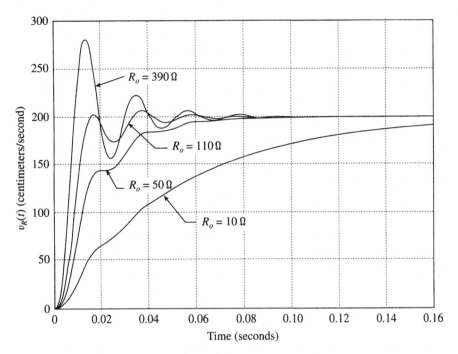

Fig. 7.23 $v_R(t)$ for four values of R_o.

quickly compared to the exponential term. Consequently, the constant term and the exponential term dominate $v_R(t)$ after the initial 30 milliseconds of the transient. An accurate plot of $v_R(t)$ is given in Fig. 7.23 as the curve labeled $R_o = 10\ \Omega$. This response, described as "heavily damped" or "overdamped," might be puzzling to someone expecting a very lightly damped response from a mechanical system with a damping ratio of 0.2. The lightly damped component is present, as we have seen, but it is overshadowed by the larger exponential term. The physical origin of the exponential term is the amplifier-magnet circuit, and the time constant $L_c/(R_o + R_c)$ depends on the physical characteristics of two separate units. Further, while R_c is simply the ohmic resistance of the magnet coil wire, R_o is not so easily identified as a separate resistor in the amplifier; it is an "equivalent resistance" of an electronic circuit, representing only the electrical nature of the amplifier as viewed from the output terminals of that device. This can be puzzling to those who are accustomed to associating each time constant of a system response with one specific device in the system, and to those for whom R_o is an obscure parameter. Control engineers must always be aware of the interactions between devices and the dynamic consequences of such interactions on overall system performance.

If the electronic design in this example is changed to increase the equivalent resistance R_o to 50 Ω, and K_A is increased to 0.6 in order to maintain a constant value for K_c, then the time constant of the exponential term will be decreased to 1/60 second, the same as that of the damped sinusoidal term. This is indicated on the pole-zero plot of Fig. 7.24. Now the residues at the complex

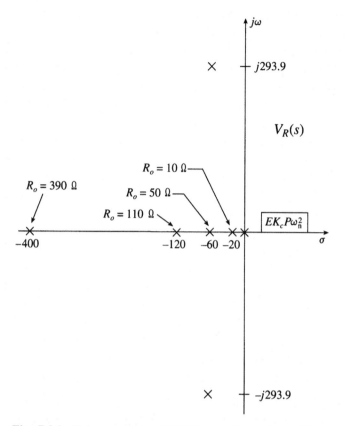

Fig. 7.24 Pole-zero plots of $V_R(s)$ for the four values of R_o.

poles have become larger in comparison to the residues at $s = 0$ and $s = -60$, and this, along with the equality of the time constants, has made the oscillatory term more pronounced in the overall response, as shown in Fig. 7.23. The overall response has become quicker as well, but it does not overshoot its final value. If R_o is further increased to 110 Ω then the exponential term dies out more quickly than does the oscillatory term, which now becomes prominent in the overall response, as indicated in Fig. 7.23; $v_R(t)$ now overshoots its final value slightly, with the peak overshoot occurring on the second local maximum of the curve rather than on the first maximum, as would occur in a second-order system. The response time is also significantly improved by this increase in R_o. If R_o is increased still further to 390 Ω then the oscillatory term dominates, causing the overall response to overshoot its final value by 40 percent and to oscillate excessively before coming to rest. The exponential term dies out quickly compared to the oscillatory component, and the overall response resembles that of a lightly damped second-order system without finite zeros in its transfer function. A system designer would likely decide on a value of R_o in the neighborhood of 100 Ω in this case, since that offers a reasonable compromise between the speed of response of $v_R(t)$ and its overshoot and terminal oscillations.

This example has illustrated two principles of control system engineering. The first is that the model of the dynamic system must include all pertinent physical parameters affecting the dynamic response of the system, and they must be modeled as they are to be used in practice. In this case we find that R_o – a parameter of the amplifier that might be chosen simply on the basis of its static interaction with the magnet resistance R_c – has more effect on the damping of the overall response than does the mechanical damping coefficient k_v of the valve.

The second principle pertains to the response that a dynamic system exhibits to a step input. If the transfer function of the system has only two complex poles and no finite zeros, then the overshoot in response to the step is correlated with the damping ratio ζ of the complex poles, as we see in Fig. 7.14 and in Appendix E. However, if the transfer function of the second-order system has a finite zero, or if the system is of higher order with a pair of complex poles, then the damping ratio of the complex pair is not directly related to the amount of overshoot to a step input. This is especially noticeable in the step response of the third-order example if the time constant of the real-valued pole is about the same as, or longer than, the decay time of its lightly damped pair. Therefore, the term "damping ratio" must be used with care in describing the dynamic nature of a physical system. Strictly speaking, the damping ratio of a system refers only to the ζ appearing in the quadratic factor of the denominator of the system's transfer function; it has a direct quantitative relationship to the overshoot only in a second-order system whose transfer function has no finite zeros.

Appendix E includes a set of normalized response curves for the third-order system, as well as a set for the second-order system having a finite zero in its transfer function.

7.5 Higher-Order Systems

We continue to focus attention on the zero-state response of a linear system having input $u(t)$ and zero-state output $y(t)$. Utilizing the transfer function method, we have the relationship

$$Y(s) = G(s) \times U(s). \tag{7.83}$$

The *steady-state gain* of the system is defined in terms of its response to a step input function of strength E, so that $U(s) = E/s$, and the output function is

$$Y(s) = \frac{E}{s} \times G(s). \tag{7.84}$$

The final-value theorem will give the steady-state value of the output, provided the system is stable:

$$\lim_{t \to \infty} [y(t)] = y(\infty) = \lim_{s \to 0} [sY(s)] = EG(0). \tag{7.85}$$

The steady-state gain is defined as the ratio of the steady-state output to the input step strength:

$$\begin{bmatrix} \text{steady-state gain} \\ \text{of the system} \end{bmatrix} = \frac{y(\infty)}{E} = G(0). \tag{7.86}$$

The *impulse response* of the system is defined as the zero-state response of the system when the input is a unit impulse function $\delta(t)$. The Laplace transform of the unit impulse is 1, so $Y(s)$ is

$$Y(s) = 1 \times G(s) = G(s), \tag{7.87}$$

making the impulse response simply the inverse Laplace transform of the transfer function $G(s)$:

$$[\text{impulse response}] = \mathcal{L}^{-1}\{G(s)\}. \tag{7.88}$$

The dynamics of a linear system are often expressed in terms of its impulse response (not including the initial-condition response), which is directly related to the physical properties of the system (Eqn. 7.88). The transfer function of an nth-order system has n poles and m zeros (usually $n \geq m$), which can always be expressed in the standard form

$$G(s) = \frac{K[s^m + a_{m-1}s^{m-1} + \cdots + a_1 s + a_0]}{[s^n + b_{n-1}s^{n-1} + \cdots + b_1 s + b_0]} = \frac{K(s+Z_1)\cdots(s+Z_m)}{(s+P_1)\cdots(s+P_n)} \tag{7.89}$$

and which can be expanded in partial fractions exactly as shown in Section 7.1. The transform of the impulse response, for the distinct-pole case, is then

$$Y_{\text{impulse response}}(s) = G(s) = \frac{K_1}{s+P_1} + \frac{K_2}{s+P_2} + \cdots + \frac{K_n}{s+P_n}, \tag{7.90}$$

and the impulse response itself is

$$[\text{impulse response}] = g(t) = \underbrace{K_1 \epsilon^{-P_1 t}}_{\substack{\text{system} \\ \text{mode 1}}} + \underbrace{K_2 \epsilon^{-P_2 t}}_{\substack{\text{system} \\ \text{mode 2}}} + \cdots + \underbrace{K_n \epsilon^{-P_n t}}_{\substack{\text{system} \\ \text{mode } n}}. \tag{7.91}$$

Each term in $g(t)$ is called a *system mode* because $g(t)$ is characterized by the physical properties of the system itself, not by an input or by initial conditions. Because the exponential constant of each mode is a pole of $G(s)$, it has often been said that "the poles of $G(s)$ characterize $g(t)$." But, as we have seen before, the sign and magnitude of each mode depend on the residues, and the residues are determined not only by the poles of $G(s)$ but also by the zeros of $G(s)$ and by the flag constant K. The *relative* values of the residues do not depend on K, so a more accurate statement is that "the poles *and zeros* of $G(s)$ characterize $g(t)$."

Consider a fifth-order system whose zero-state transfer function is displayed in the pole-zero plot in Fig. 7.25:

$$G(s) = \frac{10{,}800(s+8)}{(s+4)(s+9)(s+30)[s^2+8s+80]}. \tag{7.92}$$

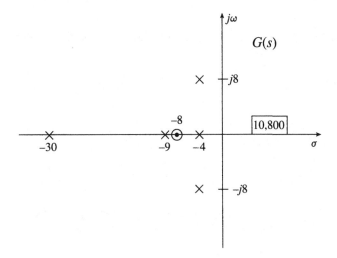

Fig. 7.25 Pole-zero plot for $G(s)$ (Eqn. 7.92).

The flag constant 10,800 is chosen to make the steady-state gain of this system 1. A partial-fraction expansion of this impulse response function is

$$G(s) = \frac{5.1923}{s+4} + \frac{1.1557}{s+9} - \frac{0.58806}{s+30} + \frac{-2.881 - j0.5944}{s+4+j8} + \frac{-2.881 + j0.5944}{s+4-j8},$$

$$(7.93)$$

and the time response $g(t)$ obtained by the inverse transformation of $G(s)$ is the sum of four real-valued components, one of which is a well-damped sinusoid oscillating at 8 rad/s. The response $g(t)$ is calculated and plotted in Fig. 7.26.

We note that the residue at $s = -30$ is much smaller in magnitude than the residue at $s = -4$. The reason for this is apparent from the pole-zero plot of $G(s)$. The pole at $s = -30$ is remote from the other poles and zeros. Four poles

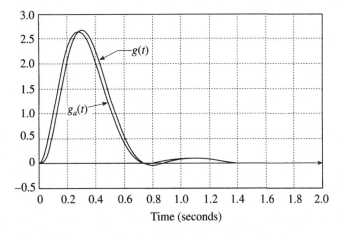

Fig. 7.26 $g(t)$ and its approximation $g_a(t)$.

and one zero are grouped near the origin; therefore, if we make an "arrow" cal-culation of residues (cf. Figs. 7.3 and 7.7), we will have by inspection the quali-tative comparison:

$$
\begin{bmatrix} \text{magnitude of} \\ \text{residue at } s = -30 \end{bmatrix} \approx \frac{10,800 \ (\text{long})}{(\text{long})(\text{long})(\text{long})(\text{long})} \approx \frac{10,800}{(\text{long})^3},
$$

$$
\begin{bmatrix} \text{magnitude of} \\ \text{residue at } s = -4 \end{bmatrix} \approx \frac{10,800 \ (\text{short})}{(\text{short})(\text{medium})(\text{medium})(\text{long})}
$$
(7.94)

$$
\approx \frac{10,800}{(\text{medium})^2 (\text{long})}.
$$

Further, the residue at $s = -9$ is noticeably smaller than the residue at $s = -4$ because of the zero at $s = -8$. The qualitative expression for this residue is

$$
\begin{bmatrix} \text{magnitude of} \\ \text{residue at } s = -9 \end{bmatrix} \approx \frac{10,800 \ (\text{very short})}{(\text{short})(\text{medium})(\text{medium})(\text{long})}
$$

$$
\approx \frac{10,800 \ (\text{very short})}{(\text{medium})^3}.
$$
(7.95)

The time constants of the modes associated with these two poles, $1/9$ s and $1/30$ s, are also noticeably shorter than the time constant corresponding to the pole at $s = -4$. The combination of short time constants and small resi-dues indicates that these two modes contribute very little to the total response $g(t)$ as compared to the contribution of the other three modes. This suggests the possibility of approximating the transfer function $G(s)$ with a lower-order transfer function by somehow eliminating the two modes that contribute only a minor part to the impulse response. This can be accomplished by factoring $G(s)$ in the following way:

$$
G(s) = \frac{((30 \times 9)/8)(s+8)}{(s+9)(s+30)} \times \frac{320}{(s+4)[s^2 + 8s + 80]},
$$
(7.96)

and by defining an approximation to $G(s)$ to be the second factor

$$
G_a(s) = \frac{320}{(s+4)[s^2 + 8s + 80]}.
$$
(7.97)

The inverse transform $g_a(t)$ of $G_a(s)$ is also plotted in Fig. 7.26 for compar-ison with the original $g(t)$. This approximation would be considered very good in most control engineering applications for the following reason. Uncertain-ties in the physical parameters on which the original transfer function is based introduce modeling errors that make the original $g(t)$ itself only an approxi-mation to the impulse response found in the real system. In other fields (such as communication systems) where physical parameter values may be known with precision and where the real system behaves linearly, this approximation might be considered to be unsatisfactory.

Insight into how $G(s)$ is factored to obtain the approximation may be gained by thinking of the original system as being composed of a hypothetical

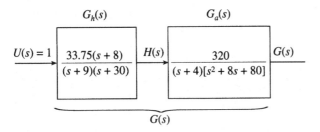

Fig. 7.27 Illustration of how $G(s)$ is approximated by $G_a(s)$.

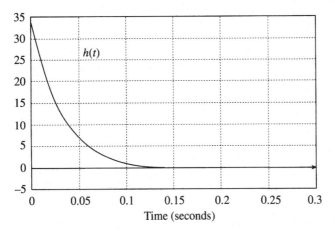

Fig. 7.28 $h(t)$ is approximately a unit impulse for the system represented by $G_a(s)$.

system having a transfer function $G_h(s)$ cascaded with $G_a(s)$, as shown in Fig. 7.27. The response of the hypothetical system is represented by $H(s)$, which for a unit impulse input equals $G_h(s)$:

$$H(s) = 1 \times G_h(s) = \frac{33.75(s+8)}{(s+9)(s+30)}. \tag{7.98}$$

The corresponding time response $h(t)$ is plotted in Fig. 7.28. The area under the $h(t)$ curve is $\int_0^\infty h(t)\,dt$. This area may be calculated by applying the final-value theorem to the Laplace transform of $\int_0^t h(t)\,dt$:

$$\int_0^\infty h(t)\,dt = \lim_{t \to \infty} \int_0^t h(t)\,dt = \lim_{s \to 0} s\left(\frac{1}{s}H(s)\right) = \frac{33.75(8)}{(9)(30)} = 1. \tag{7.99}$$

The response $h(t)$ resembles a unit impulse because it is a large pulse of strength 1, and it has a short duration compared to the time constants of the approximate system. Consequently, the response of the approximate system to a precise unit impulse, $g_a(t)$ (Fig. 7.26), is close to the impulse response of the original system.

A significant analytical difference between the impulse response $g(t)$ and its approximation $g_a(t)$ lies in initial values of the derivatives of the two functions.

These can be compared by applying the initial-value theorem to both $g(t)$ and $g_a(t)$, as follows.

	$g(t)$	$g_a(t)$
$g(0^+)$	0	0
$\dot{g}(0^+)$	0	0
$\ddot{g}(0^+)$	0	320
$\dddot{g}(0^+)$	10,800	∞

A discrepancy between the two initial values of the second derivatives occurs because the approximation was made by eliminating one zero and two poles from the original transfer function. If an equal number of poles and zeros are eliminated, this discrepancy would not exist. Whether this discrepancy would be seriously misleading in the practical setting depends on the particular application at hand.

An important technique of experimental engineering analysis is based on the principle illustrated by the examples in Section 7.2 – namely, that the dynamic response of a linear system is the sum of exponential modes, each corresponding to a pole of the s-domain representation of the response. It is often necessary for an engineer to deduce the physical properties of a system by studying experimental data derived from dynamic records. The data on the system that are available for this study are often incomplete, redundant, or incorrect. Nevertheless, knowing the structure of linear system responses, the engineer is frequently able to use the available data to formulate an accurate picture of what occurred during the reported test and to make useful quantitative estimates of physical parameters in the system.

A key element in this technique is the determination of the pole and zero locations of $Y(s)$ from a study of $y(t)$. This is called *decomposition of $y(t)$*. A simple example will illustrate the basic idea, and further examples are available in the problems in Section 7.9.

Assume that the curve in Fig. 7.29 labeled $y(t)$ is known to be the response of a system (with initial conditions known to be zero), the input being an impulse of strength α; that is, $u(t) = \alpha\delta(t)$. Possibly α is unknown, and we must estimate the pole and zero locations of the transfer function. We know that the transfer function has the form

$$G(s) = \frac{K(s^m + a_{m-1}s^{m-1} + \cdots + a_0)}{s^n + b_{n-1}s^{n-1} + \cdots + b_0}. \tag{7.100}$$

Therefore $Y(s)$ has the form

$$Y(s) = \frac{\alpha K(s^m + a_{m-1}s^{m-1} + \cdots + a_0)}{s^n + b_{n-1}s^{n-1} + \cdots + b_0}. \tag{7.101}$$

The data in Fig. 7.29 show the initial value of $y(t)$ to be zero, while that of the derivative of $y(t)$ is a positive number. When the initial-value theorem is applied to $Y(s)$, these two facts reveal that

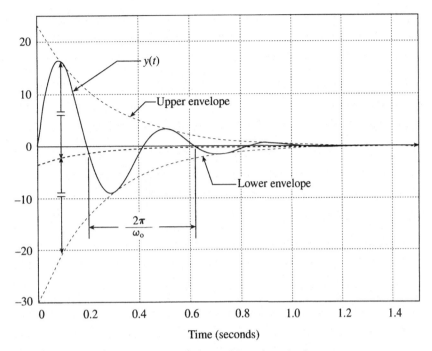

Fig. 7.29 $y(t)$ for an unknown system with an impulse input.

$$\lim_{s \to \infty} \frac{\alpha K s^{m+1}}{s^n} = 0 \quad \text{so that } n > m+1;$$

$$\lim_{s \to \infty} \frac{\alpha K s^{m+2}}{s^n} > 0 \quad \text{so that } n = m+2 \quad \text{and } \dot{y}(0^+) = \alpha K. \tag{7.102}$$

The positive value of αK could be estimated by measuring the slope of the curve $y(t)$ at $t=0$, which appears to be in the neighborhood of 25 amplitude units for 0.05 s; this gives a numerical value of roughly 500. Furthermore, we know that α and K are either both positive or both negative, since their product is positive. From the final-value theorem, which is obviously valid for this stable function, we see that $b_0 \neq 0$ because $y(\infty) = 0$, so $Y(s)$ cannot have a pole at $s = 0$.

We will now attempt to determine the modes of $y(t)$ by inspecting it in detail. We will try to find the least number of modes that can provide a reasonable fit to the given curve. From the damped oscillations in $y(t)$, it is clear that a complex pair of poles lies in the left-half s plane, and that these poles can be located by measuring the frequency of the oscillations and the time constant of the *envelope*. A careful examination of $y(t)$ reveals that the curve does not cross the $y = 0$ axis at uniform intervals, so $y(t)$ must have at least one mode in addition to the damped sinusoidal mode. A sketch of the envelope of $y(t)$ (Fig. 7.29) is useful in decomposing $y(t)$. We locate for each instant the midpoint of the segment between the upper and the lower envelope. This can be done approximately by sketching and checking several points along the sketch, as indicated at the point $t \cong 0.1$ s. The locus of midpoints is shown as the dashed curve, which turns out to have an exponential shape.

We next check to see whether $y(t)$ crosses this midpoint curve at uniform intervals, which it does in this case. The period of the sinusoid, $2\pi/\omega_0$ s, is then measured as indicated; since it is approximately 0.417 s, $\omega_0 \cong 15.1$ rad/s. This establishes the $j\omega$-axis coordinates of the complex poles. The σ-axis coordinate is the inverse of the time constant of the envelope of the damped sinusoid. This time constant is the time required for the amplitude of the envelope, measured from the midpoint curve, to diminish from its value at any starting point to $1/\epsilon$ ($\cong 0.3679$) times that starting value. In this case that time constant is about 0.251 s, so the complex poles are located at approximately $-4 \pm j15$ on the s plane. The midpoint curve is an exponential having a time constant of about 0.2 s and an initial value of about -3.6. Thus there is a pole in the transfer function at $s = -5$, and the residue of $Y(s)$ at that pole is about -3.6. It appears that these three poles are sufficient to characterize the response, but the negative sign of the residue at the real pole indicates that a zero in $Y(s)$ must lie somewhere to the right of the pole, since the flag constant αK is known to be positive. This zero may be located by considering the relative values of the residues at the real pole and at one of the complex poles. The intersection of the envelopes with the y axis at $t = 0$ gives the value of approximately 26.6, which is twice the absolute value of the residue of $Y(s)$ at $s = -4 - j15$ (see pair 16 in the table of transforms). A short series of trials shows that a zero located at approximately $s = -3$ will provide the ratio of residue values corresponding to this analysis of the experimental data. The complete form of $Y(s)$ is thus

$$Y(s) \cong \frac{\alpha K(s+3)}{(s+5)[s^2+8s+240]}. \tag{7.103}$$

An accurate estimate of the flag constant, known to be roughly 500 from the estimated initial value of $\dot{y}(t)$, can be obtained by calculating the residue of $Y(s)$ at $s = -4 - j15$. Using the graphical technique for this, we have

$$|K_{-4-j15}| = \frac{\alpha K|-4-j15+3|}{|-4-j15+5||-j30|} = \frac{\alpha K}{30} \cong 13.3, \tag{7.104}$$

from which we see that $\alpha K \cong 400$. This estimate for $Y(s)$ can be quickly checked by making a numerical calculation of $y(t)$ on a control system computer program and comparing it with the experimental data. Once the pole and zero locations have been verified, the values of the physical parameters of the system can be determined from the physical principles upon which the transfer function is originally derived.

7.6 Series Compensation

Pole-zero analysis is useful in improving the dynamic performance of a linear system. Consider the plant represented by its transfer function relating the input $u(t)$ to the zero-state response $y(t)$ in Fig. 7.30. (The term "plant"

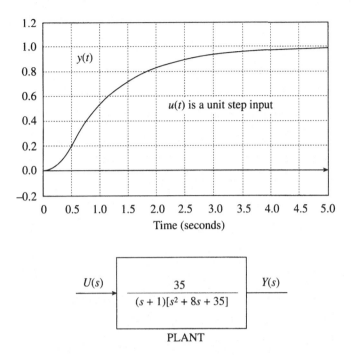

Fig. 7.30 $y(t)$, the zero-state response of the plant to a unit step input.

often implies that the system is fixed and that the control engineer does not have authority over its design.) The response of the plant to a unit step input is also plotted to show that it is rather sluggish, due to the pole at $s = -1$. To overcome the undesirable effects of this pole it may be possible to install an auxiliary system, called a *series compensator* or a *controller,* at the input of the original system; this is shown in Fig. 7.31. The input to the composite system is denoted $\hat{u}(t)$, and the overall transfer function is the product of the two transfer functions. If the compensator is designed with a zero where the original system has the troublesome pole, that pole will not appear in the overall transfer function. This design technique is called *pole-zero cancellation*. A less troublesome pole, in this example at $s = -10$, can also be designed into the compensator. The overall transfer function is

$$\frac{Y(s)}{\hat{U}(s)} = \frac{350}{(s+10)[s^2+8s+35]}. \tag{7.105}$$

If a unit step is applied at the input to the series compensator, the response $y(t)$ of the plant is that shown in Fig. 7.31. For direct comparison, the step response of the uncompensated plant is also plotted. The comparison reveals an improvement in the speed of response of the plant by a factor of five or more. This remarkable improvement in performance, achieved by the addition of a simple signal conditioning device at the input to the plant, warrants some further remarks about the pole-zero cancellation technique.

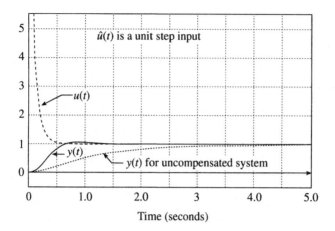

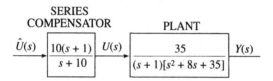

Fig. 7.31 Series compensation of stable plant by pole-zero cancellation.

Remark 1   The physical reason behind the improvement in the speed of response is simply that the input to the plant, $u(t)$, has been changed by the compensator into the large initial pulse shown in Fig. 7.31. The plant, which itself remains unaltered, is now driven by a larger input that fluctuates (or is "shaped"), so that the response $y(t)$ is improved as shown. The compensator design by pole-zero cancellation is a convenient way to achieve this input signal modification. It has the effect of replacing the troublesome plant pole at $s = -1$ with a more suitable pole at $s = -10$.

Remark 2   In most cases the cost of the compensator is trivial compared to the cost of the plant, so that series compensation is a very popular design technique. In many cases it provides improvement in plant performance comparable to, or even superior to, that demonstrated in this example. However, the negative aspects of this design approach are often not apparent from the pole-zero analysis of linear systems. These negative aspects are described as follows.

Remark 3   Transfer function analysis will accurately predict the dynamic behavior of the plant only if the input signal amplitude remains small enough to keep the plant in its linear range of operation. If this circumstance prevails, series compensation can be successful. But all physical plants are nonlinear when the driving signals exceed the linear range. Consider the electric amplifier–motor "plant" described in Fig. 4.9. The amplifier is unable to deliver more than a prescribed current to the motor even if the input voltage $e(t)$ exceeds the design level. Consequently, the driven load cannot accelerate beyond the limit provided by the driving current, so the performance one would expect from a linear analysis of the system will not be realized when the compensator calls for an excessive voltage $e(t)$, even though the signal level at the input to the compensator might be modest. Attempted pole-zero cancellation in this case would yield disappointing results.

Remark 4 A second cause of disappointing results can be the noise that is invariably superimposed on the control signal at the input to the system. This problem is discussed briefly in Section 3.4. The transfer function of a compensator designed via pole-zero cancellation usually has at least one zero close to the origin of the s plane that is critically important to the compensator design. The poles of the compensator, which are normally included to alleviate the noise problem, are located farther out in the s plane. We have seen that the presence of a zero near the origin has the signal-conditioning effect of introducing a significant term in the output proportional to the *derivative* of the input signal. Because the noise component of the input signal usually fluctuates much more rapidly than does the desired component, and because the compensator can act only on the sum (signal + noise), the signal-to-noise ratio of its output signal will be degraded as compared to that of its input signal. This degradation can negate the benefits of pole-zero cancellation expected from an analysis that ignores the noise problem. A quantitative analysis of the influence of noise in dynamic systems is necessary to evaluate this situation accurately, but such an analysis is beyond the scope of this book.

Remark 5 A third possibility for disappointment lies in the inherent uncertainty regarding the value of the physical parameters of the plant. These values are obtained by calculations based on expected conditions that may not actually prevail in practice, or on experimental measurements that contain undetermined errors, such as wind-tunnel measurements on scale models of aircraft. Also, the physical parameters of the plant may not remain fixed during operation. Consequently, the locations of the troublesome poles and zeros of the plant may be uncertain and subject to variation. Although it is usually possible to design a compensator with precise locations of its poles and zeros, the cancellation may be inexact owing to these uncertainties of the plant. The compensator design may be less successful than expected, and it may introduce serious problems that become apparent when feedback control is attempted.

Remark 6 A corollary to Remark 5 pertains to plants that are inherently unstable. This very important class of plants can be satisfactorily analyzed only with the full extent of the theory presented in this book. Linear unstable plants have transfer functions with one or more poles in the right-half s plane, and these pole values always have some degree of uncertainty. Therefore, an attempt to stabilize the plant using series compensation by pole-zero cancellation will always result in inexact cancellation of the right-half–plane poles. The

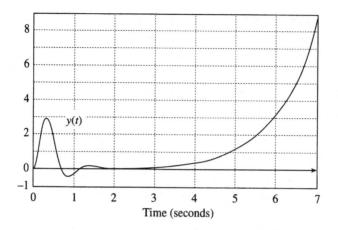

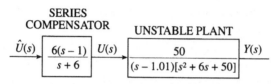

Fig. 7.32 Series compensation of unstable plant, inexact pole-zero cancellation.

composite system will still have the unstable mode, and even though the residue associated with this pole might be very small owing to the nearby zero, the unstable mode – once excited – will grow without bound. Figure 7.32 illustrates such a case, showing the response obtained for a unit impulse $\hat{u}(t)$. Inexact but close cancellation of *left*-half-*s*-plane poles is often acceptable because the troublesome mode will be very small in the beginning and will diminish with time. But attempted cancellation of *right*-half-plane poles invariably produces the result illustrated in Fig. 7.32. The only circumstance under which such a design approach could be successful is if the life of the system is short compared to the time constant of the unstable mode, as occurs in some short-term navigation and steering problems. Even in these cases, though, misunderstandings as to when $t = 0$ occurs can cause loss of control of the plant. Unstable plants are successfully controlled by feedback techniques, and this should be the primary design approach for these systems. We begin the study of feedback systems in Chapter 8.

Remark 7 Another circumstance where design by the pole-zero cancellation approach fails to achieve its intended result occurs when nonzero initial conditions exist on the plant. The reason for this is not apparent from our analysis because we focused on the zero-state response, where initial conditions are ignored. All transfer functions relating the initial conditions to the output function $Y_{ZIR}(s)$ have the same poles as the ZSR transfer functions, and the series compensator has no influence on these transfer functions. Consequently, even if perfect cancellation of right-half-plane poles could be achieved, the output of the plant will exhibit an unstable mode if one of the initial conditions is different from zero, even by a very small amount. This phenomenon occurs when an otherwise linear system is temporarily driven into a nonlinear domain. As it comes back into the linear range of operation a new episode of the initial-value problem begins, and not all the initial conditions for this new episode will be zero.

7.7 Nonminimum Phase Systems

A linear system that has a transfer function with one or more poles or zeros in the right-half s plane is called a *nonminimum phase* system. (This terminology comes from the frequency response characteristics of the system, discussed in Chapter 11.) An unstable system is a nonminimum phase system because its transfer function has one or more poles in the right-half plane. We discuss unstable systems thoroughly in later chapters, but here we consider nonminimum phase systems having only right-half–plane zeros in the transfer function. These systems can exhibit dynamic properties that seem peculiar to the reader who has studied only minimum phase systems. Calculating the response of such a system using the pole-zero methods of this chapter requires no new techniques, so that any special effect that right-half–plane zeros have on the dynamic response can be explained by the influence which the zero locations have on the magnitudes and signs of the residues. One such effect is illustrated in Fig. 7.33. The input in this example is a positive unit step, so that

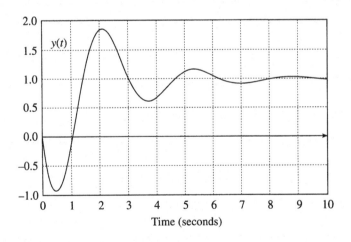

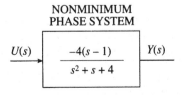

Fig. 7.33 Response of nonminimum phase system to unit step input.

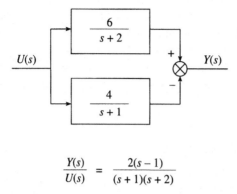

$$\frac{Y(s)}{U(s)} = \frac{2(s-1)}{(s+1)(s+2)}$$

Fig. 7.34 Constructing a nonminimum phase system from two minimum phase systems.

$U(s) = 1/s$, and the resultant time response is $y(t)$, as shown. The steady-state gain of this system is $+1$, opposite to the sign of the constant factor of $Y(s)$, but the transient starts out in the negative direction before reversing direction and settling at the positive final value. If the transfer function had two zeros in the right-half plane, then the final value would be negative (the same sign as the constant factor of $Y(s)$) and the initial part of the transient could have two reversals.

Nonminimum phase systems can be easily constructed by interconnecting minimum phase systems; an example is given in Fig. 7.34. Nonminimum phase systems also occur naturally. A response like the one in Fig. 7.33 can be observed in the motion of a conventional aircraft that has its pitch-axis control surface aft of the center of mass.

7.8 Summary

In this chapter we have concentrated on the analysis of the standard input–output problem, with the transfer function for the zero-state response playing the central role represented in Fig. 7.1. The ideas emphasized here that are essential to the study of feedback control systems are as follows.

(a) The poles, zeros, and flag constant of the transfer function depend on the physical parameters of the system.
(b) The modes of $y(t)$ are defined by the poles of $Y(s)$, and may be calculated almost by inspection of $Y(s)$ using the partial-fraction expansion.
(c) The response $y(t)$ is itself a linear combination of its modes.
(d) Given a response function $y(t)$, quantitative values for the physical parameters of the system may be deduced by the art of decomposition of $y(t)$ in order to determine the poles and zeros of $Y(s)$.
(e) The apparent dynamic properties of a plant may be altered using series compensation, but this technique has severe practical limitations.

It is important to keep in mind that the output of a linear system will contain terms due to nonzero initial conditions. This is clearly shown in Section 6.5. The analysis of these terms is conducted using the same techniques as those applied to the zero-state response.

Although the role of the computer in the study of linear system dynamics is not emphasized in this chapter, computer assistance is necessary for studying the principles of analysis. In dealing with specific numerical examples the computer can be used to factor polynomials, to obtain partial-fraction expansions, to calculate $y(t)$ from the transfer function and the given $u(t)$, and to plot $y(t)$. These tasks, none of which can be done in a reasonable time by hand calculations for systems of even modest order, can all be accomplished within mere seconds by the computer. The numerical results obtained are usually correct to a degree of precision greater than is necessary in engineering problems.

The computer is indeed an indispensable aid to the engineer. However, the computer cannot perform the ratiocination that translates an engineering problem from physical data into a numerical problem, and it cannot assess the meaning of the numerical output to the engineering problem at hand. Occasionally the algorithms within the computer program provide incorrect numerical results, and part of the task of the engineer is to recognize when this happens and to find the right answer. Such problems are usually resolved by applying analytical techniques combined with physical principles.

The computer is also used to simulate dynamic systems, often those having nonlinear components or random inputs, and to produce graphical representations of the state-variable or output-variable responses. Simulation is an important facet of control engineering, but it is one of many topics that are beyond the scope of this introductory book.

7.9 Problems

Problem 7.1

An accurate pole-zero plot of a function $Q(s)$ appears in Fig. P7.1.

(a) Express $Q(s)$ in analytical form.

(b) Find the initial values $q(0^+)$, $\dot{q}(0^+)$, and $\ddot{q}(0^+)$.

(c) Is $q(t)$ a stable function? If so, find the numerical value of K for which the final value of $q(t)$ will be 1. If not, explain why $q(t)$ is unstable.

Problem 7.2

Given

$$Y(s) = \frac{K(s+Z)}{s(s+P)(s+\alpha s+\beta)},$$

find the conditions on K, P, α, and β that must be satisfied in order that $y(\infty) = 1$.

Problem 7.3

The general form for $Y(s)$ given in Eqn. 7.1 is

$$Y(s) = \frac{K[s^m + a_{m-1}s^{m-1} + \cdots + a_1 s + a_0]}{s^n + b_{n-1}s^{n-1} + \cdots + b_1 s + b_0} = \frac{K(s+Z_1)(s+Z_2)\cdots(s+Z_m)}{(s+P_1)(s+P_2)\cdots(s+P_n)}.$$

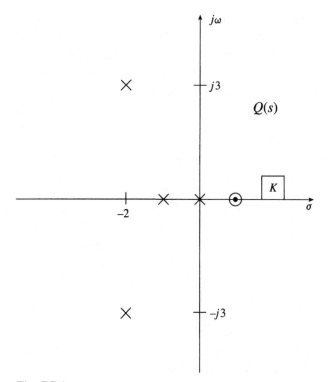

Fig. P7.1

(a) Show that $Z_1 + Z_2 + \cdots + Z_m = a_{m-1}$ and that $P_1 P_2 \cdots P_n = b_0$.

(b) Using the final-value theorem, determine how the terms in $Y(s)$ must be related to one another in order that $y(\infty) = A$.

(c) Determine the relationships which must exist in order that $y(\infty) = 0$.

(d) Which relationships must exist in order that the rth derivative and all lower-order derivatives of $y(t)$ have initial values equal to zero?

Problem 7.4

The function $m(t)$ has the Laplace transform

$$M(s) = \frac{10(s-2)^2}{s^5 + 8s^4 + 19s^3 + 12s^2 + \beta s}.$$

(a) Can you find a value of β for which the initial value of $m(t)$ is 1?

(b) Can you find a value of β for which the final value of $m(t)$ is 1? If so, specify that value. If not, explain why not.

Problem 7.5

A linear system has a zero-state transfer function $G(s)$. The input is a unit step function $\sigma(t)$. The output response $y(t)$ is plotted accurately in Fig. P7.5. Find $G(s)$.

Problem 7.6

Expand each of the following functions into partial fractions:

(a) $\dfrac{2[3s^2 + 11s + 9]}{s^3 + 6s^2 + 11s + 6}$;   (b) $\dfrac{16(s+1)}{s^4 + 4s^3 + 8s^2 + 8s}$;   (c) $\dfrac{3s^2 - s + 2}{s^3 + 2s^2 + 2s + 1}$;

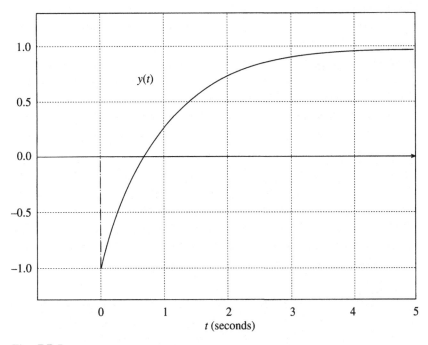

Fig. P7.5

(d) $\dfrac{2(s+15)}{s^2+10s+50}$; **(e)** $\dfrac{40(s+1)^2}{(s+2)^3(s+5)}$; **(f)** $\dfrac{625}{s^4+12s^3+86s^2+300s+625}$.

Problem 7.7

Use the formulas in Eqn. 7.31 to prove the assertion of Eqn. 7.38.

Problem 7.8

Construct a pole-zero map of

$$F(s) = \frac{250(s+3)}{(s^2+14s+24)(s^2+14.1s+49.7)}.$$

Note that the residues of $F(s)$ at the two poles near $s = -7$ are both very much larger than the residues at the other two poles. Does this mean that the modes in $f(t)$ corresponding to the other two poles may be neglected? Explain your answer.

Problem 7.9

Refer to Problem 2.5. Use the notation $v_m(t) = \dot{z}_m(t)$. Derive the transfer function $V_m(s)/F(s)$. If $M = 1$ kg, $k = 144$ N/m, $\alpha = 12$ N/(m/s), and $f(t)$ is a step function of 8 N applied at $t = 0$, find the following quantities using the normalized time response curves in Appendix E:

(a) the final velocity of the piston, $v_m(\infty)$;
(b) the instant at which the piston first reaches its final velocity;
(c) the peak velocity of the piston; and
(d) the instant at which the peak velocity of the piston is reached.

Problem 7.10

Let

$$Q(s) = \frac{K}{s(s+P)[s^2+2\zeta\omega_n s+\omega_n^2]},$$

where P, ζ, and ω_n are all positive.

(a) What value must K have in order that $q(\infty) = 1$?

(b) Prove that $q(t)$ will not overshoot its final value if $P < \zeta\omega_n$.

Problem 7.11

Refer to the diagram for the electrohydraulic valve and ram system in Fig. 7.21. The following parameter values are known: $M = 5$ g, $R_c = 180$ Ω, $R_o = 500$ Ω, $K_A = 13.6$, $K_F = 10^6$ dynes/A, and $K_q = 3,000$ (cm^3/s)/cm. A one-volt step is applied to the input of the amplifier, and the resultant response of the ram velocity $v_R(t)$ is recorded in Fig. P7.11. Find the numerical values of A_R, L_C, k_e, and k_v. *Hint:* use the normalized time response curves in Appendix E.

Problem 7.12

Refer to Problem 3.8. Derive the transfer function $E_{out}(s)/E_{in}(s)$ and express it in terms of K_P, K_I, and K_D.

Problem 7.13

Refer to Problem 3.4. Derive the transfer function $Y(s)/U(s)$. Let the physical parameters have the following values: $R_1 = R_2 = 10$ Ω, $L = 1$ H, and $C_1 = 100$ μF. K can be varied over the range $-50 < K < 50$. Draw a root-locus plot showing how the poles of

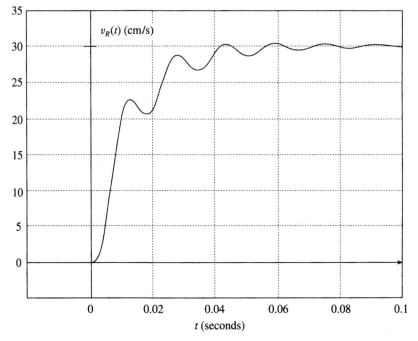

Fig. P7.11

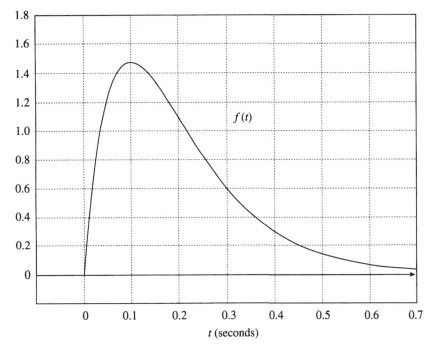

$f(t)$

t (seconds)

Fig. P7.14

$Y(s)/U(s)$ migrate in the s plane as K is varied from -50 to 50. Identify the range of values of K for which this circuit will be unstable.

Problem 7.14
An accurate plot of $f(t)$ appears in Fig. P7.14. Find $F(s)$ and express it in the standard factored form indicated in Eqn. 7.2.

Problem 7.15
An accurate plot of $g(t)$ appears in Fig. P7.15. Find $G(s)$.

Problem 7.16
An accurate plot of $r(t)$ appears in Fig. P7.16. Find $R(s)$.

Problem 7.17
How would the response of the system in Fig. 7.32 be changed if the right-half-plane pole were at $s = +0.99$ instead of at $s = +1.01$?

Problem 7.18
Let

$$F(s) = \frac{K(s+Z_1)(s+Z_2)}{s(s+3)(s^2+4s+25)}.$$

Write $F(s)$ in the form $F(s) = Q(s) + k_1 s Q(s) + k_2 s^2 Q(s)$, where $Q(s)$ has no finite zeros and k_1 and k_2 are expressed in terms of K, Z_1, and Z_2. Note that $f(t) = \mathcal{L}^{-1}\{F(s)\}$ can be expressed in the form $f(t) = q(t) + k_1 \dot{q}(t) + k_2 \ddot{q}(t)$.

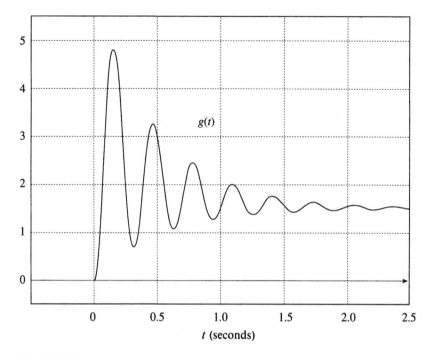

Fig. P7.15

(a) Specify the conditions on K, Z_1, and Z_2 that must hold for $f(t)$ to have a final value of 1 and for the zeros of $F(s)$ to be distinct from its poles.

(b) Assume that the conditions of (a) are satisfied. Find further conditions on the three parameters under which k_1 and k_2 satisfy each of the following conditions:

1. k_1 and k_2 are both positive;
2. k_1 and k_2 are both negative;
3. $k_1 > 0$ and $k_2 < 0$;
4. $k_1 < 0$ and $k_2 > 0$.

(c) In which of the four cases in part (b) is it possible to have a pair of complex conjugate zeros in $F(s)$?

(d) Use a computer program to plot $f(t)$ versus t for examples of each of the four cases in part (b).

Problem 7.19

A sixth-order system has the zero-state transfer function

$$G(s) = \frac{560(s^2 + 3.05s + 2.1)}{(s+12)(s^2 + 4s + 25)(s^3 + 5.4s^2 + 8.4s + 4)}.$$

Find a third-order transfer function that is a reasonable approximation to $G(s)$.

Problem 7.20

The zero-state transfer function of a linear system is known to have the form

$$\frac{Y(s)}{U(s)} = \frac{K}{(s+50)(s^2 + 50s + \omega_n^2)},$$

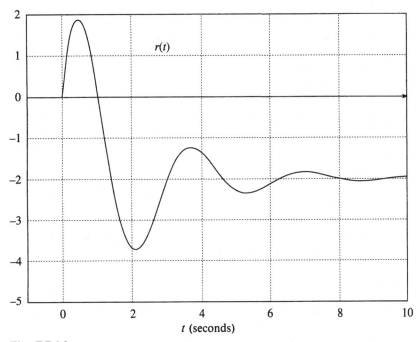

Fig. P7.16

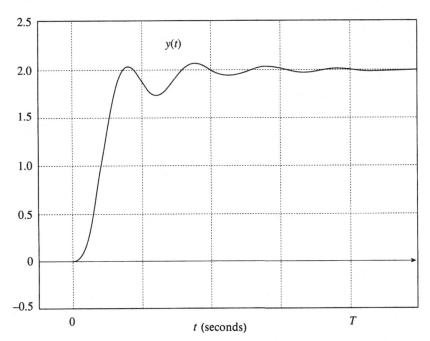

Fig. P7.20

where the numerical values of K and ω_n are not known. A dynamic test is conducted on this system by exciting it with a unit step input, $u(t) = \sigma(t)$. The resultant output, $y(t)$, is plotted in Fig. P7.20. Unfortunately, the time scale was lost in the recording process. Find the numerical values of K and ω_n and establish the time scale on the recording by specifying T.

Problem 7.21

Consider the transfer function of a stable second-order system,

$$\frac{Y(s)}{U(s)} = \frac{K(s+Z)}{(s^2+2\zeta\omega_n s+\omega_n^2)},$$

where K and Z are adjustable parameters. If the input is a ramp function $u(t) = At$, determine what values K and Z must have if $y(t)$ is to approach $u(t)$ for large t, that is, $\lim_{t \to \infty}[y(t) - u(t)] = 0$.

Problem 7.22

With K and Z adjusted to satisfy the condition stated in Problem 7.21, show that if $u(t)$ is a step input $u(t) = B\sigma(t)$ then:

   (a) $\lim_{t \to \infty} y(t) = B$; and
   (b) the output response $y(t)$ must overshoot its final value of B; that is, there must be an instant t_1 at which $y(t_1) > B$.

Problem 7.23

Prove or disprove the following statement: Any stable nth-order linear system satisfying the "zero ramp error" property described in Problem 7.21 will also satisfy the step input properties (a) and (b) described in Problem 7.22.

Problem 7.24

An accurate plot of $h(t)$ appears in Fig. P7.24. Find $H(s)$.

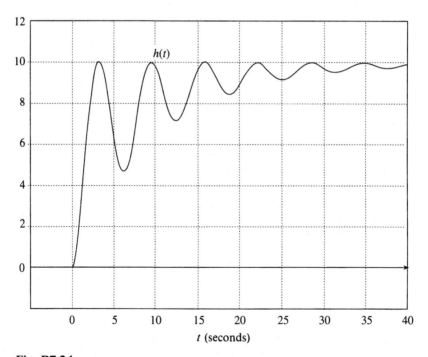

Fig. P7.24

8 Automatic Feedback Control

8.1 Automatic Control of Dynamic Systems

In this chapter we restrict our attention to single-input–single-output (SISO) plants, reserving consideration of the multi-input–multi-output (MIMO) plant for Chapter 15. The basic control problem for the SISO dynamic plant is defined in the block diagram of Fig. 8.1, where $y(t)$ is either the physical quantity (of the plant) that is to be controlled or a signal produced by a sensor measuring the controlled variable. The energy source required to operate the plant must be included in the plant model if that source significantly affects the dynamic performance of the plant. The input variable $u(t)$ of the plant is usually provided by an actuator such as an electromagnetic or electrohydraulic energy converter. Both the actuator and the energy source required to operate it may be included in the model for the plant, or they may be considered as part of the controller, as in Fig. 8.1. The controller is designed to accept an input command $r(t)$, which is a signal furnishing information only, not energy, to a computing device termed the *control law*. The control law issues the driving signal $u_i(t)$ to the actuator.

In a typical control problem, the control designer cannot change the fixed plant. Sometimes the actuator is also unchangeable, in which case the actuator

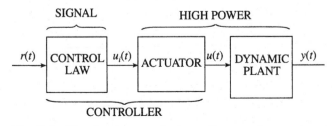

Fig. 8.1 Signal and power elements of a control system.

189

would be considered part of the fixed plant. Enough information must be available about the physical properties of the plant to construct a mathematical model (transfer function) of the plant. The control designer must know the design specifications on the desired output of the plant over a specified time interval. These specifications may be stated in terms of the response exhibited by $y(t)$ to a specified control signal $r(t)$. For example, a step input function $r(t)$ of specified magnitude must result in a stable response $y(t)$ that settles promptly toward a specified final value. Other specifications that may be placed on detailed dynamic features of the response are: the largest overshoot of its final value permitted $y(t)$, the number of oscillatory cycles that $y(t)$ can exhibit before a specified settling time, restrictions on the peak values of $\dot{y}(t)$ and $\ddot{y}(t)$, and other limitations on the trajectory that $y(t)$ makes in response to the step input.

Test inputs other than the unit step – a ramp function, a power series of a few terms, a sinusoidal function of specified amplitude and frequency or a spectrum of such functions, a random function $r(t)$ specified in terms of its statistical parameters – may also be specified, with corresponding requirements placed on $y(t)$. Because some of these test functions are redundant, one must be wary of multiple performance specifications that may be inconsistent. Most common dynamic performance requirements on $y(t)$ can be expressed as specifications on the step response, and therefore much use is made of the step function. The specifications themselves come from the intended use of the system. Because the system is often a component in a larger system, its performance specifications must be met to ensure acceptable performance of the larger system.

The control engineer designs the controller so that the composite controller–plant system provides an acceptable $y(t)$ in response to the specified test input $r(t)$. The engineer must find a transfer function for the controller $G_{\text{cont}}(s)$ so that the following relationship holds:

$$Y_{\text{acceptable}}(s) = R_{\text{test}}(s) \times G_{\text{cont}}(s) \times G_{\text{plant}}(s). \tag{8.1}$$

Equation 8.1 can be re-arranged to provide an immediate solution to the design problem:

$$G_{\text{cont}}(s) = \frac{Y_{\text{acceptable}}(s)}{R_{\text{test}}(s) \times G_{\text{plant}}(s)}. \tag{8.2}$$

Since the poles, zeros, and gain constants of all the terms on the right-hand side are known from the specification of the problem, the poles, zeros, and gain constant of $G_{\text{cont}}(s)$ are apparently determined! Furthermore, Eqn. 8.2 has within it the possibility of providing multiple solutions to the design problem, because many functions $Y_{\text{acceptable}}(s)$ will normally satisfy the desired response specifications. However, this mathematical approach to the design problem warrants more careful consideration. In most practical applications it is unsatisfactory for the following reasons, which are not obvious from Eqn. 8.2.

1. Usually the plant parameters are not precisely known, and they may drift with use. Since the transfer function poles and zeros are not precisely known, Eqn. 8.2 could yield a design for the controller that would prove unsatisfactory when used with the real plant. In fact, such a design is certain to be unsatisfactory if the plant is unstable, because the slightest uncertainty in the right-half–plane pole of $G_{\text{plant}}(s)$ would result in an unstable mode in $y(t)$.

2. Equation 8.1 gives only the zero-state response. Nonzero initial conditions on the plant state variables are not accounted for in this analysis, and they will produce a component of $y(t)$ which does not appear in Eqn. 8.1 and which could render $y(t)$ unacceptable.

3. Most plants have a second input, called a *disturbance input,* which drives the plant in the same way as does the control input. This (usually random) disturbance, which comes from outside the system, cannot be directly measured or suppressed. The flight-control system of an aircraft in cruising flight provides an example of this situation. The natural turbulence of the atmosphere places undesired random aerodynamic forces on the airframe. The solution offered by Eqn. 8.2 cannot satisfactorily provide control forces to counteract the disturbing forces.

8.2 Feedback Control of Dynamic Plants

The practical difficulties of controlling a dynamic plant using the configuration in Fig. 8.1 are largely overcome, or at least minimized, by using *feedback control.* The basic idea behind feedback control is to make the control input $u(t)$ dependent on what the plant output is actually doing, and *not* on what it would be doing were there no uncertainties in the physical parameters, unknown initial conditions, or plant disturbances. This control strategy in a SISO system requires a sensor or instrument to measure $y(t)$ instantly and to provide a feedback signal to the controller, as shown in Fig. 8.2. The disturbances and initial conditions, if any, will upset $y(t)$ just as they do in the open-loop case of Fig. 8.1, but the feedback signal gives the controller current information on the

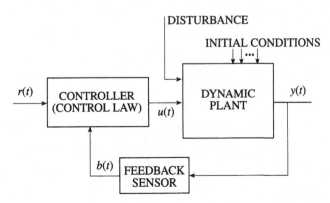

Fig. 8.2 General feedback configuration for a control system.

actual condition of the plant output. In addition to the current value $y(t)$ of the plant output, the control designer also has $r(t)$ available to represent the desired value of $y(t)$. These two signals can be used, somehow, to generate a control input to the plant $u(t)$ capable of driving $y(t)$ in a manner superior to that which is possible without feedback.

It is important to note that the fixed plant in Fig. 8.1 is exactly the same as the fixed plant in Fig. 8.2, but the closed-loop configuration of Fig. 8.2 renders the dynamic relationship between $r(t)$ and $y(t)$ very different from that in the open-loop configuration of Fig. 8.1. In practice, the potential for altering the apparent dynamic behavior of the fixed plant by feedback through proper design of the controller often leads to remarkable enhancements in the performance of the plant. Unfortunately, the feedback connection also has the potential to destroy the performance of the plant (and even, in some cases, the plant itself) when the controller is not properly designed. The purpose of the remainder of this book is to analyze the dynamics of the feedback connection and to develop some rudimentary techniques for controller design that will realize the former potential while avoiding the latter.

Figure 8.2 does not show just how the controller processes the two input signals $r(t)$ and $b(t)$ to form the controller output $u(t)$. In some situations a person is the controller. For example, a skilled crane operator observes the progress of a process ($b(t)$), mentally compares that progress with the desired progress ($r(t)$), and steers the crane accordingly, thus completing the control loop. This is an example of *manual feedback control.* However, in our study of *automatic* feedback control, the controller is usually a computer programed to realize the desired control objective.

The terms $r(t)$ and $b(t)$ may be combined and processed in an infinite variety of ways to produce the control input to the plant $u(t)$. Here we must limit our attention to a severely restricted class of strategies for controller design. The class of control laws upon which we concentrate matches the analytical form of the plant model – namely, ordinary differential equations with $u(t)$ as the dependent variable and $r(t)$ and $b(t)$ as driving functions. This approach permits us to form a set of *simultaneous differential equations* (represented by transfer functions) from those of the plant plus those of the controller, so that the solution $y(t)$ to this set may be found when the appropriate input functions and initial conditions are supplied. This approach has been used with outstanding success in automatic control engineering since the middle of the nineteenth century, when Maxwell performed the first mathematical analysis of feedback systems. In Chapter 15 the question of a more general approach to controller design is addressed, but in the intervening chapters a major effort will be required to learn the mature science called *classical control theory.*

We further restrict the form of the controller to that which operates on the difference between $r(t)$ and $b(t)$, as illustrated in Fig. 8.3. The differential symbol represents the algebraic operation

$$e(t) = r(t) - b(t). \tag{8.3}$$

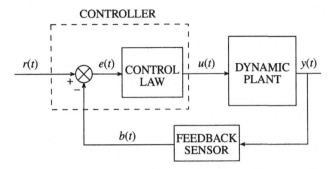

Fig. 8.3 Configuration for an error-actuated automatic feedback control system.

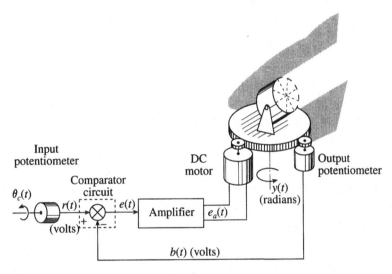

Fig. 8.4 Searchlight servomechanism.

The difference $e(t)$ is frequently called the *error,* and the feedback system is sometimes called an *error-actuated* system to distinguish it from the open-loop control scheme in Fig. 8.1. Because we will first study the zero-state response in the disturbance-free case, returning to the general case later, we omit the plant disturbance and initial conditions from the block diagram in Fig. 8.3.

Figure 8.4 shows a physical example of such a system. The controlled variable in this electromechanical servomechanism is the azimuth angle of the searchlight platform $y(t)$, which is measured by the output potentiometer to provide the feedback signal $b(t)$ volts. (A typical potentiometer circuit is diagramed in Fig. 3.14.) The error signal is generated by a summing circuit that could be similar to those diagramed in Figs. 3.10 and 3.11. When the searchlight is stationary and the platform angle $y(t)$ is equal to the input potentiometer angle $\theta_c(t)$, the error is zero and the voltage at the motor is also zero. If the input command $\theta_c(t)$ is suddenly changed to a new setting then the suddenly increasing error will be amplified, and the motor will accelerate the platform in the direction corresponding to the input command in order to reduce the error. If the whole system is dynamically stable, the error will eventually be driven to zero, with the searchlight pointing in the newly commanded direction. It is essential to the success of this system that the electrical connections in the potentiometer circuits, the summing circuit, and the amplifier be coordinated so that the motor will accelerate in the proper direction when excited, and also so that the *negative feedback* relationship indicated by the summing circuit is realized. Another practical consideration is to make the *gain* of the amplifier (the ratio of the voltage at the motor terminals to the error voltage) as high as possible, so that a small error will produce a large driving torque on the platform. The friction in the drive mechanism, which has been minimized but not eliminated, will thus be overcome by a very small error.

Figure 8.4 also pictures two typical properties found in automatic control systems. The first is power amplification. The driving motor can expend much power as it moves the searchlight back and forth to track the command that an operator applies with a light twist of the input potentiometer shaft. The second is remote control. The searchlight platform may be located in a desolate and hostile environment, but the operator may be located some distance away in a comfortable office. Electrical communication between the two sites, possibly using wireless techniques, makes remote control possible.

As a first step in the dynamic analysis of negative feedback systems, we simplify the configuration of Fig. 8.3 to that of Fig. 8.5, retaining the necessary analytical features of the system. In this diagram the transfer functions of the control law and the plant are combined into the single transfer function, denoted $G(s)$. This is called the *forward-path* transfer function or the *open-loop* transfer function:

$$\left[\begin{array}{c} \text{open-loop} \\ \text{transfer function} \end{array} \right]_{\text{ZSR}} = \frac{Y(s)}{\epsilon(s)} = G(s). \tag{8.4}$$

Note the use of the nonstandard notation $\mathcal{L}\{e(t)\} = \epsilon(s)$.

Assume that the feedback sensor has a scale factor of 1, so that the output itself is fed back to the summing device. (Ignore, temporarily, the difference in physical units that usually exists between $y(t)$ and $b(t)$.) The s-domain model

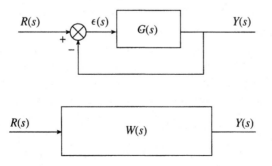

Fig. 8.5 Transfer function block diagrams for unity negative feedback system.

of the summing device is obtained by taking the Laplace transform of Eqn. 8.3, with $y(t) = b(t)$:

$$\epsilon(s) = R(s) - Y(s). \tag{8.5}$$

From Eqn. 8.4 we have

$$Y(s) = \epsilon(s)G(s). \tag{8.6}$$

Substitute $\epsilon(s)$ from Eqn. 8.5 into Eqn. 8.6, and re-arrange the result to obtain the formula upon which most of classical control theory is based.

$$\begin{bmatrix} \text{closed-loop} \\ \text{transfer function} \end{bmatrix}_{\text{ZSR}} = \frac{Y(s)}{R(s)} = \frac{G(s)}{1+G(s)} = W(s). \tag{8.7}$$

This important formula shows the effect of the physical properties of the system components (modeled in the transfer function $G(s)$) upon the dynamic behavior of the overall system (modeled in the transfer function $W(s)$). This relationship guides the control engineer in the design of $G(s)$ so that performance specifications on $W(s)$ are satisfied. The relationship is easily inverted to show that

$$G(s) = \frac{W(s)}{1-W(s)}. \tag{8.8}$$

It is very useful to represent $G(s)$ in the standard form – showing its poles, zeros, and flag constant – and to use the notation $N(s)$ for numerator and $D(s)$ for denominator.

$$\begin{aligned} G(s) = \frac{N(s)}{D(s)} &= \frac{K(s^m + a_{m-1}s^{m-1} + \cdots + a_1 s + a_0)}{s^n + b_{n-1}s^{n-1} + \cdots + b_1 s + b_0} \\ &= \frac{K(s+Z_1)(s+Z_2)\cdots(s+Z_m)}{(s+P_1)(s+P_2)\cdots(s+P_n)}. \end{aligned} \tag{8.9}$$

Substitute this detailed expression for $G(s)$ into Eqn. 8.7 to observe greater detail on the closed-loop transfer function:

$$\begin{aligned} W(s) &= \frac{N(s)}{N(s)+D(s)} \\ &= \frac{K(s^m + a_{m-1}s^{m-1} + \cdots + a_1 s + a_0)}{[K(s^m + a_{m-1}s^{m-1} + \cdots + a_1 s + a_0)] + [(s^n + b_{n-1}s^{n-1} + \cdots + b_1 s + b_0)]} \\ &= \frac{K(s+Z_1)(s+Z_2)\cdots(s+Z_m)}{[K(s+Z_1)(s+Z_2)\cdots(s+Z_m)] + [(s+P_1)(s+P_2)\cdots(s+P_n)]} \\ &= \frac{K_{\text{CL}}(s+Z_1)(s+Z_2)\cdots(s+Z_m)}{(s+Q_1)(s+Q_2)\cdots(s+Q_p)}. \end{aligned} \tag{8.10}$$

This expression for $W(s)$ reveals two key facts about the relationship between the dynamics of the open-loop system, as characterized by $G(s)$, and those of the closed-loop system, as characterized by $W(s)$. First, the denominator polynomial of $W(s)$ is the sum of the numerator polynomial of $G(s)$ and the denominator polynomial of $G(s)$, which has this form:

$$(a_0 + b_0) + (a_1 + b_1)s + (a_2 + b_2)s^2 + \cdots + (c_p)s^p, \tag{8.11}$$

where:

for $n > m$, $p = n$ and $c_p = 1$;
for $n = m$, $p = n = m$ and $c_p = (K+1)$; and
for $n < m$, $p = m$ and $c_p = K$.

p, the dynamic order of $W(s)$, is an important parameter in the closed-loop transfer function. $p = n$ if $G(s)$ has more finite poles than finite zeros ($n > m$); otherwise, $p = m$. The roots of polynomial Eqn. 8.11 are the poles of $W(s)$. These roots depend on the coefficients of both the numerator polynomial and the denominator polynomial of $G(s)$. These roots can be determined only by factoring the polynomial, a formidable task that makes a computer with a program tailored for control system analysis an absolute necessity. Second, the flag constant of $W(s)$, designated as K_{CL} in Eqn. 8.10, also depends on n and m:

$$K_{CL} = \begin{cases} K & \text{if } n > m, \\ K/(K+1) & \text{if } n = m, \\ 1 & \text{if } n < m. \end{cases}$$

The zeros of $W(s)$ coincide with the zeros of $G(s)$ in this unity feedback case. An example of a typical open-loop transfer function is

$$G(s) = \frac{40(s+1)}{s(s+4)(s+5)(s+6)}, \tag{8.12}$$

so that the closed-loop transfer function is

$$\begin{aligned} W(s) &= \frac{40(s+1)}{[s^4 + 15s^3 + 74s^2 + 120s] + [40s + 40]} \\ &= \frac{40(s+1)}{s^4 + 15s^3 + 74s^2 + 160s + 40} \\ &= \frac{40(s+1)}{(s+0.286)(s+8.38)[s^2 + 6.335s + 16.71]}. \end{aligned} \tag{8.13}$$

Pole-zero plots of $G(s)$ and $W(s)$ are superimposed in Fig. 8.6.

8.3 Block Diagrams for Control System Analysis

Many basic principles of feedback control theory can be illustrated through analysis of the unity feedback system in Fig. 8.5. The more comprehensive block diagram in Fig. 8.7 permits us to analyze the effects of plant disturbances, initial conditions, and command input on the system response. It also permits us to include dynamic elements in the feedback path. In this diagram the transfer function $H(s)$ represents the dynamics (if any) associated with the feedback

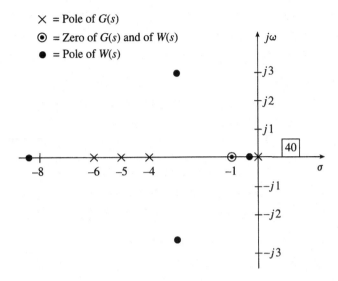

X = Pole of $G(s)$
⊙ = Zero of $G(s)$ and of $W(s)$
● = Pole of $W(s)$

Fig. 8.6 Pole-zero plots for $G(s)$ and $W(s)$ (Eqns. 8.12 and 8.13).

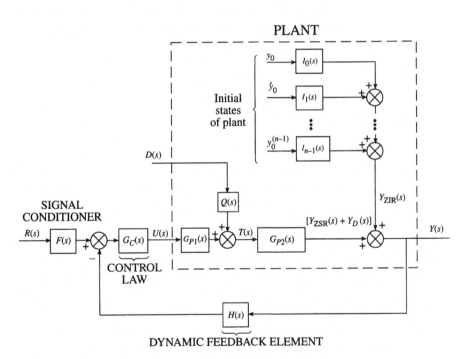

Fig. 8.7 Comprehensive block diagram of feedback control system.

sensor devices. This might include electronic filters designed to reduce noise on the feedback signal coming from the disturbance (shown as $D(s)$) or from other noise sources not shown. $H(s)$ might also be designed to modify the feedback signal in order to improve system performance in ways not related to noise. $H(s)$ is taken to be 1 when the system has unity feedback.

The zero input response of the plant is represented by the n initial-state transfer functions, as derived in Chapter 6. The plant transfer function is divided into blocks to delineate the effect of the disturbance. For example, say the disturbance $d(t)$ m/s is the wind blowing on the searchlight shown in Fig. 8.4. This places an undesired load torque, expressed in units of N-m, on the platform. The transfer function $Q(s)$ in Fig. 8.7 represents the aerodynamic properties of the searchlight mechanism that converts wind velocity in m/s to disturbance torque in N-m. The quantity labeled $T(s)$ represents the sum of the disturbance torque and the control torque. $G_{P1}(s)$ represents that part of the plant relating the input signal $u(t)$ to the control torque. The total response function $y(t)$ is measured by the sensor, which cannot distinguish among the three components of motion caused by control, wind, and initial conditions. Its Laplace transform is

$$Y(s) = Y_{ZSR}(s) + Y_{ZIR}(s) + Y_D(s). \tag{8.14}$$

The input signal $r(t)$ may also contain unwanted noise, or it may need to be modulated in some sense to make it acceptable to the control system. In this case a filter denoted by $F(s)$, sometimes called a *signal conditioner,* would be installed at the input as shown. The mathematical definition of system error, which in the unity feedback case is

$$[\text{system error}] = e(t) = r(t) - y(t), \tag{8.15}$$

continues to be valid, and remains useful in performance specifications even though in this system the signal at the input to the control law is not $e(t)$; $e(t)$ is *not available as a measurable quantity* anywhere in this system.

In subsequent chapters much of the analysis and design work assumes the disturbances to be negligible, the initial conditions to be zero, and the signal conditioner to be unnecessary. Under these circumstances the block diagram of Fig. 8.7 reduces to that of Fig. 8.8, which is almost as simple as the unity feedback diagram treated in Section 8.2. All the forward-path elements are collected into the transfer function $G(s)$, and the feedback elements have the transfer function $H(s)$. The system error, as defined in Eqn. 8.15, is not available as a measurable variable, so the input to $G(s)$ is designated as $\Delta(s)$. The three constituent equations for this configuration are:

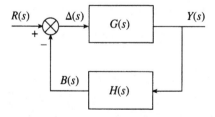

Fig. 8.8 Block diagram for a non-unity negative feedback system.

$$Y(s) = \Delta(s)G(s),$$
$$B(s) = Y(s)H(s),$$
$$\Delta(s) = R(s) - B(s).$$

(8.16)

To obtain the closed-loop transfer function, again designated by $W(s)$, combine these three equations by eliminating the two "internal" variables $\Delta(s)$ and $B(s)$, and re-arrange the result to yield

$$\begin{bmatrix} \text{closed-loop} \\ \text{transfer function} \end{bmatrix} = \frac{Y(s)}{R(s)} = W(s) = \frac{G(s)}{1 + G(s)H(s)}.$$

(8.17)

Note that if $H(s)$ is set equal to 1 then this expression reduces to that for the unity feedback case in Eqn. 8.7. Because we are dealing only with the zero-state response in this example, we drop the subscript ZSR appearing in Eqn. 8.7. Nevertheless, it is prudent to indicate somehow in the notation that this is the special case of zero initial conditions. This precaution against ambiguity is frequently overlooked in practice, with the result that the analyst may be confused by experimental data which is due in part to nonzero initial conditions but which cannot be explained from the analysis of the ZSR only. If $G(s)$ and $H(s)$ are written in more detail as

$$G(s) = \frac{N_1(s)}{D_1(s)} \quad \text{and} \quad H(s) = \frac{N_2(s)}{D_2(s)},$$

(8.18)

then the expression for $W(s)$ also shows a more detailed structure,

$$W(s) = \frac{\dfrac{N_1(s)}{D_1(s)}}{1 + \dfrac{N_1(s)N_2(s)}{D_1(s)D_2(s)}} = \frac{N_1(s)D_2(s)}{N_1(s)N_2(s) + D_1(s)D_2(s)},$$

(8.19)

which reveals the following useful information.

(a) The zeros of $W(s)$ are the zeros of $G(s)$ and the poles of $H(s)$.

(b) The poles of $W(s)$ are roots of the polynomial sum $N_1(s)N_2(s) + D_1(s)D_2(s)$. This shows that, insofar as the poles of $W(s)$ are concerned, the loop transfer function $[G(s)H(s)]$ in this system plays the same role that $G(s)$ plays in the unity feedback case.

Our next example, closely related to that of Eqns. 8.12 and 8.13, is useful. Let $G(s)$ and $H(s)$ be such that their product is the same as the $G(s)$ in Eqn. 8.12:

$$G(s) = \frac{8}{s(s+4)(s+6)}, \qquad H(s) = \frac{5(s+1)}{s+5}.$$

(8.20)

The closed-loop transfer function is then

$$W(s) = \frac{8(s+5)}{[8 \times 5(s+1)] + [s(s+4)(s+6) \times (s+5)]}$$

$$= \frac{8(s+5)}{s^4 + 15s^3 + 74s^2 + 160s + 40}.$$

(8.21)

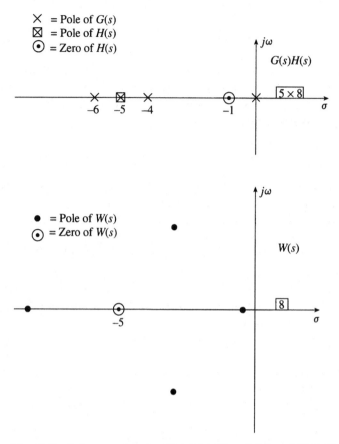

Fig. 8.9   Pole-zero plots for $G(s)H(s)$ (Eqn. 8.20) and $W(s)$ (Eqn. 8.21).

Here the denominator is the same as that in Eqn. 8.12 but the numerator is different, indicating that although the product $G(s)H(s)$ in this example plays a role identical to that played by $G(s)$ in the unity feedback case with respect to the *poles* of $W(s)$, the same is not true with respect to the *zeros* of $W(s)$. It is noteworthy that 5 and 8, the two flag constants in this example, were chosen so that their product matches the flag constant of $G(s)$ in the previous example, and so that the steady-state gain of the closed-loop system $W(0)$ is 1 in both examples. Pole-zero plots for $G(s)H(s)$ and for $W(s)$ are drawn in Fig. 8.9.

8.4   Control System Design

The major problems faced by control system design engineers were outlined in Section 8.1. These problems, which are similar to those faced by designers in other branches of engineering, reflect the economic necessity of making a product that satisfies given performance specifications while minimizing costs of time and material resources in the process.

One aspect of the design process that distinguishes control system engineering is that the design specifications invariably emphasize *dynamic* performance requirements for accuracy of control. This accounts for the emphasis on the dynamics of physical devices and processes in automatic control textbooks. Another distinguishing aspect of control engineering is the attention paid to the *robustness* of a design. Robustness refers to the ability of the finished system to function satisfactorily under the following typical operating conditions.

(a) The physical parameters of the plant are different from the nominal values assumed in the design, but they lie within known limits of uncertainty.

(b) Some existing properties of the physical system are ignored in the design. Examples include: incipient friction in mechanisms; nonlinear characteristics in amplifiers, power supplies, actuators, and sensors; and characteristics of signal-processing devices that are too difficult to model correctly.

(c) A monitoring subsystem may be used to detect failures of sensors, actuators, computers, or other components of the plant.

Robustness is one facet of the design approach called *optimal control*. In this approach, the goal is to satisfy the dynamic response specifications while at the same time minimizing a *cost function* (or *figure of merit*). The cost function depends on the energy expended to accomplish a given control task, on the time required for the task, or on other important aspects of the design. The analytical techniques of optimal control (including statistical analysis of random processes) are more advanced than those discussed here.

A problem in *design* is normally distinguished from a problem in *analysis* as follows: the well-posed analysis problem has only one solution, but a well-specified design problem can have many satisfactory solutions. Furthermore, a design problem that appears to be well-specified can turn out to have no solution at all. While the design task can require a broad class of talents (often characterized by experience and judgment) in addition to a facility at analysis, the most elementary and frequently used approach to design is the trial-and-error procedure, which is largely analysis. The starting point is a trial design for the controller, which might be a modification of a design obtained previously for a similar problem or a design obtained from an engineering handbook. This trial design is then analyzed to see whether it satisfies the required performance specifications. In most cases it will not, so a second trial is called for. The results of the first trial should indicate how a modification of the first trial design might improve the system performance; normally the modification involves the shifting of poles and zeros in the controller. This modification is made, and the second trial design is analyzed. This sequence of trial–analysis–assessment–alteration cycles is continued until a satisfactory design results. Often, a satisfactory solution is found after a few trials. More difficult problems are successfully solved using computer programs, which have accelerated the trial-and-error process far beyond what was possible as late as the mid-1980s.

Computers used as *simulators* are especially useful for the analysis of non-linear systems. Operational hardware components of a large system can be tested by connecting them to a simulation of the other parts of the system. The answers from a simulator are usually time trajectories of critical variables of the system, as opposed to pole and zero values or to other analytical properties of the design.

Once a satisfactory controller design has been identified by its transfer function, the right hardware must be found to realize the design. For example, in the searchlight servomechanism design it may be that the gain of the power amplifier and the drive-gear ratio appear as a product in the final system model, and that the product is specified by the system engineer to be, say, 100. A gear ratio in the range from 4 to 30 with a corresponding amplifier gain in the range from 25 to 3.3 could satisfy this product, and the component designer must choose the combination. The gear ratio is selected on the basis of speed requirements, friction levels, size, weight, and cost, while considerations of electrical noise levels and output impedance bear on the choice of amplifier gain. Obviously, the system engineer must understand some of the basic physical properties of the components to be used in the system if reasonable specifications concerning those components are to be made. Only the salient physical features of actuators, gear trains, electric circuits, and simple mechanisms are described in this book. Detailed constructional features of these devices, as well as other control system hardware such as sensors and computers, are described in the referenced books.

8.5   Historical Background and Full-State Feedback

Various forms of feedback control have been practiced since ancient times, in many cases for the regulation of water levels. An early example is the use of a float and orifice valve to control of the water level in the metering chamber of a water clock. Beginning in the eighteenth century, attempts were made to apply feedback techniques to manufacturing processes that required position control of mechanisms and speed control of prime movers. Many of these feedback controls, such as Watt's speed governor for steam engines, met with remarkable success. But many attempts floundered because the feedback caused unwanted oscillations in the regulated variable. Determining the cause of these unstable motions was impossible in terms of the *static* behavior of the feedback process, which was the only quantitative approach available to most experimenters in those early times.

The scientific approach to the study of such motions began in the mid-nineteenth century. Maxwell's paper of 1868 is the earliest publication to treat the problem from the *dynamic* point of view, using differential equations. This paper initiated a 100-year period of evolution of *classical control theory,* the dynamic study of SISO linear systems based on differential equations.

A simple example shows first how feedback may be applied to modify the dynamic performance of physical systems in very useful and very dramatic ways, and second, how such feedback can also induce unwanted oscillations. Consider the second-order spring–mass–damper system in Fig. 8.10. Let the

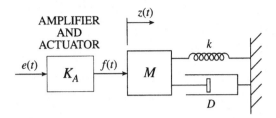

Fig. 8.10 Spring, mass, and damper system.

driving force on the mass be applied by an amplifier–actuator combination excited by a control voltage $e(t)$. Neglect the dynamics of the actuator and assume that the actuating force on the mass $f(t)$ is simply proportional to the input voltage:

$$f(t) = K_A e(t). \tag{8.22}$$

Combine this with the dynamics of the mass–spring–damper to obtain the familiar second-order equation

$$K_A e(t) = M\ddot{z}(t) + D\dot{z}(t) + kz(t). \tag{8.23}$$

The two roots of the characteristic equation for this system are

$$\frac{-D \pm \sqrt{D^2 - 4kM}}{2M}. \tag{8.24}$$

If the input voltage is a step or an impulse, the response $z(t)$ will be overdamped, critically damped, or underdamped, depending on the values of the three parameters M, D, and k. Normalized response curves showing the exact forms that $z(t)$ can have are given in Appendix E. Because M, D, and k are all positive real numbers, it is impossible for this system to be unstable; at worst, an undamped response bordering on instability would occur if $D = 0$.

Now, let us introduce two sensors to measure the motion of the mass. One of these is a position transducer that puts out a voltage proportional to $z(t)$, and the other is a velocity transducer measuring $\dot{z}(t)$. These sensors have scale factors k_p V/m and k_v V/(m/s):

$$e_p(t) = k_p z(t), \qquad e_v(t) = k_v \dot{z}(t). \tag{8.25}$$

Figure 8.11 depicts this arrangement. The two sensor voltages are fed back to the input of the amplifier and are summed with a new control voltage $e_c(t)$, as shown. Note that the original physical system has not been modified. The driving voltage at the input to the amplifier is still $e(t)$, the driving force on the mass $f(t)$ is still given by Eqn. 8.22, and the physical dynamics of the mass, spring, and damper are still governed by Newton's law as expressed in Eqn. 8.23. But now, because of the feedback arrangement, the driving voltage is a function of the position and velocity of the mass as well as of the newly introduced control voltage $e_c(t)$:

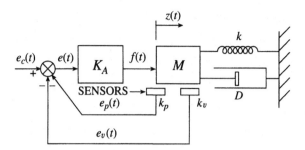

Fig. 8.11 Spring–mass–damper system with instrumentation for full-state feedback.

$$e(t) = e_c(t) - e_p(t) - e_v(t)$$
$$= e_c(t) - k_p z(t) - k_v \dot{z}(t). \qquad (8.26)$$

The behavior of the system with the feedback configuration shown in Fig. 8.11 is described by combining the two simultaneous Eqns. 8.23 and 8.26:

$$K_A e_c(t) = M\ddot{z}(t) + (D + K_A k_v)\dot{z}(t) + (k + K_A k_p)z(t). \qquad (8.27)$$

Compare Eqn. 8.27 with Eqn. 8.23. They have exactly the same form, but now the damping coefficient has been augmented by the velocity feedback, and the spring coefficient by the position feedback. We can express this as

$$\begin{bmatrix} \text{new damper} \\ \text{coefficient} \end{bmatrix} = [D + K_A k_v], \qquad \begin{bmatrix} \text{new spring} \\ \text{coefficient} \end{bmatrix} = [k + K_A k_p]. \qquad (8.28)$$

The possibilities now available for the dynamic response are exhibited, as before, by the two roots of the characteristic equation corresponding to Eqn. 8.27:

$$\frac{-(D + K_A k_v) \pm \sqrt{(D + K_A k_v)^2 - 4(k + K_A k_p)M}}{2M}. \qquad (8.29)$$

The terms "overdamped," "critically damped," and "underdamped" are sufficient to indicate qualitatively the general nature of the dynamic response to be expected in the passive mass–damper–spring system, that is, when the coefficients M, D, and k can have only positive values. But in the active (feedback) system, these three qualitative terms have meaning only when the *new* coefficients have positive values. We note that since the scale factors of the sensors can have either a positive or negative sign, the signs of both the new damper and the new spring coefficients can be made negative as well as positive. Further, the feedback signals can be summed negatively (as shown in Fig. 8.11) or positively, and the amplifier gain K_A can also be set at any desired positive or negative value. The apparent values of the spring and damper coefficients are easily tailored by adjustments to the feedback elements, and this makes a remarkable range of dynamic responses possible, including some that are not possible in the passive system.

If high-viscosity fluid in the damper causes the mechanical damping coefficient D to be too large, the dynamic response may be too overdamped for the application at hand. The viscosity can be effectively reduced by adjusting the velocity feedback circuit, and the response can be made as brisk as desired. In this case the velocity feedback signal simply causes an increase in the applied force $f(t)$ proportional to the velocity. This increment of applied force, which overcomes the excess viscous drag on the mass, effectively reduces the viscosity of the fluid. (The amplifier–actuator must have the force capability to overcome the viscous drag.) If the fluid viscosity should change with the temperature of the operating environment, compensating adjustments (guided by the prevailing temperature) can be made in the velocity feedback circuit. The velocity feedback voltage can also be tailored to obtain nonlinear damping characteristics for the overall system, which are desirable in some applications.

The velocity feedback can also make the effective damping negative, causing an unstable oscillation which will destroy the control authority of the input voltage $e_c(t)$ and which could damage the system. Such unstable oscillations increase exponentially, as described by the linear analysis, only until the physical limit of the actuating force is reached; then linear analysis no longer describes the motion. In many cases the oscillations reach a steady state, called a *limit cycle,* in which $z(t)$ has a periodic wave form that is not sinusoidal. This phenomenon is discussed in Chapter 15.

The effective spring constant can also be increased, decreased, and even made negative by adjusting the position feedback circuit. In some designs the mechanical spring is eliminated entirely, and the elastic property of the overall system is provided by the position feedback. In some applications where a special mechanical device is provided to damp the mass, the damper can also be replaced using velocity feedback. In this case the energy loss associated with mechanical damping will be eliminated.

The feedback arrangement depicted in Fig. 8.11 provides what is known as *full-state feedback* in this simple system. (Full-state feedback is discussed further in Chapter 15.) Linear analysis of this example suggests that such feedback control applied to the basic M, k, D system has almost limitless potential for improving the dynamic performance of that system. While this potential can be realized to a remarkable extent in practical applications, a limit to its utilization lies in the force bounds of the actuator. A design based on a linear analysis that predicts acceleration and velocity levels requiring $f(t)$ to exceed the force capability of the actuator will fail to meet the performance objectives.

8.6 Problems

Problem 8.1
The diagram for a nondynamic feedback system in Fig. P8.1 is useful for illustrating the robustness of feedback systems, and it is a practical representation of some electronic circuits. Let the ratio of output to input be called the gain of the system: $[y/r] = M$.

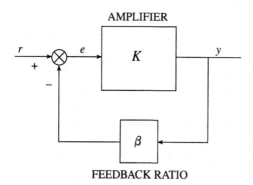

Fig. P8.1

(a) Express M as a function of the amplifier gain K and the feedback ratio β.
(b) Assume that β is known to be precisely 0.1000 and that it remains constant in spite of expected variations in its operating environment. K is a very high gain but is subject to severe fluctuations due to aging and changes in the environment, which cause K to vary randomly in the range from 10^5 to 10^7. (These are typical operating conditions for ordinary feedback amplifiers.) To illustrate the robustness of the feedback amplifier to fluctuations in K, calculate the fluctuations in M due to those in K.

Problem 8.2

The diagram in Fig. P8.2 is useful for defining the zero-state transfer function from the system input to the system error and to the system output. It permits a model transfer function to be used in the analysis, and it separates the plant, the controller, and the dynamic feedback element. Derive $Y(s)/R(s)$ and $\epsilon(s)/R(s)$ in terms of the four individual transfer functions.

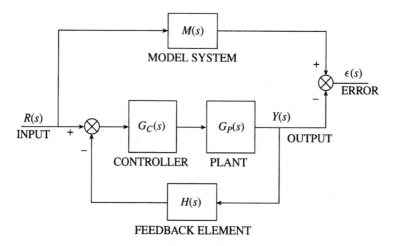

Fig. P8.2

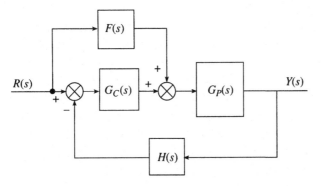

Fig. P8.3

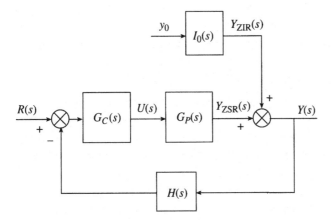

Fig. P8.4

Problem 8.3

The diagram in Fig. P8.3 also involves only zero-state transfer functions. Find the system transfer function $W(s) = Y(s)/R(s)$ in terms of the four individual transfer functions.

Problem 8.4

The diagram in Fig. P8.4 incorporates one initial-condition transfer function, $I_0(s)$. The other three are zero-state transfer functions. The system output can be expressed as $Y(s) = R(s)W_R(s) + y_0 W_0(s)$. Find $W_R(s)$ and $W_0(s)$ in terms of the four individual transfer functions shown in the diagram.

Problem 8.5

The steady-state gain of a transfer function $M(s)$ is defined as $M(0)$. Refer to the diagram for a single-loop feedback system in Fig. 8.8. Let

$$G(s) = \frac{136}{s(s^2 + 14s + 100)} \quad \text{and} \quad H(s) = 2.$$

(a) Calculate the closed-loop transfer function $Y(s)/R(s)$ and put it in factored form.

(b) Is the closed-loop system stable?

(c) Calculate the steady-state gain of $G(s)$ and that of $Y(s)/R(s)$. Compare the steady-state configuration of this system with the nondynamic system in Problem 8.1.

(d) Will the closed-loop system be stable if the flag constant of $G(s)$ is increased from 136 to 800?

Problem 8.6

Refer to the diagram for a single-loop feedback system in Fig. 8.8. Let the forward-path transfer function be

$$G(s) = \frac{K_1(s+6)}{s(s+2)(s^2+8s+25)}$$

and the feedback transfer function be $H(s) = 1$. Calculate the closed-loop transfer function $W(s) = Y(s)/R(s)$, showing K_1 as a parameter.

(a) Evaluate the steady-state closed-loop system gain $W(0)$. Does this gain depend on K_1? Why?

(b) Let $K_1 = 50$, evaluate $W(s)$, and show it in the standard factored form.

(c) Is the closed-loop system stable?

(d) Repeat (b) and (c) for $K_1 = 100$.

Problem 8.7

Refer to the diagram for a single-loop feedback system in Fig. 8.8. Let the forward-path transfer function be

$$G(s) = \frac{K_1(s+6)}{s(s+2)(s^2+8s+25)}$$

as in Problem 8.6, but now let the feedback transfer function be $H(s) = 10(s+3)/(s+30)$. Calculate the transfer function $W(s) = Y(s)/R(s)$, showing K_1 as a parameter.

(a) Evaluate the steady-state closed-loop system gain $W(0)$.

(b) Let $K_1 = 100$, evaluate $W(s)$, and show it in the standard factored form.

(c) Is the closed-loop system stable? Compare this result with that of part (d) of Problem 8.6.

(d) Repeat (b) and (c) for $K_1 = 200$.

Problem 8.8

Use the same system diagram and the same $G(s)$ as those in Problems 8.6 and 8.7. Let $H(s) = 0.2(s+30)/(s+6)$.

(a) Evaluate the steady-state closed-loop system gain $W(0)$.

(b) Let $K_1 = 20$, evaluate $W(s)$, and show it in a completely rationalized standard factored form.

(c) Repeat (b) for $K_1 = 50$, and determine the stability of the closed-loop system.

(d) Compare the results of this problem with those of Problems 8.6 and 8.7.

Problem 8.9

Refer to the M, k, D system described in Section 8.5 and Figs. 8.10 and 8.11. Adopt the notation $\dot{z}(t) = v(t)$.

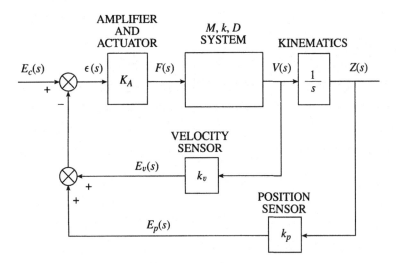

Fig. P8.9

(a) Verify that the block diagram in Fig. P8.9 is an accurate analytical portrayal of the physical system sketched in Fig. 8.11.

(b) Derive the transfer function $V(s)/F(s)$.

9 Dynamic Analysis of Feedback Control Systems

9.1 The Stability Problem in Feedback Systems

The following example illustrates the most important problem confronting automatic control engineers, that of ensuring the stability of a feedback system. Consider the searchlight servomechanism previously discussed (Section 8.2). Figure 9.1 shows a block diagram of this feedback system, where the feedback gain is taken to be 1 in order to simplify the analysis. The transfer function of the amplifier is shown to have a pole that represents the dynamic interaction between the amplifier and the motor, as discussed in Section 6.6. We use Eqn. 8.10, in which $N(s) = K_1 K_2$ and $D(s) = s(s + P_1)(s + P_2)$, to write the closed-loop transfer function for the zero-state response for this system:

$$W(s) = \frac{Y(s)}{R(s)} = \frac{K_1 K_2}{s^3 + (P_1 + P_2)s^2 + (P_1 P_2)s + K_1 K_2}. \tag{9.1}$$

We now study the dynamic response of this system in terms of its zero-state response to a unit step input, that is, $r(t) = \sigma(t)$, so that $R(s) = 1/s$. This yields the following:

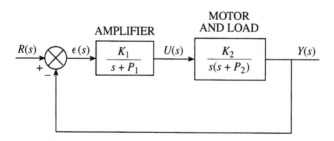

Fig. 9.1 Block diagram for searchlight servomechanism.

210

$$Y(s) = \frac{K_1 K_2}{s[s^3 + (P_1 + P_2)s^2 + (P_1 P_2)s + K_1 K_2]}$$

$$= \frac{K_1 K_2}{s(s + Q_1)(s + Q_2)(s + Q_3)}. \tag{9.2}$$

The three poles $-Q_1, -Q_2, -Q_3$, which are the roots of the bracketed polynomial, depend on the four coefficients K_1, K_2, P_1, and P_2, which in turn are determined by the physical parameters of the searchlight motor, platform, sensors, and amplifier. To make a specific case, assume that all the physical parameters are fixed except for the gain constant K_1 of the amplifier, which can be varied at will. Take $K_2 = 20$, $P_1 = 30$, and $P_2 = 10$; $Y(s)$ now depends only on K_1:

$$Y(s) = \frac{20K_1}{s[s^3 + 40s^2 + 300s + 20K_1]}. \tag{9.3}$$

A static analysis of the error channel reveals that

$$[u(t)]_{\text{static}} = \frac{K_1}{30}[e(t)]_{\text{static}}, \tag{9.4}$$

which shows that if K_1 is large then the static error will be small in order to support a given static voltage on the motor. Because static error is an important measure of system performance, high amplifier gain is normally desired. But since static analysis tells nothing about the dynamic performance of the system, we now look at the dynamic performance by calculating $y(t)$ for three values of K_1. We display the poles of the transfer function for each of the test values of K_1 as follows.

$K_2 = 20, P_1 = 30, P_2 = 10$			
K_1	Q_1	Q_2	Q_3
80	$-3.884 + j5.876$	$-3.884 - j5.876$	-32.23
300	$-1.855 + j12.72$	$-1.855 - j12.72$	-36.29
700	$+0.5046 + j18.47$	$+0.5046 - j18.47$	-41.01

These pole distributions are shown on the s-plane plot in Fig. 9.2. Here we see that, as K_1 is increased, the poles of the closed-loop transfer function migrate into the right half of the s plane. The system step response corresponding to each of these three test values for K_1 is shown in Fig. 9.3. When $K_1 = 80$ the response is reasonably stable (with an overshoot of about 12%), and the error first reaches zero in about 0.4 s. When $K_1 = 300$ the error first reaches zero in less than 0.2 s, but the overshoot is excessive and the oscillations persist longer than for $K_1 = 80$. When $K_1 = 700$ the response is even more prompt than in the previous two cases, but the oscillations are unstable. The two complex poles of $Y(s)$ lie in the right-half s plane, which signifies instability.

The cause of the unstable behavior in the last case is a combination of the dynamics of the amplifier and its high gain. The high gain yields a large acceleration torque on the load for a small error, driving the error to zero in about

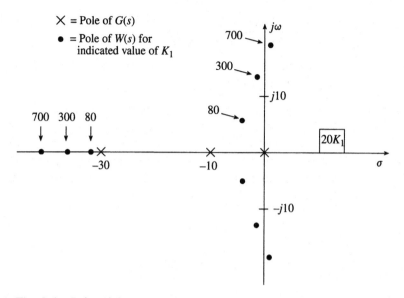

Fig. 9.2 Poles of $G(s)$ and poles of $W(s)$ for three values for K_1.

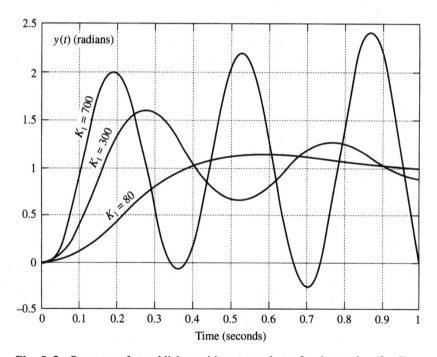

Fig. 9.3 Response of searchlight position to step input for three values for K_1.

0.1 s. Because the load velocity is high at the instant the error becomes zero, the load continues past the zero error point. It continues to accelerate temporarily because the dynamics of the amplifier causes its output to lag behind its input. Consequently, the torque changes sign not at the instant of zero error but rather at a slightly later instant, exactly when the load velocity reaches its peak. This phenomenon recurs as the error returns to zero at about 0.28 s, but since at this instant the velocity is higher in magnitude than it was at the first zero crossing, the second peak of the oscillation, at about 0.36 s, exceeds the first peak. The story continues in this fashion, each peak of the oscillation exceeding its predecessor in an exponentially growing pattern. The system can be stabilized either by reducing the amplifier gain or by reducing the time constant of the amplifier, $1/P_1$, so that the oscillations will diminish rather than grow.

The roles of amplifier gain and amplifier time constant in the dynamic response are clear in this simple system. However, for systems where $G(s)$ has either some finite zeros or more than three poles, qualitative analysis of the stability problem based on experience with simple systems seldom produces useful guidance in design, and is sometimes simply wrong. Quantitative analysis, using the pole-zero methods developed in Chapter 7 coupled with labor-saving computer aids for factoring polynomials, is the most reliable approach to analysis in control engineering.

We now consider the application of quantitative analysis techniques to feedback systems of the nth order, starting with the unity feedback model. The closed-loop transfer function for the unity feedback system $W(s)$ is expressed in terms of the open-loop transfer function $G(s)$, as derived in Chapter 8:

$$W(s) = \frac{G(s)}{1+G(s)} = \frac{N(s)}{D(s)+N(s)}, \quad \text{where}$$

$$G(s) = \frac{N(s)}{D(s)} = \frac{\text{a polynomial in } s}{\text{a monic polynomial in } s}.$$

(9.5)

The denominator of $W(s)$ is an nth-order polynomial formed by the sum of the two polynomials $N(s)$ and $D(s)$. Its coefficients therefore depend on the coefficients of both $N(s)$ and $D(s)$, as shown in Eqn. 8.11. n is the order of $D(s)$ or the order of $N(s)$, whichever is greater; in most cases, the order of $D(s)$ is greater. This characteristic polynomial of the feedback system has the following form:

$$\begin{bmatrix} \text{characteristic} \\ \text{polynomial of } W(s) \end{bmatrix} = D(s) + N(s)$$

$$= c_n s^n + c_{n-1} s^{n-1} + c_{n-2} s^{n-2} + \cdots + c_1 s + c_0$$

$$= c_n (s+Q_1)(s+Q_2)(s+Q_3)\cdots(s+Q_n). \quad (9.6)$$

The poles of $W(s)$ are the n roots of this polynomial, $-Q_1, -Q_2, ..., -Q_n$, some of which are real-valued and some of which appear in complex conjugate pairs. The system is called *stable* if all n roots have negative real parts, that is, if the poles of $W(s)$ lie in the left-half s plane. In a stable system all n modes of the impulse response of the system (or of the zero-input response) will *diminish* exponentially with time. The time constants and oscillating frequencies of the modes are defined by the locations of the corresponding poles in the s plane. But if one or more of the poles lies in the right-half s plane the system will be *unstable,* because in the impulse response (or in the zero-input response) one or more modes corresponding to the right-half–plane pole(s) will *grow* exponentially with time.

Maxwell was the first to point out that the roots of the characteristic equation of a dynamic system determine the stability of that system's motion. The task of relating the coefficients $c_0, c_1, ..., c_n$ to the roots then became an essential part of stability analysis. Routh, who followed Maxwell by a few years, discovered an algebraic procedure for determining the necessary and sufficient conditions on the coefficients which guarantees that none of the n roots will lie in the right-half s plane. This procedure, known as Routh's stability criterion, has been an important tool in the dynamic analysis of linear systems since 1877. It is described in detail in Appendix D.

Routh's test applied to the cubic polynomial in Eqn. 9.1 reveals that the condition that must be satisfied for stability is

$$K_1 K_2 < P_1 P_2 (P_1 + P_2). \tag{9.7}$$

In the example treated in Eqn. 9.3 (for $P_1 = 30$, $P_2 = 10$, and $K_2 = 20$), Routh's stability test imposes the following constraint on the amplifier gain K_1:

$$K_1 < \frac{P_1 P_2 (P_1 + P_2)}{K_2} = 600. \tag{9.8}$$

This is consistent with the results displayed in Figs. 9.2 and 9.3, which show a stable response for $K_1 = 300$ and an unstable response for $K_1 = 700$.

Routh's test shows that the four roots of the polynomial $c_4 s^4 + c_3 s^3 + c_2 s^2 + c_1 s + c_0$ will lie in the left-half plane provided that the following conditions are satisfied:

all c_i must have the same sign, $i = 0, 1, 2, 3, 4$; $\tag{9.9}$

$$c_1 (c_2 c_3 - c_1 c_4) > c_3^2 c_0. \tag{9.10}$$

For polynomials of order 5 and higher, the Routh conditions require that two or more inequalities among the coefficients be satisfied.

The availability of computer programs that quickly produce the roots of a high-order polynomial from the coefficients has made Routh's stability rule unnecessary in most routine engineering work on feedback control systems. Nevertheless, Routh's rule is still useful in analysis for making quick evaluations of stability in systems of fourth order and lower when a computer is not immediately available. It is also a convenient check on a computer result that seems to be incorrect. This often happens, either because a wrong number for one of the coefficients was entered into the program, or (less frequently) because of a failure of the numerical algorithm within the program.

9.2 Root-Locus Method for Factoring $1 + G(s)$

In 1948, W. R. Evans developed the root-locus method for factoring the denominator of $W(s)$ (Eqn. 9.5). This technique, now widely used in control system engineering, depicts the way in which the locations of the poles of $W(s)$ on the complex plane change as one of the physical parameters of the system is changed. An example is the amplifier gain of the servomechanism in Section 9.1. Using this method, the designer is often able to determine those values for the physical variables described in $G(s)$ that will achieve a desired dynamic response of the closed-loop system. Furthermore, the method often shows directly whether or not the desired dynamic performance can be achieved by simple adjustment of the parameters of $G(s)$; in those cases where no such adjustment is possible, the root-locus method may indicate how the open-loop components can be redesigned to satisfy the closed-loop requirements.

We start by considerng the unity feedback case in which $G(s)$ has the standard form

$$G(s) = \frac{K(s + Z_1)(s + Z_2) \cdots (s + Z_m)}{(s + P_1)(s + P_2) \cdots (s + P_n)}, \tag{9.11}$$

where the m finite zeros $-Z_1, -Z_2, \ldots, -Z_m$ and the n poles $-P_1, -P_2, \ldots, -P_n$ have known values but where the gain constant K is variable. For simplicity we temporarily assume that $n > m$ in order that the characteristic polynomial of $W(s)$, which is

$$D(s) + N(s) = (s + P_1)(s + P_2) \cdots (s + P_n) + K(s + Z_1)(s + Z_2) \cdots (s + Z_m), \tag{9.12}$$

will be a monic polynomial of order n, as in Eqn. 9.6, with $c_n = 1$. The poles of $W(s)$ are the roots of the characteristic polynomial. These can be found for a given value of K by expanding the factored forms of $D(s)$ and $N(s)$ into polynomials, combining the like-powered terms from the two expanded forms to obtain the composite polynomial $D(s) + N(s)$, and then factoring the composite polynomial to obtain the roots.

If n exceeds 4, the task of extracting the roots of the polynomial becomes so tedious that it is wise to use a computer-based control system analysis program. These programs provide convenient subroutines, usually called *root-locus* commands, for computing the n roots of Eqn. 9.12 for a range of values of K and for plotting the roots for each value of K on the s plane. The root-locus plot thus reveals the locations for the poles of $W(s)$ that are attainable for K lying in the designated range.

An example of such a root-locus plot for a unity feedback system having negative feedback is shown in Fig. 9.4. In this plot, the open-loop transfer function is

$$G(s) = \frac{K[s^2 + 2s + 16]}{s(s + 0.5)(s + 8)[s^2 + 12s + 38]}; \tag{9.13}$$

K ranges from 0 to 3000. The arrows indicate the direction of increasing K. The loci pass into and out of the right half of the s plane as K traverses the designated range of values, indicating that this feedback system can be either stable

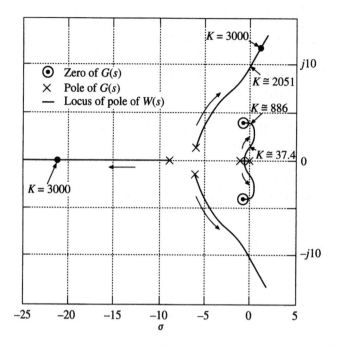

Fig. 9.4 Root-locus plot for unity negative feedback system, $G(s)$ in Eqn. 9.13.

or unstable, depending upon K. From the root-locus plot we identify the stability ranges as follows.

Approximate K Range	Closed-Loop Stability
< 0	unstable
0 to 37.4	stable
37.4 to 886	unstable
886 to 2051	stable
> 2051	unstable

Root-locus plots like this one are easily obtained by computer methods, even for systems of high order.

Although using the computer for calculations in most routine analysis work is the only practical way to obtain information on pole placement possibilities and stability margins, it is nevertheless essential to study the analytical properties of the root locus. This enables one not only to gain an understanding of the geometrical origins of the attractive shapes that frequently occur in the plots, but also to develop diagnostic tools for use when mistakes occur somewhere in the computing–plotting process. We therefore now turn our attention to the basic geometry of the s plane, which is implied by the characteristic equation $1 + G(s) = 0$.

The characteristic polynomial $D(s) + N(s)$, set equal to zero, establishes the *characteristic equation* for the unity feedback system. The solutions to this equation are the values of s for which the polynomial is zero, namely the poles of $W(s)$. The characteristic equation may be re-arranged to form a useful equivalent as follows:

$$D(s) + N(s) = 0 \;\Rightarrow\; 1 + \frac{N(s)}{D(s)} = 0 \;\Rightarrow\; 1 + G(s) = 0 \;\Rightarrow\; G(s) = -1. \qquad (9.14)$$

The complex variable s is often denoted as the sum of its real and imaginary parts, $s = \sigma + j\omega$. Since $G(s)$ is a function of the complex variable s, it is also complex-valued. We sometimes also denote $G(s)$ as the sum of its real and imaginary parts,

$$G(s) = \mathrm{Re}[G(s)] + j\,\mathrm{Im}[G(s)] \qquad (9.15)$$

where both $\mathrm{Re}[G(s)]$ and $\mathrm{Im}[G(s)]$ are real-valued. This *rectangular* expression for $G(s)$ is a convenient representation for some computations. But for other computations, including the study of the root locus, the following *polar* representation of $G(s)$ is more convenient:

$$G(s) = |G(s)| \epsilon^{j \angle G(s)} \qquad (9.16)$$

where $|G(s)|$ is a positive real function called the *magnitude* of $G(s)$; $\angle G(s)$, which may be expressed in degrees or in radians, is called the *argument* of $G(s)$. This argument is real-valued, and it can be either positive or negative. As an example of these two representations, take

$$G(s) = \frac{20(s+2)}{(s+5)(s+9)} \quad \text{for } s = -3 + j4. \qquad (9.17)$$

The rectangular form for $G(-3 + j4)$ is calculated as

$$G(-3 + j4) = \frac{20(-3 + j4 + 2)}{(-3 + j4 + 5)(-3 + j4 + 9)} = \frac{20(-1 + j4)}{(-4 + j32)}$$

$$= \frac{20(132 + j16)}{1040} = \underbrace{2.5385}_{\mathrm{Re}\,G(-3+j4)} + \underbrace{j0.30769}_{\mathrm{Im}\,G(-3+j4)}. \qquad (9.18)$$

The polar form is calculated as

$$G(-3 + j4) = \frac{20(-1 + j4)}{(-4 + j32)} = \frac{20\sqrt{17}\,\epsilon^{j\theta}}{\sqrt{1040}\,\epsilon^{j\phi}} = 2.5570\,\epsilon^{j(\theta - \phi)}, \quad \text{where}$$

$$\theta = \tan^{-1}\left[\frac{4}{-1}\right] = 104.036° \quad \text{and} \quad \phi = \tan^{-1}\left[\frac{32}{-4}\right] = 97.125°; \quad \text{thus} \qquad (9.19)$$

$$|G(-3 + j4)| = 2.5570 \quad \text{and} \quad \angle G(-3 + j4) = 6.911°.$$

Euler's formula can be used to convert complex numbers from their polar representation to their rectangular representation (see Appendix B).

When $G(s)$ has the standard form, the characteristic equation becomes

$$G(s) = \frac{N(s)}{D(s)} = \frac{K(s + Z_1)(s + Z_2) \cdots (s + Z_m)}{(s + P_1)(s + P_2) \cdots (s + P_n)} = -1. \qquad (9.20)$$

Assume for the moment that $n > m$, so that s has n values satisfying this characteristic equation. For each of these n values of s, $G(s)$ will be complex-valued, which means that it must embody two properties:

(a) $\angle G(s) = \pm 180°, \pm 540°, \pm 900°, \dots$;

(b) $|G(s)| = 1.$ (9.21)

We call (a) the *angle requirement* and (b) the *magnitude requirement*. The angle requirement applied to the general form of $G(s)$ (Eqn. 9.20) shows us that

$$\angle G(s) = \angle N(s) - \angle D(s)$$
$$= \angle K + \angle(s+Z_1) + \angle(s+Z_2) + \cdots + \angle(s+Z_m)$$
$$- \{\angle(s+P_1) + \angle(s+P_2) + \cdots + \angle(s+P_n)\}$$
$$= \angle K + \sum_{i=1}^{m} \angle(s+Z_i) - \sum_{j=1}^{n} \angle(s+P_j) = \pm 180°, \pm 540°, \dots . \quad (9.22)$$

We note that $\angle K$ is 0 if K is positive, but $\angle K$ is $\pm 180°$ if K is negative. In many cases, K is positive and so the term $\angle K$ is omitted from the expression for $\angle G(s)$. However, we will carry it along to avoid the error of overlooking it in those cases where K is negative. The magnitude requirement of Eqn. 9.21(b) applied to the general form of $G(s)$ is

$$|G(s)| = \frac{|N(s)|}{|D(s)|} = \frac{|K||s+Z_1||s+Z_2|\cdots|s+Z_m|}{|s+P_1||s+P_2|\cdots|s+P_n|}$$
$$= \frac{|K|\prod_{i=1}^{m}|s+Z_i|}{\prod_{j=1}^{n}|s+P_j|} = 1. \quad (9.23)$$

We find the n roots of the characteristic equation in two steps. First, we find all values of s satisfying the angle requirement of Eqn. 9.21(a). This eliminates nearly all the points in the s plane as potential roots, but an infinite number of points satisfying the angle requirement will remain! These points form the root-locus curves. The second step in our search is to determine which n of those points on the root loci satisfy the magnitude requirement of Eqn. 9.21(b). This trial-and-error procedure is very much akin to the procedure used in computer algorithms to find roots of polynomials. The transcendental equations that must be satisfied in the root-finding problem can be solved only by trial and error. The computer has the overwhelming advantages of speed and dependable memory. A simple example illustrates this procedure and shows some of the geometrical properties of the root locus.

Let the open-loop transfer function of the unity feedback system be

$$G(s) = \frac{K(s+2)}{s(s+1)(s+4)(s+10)}. \quad (9.24)$$

The angle requirement of Eqn. 9.21(a) applied to this example is

$$\angle K + \angle(s+2) - \{\angle(s) + \angle(s+1) + \angle(s+4) + \angle(s+10)\}$$
$$= \pm 180°, \pm 540°, \dots . \quad (9.25)$$

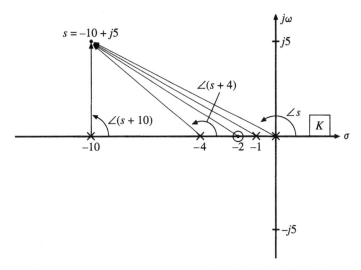

Fig. 9.5 Measuring $\angle G(s)$ for $s = -10 + j5$, $G(s)$ in Eqn. 9.24.

We select an arbitrary value for s, say $s = -10 + j5$, to see whether it satisfies the angle requirement. Figure 9.5 is a pole-zero plot of $G(s)$ showing the geometry implied by Eqn. 9.25. We note that an arrow drawn from the pole at $s = -4$ to the point $s = -10 + j5$ is a graphical representation of the complex number $(s+4)$, for $s = -10 + j5$.

$$(s+4)|_{s=-10+j5} = -6+j5. \tag{9.26}$$

The argument of this complex number, indicated as $\angle(s+4)$ in the pole-zero plot, is calculated as

$$\angle(s+4) = \angle(-6+j5) = \tan^{-1}\left(\frac{5}{-6}\right) = 140.19°. \tag{9.27}$$

The same geometry applies to the other poles and to the zero. The arguments of these terms could be approximately determined simply by measuring the angles with a protractor on the pole-zero plot, or they can be calculated as

$$\angle K = 0° \text{ or } \pm 180° \text{ (as noted previously)},$$

$$\angle(s) = \angle(-10+j5) = \tan^{-1}\left(\frac{5}{-10}\right) = 153.43°,$$

$$\angle(s+1) = \angle(-9+j5) = \tan^{-1}\left(\frac{5}{-9}\right) = 150.95°,$$

$$\angle(s+10) = \angle(j5) = \tan^{-1}\left(\frac{5}{0}\right) = 90°, \tag{9.28}$$

$$\angle(s+2) = \angle(-8+j5) = \tan^{-1}\left(\frac{5}{-8}\right) = 147.99°.$$

The sum of these five pole and zero angles is taken as indicated in Eqn. 9.25. $\angle G(s)|_{s=-10+j5}$ is $-386.58°$ if K is positive and $-206.58°$ if K is negative. In

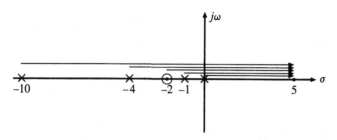

Fig. 9.6   Measuring ∠*G*(*s*) for *s* = 5, *G*(*s*) in Eqn. 9.24.

neither case is the angle requirement of Eqn. 9.21(a) satisfied, so the point *s* = −10 + *j*5 cannot be a pole of *W*(*s*). (We note that the angle −386.58° can also be denoted as −26.58° or as +333.42°.) We must continue to search the *s* plane for points that do satisfy the angle requirement, and it should somehow be a systematic search.

A search along the *σ* and *jω* axes is one systematic approach that might be productive. Consider the test point *s* = 5, as shown on the pole-zero plot in Fig. 9.6. All of the arrows drawn to this point have zero angles, which makes the calculation of ∠*G*(*s*) possible by inspection. All the terms in Eqn. 9.25 are zero except for ∠*K*, which is also zero for positive *K* but is ±180° for negative *K*. The point *s* = 5 therefore satisfies the angle requirement of Eqn. 9.21(a) for negative *K* but not for positive *K*. Furthermore, the arrow diagram makes it clear that the same is true for all the points on the *σ* axis to the right of the pole at the origin.

We have found results for negative *K* that are opposite to those for positive *K*. In order to focus on the geometrical properties of the root locus without dividing our attention unnecessarily, we will examine the case of positive *K* first. We will return to the less common case of negative *K* in Section 9.6.

We continue the search along the *σ* axis for points that satisfy the angle requirement of Eqn. 9.21(a), assuming that ∠*K* = 0. The point *s* = −0.5, for example, has an arrow diagram with four arrows having zero angles and one arrow, drawn from the pole at the origin, with an angle of −180°, making ∠*G*(*s*) = −180° (or +180°, which is equivalent). This point, and in fact all the points on the *σ* axis lying between *s* = 0 and *s* = −1, therefore satisfy the angle requirement. Continuing along the *σ* axis, we discover that any point having an odd number of arrows coming from its right side will satisfy the angle requirement; it makes no difference whether the arrows come from zeros of *G*(*s*) or from poles. We account for all points on the *σ* axis by denoting those points for which ∠*G*(*s*) = ±180°, ±540°, ... with heavy lines in Fig. 9.7.

Next, we search the *jω* axis for possible points satisfying the angle requirement of Eqn. 9.21(a). For points on this axis close to the origin, all the arrows will have very small angles, except for the short arrow from the pole at the origin, whose angle is 90°. ∠*G*(*s*) will be close to −90°, far short of the required −180°, for points low on the *jω* axis. Now move to points far up on the *jω* axis, say to points beyond *j*100. The arrows from the poles and zeros at

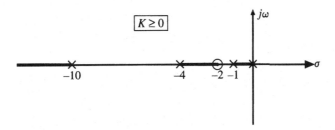

Fig. 9.7 Points on the real axis satisfying $\angle G(s) = \pm 180°$, $G(s)$ in Eqn. 9.24.

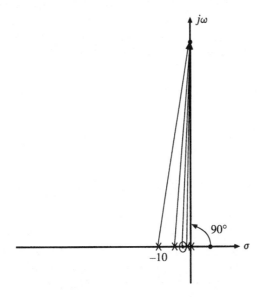

Fig. 9.8 Measuring $\angle G(s)$ for s lying high on the $j\omega$ axis, $G(s)$ in Eqn. 9.24.

the points $s = 0, -1, -2, -4$, and -10 now all have angles of approximately $90°$, as shown in Fig. 9.8. Since four of these arrows come from poles and one from a zero, we have

$$\begin{bmatrix} \angle G(s) \text{ at points on } j\omega \text{ axis} \\ \text{far above the origin} \end{bmatrix} \approx \underset{\underset{\text{number of zeros}}{\uparrow}}{[1]} \times 90° - \underset{\underset{\text{number of poles}}{\uparrow}}{[4]} \times 90° = -270°. \qquad (9.29)$$

Thus for points that are low on the $j\omega$ axis the total angle is about $-90°$, and for points high on the axis the total angle is about $-270°$. Now $\angle G(s)$ is a continuous function of s at all points in the s plane except at the poles and zeros of $\angle G(s)$. This means that a small shift in a test point will produce only a small change in each of the arrow angles and therefore only a small change in $\angle G(s)$, unless the test point is at one of the singularities of $G(s)$. Because in this example no singularities of $G(s)$ exist on the $j\omega$ axis, we know from the continuity argument that $\angle G(s) = -180°$ for at least one point on the $j\omega$ axis between low and high values of $j\omega$. (The same is true for any continuous path in the s

plane connecting a low region to a high region and not passing through poles or zeros.)

However, it is not necessary to invoke mathematical properties to see that a point on the $j\omega$ axis in the neighborhood of $j5$ satisfies the angle requirement of $\pm180°$. A sketch on paper roughly to scale shows that the arrow from $s = 0$ to all points on the $j\omega$ axis has an angle of $90°$, and so the other four arrows to the solution point must contribute a net angle of only $90°$. At roughly $j5$, the arrow angle from the zero at $s = -2$ and that from the pole at $s = -1$ are approximately the same, so the net angle contribution from these two is positive but small. This leaves only the two arrows from the poles at $s = -4$ and $s = -10$ to contribute the angle balance of about $90°$. That contribution would be less than $90°$ for a test point $j4$, and it would be greater than $90°$ at $j10$. A search on the $j\omega$ axis near the point $j5$ shows the following result.

Test Point	$\angle G(s)$
$j4.8$	$-176.69°$
$j4.9$	$-177.55°$
$j5.0$	$-178.40°$
$j5.1$	$-179.23°$
$j5.2$	$-180.06°$

A finer search shows that $\angle G(j5.193) = -180.00°$. Because for points above $j5.193$ the angle grows asymptotically toward $-270°$, only one point on the upper $j\omega$ axis satisfies the angle requirement of Eqn. 9.21(a) for the root locus. But the mirror image of this positive $j\omega$ axis point, $-j5.193$, must show that $\angle G(-j5.193) = +180.00°$, because the angle of each arrow drawn from the poles and the zero to $s = -j5.193$ is the negative of its counterpart corresponding to $s = +j5.193$. So for each solution point on the positive $j\omega$ axis there is also a solution on the negative $j\omega$ axis at the mirror-image point. In fact, for any solution point in the upper half of the s plane, a mirror-image solution point exists in the lower half of the s plane.

The two solution points we have located on the $j\omega$ axis at $\pm j5.193$ are not isolated solutions to the angle requirement of Eqn. 9.21(a) for the root locus. They are simply the points on the $j\omega$ axis that also lie on the root loci where those loci cross the axis. We must now find all the remaining points in the s plane that satisfy the angle requirement. We can do this in a systematic way by looking next at points lying in the remote regions of the s plane. Let $s_A = \sigma + j\omega$ (where both σ and ω are much larger than 10) be such a point, as depicted in Fig. 9.9. The arrows drawn from the finite poles and zeros of $G(s)$ to this remote point all have very nearly the same angle, shown as θ_A. Therefore Eqn. 9.22 gives us

$$\angle G(s_A) \approx [1-4] \times \theta_A = -3\theta_A. \tag{9.30}$$

Are there any remote points for which the angle $[-3\theta_A]$ will be $\pm180°$, $\pm540°$, ... ? The answer is clearly yes. Three remote regions in which they reside are identified by three angles $\theta_A = +60°$, $-60°$, $\pm180°$. One of these regions,

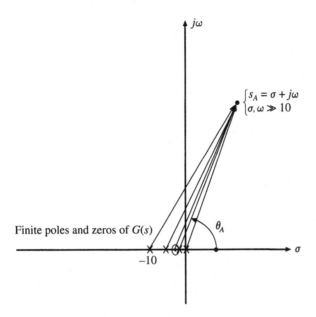

Fig. 9.9   Measuring $\angle G(s)$ for s lying in a remote region of the s plane, $G(s)$ in Eqn. 9.24.

identified by $\theta_A = \pm 180°$, contains that branch of the root locus shown in Fig. 9.7 which lies to the left of the pole at $s = -10$. We can show by continuity arguments that the other two regions also have a locus of points satisfying the angle requirement of Eqn. 9.21(a). A solution point for the angle requirement is found by searching a remote region in the direction of $\theta_A = 60°$. Since one such point is located at $s_A = 36.198 + j70$, we have

$$\angle G(36.198 + j70) = -180.00°. \tag{9.31}$$

The region near this point is shown in Fig. 9.10. If we test points slightly to the left of s_A, we find that $\angle G(s)$ will be slightly less than $-180°$; if we test points slightly to the right, we find that $\angle G(s)$ increases slightly. In Fig. 9.10 we have, for example,

$$\angle G(s_A - 2) = \angle G(34.198 + j70) = -183.7°,$$

$$\angle G(s_A) = \angle G(36.198 + j70) = -180.0°,$$

$$\angle G(s_A + 2) = \angle G(38.198 + j70) = -176.4°.$$

Any continuous curve drawn between the two points $(s_A - 2)$ and $(s_A + 2)$ (the circle of radius 2 as shown in Fig. 9.10, for example) must contain at least one point for which $\angle G(s) = -180°$. The angle measurements for twelve test points on this circle are indicated in the figure. Two of these satisfy the angle requirement, as shown. It is possible to connect these two points with a smooth curve, passing through s_A, *all* of whose points satisfy the angle requirement of Eqn. 9.21(a). This curve will be very nearly a straight line, making an angle of 60° with the σ axis. Because of the continuity of the function $\angle G(s)$, any point in

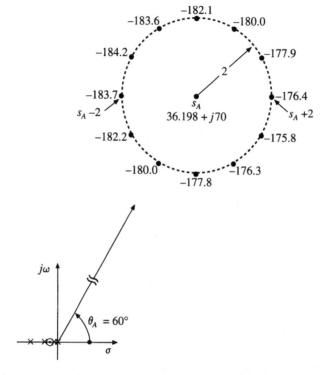

Fig. 9.10 $\angle G(s)$ for s near $s_A = 36.198 + j70$, $G(s)$ in Eqn. 9.24.

the s plane which satisfies the angle requirement will have a nearby point which also satisfies that requirement. Therefore, all points satisfying the angle requirement lie on continuous curves, called the *root loci*. Starting at point s_A in Fig. 9.10, we can work our way back toward the origin of the s plane, one adjacent point at a time, completing the root-locus sketch. Figure 9.11 shows the complete picture for points near the origin.

We must now find the four specific points on the loci that are the roots of the characteristic equation (Eqn. 9.20), which for this example is

$$G(s) = \frac{K(s+2)}{s(s+1)(s+4)(s+10)} = -1,$$

which is re-arranged into the characteristic polynomial

$$(9.32)$$

$$s^4 + 15s^3 + 54s^2 + (40+K)s + 2K = 0.$$

Each value of K will produce a unique set of four roots, or poles, of $W(s)$. These will lie at points on the loci satisfying the magnitude requirement of Eqn. 9.2(b). When K is positive in this example, the magnitude requirement is

$$|G(s)| = \frac{K|s+2|}{|s||s+1||s+4||s+10|} = 1.$$

$$(9.33)$$

We re-arrange Eqn. 9.33 to show more clearly the role played by K:

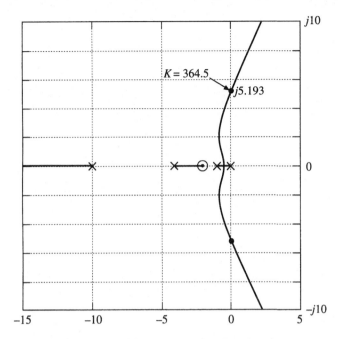

Fig. 9.11 Locus of all points near the origin for which $\angle G(s) = \pm 180°$, $G(s)$ in Eqn. 9.24.

$$K = \frac{|s||s+1||s+4||s+10|}{|s+2|}. \tag{9.34}$$

The geometrical interpretation of this magnitude requirement is that the terms $|s|$, $|s+1|$, $|s+4|$, $|s+10|$, and $|s+2|$ are the lengths of arrows drawn to a specified point on the loci from the poles and zero of $G(s)$. Equation 9.34 yields the K which will make that point a pole of $W(s)$. For example, to find the *critical* value of K, above which the closed-loop system will be unstable, we draw arrows to the point $s = j5.193$, as shown in Fig. 9.12. The absolute values in Eqn. 9.34 yield the following result:

$$K_{\text{crit}} = \frac{|s||s+1||s+4||s+10|}{|s+2|}\Bigg|_{s=j5.193}$$

$$= \frac{(5.193)(5.289)(6.555)(11.268)}{5.565} = 364.5. \tag{9.35}$$

This value for K places two poles of $W(s)$ at the points $s = \pm j5.193$, but two other poles of $W(s)$ exist for this same value of K. These are also located at points on the loci for which Eqn. 9.34 is satisfied for $K = 364.5$.

We note that the short branch of the locus on the σ axis between $s = -2$ and $s = -4$ has points very close to the pole of $G(s)$ at $s = -4$, as well as points very close to the zero of $G(s)$ at $s = -2$. For the points close to $s = -4$ the term $|s+4|$ in the numerator of Eqn. 9.34 will be very small, indicating that K will be very small. But for points on the short branch near $s = -2$ the term in the

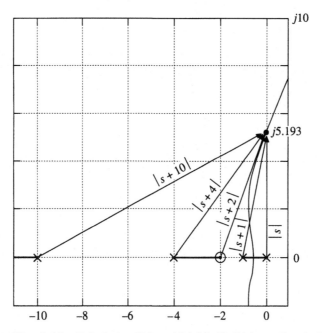

Fig. 9.12 Calculating K for which $|G(j5.193)| = 1$, $G(s)$ in Eqn. 9.24.

denominator, $|s+2|$, will be very small, indicating that K will be very large. In fact, for points on this short branch, K ranges from 0 to ∞, so for $K = 364.5$ a point somewhere between $s = -2$ and $s = -4$ satisfies the magnitude requirement (Eqn. 9.34); that point is $s = -2.095$. For points on the branch lying to the left of but close to $s = -10$, K will also be very small because of the term $|s+10|$, but for points increasingly far from $s = -10$ the four numerator terms in Eqn. 9.34 will also grow, overwhelming the single denominator term so that K increases toward infinity as points on this branch become increasingly distant from $s = -10$. It is at $s = -12.91$ that K reaches 364.5.

We can trace the trajectories of the four roots of the characteristic polynomial as K is varied over the range 0 to ∞ simply by factoring the polynomial (Eqn. 9.32) for several values of K.

K	Poles of $W(s)$				
0	0	-1	-4	-10	
5	-0.370	-0.6944	-3.863	-10.073	
50	$-0.75 \pm j1.65$		-2.865	-10.64	(9.36)
364.5	$0 \pm j5.193$		-2.095	-12.91	
500	$0.3107 \pm j5.965$		-2.068	-13.55	
∞	$\infty \pm j\infty$		-2.000	$-\infty$	

The locations of these five sets of poles are displayed as the sequence of large dots on the root loci in Fig. 9.13. As K is increased from 0 to ∞, the poles of $W(s)$ start at the four poles of $G(s)$ and migrate along the four branches of the root loci. One branch terminates on the zero of $G(s)$ at $s = -2$. The other

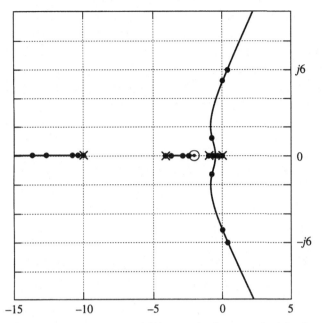

Fig. 9.13   Poles of $W(s)$ lying on the four root loci for five values of K, Eqn. 9.36.

three branches extend to infinity and approach straight-line asymptotes; one of these coincides with the negative σ axis, and the other two make angles of $\pm 60°$ to that axis. The two poles on the branches emanating from $s = 0$ and $s = -1$ meet at $s = -0.5359$ (for $K \approx 5.5692$) and then move off the σ axis, becoming complex-valued. These two poles pass into the right half of the s plane at $K = K_{crit} = 364.5$, establishing this as the limiting value of K for stability of the closed-loop system. (The operational value for K would normally be set between 60 and 120, to provide a gain margin in the range of 3 to 6.) The value of K_{crit} found by repeated factoring of Eqn. 9.32 can also be found by applying Routh's stability criterion to the characteristic polynomial, which in this case requires factoring only a second-order polynomial.

We now look at a generalization of the root-locus technique for the unity feedback case, where K is positive. Consider an open-loop transfer function having the same number of finite zeros as it has finite poles, $m = n$:

$$G(s) = \frac{K(s+Z_1)(s+Z_2)\cdots(s+Z_n)}{(s+P_1)(s+P_2)\cdots(s+P_n)},\qquad(9.37)$$

where there are no common zeros and poles. The closed-loop transfer function then has the form

$$W(s) = \frac{K(s+Z_1)(s+Z_2)\cdots(s+Z_n)}{(s+P_1)(s-P_2)\cdots(s+P_n)+K(s+Z_1)(s+Z_2)\cdots(s+Z_n)}.\qquad(9.38)$$

The zeros of $W(s)$, which are simply the zeros of $G(s)$, do not depend on the loop gain K. For $K = 0$ we note that the second term in the denominator disappears, and the poles of $W(s)$ are simply the poles of $G(s)$ (as we discovered from the root-locus plot in the previous example). Next, let $K \to \infty$ and consider the poles of $W(s)$. Now it is the first term in the denominator of $W(s)$ that becomes insignificant compared to the second term, and the poles of $W(s)$ become the zeros of $G(s)$! Hence a root-locus plot for the system would show one branch of the root locus for each pole of $G(s)$ for a total of n branches. Further, for $K \cong 0$, each of the n poles of $W(s)$ resides on a separate branch of the loci very close to one of the poles of $G(s)$. Then, as K increases, the n poles of $W(s)$ will migrate along the loci, their locations for each value of K being determined by the magnitude requirement of Eqn. 9.21(b). As $K \to \infty$, the poles of $W(s)$ approach the zeros of $G(s)$.

A simple example illustrates this principle of the root locus for the case $m = n = 3$. Let $G(s)$ have the form

$$G(s) = \frac{K(s+7)[s^2+8s+25]}{(s-1)(s+3)(s+4)}. \tag{9.39}$$

Figure 9.14 shows the root-locus plot for this system in which the loci emanate from the poles of $G(s)$ at $s = 1$, $s = -3$, and $s = -4$, and terminate on the zeros of $G(s)$ at $s = -7$, $s = -4+j3$, and $s = -4-j3$. Interestingly, $G(s)$ has a pole in the right-half s plane, indicating that it is the transfer function of an unstable mechanism. However, the closed-loop system is stable, provided K is large enough so that all three poles of $W(s)$ reside in the left-half plane. In this case the critical value of K may be determined from Eqn. 9.21 by applying the magnitude requirement of Eqn. 9.21(b) at the point where the locus, starting from $s = 1$, crosses into the left-half plane at $s = 0$:

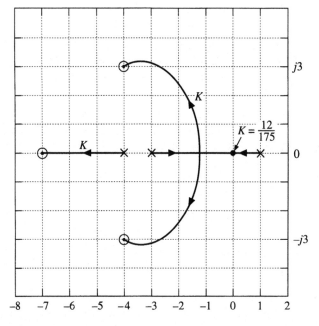

Fig. 9.14 Complete root loci for positive K, $G(s)$ in Eqn. 9.39.

$$K_{\text{crit}} = \frac{(1)(3)(4)}{(7)(5)(5)} = \frac{12}{175} \tag{9.40}$$

where the condition $K > K_{\text{crit}}$ is necessary for stability of the closed-loop system.

It has been noted (Section 7.1, Remark 6) that any transfer function has the same number of zeros as poles if those zeros (or poles) at infinity are included in the count. If $G(s)$ has the general form of Eqn. 9.11, where (for example) $m < n$, then $G(s)$ has n poles, m finite zeros, and $n - m$ zeros at infinity. As $s \to \infty$, $G(s)$ behaves like $K/s^{(n-m)}$. As K is increased from 0 toward ∞, the poles of $W(s)$ move along the root loci *toward the zeros of* $G(s)$. Because there are m finite zeros of $G(s)$, m poles of $W(s)$ will move along the loci toward these zeros as K is increased. That is, m branches of the loci connect poles of $G(s)$ to the finite zeros of $G(s)$. But the remaining $n - m$ branches of the loci lead out toward infinity in the s plane, terminating on those $n - m$ infinite zeros of $G(s)$. These infinite branches of the loci approach straight-line asymptotes that make angles with the σ axis of

$$\frac{\pm 180°}{n - m}, \frac{\pm 540°}{n - m}, \frac{\pm 900°}{n - m}, \dots .$$

The $n - m$ asymptotes may be determined by considering the general form of $G(s)$:

$$G(s) = \frac{K(s^m + a_{m-1}s^{m-1} + \cdots + a_1 s + a_0)}{s^n + b_{n-1}s^{n-1} + \cdots + b_1 s + b_0} = \frac{K(s + Z_1)(s + Z_2) \cdots (s + Z_m)}{(s + P_1)(s + P_2) \cdots (s + P_n)}. \tag{9.41}$$

Recall from Eqns. 7.5 and 7.7 the two relationships between the coefficients of the polynomials and the poles and zeros:

$$\begin{aligned} a_{m-1} &= Z_1 + Z_2 + \cdots + Z_m \quad \text{and} \quad a_0 = Z_1 Z_2 \cdots Z_m; \\ b_{n-1} &= P_1 + P_2 + \cdots + P_n \quad \text{and} \quad b_0 = P_1 P_2 \cdots P_n. \end{aligned} \tag{9.42}$$

We can say that "$-a_{m-1}$ is the sum of the zeros of $G(s)$, and $-b_{n-1}$ is the sum of the poles of $G(s)$." Now, divide the numerator of $G(s)$ into its denominator and write

$$G(s) = \frac{K}{s^{(n-m)} + (b_{n-1} - a_{m-1})s^{(n-m-1)} + \cdots}. \tag{9.43}$$

To obtain the root loci for this system we must set $G(s) = -1$. Using this form for $G(s)$, we have the characteristic equation

$$s^{(n-m)} + (b_{n-1} - a_{m-1})s^{(n-m-1)} + \cdots = -K. \tag{9.44}$$

For very large values of s, Eqn. 9.44 may be written approximately as an $(n-m)$th-order polynomial equal to zero:

$$\begin{aligned} s^{(n-m)} &+ (b_{n-1} - a_{m-1})s^{(n-m-1)} + \cdots + K \\ &= (s + R_1)(s + R_2) \cdots (s + R_{(n-m)}) = 0, \quad \text{where} \end{aligned} \tag{9.45}$$

$$\sum_{i=1}^{n-m} R_i = (b_{n-1} - a_{m-1}) \quad \text{and} \quad K \to \infty \quad \text{as} \quad s \to \infty.$$

Because we are considering the behavior of the characteristic polynomial only for $|s| \to \infty$ (i.e., where the $n - m$ branches of the root locus essentially coincide with the asymptotes), the $n - m$ roots of Eqn. 9.45, $-R_1, -R_2, ..., -R_{(n-m)}$, also lie on the asymptotes. The asymptotes are straight lines, making angles of $\pm 180°/(n - m), \pm 540°/(n - m), ...$ with the σ axis and intersecting the σ axis at a point $s = -\sigma_I$, which is yet to be determined.

To find σ_I we re-arrange Eqn. 9.45, which pertains only to the asymptotes, as

$$s^{n-m}\left[1 + \frac{(b_{n-1} - a_{m-1})}{s} + \cdots\right] = -K. \tag{9.46}$$

For $K \to \infty$ and $|s| \to \infty$, all of the terms, except for the first two in the brackets, are negligible and are therefore dropped. Now take the $(n - m)$th root of both sides:

$$s\left[1 + \frac{(b_{n-1} - a_{m-1})}{s}\right]^{1/(n-m)} = [-K]^{1/(n-m)}. \tag{9.47}$$

We approximate the bracketed term by the first two terms in its infinite series expansion to obtain

$$\left[s + \frac{(b_{n-1} - a_{m-1})}{n - m}\right] = [-K]^{1/(n-m)}. \tag{9.48}$$

Equation 9.48 gives both magnitude and argument information on the location of points on the asymptotes at the remote regions of the s plane. But since the asymptotes are straight lines, we can also find the intersection point $s = -\sigma_I$ from Eqn. 9.48. For this we introduce a new set of coordinates establishing a "β plane," with axes parallel to the σ and $j\omega$ axes and origin at the intersection point $s = -\sigma_I$. The juxtaposition of the two planes is illustrated in Fig. 9.15 for the case $n - m = 5$. The relationship between the s plane and the β plane is simply

$$\beta = s + \sigma_I. \tag{9.49}$$

For $n - m = 5$, five asymptotes originate at the intersection point (which is the origin of the β plane) with the angles given by

$$\theta_A = \frac{\pm 180°}{5}, \frac{\pm 540°}{5}, \frac{\pm 900°}{5} = \pm 36°, \pm 108°, \pm 180°. \tag{9.50}$$

Let R be a point on one of the asymptotes, as shown in Fig. 9.15. The coordinates of point R can be expressed in β-plane coordinates, using the polar form for β.

$$\vec{\beta} = |\beta|\epsilon^{j\theta_A}. \tag{9.51}$$

This expression is valid for all points on the asymptote, including those points in the remote regions. But the expression in Eqn. 9.48 for the coordinates of R is valid for R only in the remote regions. That expression can be written as

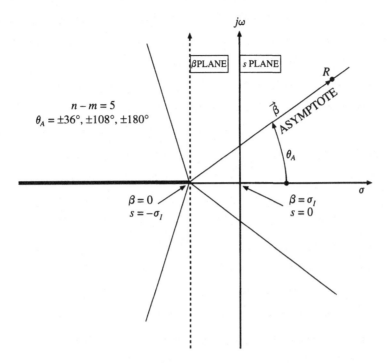

Fig. 9.15 Constructing the asymptotes of the root loci for a $G(s)$ in which $n-m=5$.

$$\left[s+\frac{(b_{n-1}-a_{m-1})}{n-m}\right]=[-K]^{1/(n-m)}=\{[K]^{1/(n-m)}\}\epsilon^{jk\pi}, \quad \text{where}$$

$$K\geq 0 \quad \text{and} \quad k=\frac{1}{n-m},\frac{3}{n-m},\frac{5}{n-m},\cdots. \tag{9.52}$$

Now substitute $s=\beta-\sigma_I$ from Eqn. 9.49 into Eqn. 9.52, and let $|\beta|=\{[K]^{1/(n-m)}\}$. Equation 9.52 then becomes

$$\left[\beta+\left(\frac{(b_{n-1}-a_{m-1})}{n-m}-\sigma_I\right)\right]=|\beta|\epsilon^{j\theta_A}, \tag{9.53}$$

which is valid for all values of β, as noted before. To find the intersection point, let $\beta=0$:

$$\sigma_I=\frac{b_{n-1}-a_{m-1}}{n-m}=\frac{\sum_{i=1}^{n}P_i-\sum_{j=1}^{m}Z_j}{n-m}. \tag{9.54}$$

The asymptotes that may be constructed by inspection of $G(s)$ aid in sketching the root locus when a computer is not available.

As an example, consider the seventh-order unity feedback system whose open-loop transfer function is

$$G(s)=\frac{K(s+2)(s+4)}{s(s+1)(s+10)[s^2+12s+52][s^2+5.6s+25]}. \tag{9.55}$$

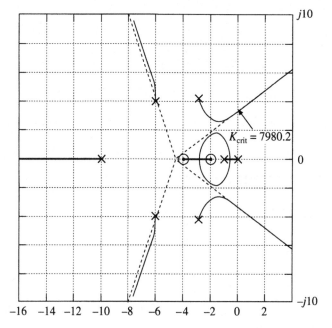

Fig. 9.16   Root loci in the vicinity of the origin for positive K, $G(s)$ in Eqn. 9.55.

Here we have $m = 2$ and $n = 7$, so $n - m = 5$. The five asymptotes have angles $\pm 36°$, $\pm 108°$, and $\pm 180°$. The intersection point of the asymptotes is calculated as

$$s = -\frac{[0 + 1 + 10 + 12 + 5.6] - [2 + 4]}{7 - 2} = -4.52. \tag{9.56}$$

Figure 9.16 shows the asymptotes and the root-locus plot drawn for K ranging from 0 to 80,000. The critical K for stability occurs at $K = 7980.2$.

The closed-loop transfer function, when written in the following form, shows an interesting property of this system:

$$W(s) = \frac{K(s+2)(s+4)}{\begin{array}{c} s^7 + 28.6s^6 + 347.8s^5 + 2353.4s^4 + 9245.2s^3 \\ + (K + 20{,}212)s^2 + (6K + 13{,}000)s + 8K \end{array}}$$

$$= \frac{K(s+2)(s+4)}{(s+Q_1)(s+Q_2)(s+Q_3)(s+Q_4)(s+Q_5)(s+Q_6)(s+Q_7)}. \tag{9.57}$$

The seven closed-loop poles lie on the root loci at points that depend on the loop gain K. But the property of polynomials that relates the roots to the coefficients (Eqn. 9.42) ensures that, however the closed-loop poles move along the loci as K is varied, the sum $Q_1 + Q_2 + \cdots + Q_7$ must remain constant at 28.6. If some roots move to the left as K increases, others must move to the right to keep the sum constant. We also note from Eqn. 9.42 that the product $Q_1 Q_2 \cdots Q_7$ must be $8K$. Therefore, as K increases, some of the roots must move outward to maintain this relationship. These properties of root-locus plots for unity feedback systems, where K is the variable parameter, hold for all systems in which $n > m$. They often aid in sketching plots where, depending on the relative locations of the poles and zeros, it is uncertain which of two possible patterns will occur.

Two further properties of the root locus are illustrated by another example. In this case we again consider a unity feedback system having a fourth-order $G(s)$, and we restrict our attention to positive values of the loop gain:

$$G(s) = \frac{K}{s(s+4)[s^2+6s+18]}. \tag{9.58}$$

The asymptotes for the root-locus plot are given as

$$\begin{bmatrix} \text{asymptote} \\ \text{angles} \end{bmatrix} = \theta_A = \frac{\pm 180°}{4-0}, \frac{\pm 540°}{4-0} = \pm 45°, \pm 135°,$$

$$\begin{bmatrix} \text{intersection} \\ \text{point} \end{bmatrix} = \sigma_I = -\frac{0+4+6-0}{4-0} = -2.5.$$

The asymptotes and the root loci are plotted in Fig. 9.17. The critical value of K for stability of the closed-loop system is 250.6. We note a common occurrence in this case, which is that two of the loci cross their own asymptotes. This plot also shows two characteristic features, the breakaway point in the vicinity of $s = -1.5$ and the departure angle of the loci from the complex poles. If a computer is not available to calculate and plot the root loci then these two features, in addition to the asymptotes, will aid in sketching the root-locus plot.

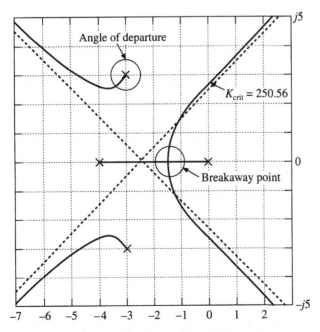

Fig. 9.17   Root loci for $G(s)$ in Eqn. 9.58 showing asymptotes, breakaway point, and angles of departure.

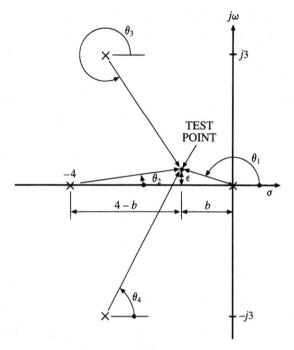

Fig. 9.18 Finding the breakaway point of root loci for $G(s)$ in Eqn. 9.58.

The *breakaway point* is the point on the locus between the two real poles at which K has its maximum value. At any point on the locus, K is given by the formula

$$K = K(\sigma) = \frac{\prod_{i=1}^{4}|\sigma + P_i|}{1} \quad \text{for } -4 \leq \sigma \leq 0. \tag{9.59}$$

If $K(\sigma)$ is expressed analytically, its maximum can in principle be found by setting the derivative $dK/d\sigma$ to zero and solving the resulting expression for the breakaway point σ_B. This is often a time-consuming task. The solution in this case shows that the maximum K occurs at $s = -1.50$ for $K = 42.19$.

Another analytical approach to calculating σ_B is illustrated in Fig. 9.18. Choose a test point on the locus an infinitesimal distance ϵ above the breakaway point, as yet unknown, at $s = -b$. The test point must satisfy the angle requirement of the root locus, $\angle G(s) = \pm 180°, \pm 540°, \dots$. For this system we therefore seek a constant b such that, for $\epsilon \to 0$, we have

$$\angle G(s) = \underbrace{\left[180° - \tan^{-1}\left(\frac{\epsilon}{b}\right)\right]}_{\theta_1} + \underbrace{\left[\tan^{-1}\left(\frac{\epsilon}{4-b}\right)\right]}_{\theta_2}$$

$$+ \underbrace{\left[360° - \tan^{-1}\left(\frac{3-\epsilon}{3-b}\right)\right]}_{\theta_3} + \underbrace{\left[\tan^{-1}\left(\frac{3+\epsilon}{3-b}\right)\right]}_{\theta_4} = \pm 180°, \dots . \tag{9.60}$$

This requirement can be reduced to

$$\tan^{-1}\left(\frac{\epsilon}{4-b}\right)-\tan^{-1}\left(\frac{\epsilon}{b}\right)=\tan^{-1}\left(\frac{2\epsilon(b-3)}{(b-2)^2+(3)^2-\epsilon^2}\right); \tag{9.61}$$

when $\epsilon\to 0$, Eqn. 9.61 becomes a nonlinear equation in b:

$$\frac{1}{4-b}-\frac{1}{b}=\frac{2(b-3)}{(b-2)^2+(3)^2}. \tag{9.62}$$

Three solutions exist, only one of which, $b=1.50$, lies on the root locus.

The angle at which the root locus departs from a complex pole can be easily calculated. Choose a test point q which is on the locus but which is an infinitesimal distance from the complex pole, as indicated in Fig. 9.19. The angle θ_3 is called the *departure angle,* and since q is on the locus we must have

$$\theta_1+\theta_2+\theta_3+\theta_4=\pm 180°,\ \pm 540°,\ \dots. \tag{9.63}$$

In this case θ_1, θ_2, and θ_4 may be determined by inspection, because point q is very close to the pole at $s=-3+j3$. Since $\theta_1=135°$, $\theta_2=71.565°$, and $\theta_4=90°$, from Eqn. 9.63 we obtain

$$\theta_3=\pm 180°-[135°+71.565°+90°]=-116.565°=243.435°. \tag{9.64}$$

Knowing the angle of departure of the locus from the complex poles aids in sketching the root-locus plot when computing aids are unavailable.

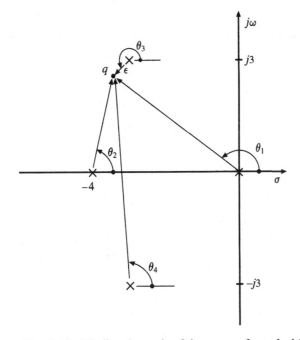

Fig. 9.19 Finding the angle of departure of root loci for $G(s)$ in Eqn. 9.58.

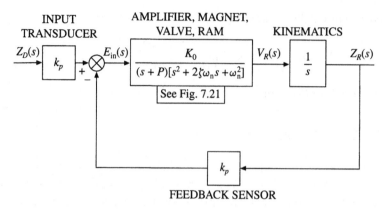

Fig. 9.20   Block diagram for electrohydraulic servomechanism.

9.3   Stability Analysis of an Electrohydraulic Servomechanism

We now apply the root-locus method to investigate the dynamic performance of an electrohydraulic positional servomechanism. The electrohydraulic amplifier, magnet, valve, and ram system depicted in Fig. 7.21 is used in the forward path of the servomechanism. The output variable in that diagram is the ram velocity $\dot{z}_R(t)$, but in a *position* control system we will measure the position of the ram $z_R(t)$ with a feedback sensor to compare it with the commanded position. We must supplement the block diagram of Fig. 7.21 to include the kinematic relationship between the velocity and the position of the ram. This is done in Fig. 9.20, where the position sensor gain, k_p V/cm, is included along with the input and error-generation elements. The system input command is now the *desired* ram position $z_D(t)$. For simplicity we use the same input transducer gain, k_p V/cm, as the feedback sensor. The notation used here is derived from that in Section 7.4:

$$K_0 = \frac{2K_A K_F K_q}{L_c m A_R} \frac{\text{cm}}{\text{Vs}^4}; \qquad P = \frac{R_o + R_c}{L_c} \frac{1}{\text{s}};$$

$$2\zeta\omega_n = \frac{k_v}{m} \frac{1}{\text{s}}; \qquad \omega_n^2 = \frac{k_e}{m} \frac{1}{\text{s}^2}. \tag{9.65}$$

To put the block diagram in Fig. 9.20 into the standard unity feedback configuration, we can move the transducer gain constants to the output of the summing junction without changing the mathematical validity of our model. This is done in Fig. 9.21, where the position error now appears at the output of the summer. The loop gain K is

$$K = K_0 k_p \frac{1}{\text{s}^4}. \tag{9.66}$$

Typical values for the physical parameters of mechanisms of this type are:

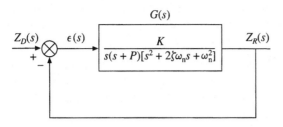

Fig. 9.21 Block diagram of Fig. 9.20 reduced to unity feedback configuration.

$$R_c = 10\ \Omega, \qquad\qquad R_o = 240\ \Omega, \quad L_c = 1\ \text{H},$$

$$K_F = 10^6\ \text{dyne/A}, \qquad m = 5\ \text{g}, \qquad k_p = 0.1\ \text{V/cm},$$

$$k_v = 600\ \text{dyne/(cm/s)}, \qquad\qquad k_e = 45 \times 10^4\ \text{dyne/cm},$$

$$K_q = 8000\ (\text{cm}^3/\text{s})/\text{cm}, \qquad\qquad A_R = 8\ \text{cm}^2,$$

so that the open-loop transfer function with these values becomes

$$G(s) = \frac{K}{s(s+250)[s^2 + 120s + 9 \times 10^4]}, \tag{9.67}$$

where the loop gain K is related to the amplifier gain by

$$K = [4 \times 10^7] K_A. \tag{9.68}$$

We consider the amplifier gain K_A to be positive but variable, and we now study the dependence of the dynamic performance of the closed-loop system on K_A, giving special attention to the stability of the system.

The first step is to draw a pole-zero plot of $G(s)$ and to construct a root-locus plot. The asymptotes for this plot are

$$\begin{bmatrix} \text{asymptote} \\ \text{angles} \end{bmatrix} = \theta_A = \frac{\pm 180°}{4-0}, \frac{\pm 540°}{4-0} = \pm 45°, \pm 135°,$$

$$\begin{bmatrix} \text{intersection} \\ \text{point} \end{bmatrix} = \sigma_I = -\frac{0 + 250 + 120 - 0}{4 - 0} = -92.5,$$

and the angle of departure of the locus from the upper complex pole is

$$\theta_D = -[101.537° + 90° + 57.122°] + 180° = -68.66° = 291.34°.$$

The transfer function of the closed-loop system as a function of K is

$$W(s) = \frac{Z_R(s)}{Z_D(s)} = \frac{N(s)}{D(s) + N(s)} = \frac{K}{s^4 + 370s^3 + 1.2 \times 10^5 s^2 + 2.25 \times 10^7 s + K}. \tag{9.69}$$

The four poles of $W(s)$, as they depend on K, are determined from the root-locus plot in Fig. 9.22. The system becomes unstable if K exceeds 3.599×10^9, which places a limit on the amplifier gain of 89.75 V/V. The engineer would probably specify an amplifier whose gain could be adjusted in the range $0 < K_A < 100$ in

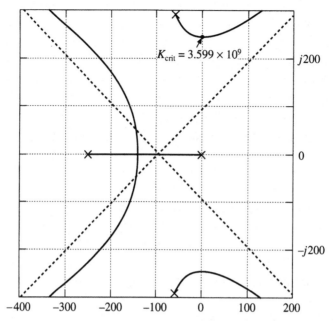

Fig. 9.22   Root-locus plot for electrohydraulic servomechanism with positive K, $G(s)$ in Eqn. 9.67.

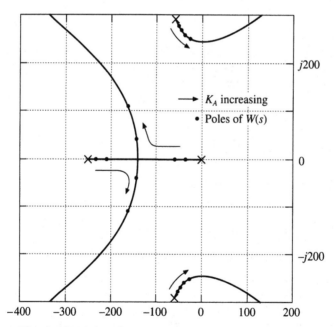

Fig. 9.23   Poles of $W(s)$ on root loci for $K_A = 15, 25, 40,$ and 60.

order to accommodate the normal tolerances expected in the values of physical parameters used in this kind of analysis.

Next we calculate the transient response of the closed-loop system for four values of amplifier gain in the design range. The input will be a unit step function, $z_D(t) = \sigma(t)$, so that $Z_D(s) = 1/s$. The system output function will be

$$Z_R(s) = \frac{[4 \times 10^7] K_A}{s\{s^4 + 370s^3 + 1.2 \times 10^5 s^2 + 2.25 \times 10^7 s + [4 \times 10^7] K_A\}}. \tag{9.70}$$

Because of the pole of $G(s)$ at the origin – that is, because of the "integrating" property of the open-loop transfer function – the final value of $z_R(t)$ will be 1, which is equal to the commanded input if the system is stable. (This "integrating" property is discussed further in Section 9.4.) The four values of amplifier gain are 15, 25, 40, and 60. The following table shows the pole locations of $W(s)$ for each of these gain settings.

K_A	Gain Margin	Poles of $W(s)$
15	6.0	$-31.5,\ -227,\ -55.8 \pm j284$
25	3.6	$-61.4,\ -204,\ -52.1 \pm j278$
40	2.25	$-141 \pm j46.1,\ -44.5 \pm j267$
60	1.50	$-156 \pm j111,\ -28.8 \pm j254$

Figure 9.23 illustrates these locations on the s plane to show how the closed-loop poles migrate along the root loci as K_A is increased from a low value of 15 to a high value of 60. The reason that these gains are designated as "low" and "high" is shown in the transient responses in Fig. 9.24.

The gain margin is defined as follows:

$$[\text{gain margin}] = \frac{K_{A\text{crit}}}{K_A} = \frac{89.96}{K_A}. \tag{9.71}$$

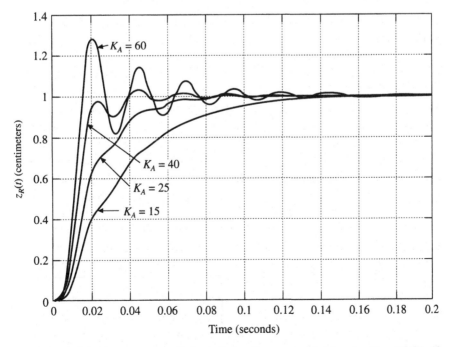

Fig. 9.24 Step response of electrohydraulic servomechanism for $K_A = 15, 25, 40,$ and 60.

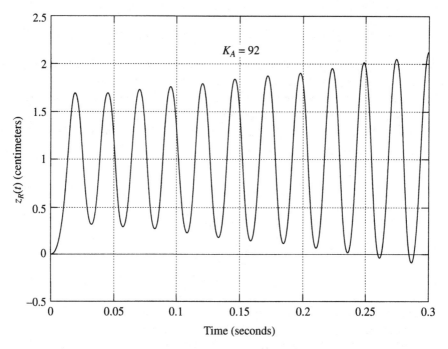

Fig. 9.25 Step response of electrohydraulic servomechanism for $K_A = 92$.

The gain margin is often used as a performance measure for a feedback system, and it is usually selected to lie between 2 and 6. The transient responses indicate that K_A between 25 and 40 would provide a response which is faster than that for $K_A = 15$ and which lacks the excessive oscillations shown in the response for $K_A = 60$.

Figure 9.25 shows the response of the hydraulic ram for an amplifier gain slightly above the critical value. In this case $K_A = 92$, putting two complex poles just across the $j\omega$ axis into the right-half plane. The oscillations of $z_R(t)$ will continue to build exponentially until a limit to the physical capability of the system is reached. For example, the peak velocity $\dot{z}_R(t)$ of the ram in the time interval around $t \approx 0.15$ s is about 200 cm/s, and the peak acceleration $\ddot{z}_R(t)$ of the ram is about 49,000 cm/s^2. During this interval an hydraulic pump that can deliver fluid at a peak rate of about 1.6 l/s is required to drive the ram. The peak displacement of the valve spool during these oscillations is about 2 mm. In establishing the model for this system, the mass of the ram (plus the mass of any device driven directly by the ram) is ignored. If the mass of the ram plus its load is 1 kg, for example, the peak force required to accelerate the ram during this interval is about 500 N. This force requires a peak differential pressure of about 625,000 Pa (about 90 psi) across the ram piston.

These peak-flow, displacement, and pressure requirements increase exponentially as the amplitude of the oscillations increase, and they soon approach a level that cannot be sustained by the flow capability of the hydraulic supply, the mechanical limits on the valve spool displacement, or the pressure capability of the hydraulic supply. The linear model will then fail to describe the motion, and in this high-powered hydraulic servomechanism the peak acceleration may be high enough to destroy the mechanism being controlled, or the ram may continue its nonlinear oscillation at a fixed amplitude but with a nonsinusoidal waveform.

9.4 Control of Non-Integrating Processes

Many controlled processes have the integrating property exhibited by the actuators in positional servomechanisms, such as the searchlight system described in Sections 8.2 and 9.1 and the electrohydraulic system described in Section 9.3. The transfer functions relating the actuator input signal to the position of the actuated member ($Y(s)/U(s)$ in Fig. 9.1 and $Z_R(s)/E_{in}(s)$ in Fig. 9.20) have a pole at the origin that represents the kinematic relationship between the velocity of the actuated member and its position.

However, many controlled processes lack this integrating property. A common example is the DC motor used in a speed control (not position control) application. Figure 9.26 is a block diagram of such a system, where the input $r(t)$ is the reference control voltage and the output $y(t)$ is the velocity of the load shaft. Consequently, the transfer function $Y(s)/R(s)$ has units (rad/s)/V, and the feedback sensor gain k_V has units V/(rad/s). Assume for the moment that the control law is a simple gain, $G_C(s) = K_1$ V/V.

To study the performance of this system, we calculate its response to a step input of reference control voltage and observe the system error. The transfer function for this assessment is

$$\frac{\epsilon(s)}{R(s)} = \frac{(s+P_1)(s+P_2)}{K_1 K_2 k_V + (s+P_1)(s+P_2)}. \tag{9.72}$$

For $r(t) = E\sigma(t)$ V, the Laplace transform of the error response is

$$\epsilon(s) = \frac{E(s+P_1)(s+P_2)}{s[s^2+(P_1+P_2)s+(K_1K_2k_V+P_1P_2)]}. \tag{9.73}$$

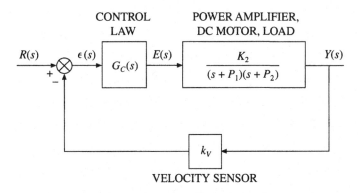

CONTROL   POWER AMPLIFIER,
LAW   DC MOTOR, LOAD

Fig. 9.26 Block diagram for a speed control system for a DC motor.

If we assume that K_1, K_2, k_V, P_1, and P_2 are all positive then the system is stable, and the final-value theorem may be applied to $\epsilon(s)$ to find the steady-state error:

$$\begin{bmatrix} \text{steady-state} \\ \text{error} \end{bmatrix} = \hat{\epsilon}(\infty) = \lim_{t \to \infty}[\hat{\epsilon}(t)] = \lim_{s \to 0}[s\epsilon(s)] = E\left(\frac{P_1 P_2}{P_1 P_2 + K_1 K_2 k_V}\right) \text{V},$$

(9.74)

where $\hat{\epsilon}(t) = \mathcal{L}^{-1}\{\epsilon(s)\}$.

Typical numerical values for the parameters in a system of this type are

$$k_V = 1, \quad K_1 = 2.6, \quad K_2 = 2000, \quad P_1 = 10, \quad P_2 = 100,$$

which yields a steady-state error of $0.161E$ V. The steady-state output velocity is

$$y(\infty) = \frac{0.839E}{k_V} \frac{\text{V}}{\text{V/(rad/s)}}$$

$$= 0.839E \text{ rad/s}, \tag{9.75}$$

and the steady-state voltage at the input of the power amplifier is $0.419E$ V.

The large steady-state error may be eliminated by introducing an integrator into the control law. The simplest transfer function for such a control law is

$$G_C(s) = \frac{K_1}{s}, \tag{9.76}$$

which increases the dynamic order of the system to 3 and places a limit on the loop gain to ensure stability. The transfer function relating $R(s)$ to $\epsilon(s)$ is now

$$\frac{\epsilon(s)}{R(s)} = \frac{s(s+P_1)(s+P_2)}{s^3 + (P_1 + P_2)s^2 + (P_1 P_2)s + K_1 K_2 k_V}. \tag{9.77}$$

Assuming that $k_V = 1$, $K_2 = 2000$, $P_1 = 10$, and $P_2 = 100$, and assuming that the control loop is designed to be well-damped, the control law gain will be $K_1 = 3$. The steady-state error in response to a step input will now be zero. This implies that the steady-state velocity will be E rad/s and thus the steady-state voltage at the input of the power amplifier will be $E/2$ V. Figure 9.27 illustrates the

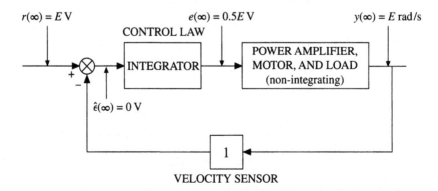

Fig. 9.27 Steady-state situation for speed control system using integral control.

steady-state situation. Note that the voltage at the input to the integrator is zero, but the voltage at the output is $E/2$ V. This important dynamic characteristic of the integrator was described in Section 3.3.

9.5 Non-Unity Feedback Systems

It is not always possible, nor is it always desirable, to use a constant gain element in the feedback path of a control system. Figures 8.7 and 8.8 showed the general configuration for a single-loop control system having a dynamic element with a transfer function, denoted $H(s)$, in the feedback path. The initial-condition excitations, as well as the disturbance input and a conditioned command input, were also shown. A simplified version of these figures is repeated here as Fig. 9.28. We note again that the input to the forward-path element (denoted $\Delta(s)$ on the diagram) is not the system error, if that error is defined as the difference between the input and the output of the overall system. Following Eqn. 8.17, we can define the transfer function relating the system error to the input, neglecting the initial conditions and disturbance:

$$W(s) = \frac{Y(s)}{R(s)} = \frac{G(s)}{1+G(s)H(s)} \quad \text{and}$$

$$\epsilon(s) = R(s) - Y(s), \quad \text{so that} \tag{9.78}$$

$$\frac{\epsilon(s)}{R(s)} = \frac{1-G(s)+G(s)H(s)}{1+G(s)H(s)}.$$

We express $G(s)$ and $H(s)$ in terms of their numerator and denominator polynomials:

$$G(s) = \frac{N_1(s)}{D_1(s)} \quad \text{and} \quad H(s) = \frac{N_2(s)}{D_2(s)}. \tag{9.79}$$

The transfer functions relating the input to the output and to the error are

$$\frac{Y(s)}{R(s)} = W(s) = \frac{N_1(s)D_2(s)}{[D_1(s)D_2(s)+N_1(s)N_2(s)]},$$

$$\frac{\epsilon(s)}{R(s)} = \frac{[D_1(s)D_2(s)+N_1(s)N_2(s)]-N_1(s)D_2(s)}{[D_1(s)D_2(s)+N_1(s)N_2(s)]}. \tag{9.80}$$

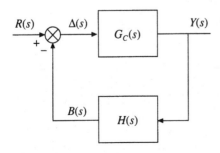

Fig. 9.28 Block diagram for non-unity negative feedback system.

Both of these transfer functions have the same denominator, whose roots – the poles of the closed-loop transfer functions – can be found using the root-locus technique. The characteristic equation for this system configuration is

$$D_1(s)D_2(s) + N_1(s)N_2(s) = 0, \tag{9.81}$$

which is re-arranged to yield

$$G(s)H(s) = -1. \tag{9.82}$$

This shows once again that to determine the poles of $W(s)$ we must treat the loop transfer function $G(s)H(s)$ of the non-unity feedback system in the same manner as we treated the loop transfer function $G(s)$ in the unity feedback case. The zeros of $W(s)$ are the zeros of $G(s)$ and the poles of $H(s)$. As an example, take the case

$$G(s) = \frac{K_1(s+3)}{s[s^2+6s+25]} \quad \text{and} \quad H(s) = \frac{K_2}{s+2}, \tag{9.83}$$

where we must find values for K_1 and K_2 such that the response to a step input will be stable, having an error that approaches zero as $t \to \infty$. Take the input to be a unit step function $\sigma(t)$, so that the error is given by

$$\epsilon(s) = \frac{1}{s}\left[\frac{s(s+2)[s^2+6s+25]+K_1K_2(s+3)-K_1(s+3)(s+2)}{s(s+2)[s^2+6s+25]+K_1K_2(s+3)}\right]. \tag{9.84}$$

The final-value theorem gives the steady-state error as

$$\lim_{t \to \infty} \hat{\epsilon}(t) = \lim_{s \to 0}[s\epsilon(s)] = \lim_{s \to 0}\left\{\frac{K_1K_2(s+3)-K_1(s+3)(s+2)}{K_1K_2(s+3)}\right\} \tag{9.85}$$

where $\hat{\epsilon}(t) = \mathcal{L}^{-1}\{\epsilon(s)\}$, provided the system is stable. The conditions for stability will be determined by a root-locus analysis, but Eqn. 9.85 gives the result that K_2 must be 2. Figure 9.29 is the root-locus plot, with the product K_1K_2 as the variable along the loci. The critical gain for stability is found to be $K_1K_2 = 112.9$, which, combined with the requirement that $K_2 = 2$, gives the stability limit as $K_1 < 56.45$. If K_1 is selected as 10 then the poles of $W(s)$ will be located as shown in Fig. 9.29, the gain margin will be 5.6, and the step response will be prompt, with an overshoot of less than 10 percent.

A second example will illustrate a special case that occurs when $H(s)$ has a zero at the same location where $G(s)$ has a pole. Let

$$G(s) = \frac{K}{s[s^2+4s+8]} \quad \text{and} \quad H(s) = \frac{s}{s+2}, \tag{9.86}$$

so that the closed-loop transfer function will be

$$W(s) = \frac{K(s+2)}{s(s+2)[s^2+4s+8]+Ks} = \frac{K(s+2)}{s((s+2)[s^2+4s+8]+K)}. \tag{9.87}$$

Notice that s is a factor of the denominator because it appears both in the numerator of $H(s)$ and in the denominator of $G(s)$. But because this factor is

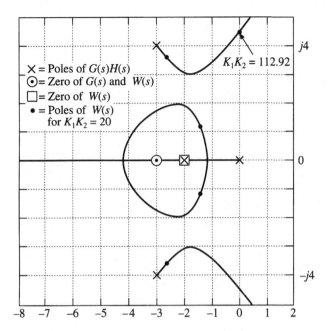

Fig. 9.29 Root-locus plot for $G(s)H(s)$ in Eqn. 9.83.

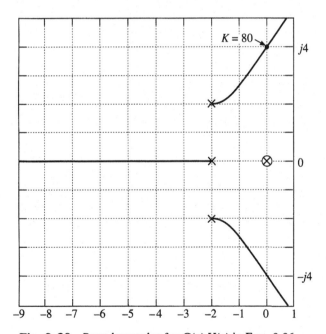

Fig. 9.30 Root-locus plot for $G(s)H(s)$ in Eqn. 9.86.

canceled in the product $G(s)H(s)$, it might be overlooked as a pole of $W(s)$ on the root-locus plot, as shown in Fig. 9.30. Because the pole-zero pair at the origin generates a branch of the root locus of zero length, $W(s)$ has a pole at the origin. Since the zero at the origin comes from $H(s)$ and is therefore not a zero of $W(s)$, no pole-zero cancellation in $W(s)$ occurs at the origin, as is clear from the algebra in Eqn. 9.87.

In summary, the system transfer function $W(s)$ for the non-unity feedback configuration in Fig. 9.28, where $G(s)$ and $H(s)$ are written in standard form, is determined as follows.

1. The zeros of $W(s)$ occur at:
   (a) the zeros of $G(s)$; and
   (b) the poles of $H(s)$.
2. The poles of $W(s)$ occur at:
   (a) the solution points of the characteristic equation $G(s)H(s) = -1$, which may be determined by a root-locus plot; and
   (b) where a pole of $G(s)$ and a zero of $H(s)$ coincide, if such points exist.
3. The flag constant of $W(s)$ is the flag constant of $G(s)$.

A technique for designing a feedback transfer function $H(s)$ that will improve the dynamic behavior of a control system is discussed in Chapter 10.

9.6 Negative-Gain Feedback Systems

We now return to the unity feedback configuration defined by Eqn. 9.5. We derive the principles of the root locus for the case where K, the flag constant of $G(s)$, is negative. The characteristic equation is still $G(s) = -1$, and the two requirements on $G(s)$ given in Eqn. 9.21 still apply. But now the angle requirement of Eqn. 9.21(a) is affected by the negativity of K. Although the angle requirement is still $\angle G(s) = \pm 180°, \pm 540°, \pm 900°, \ldots$, now $\angle K = -180°$. The angle requirement, as expressed in Eqn. 9.22, can therefore be written as

$$\sum_{i=1}^{m} \angle(s+Z_i) - \sum_{j=1}^{n} \angle(s+P_j) = \pm 0°, \pm 360°, \pm 720°, \ldots . \tag{9.88}$$

The magnitude requirement of Eqn. 9.21(b), which has the absolute value of K as a factor, remains unchanged, as expressed in Eqn. 9.23. Two examples will illustrate the differences in the root loci between the negative-K and the positive-K case. In each of these examples we use a standard unity, negative, feedback connection.

Example 1

$$G(s) = \frac{K(s+8)}{(s+3)(s+4)[s^2+10s+30]}. \tag{9.89}$$

The poles and zero of $G(s)$ are plotted in Fig. 9.31. In searching the real axis for points on the root locus – that is, for those points satisfying the angle requirement of Eqn. 9.88 – we find that all points to the right of the rightmost singularity (the pole at $s = -3$ in this case) and all points lying to the left of an even number of singularities are on the locus. In the remote regions of the s

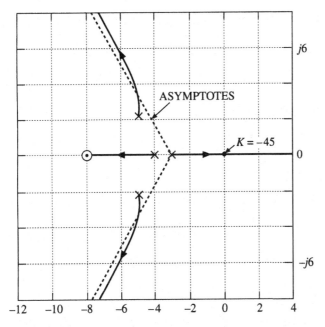

Fig. 9.31 Root-locus plot for $G(s)$ in Eqn. 9.89 with K negative.

plane, the angle requirement shows that roots lie on asymptotes which make angles with the real axis of

$$\theta_A = \frac{\pm 0°}{n-m}, \frac{\pm 360°}{n-m}, \dots = \frac{\pm 0°}{4-1}, \frac{\pm 360°}{4-1} = 0°, +120°, -120°. \tag{9.90}$$

The intersection point of the asymptotes is unaffected by the sign of K. For this example,

$$\sigma_I = -\frac{(3+4+10)-(8)}{4-1} = -3, \tag{9.91}$$

and the angle of departure from the upper complex pole is

$$\theta_D = [36.70°] - [90° + 114.1° + 131.8°] = -299.2° = 60.79°. \tag{9.92}$$

In Fig. 9.31, the locus is drawn to show the possible locations for the poles of $W(s)$. We note that the branch of the locus that starts at $s = -3$ for $K = 0$ moves toward the right-half plane as K increases, and crosses into the right-half plane when $K = -45$. This unfavorable characteristic occurs in many feedback systems when K is negative. It is especially bad for those systems in which $G(s)$ has an "integrating" nature – that is, a pole at $s = 0$ and no poles or zeros in the right-half plane. The closed-loop system will then be unstable for any nonzero negative K. Since many such systems exist, a natural prejudice arises in the control engineering profession against negative-gain systems. However, our second example is a case where negative gain offers a striking advantage over positive gain.

Example 2   In this case $G(s)$ is a *nonminimum phase* transfer function (a term designating a transfer function having one or more poles or zeros in the right-half plane). The origin of the term is made clear in Chapter 11. This particular form for $G(s)$ is found in the analysis of automatic flight-control systems for aircraft:

$$G(s) = \frac{K(s+1)(s-2)}{s(s+3)(s+5)[s^2+4s+13]}. \tag{9.93}$$

When K is negative, the asymptote angles for this root-locus plot are

$$\theta_A = \frac{\pm 0°}{5-2}, \frac{\pm 360°}{5-2} = 0°, +120°, -120°, \tag{9.94}$$

and the point of intersection of the asymptotes is

$$\sigma_I = -\frac{(0+3+5+4)-(1-2)}{5-2} = -4.33. \tag{9.95}$$

The departure angle of the locus from the upper complex pole is

$$\theta_D = [116.57° + 153.43°] - [33.69° + 63.43° + 90° + 135°] = -52.13°. \tag{9.96}$$

The pole-zero plot of $G(s)$ in Fig. 9.32 clearly shows that the closed-loop system would be unstable for any *positive* value of K because, if K were positive, a branch of the root locus would have to lie in the right-half plane between $s = 0$ and $s = 2$. Yet when K lies in the range from -74.7 to 0, the root-locus plot shows the closed-loop system to be stable.

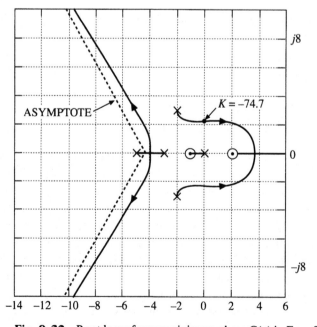

Fig. 9.32 Root locus for nonminimum phase $G(s)$ in Eqn. 9.93 with K negative.

9.7 Positive Feedback Systems

A block diagram for a unity feedback system with positive feedback is shown in Fig. 9.33. Positive feedback occurs either when the feedback signal is connected to the summing device with a polarity opposite to that used for negative feedback, or when the summing point is an intrinsic part of the system as in the geometry of some navigation systems. The closed-loop transfer function is

$$W(s) = \frac{Y(s)}{R(s)} = \frac{G(s)}{1 - G(s)} = \frac{N(s)}{D(s) - N(s)}. \tag{9.97}$$

A minus sign in the denominator is an immediate indication of potential instability of the closed-loop system, prompting a thorough investigation of the stability question. The characteristic equation for this system is obtained in the usual way:

$$1 - G(s) = 0 \;\Rightarrow\; G(s) = 1 \;\Rightarrow\; \textbf{(a) } \angle G(s) = \pm 0°, \pm 360°, \ldots \quad \text{and} \tag{9.98}$$
$$\textbf{(b) } |G(s)| = 1.$$

Written in standard form, the angle requirement (a) for $G(s)$ is

$$\angle K + \sum_{i=1}^{m}(s + Z_i) - \sum_{j=1}^{n}(s + P_j) = \pm 0°, \pm 360°, \ldots. \tag{9.99}$$

If K is restricted to negative values, this requirement becomes

$$\begin{bmatrix} \text{positive feedback} \\ \text{using negative } K \end{bmatrix} \;\Rightarrow\; \sum_{i=1}^{m}(s + Z_i) - \sum_{j=1}^{n}(s + P_j) = \pm 180°, \pm 540°, \ldots. \tag{9.100}$$

The angle requirement on the poles and zeros of $G(s)$ is the same as that in the negative feedback case using positive gain. The root-locus plots for the two cases will be identical, as will the poles of $W(s)$ for the same absolute values of K. However, since K is a factor in the numerator of $W(s)$, the numerators of $W(s)$ in the two cases will have opposite signs. An example illustrates this important distinction.

Let the feedback be positive, and let $G(s)$ have the form

$$G(s) = \frac{K}{s[s^2 + 12s + 40]}. \tag{9.101}$$

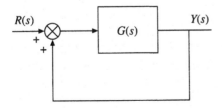

Fig. 9.33 Block diagram for a unity positive feedback system.

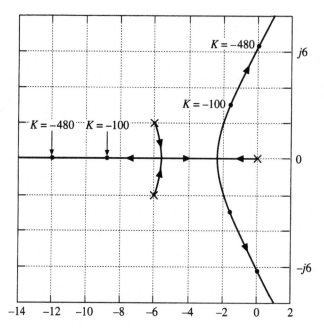

Fig. 9.34 Root-locus plot for $G(s)$ in Eqn. 9.101 with positive feedback and K negative.

Figure 9.34 shows the root-locus plot for negative K. If K is set at -100 to provide a gain margin of 4.8, the closed-loop transfer function will be

$$\begin{bmatrix} \text{positive feedback,} \\ \text{negative gain} \end{bmatrix} \Rightarrow W_{\text{PFNG}}(s) = \frac{-100}{s^3 + 12s^2 + 40s + 100}. \qquad (9.102)$$

If the same $G(s)$ as that in Eqn. 9.101 is used in a negative feedback system but with K restricted to positive values, the root-locus plot will be the same as that in Fig. 9.34. For $K = +100$, the gain margin is 4.8 and the closed-loop transfer function is

$$\begin{bmatrix} \text{negative feedback,} \\ \text{positive gain} \end{bmatrix} \Rightarrow W_{\text{NFPG}}(s) = \frac{100}{s^3 + 12s^2 + 40s + 100}. \qquad (9.103)$$

The analytical difference between $W_{\text{NFPG}}(s)$ (Eqn. 9.103) and $W_{\text{PFNG}}(s)$ (Eqn. 9.102) is simply the algebraic sign, but the practical difference between the two is profound. For example, if these transfer functions represent the automatic steering mechanism for a high-speed automobile, a command to the NFPG system to turn right will produce a right turn of the vehicle, whereas the same command to the PFNG system will cause the vehicle to turn left! One cannot underestimate the importance of understanding the polarities of the input and output quantities of all the system components, and of ensuring that they are connected to achieve the desired dynamic performance of the closed-loop system.

9.8 Summary

The problem of relating the dynamic behavior of a feedback control system to the parameters of the individual components in the control loop consists of determining the poles, zeros, and flag constant of the closed-loop transfer function $W(s)$ from those properties of the open-loop transfer function $G(s)H(s)$. This problem is essentially that of factoring polynomials in a way that shows how the changes in $W(s)$ depend on the component parameters. The most practical way to perform this task is to use a computer with an application program especially designed for control system analysis. Nevertheless, it is essential for the engineer to understand the underlying geometric relationships between the parameters of a system and the roots of the characteristic equation. The root-locus method has proved to be a significant aid to this understanding.

We shall list a useful set of rules for constructing the root locus for the unity, negative feedback system, where the poles and zeros of $G(s)$ are known and the gain constant is positive and variable. These rules are easily modified to accommodate both the non-unity feedback case and the positive feedback case, as demonstrated in Sections 9.5 and 9.7, and they may be adapted for calculating the locus of roots with parameters other than the gain constant as the variable. In the following we assume that $n \geq m$, but the rules are also easily modified to handle the less common case $m > n$.

1. Plot the poles and zeros of $G(s)$ accurately on the complex plane, using the same numerical scale on the σ and $j\omega$ axes.
2. Sketch the $n - m$ asymptotes by computing their point of intersection with the real axis (Eqn. 9.54) and the angles they make with the real axis (preceding Eqn. 9.41).
3. Construct each of the n branches of the root locus emanating from a pole of $G(s)$ (for $K = 0$) and terminating on a zero of $G(s)$ (for $K = \infty$); $n - m$ of the zeros of $G(s)$ lie at infinity in the directions indicated by the asymptotes, and the other m zeros lie in the finite portions of the s plane. The branches of the locus on the real axis lie to the left of an odd number of real-axis singularities.
4. Determine the angle of departure of the locus from a complex pole by the method demonstrated in Fig. 9.19.
5. Branches of the locus on the real axis lying between two poles or between two zeros must have a breakaway point at which they depart or join the real axis. Calculate this point as shown in Fig. 9.18, or estimate it from the positions of the nearby poles and zeros.
6. For a given value of K, use the magnitude requirement (Eqn. 9.21(b)) to determine the location of the n poles of $W(s)$ that lie on the n branches of the locus.

In addition to describing the root-locus techniques, this chapter has illustrated several other important ideas of control system engineering to be pursued in later chapters. Figure 9.14 shows that a plant which is unstable when standing alone can be rendered stable by using a properly designed feedback connection. The feedback connection automatically provides the proper driving function to the plant to keep its response stable. At the same time, input commands directing the plant to perform desired motions can be superimposed on the driving function. This is somewhat analogous to a human acrobat using visual feedback to balance an unstable load while moving about in a prescribed path. The acrobat is part of a manual feedback control system, whereas the controllers in our examples are parts of automatic feedback control systems.

A first step in the process of control system design was taken in Section 9.3, where the transient response of an electrohydraulic servomechanism was related quantitatively to the amplifier gain. The root-locus plot revealed the possible pole and zero locations for $W(s)$ as a function of amplifier gain, and the pole-zero analysis of Chapter 7 was applied to $W(s)$ to select the amplifier gain providing the fastest possible response consistent with a modest overshoot. The limit on amplifier gain beyond which the feedback system is unstable was also established from the root-locus analysis. This information guides the detailed design of an amplifier. Chapter 10 discusses further design techniques that are used when this first step in analysis reveals that the performance specifications cannot be satisfied simply by adjusting the loop gain.

A second important principle of control loop design, that of using an integrating element in the forward path of the system to eliminate a static error, was introduced in Section 9.4. This idea is also explored further in Chapter 10.

9.9 Problems

Problem 9.1
Use Routh's method (See Appendix D) to determine how many roots of each of the following polynomials lie inside the left-half plane:

(a) $s^4 + 8s^3 + 32s^2 + 80s + 100$;
(b) $s^5 + 10s^4 + 30s^3 + 80s^2 + 344s + 480$;
(c) $s^4 + 2s^3 + 7s^2 - 2s + 8$;
(d) $s^3 + s^2 + 20s + 78$;
(e) $s^4 + 6s^2 + 25$.

Problem 9.2
The fourth-order polynomial $P(s) = s^4 + s^3 - s^2 + s - 2$ has four roots.

(a) How many roots of $P(s)$ lie inside the left-half plane?
(b) How many roots of $P(s)$ lie inside the right-half plane?

Problem 9.3
In Problem 9.1(a), replace the coefficient 32 with a parameter k. Determine the range over which k can be varied while keeping all four roots in the left-half plane.

Problem 9.4
Use Routh's array to determine the nonredundant relationships that must exist among the five coefficients of the following monic polynomial to ensure that all of its roots lie in the left-half plane:

$$s^5 + a_4 s^4 + a_3 s^3 + a_2 s^2 + a_1 s + a_0.$$

Problem 9.5
Use Routh's stability criterion to verify the ranges of K for which the system described in Eqn. 9.13 and Fig. 9.4 will be stable.

Problem 9.6
Let $Q(s) = s^4 + Ks^3 + 90s^2 + 350s + 600$. Use the root-locus method to show how the roots of the polynomial $Q(s)$ depend on the parameter K.

Problem 9.7
The closed-loop transfer function for the configuration of Fig. 8.8, $Y(s)/R(s)$, is denoted as $W(s)$, as shown in Eqn. 8.17. For each of the systems (a)–(c), write $W(s)$ in the standard form $W(s) = $ [a polynomial in s]/[a monic polynomial in s]. State whether the closed-loop system is stable or unstable.

(a) $G(s) = \dfrac{30(2s+4)}{2s^3 + 12s^2 + 30s}$, $\quad H(s) = \dfrac{10}{2s+10}$;

(b) $G(s) = \dfrac{1000}{0.2s^5 + 5s^4 + 50s^3 + 250s^2 + 625s + 625}$, $\quad H(s) = 2$;

(c) $G(s) = \dfrac{150(2s+2)}{(0.5s+2)(2s^3 + 16s^2 + 50s)}$, $\quad H(s) = \dfrac{2s+4}{3(0.333s+1.333)}$.

Problem 9.8
Let

$$G(s) = \frac{K(s+1)}{s(s+2)(s+4)(s^2+2s+9)}$$

be the forward-path transfer function of a unity feedback control system. Make a root-locus diagram for this system and determine the range over which K can be varied with the system remaining stable. Consider both positive and negative K. Determine the points at which the root loci cross the $j\omega$ axis.

Problem 9.9
Let

$$G(s) = \frac{K}{s(s+1)(s+P)}$$

be the forward-path transfer function of a unity feedback control system. It is found that the system is stable only for $0 < K < 2$. Find P from a root-locus diagram. Check your answer, using Routh's criterion.

Problem 9.10
The input to a unity feedback system is a step, $r(t) = 5\sigma(t)$, and the output $y(t)$ is shown in Fig. P9.10. Find the forward-path transfer function $G(s)$.

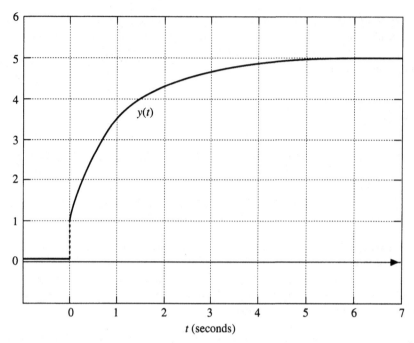

Fig. P9.10

Problem 9.11

Let

$$G(s) = \frac{K}{s(s+\alpha)(s+\beta)}$$

be the open-loop transfer function for a linear feedback control system. Show that the root loci for this system cross the $j\omega$ axis at the points $s = \pm j\sqrt{\alpha\beta}$.

Problem 9.12

A non-unity feedback positional servomechanism is diagramed in Fig. P9.12. Neglect the moments of inertia of the gears, the inductance of the armature circuit, the output resistance of the amplifier, and the loading effect of the feedback filter on the output potentiometer. The active winding on both potentiometers is 360°. E_i is 6.28 V, and E_o is 12.56 V. The gear ratio is 5:1, and J_L is 0.75 slug-ft². The manufacturer's speed–torque data for the bare motor is

$J_M = 0.01$ slug-ft²;
stall torque ($e = 100$ V) = 1 ft-lb;
no-load speed ($e = 100$ V) = 2.5 rad/s.

The system is at rest when the input potentiometer shaft is suddenly rotated 5°. In response to this step input, the output is $\theta_o(t) = A_o + A_i \sin(10t + \phi) + $ [decaying transient]. Find A_o, A_i, ϕ, K_A, and C.

Problem 9.13

Refer to the seventh-order example defined in Eqn. 9.55.

(a) Find the two breakaway points on the negative real axis.
(b) Calculate the angles of departure of the root loci from the complex poles.

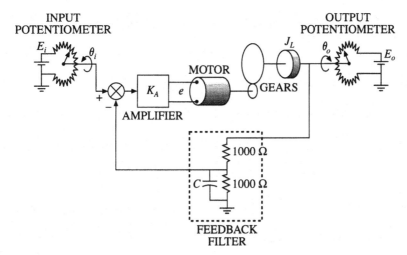

Fig. P9.12

(c) It is known that one pole of $W(s)$ lies at $s = -11$. Find the locations of the other six poles and the value of K.

(d) Can you find a value of K that will place a pole of $W(s)$ at $s = +2$? If so, find that value. If not, explain why not.

Problem 9.14

Let

$$G(s) = \frac{K}{s(s+P)(s^2+4s+16)}$$

be the transfer function of the forward path of a unity feedback system.

(a) Draw an accurate root-locus diagram for this system for $P = 4$ and $K \geq 0$.

(b) Draw root-locus diagrams for $P = 3.8$ and $P = 4.2$, noting the trend in the root-locus patterns when P is in the neighborhood of 4.

(c) If $P = 3$, calculate the angles of departure of the loci from the complex poles and the location of the breakaway point on the negative real axis.

Problem 9.15

In the electrohydraulic servomechanism described in Figs. 9.20–9.25, assume that the amplifier gain K_A is variable and that all the parameters except k_e have the values given following Eqn. 9.66. As K_A is increased from its normal operating point, the system breaks into oscillation at the frequency 26.2 Hz. Find the value of k_e and the critical setting (stability limit) of K_A.

Problem 9.16

Use the root-locus method and Routh's criterion to verify the qualitative and quantitative assertions made in Section 9.4 concerning the use of an integrating control law in the speed control system of the DC motor.

Problem 9.17

In a non-unity feedback system we have

$$G(s) = \frac{K_1(s+1)}{s(s+3)(s+5)^2} \quad \text{and} \quad H(s) = \frac{K_2(s+3)}{s+1}.$$

(a) Find K_1 and K_2 if the steady-state gain of the closed-loop system is to be 1.

(b) Find $W(s)$ for $K_1 = 600$ and $K_2 = 1/3$.

Problem 9.18

In a non-unity *positive* feedback system we have

$$G(s) = \frac{K(s+1)}{s^2+4s+9} \quad \text{and} \quad H(s) = \frac{3}{s+3}.$$

Make a root-locus plot for this system and find the number of poles of $W(s)$ that occur in the right-half plane for: (a) $K = 6$; (b) $K = 7$; and (c) $K = 10$.

Problem 9.19

A block diagram for an electrohydraulic servomechanism is shown in Fig. P9.19. Find the gain of the preamplifier if the response $x_r(t)$ to a step input $e_{in}(t) = \sigma(t)$ is to be as fast as possible but without any overshoot.

Problem 9.20

The unstable plant in the system shown in Fig. P9.20 has the transfer functions

$$G_P(s) = \frac{Y_{ZSR}(s)}{U(s)} = \frac{50}{s^3+14s^2+35s-50} \quad \text{and} \quad I_0(s) = \frac{Y_{ZIR}(s)}{y_0} = \frac{s^2+14s+35}{s^3+14s^2+35s-50}.$$

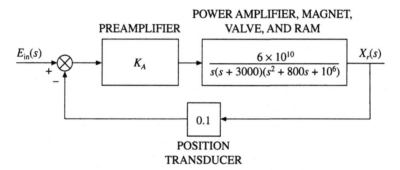

POWER AMPLIFIER, MAGNET,
PREAMPLIFIER   VALVE, AND RAM

Fig. P9.19

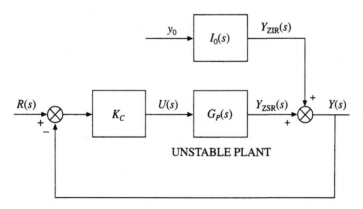

UNSTABLE PLANT

Fig. P9.20

$Y(s)$ may be expressed as $Y(s) = R(s)W_R(s) + y_0 W_0(s)$. (We assume that $\dot{y}_0$ and $\ddot{y}_0$ are both zero.)

(a) Calculate the two transfer functions $W_R(s)$ and $W_0(s)$. Note that these two transfer functions have the same denominator.

(b) Find the range of values of K_c for which the system will be stable.

(c) If one of the poles of $W_R(s)$ is -12, find the other two poles (whose sum is obviously -2). Find K_c.

(d) For $K_c = 3.64$, a unit step input is applied at $t = 0$. The initial condition is $y_0 = -0.5$. Plot $y(t)$ for $0 \le t \le 10$ s.

Problem 9.21

Let

$$G(s) = \frac{K(s+3)}{s(s+2)(s^2+\beta s+13)}$$

be the open-loop transfer function of a unity feedback system; K and β are adjustable parameters.

(a) For $\beta = 4$, construct a root-locus plot showing how the poles of the closed-loop transfer function depend on the variations in K, and determine the range of K for which the closed-loop system will be stable.

(b) For $K = 40$, construct a root-locus plot showing how the poles of the closed-loop transfer function depend on the variations in β, and determine the range of β for which the closed-loop system will be stable.

(c) Choose a combination $\{K, \beta\}$ for which the closed-loop system output $y(t)$ to a step input at $r(t)$ will respond as promptly as possible without overshooting its final value.

Problem 9.22

For the system diagramed in Fig. P9.22, use the negative feedback connection.

(a) Plot the root locus for both positive and negative values of K.

(b) Find the breakaway points and angles of departure of the root loci.

(c) Select the value of K that provides a stable closed-loop system with a gain margin of 3, and evaluate $Y(s)/R(s)$ for this gain setting.

(d) For a unit step input $r(t) = \sigma(t)$, plot the zero-state response $y(t)$.

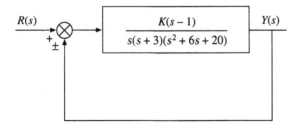

Fig. P9.22

Problem 9.23

For the system diagramed in Fig. P9.22, use the positive feedback connection.

   (i) Repeat parts (a), (c), and (d) of Problem 9.22.
   (ii) Compare the results obtained in (i) with their counterparts in Problem
   9.22.

Problem 9.24

Let

$$G(s) = \frac{K(s^2 - 2s + 6)}{s(s+3)(s+5)(s^2 + 6s + 20)}$$

be the forward-path transfer function of a unity negative feedback control system. Use a root-locus plot to find a value for K for which the closed-loop system will be stable with a gain margin of 3. Calculate and plot the system output if the system input is a unit step. Compare this problem with Problems 9.22 and 9.23.

Problem 9.25

A rudimentary scheme for the automatic control of the altitude of an aircraft during cruise conditions is depicted in Fig. P9.25. Normally $\Delta h_{com}(t)$ is zero. The system is intended to regulate the altitude $h(t)$ about its set point h_0 in presence of disturbances caused by atmospheric conditions. Here $u(t)$ is the elevator deflection from its trim value, which is frequently denoted $\Delta \delta_E(t)$. For the Navion aircraft at the cruise conditions defined in Problem 2.10, the transfer function relating $u(t)$ in degrees to $\Delta h(t)$ in feet is approximately

$$\frac{\Delta H(s)}{U(s)} = \frac{0.45(s+0.02)(s+13)(s-10)}{s(s^2+5s+13)(s^2+0.04s+0.05)} \frac{\text{feet altitude}}{\text{degree elevator}}.$$

Assume that the altimeter scale factor is 0.01 V/ft. The control law gain is then K [degrees elevator]/volt. Can you find a value of K for which the feedback system will

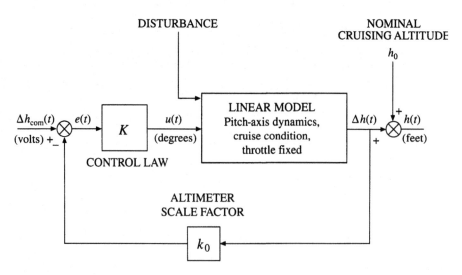

Fig. P9.25

satisfy the following two requirements? If so, show that your K will be satisfactory. If not, explain why not.

(a) The closed-loop system must be stable and must have a reasonable gain margin.

(b) In response to a step command $\Delta h_{com}(t) = 1\sigma(t)$ V, $\Delta h(t)$ should respond by rising and settling to within 3 feet of its final value of 100 feet within 30 seconds without excessive overshoot.

10 Design of Feedback Control Systems

10.1 Introduction

We now explore four problem areas in the design of single-input–single-output (SISO) feedback systems. The common theme in these problems is the control of the system error in response to input commands and to undesired disturbances. These problem areas are:

1. series compensation to satisfy design specifications;
2. parallel compensation to satisfy design specifications;
3. disturbance rejection; and
4. control of nonminimum phase plants.

The system error is defined as the difference between an ideal response of the system to these inputs and the actual response, as depicted in Fig. 10.1. This

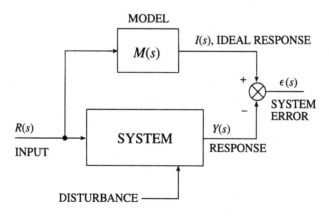

Fig. 10.1 System error defined.

definition permits the ideal response to be defined by the model transfer function $M(s)$ if the ideal response is not equal to the input.

We begin in Section 10.2 by finding simple series compensators for a small third-order servomechanism powered by a DC motor. The design of the compensators must satisfy both a dynamic response requirement and a static gain requirement. The trial-and-error method, using iterative root-locus analysis, is the design approach. This approach is effective in most SISO systems. In Section 10.3 we introduce a special type of series compensation, the PID controller, which is widely used in industrial control processes and is remarkably versatile in other technical settings as well. Pole-zero cancellation as a compensation technique is discussed in Section 10.4, and the disadvantages of that technique – described in Chapter 7 for open-loop systems – are illustrated in closed-loop systems.

Parallel compensation takes two forms in Section 10.5. The first is the single-loop form, where the compensating element is placed in the feedback path. The same trial-and-error approach as that used for series compensation design is applied in this case, and the unique performance results obtainable with feedback compensation are emphasized. The second form of parallel compensation is the multi-loop configuration, which offers a wider scope of performance enhancement; however, it introduces an important question of robustness that does not occur in single-loop systems. A special requirement for compensation, the tracking problem, is introduced in Section 10.6.

An example showing how controller design enhances the robustness of a system with respect to unwanted disturbances at the plant is considered in Section 10.7. A steady load-torque disturbance on a servomechanism induces a misalignment proportional to the torque. The misalignment (or "droop") is minimized by the controller design, which also maintains the dynamic performance of the system.

Control of nonminimum phase plants includes those whose transfer functions have poles or zeros in the right-half plane. Each of these types offers special control problems that do not occur with minimum phase plants. These problems are explored by means of examples in Sections 10.8 and 10.9.

The simple examples treated in detail in this chapter illustrate the rudiments of controller design, which may often be applied successfully to more complicated systems. These examples also reveal the need for a more comprehensive analytical approach to design for higher-order systems and multi-loop systems. Some of the features of these advanced methods of design are cited in Section 10.10.

10.2 Series Compensation

The diagram of a third-order servomechanism in Fig. 10.2 illustrates the principles of series compensation. The power amplifier–servomotor combination is detailed in Fig. 4.9. The manufacturer of the motor furnishes the steady-state speed-torque data shown in Fig. 10.3, along with the following parameter values:

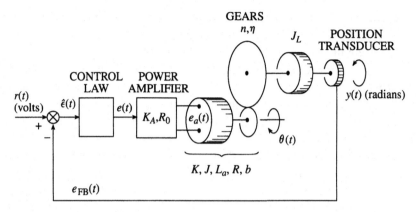

Fig. 10.2 Schematic diagram for electromechanical servomechanism.

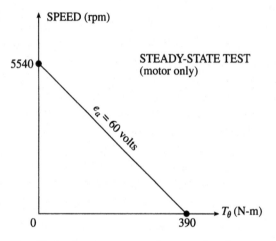

Fig. 10.3 Steady-state speed-torque data for motor only.

$$K = 13.7 \text{ oz-in}/\text{A},$$
$$J = 0.003 \text{ oz-in-s}^2, \tag{10.1}$$
$$L_a = 0.01 \text{ H}.$$

The amplifier parameters are:

$$K_A = 4 \text{ V/V}, \qquad R_o = 1 \text{ }\Omega. \tag{10.2}$$

The gear train, which has a single mesh, is characterized by the following parameters:

PINION:	20 teeth,	
LOAD GEAR:	210 teeth,	(10.3)
EFFICIENCY:	$\eta = 0.88$.	

The moments of inertia of the gears are negligible. The load is a rotating mass with a moment of inertia about the load axis of 0.002 kg-m^2. The position transducer on the output shaft has a scale factor of 1 V/rad.

The given motor data must be analyzed to determine the motor resistance R and the viscous damping coefficient b. Note that the moment of inertia of the motor is given as the ratio of torque to acceleration. It is advisable to convert the given data into a consistent set of units before making calculations. The SI system of units is recommended here. The transfer function relating the amplifier input voltage to the output shaft position for these parameter values is

$$\frac{Y(s)}{E(s)} = \frac{90{,}100}{s(s+140)(s+180)}, \tag{10.4}$$

where the numerical values have been rounded slightly.

An important parameter in the design process is the torque gain of the amplifier–motor–gear combination. This is the static torque delivered to the output shaft per volt at the amplifier input:

$$\begin{bmatrix} \text{static torque gain of} \\ \text{amplifier–motor–gears} \end{bmatrix} = \frac{T_y}{e}\bigg|_{\text{static}} = \frac{\eta n K K_A}{R+R_o} = 1.15 \text{ N-m}/\text{V}. \tag{10.5}$$

We combine this with the static gain of the control law to form the static torque gain of the forward path K_T:

$$\begin{bmatrix} \text{static torque gain} \\ \text{of the forward path} \end{bmatrix} = K_T = \frac{T_y}{\hat{\epsilon}}\bigg|_{\text{static}} = 1.15 \times G_C(0) \text{ N-m}/(\text{V error}). \tag{10.6}$$

We begin the design process by establishing a baseline system defined by the block diagram in Fig. 10.4, in which $\epsilon(s) = \mathcal{L}\{\hat{\epsilon}(t)\}$. We take the simplest form of the control law, $G_C(s) = K_C$, to determine how the transient response of the load depends on K_C. A root-locus plot is useful in this determination. Figure 10.5 shows the three loci for the poles of the closed-loop transfer function, with the root-locus gain $K_{RL} = 90{,}100 \times K_C$ as the parameter along the loci. The limit of K_{RL} for stability is 8.064×10^6, corresponding to a control

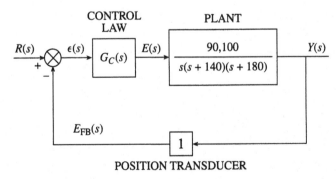

Fig. 10.4 Transfer function block diagram corresponding to schematic diagram of Fig. 10.2.

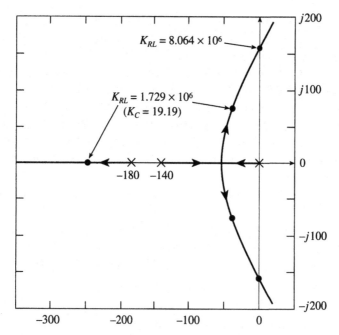

Fig. 10.5   Root-locus diagram for baseline system, $G_C(s) = K_C$, $K_{RL} = 90,100K_C$.

law gain $K_C = 89.5$. If K_C is set at 19.19, the gain margin will be 4.66 and the closed-loop poles will lie at -246.2 and $-36.92 \pm j75.24$, as indicated on the root-locus plot. The transient response $y(t)$ for a unit step input $r(t)$ is plotted in Fig. 10.6, where $K_C = 19.19$. This value of K_C is chosen so that the overshoot in $y(t)$ is exactly 20 percent, which occurs at $t = 0.0463$ s. The transient response is further characterized by its *rise time* t_R, which for our purpose is defined as the instant following the application of the step input at which $y(t)$ first reaches its final value. In this baseline case $t_R = 0.0315$ s and the static torque gain of the forward loop is $K_T = 22.07$.

We now have four significant design criteria on which to judge the performance of our servomechanism system:

1. percent overshoot in response to a unit step input;
2. rise time in response to a unit step input;
3. torque gain K_T; and
4. gain margin.

We can expect to improve the baseline system performance for these four criteria by increasing the complexity of our control law.

Suppose that the load shaft in our application is subjected to extraneous torque due to environmental conditions. This torque will displace the load shaft from its desired position. Assume that an assessment of the position error due to this disturbance torque shows that a new control law design providing a torque gain K_T of 240 N-m/(V error) will reduce the error to an acceptable level. The transient response and gain margin achieved in the new design should

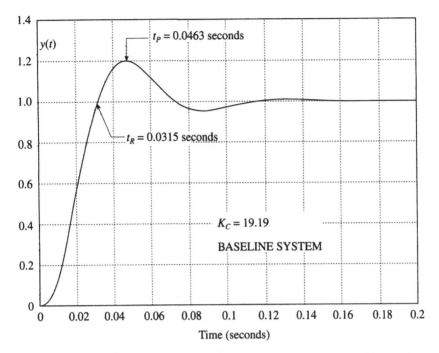

Fig. 10.6 Unit step response of baseline system, $K_C = 19.19$, with overshoot of 20 percent.

be approximately as favorable as those in the baseline system. Therefore, the new control law must have a static gain of at least 210, and it must incorporate some dynamics to maintain the stability of the system. It is prudent to try a control law transfer function with a single pole and a single zero, since this will increase the dynamic order of the system to 4, only one more than that of the baseline system:

$$G_C(s) = \frac{K_C(s+Z)}{s+P}. \tag{10.7}$$

The static gain of this control law is

$$G_C(0) = \frac{K_C Z}{P}, \tag{10.8}$$

which indicates that the required value of 210 can be achieved either by a large product $K_C Z$ or by a small value for P. The pole of $G_C(s)$ placed near the origin will produce two branches of the root locus close to the origin. These branches will migrate into the right-half s plane for very low values of K_C unless the zero is placed close enough to the origin to attract those two branches into the left-half plane, as shown in the unscaled sketch in Fig. 10.7. This first-order control law is called a *lag compensator* because $Z > P$. (The origin of this term is made clear in Chapter 11.) Many trial combinations of P, Z, and K_C can be tested using computer-aided analysis to calculate both the root-locus

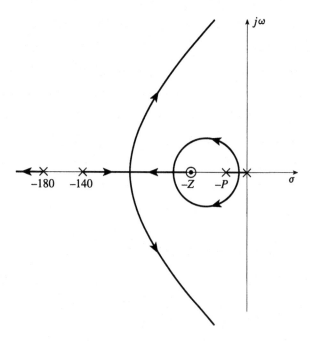

Fig. 10.7   Root-locus plot for system using lag compensation (not drawn to scale).

plot and the transient response $y(t)$. Several combinations are found to satisfy the design specifications. (In the general case the specifications may be so severe that *no* satisfactory combination could be found; a solution would then be sought using a higher-order $G_C(s)$.) A satisfactory solution in this case is

$$G_C(s) = \frac{13.5(s+5)}{s+0.2}.$$ (10.9)

Figure 10.8 shows a plot of the root locus in the region of the origin. The pole of $W(s)$ not on the plot is at $s = -234.1$. The transient response is shown in Fig. 10.9. The gain is adjusted to give an overshoot of 20 percent, which is the same as that in the baseline system. The other performance figures for this solution are:

$K_T = 388.1$ N-m/(V error);
gain margin $= 6.23$; and
$t_R = 0.0391$ s.

The torque constant achieved here, which exceeds the required level by 60 percent, is 17 times greater than that of the baseline system. The gain margin is also superior to that of the baseline system, although the response time is slightly slower in the compensated system. The pole of $Y(s)$ at $s = -5.58$ has a positive residue that contributes a relatively slow exponential term to $y(t)$, prolonging the convergence of $y(t)$ to its final value. The error at $t = 0.2$ s is about 5 percent of the final value, requiring another three quarters of a second to decay to

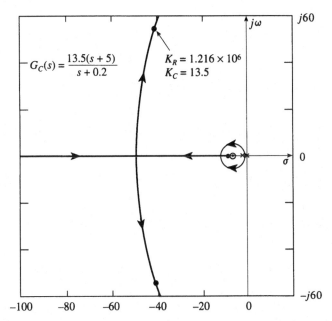

Fig. 10.8 Root-locus plot for system using lag compensation of Eqn. 10.9.

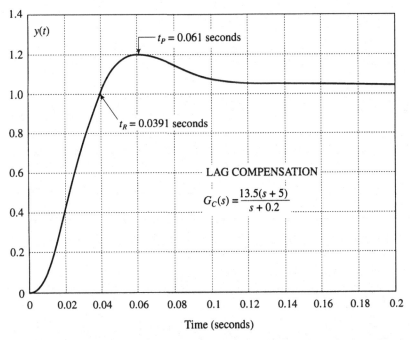

Fig. 10.9 Unit step response of lag-compensated system, with overshoot of 20 percent.

less than 0.09 percent. This "long-tailed" response could be unsatisfactory in some applications.

Suppose that another control law design is necessitated by a change in th speed of response specification, which now demands a rise time of no greate

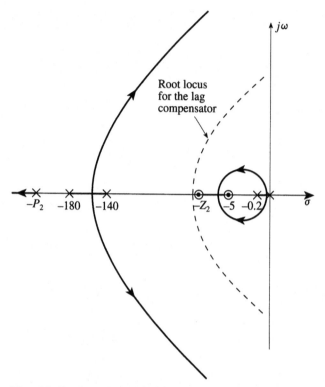

Fig. 10.10   Root-locus plot for lag-lead–compensated system compared to lag-compensated system (not drawn to scale).

than 0.015 s. The new control law must also provide a torque gain equal to or better than 240 N-m/V and an overshoot no greater than 20 percent. Because the lag compensator provided a remarkable increase in the torque gain as compared to the baseline system, we wish to retain that feature in the design of the new control law. We attempt to supplement that feature with a second pole-zero combination that will give a speed of response equal to or better than the new requirement of 0.015 s.

The unscaled root-locus sketch in Fig. 10.10 indicates that a zero properly placed between the compensator zero at $s = -5$ and the plant pole at $s = -140$ will establish a section of the root locus on the real axis between the two zeros, attracting both branches of the locus that emanate from the poles near the origin. These branches will terminate on the two zeros. The branches that lead into the right-half s plane will come from the plant poles at $s = -140$ and $s = -180$. The poles of $W(s)$ lying on these complex branches produce the oscillatory component in $y(t)$, which here has a higher frequency than that of the baseline system. A second pole is included in the control law, far to the left of the plant pole at $s = -180$, to alleviate the transmission of rapidly fluctuating noise through the compensator. A series of computer-aided trial-and-error analyses, based on the root-locus sketch of Fig. 10.10, results in a satisfactory design:

$$G_C(s) = \frac{340(s+5)(s+50)}{(s+0.2)(s+500)}. \tag{10.10}$$

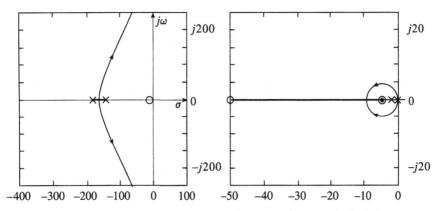

Fig. 10.11 Root-locus plot for system using lag-lead compensation of Eqn. 10.10.

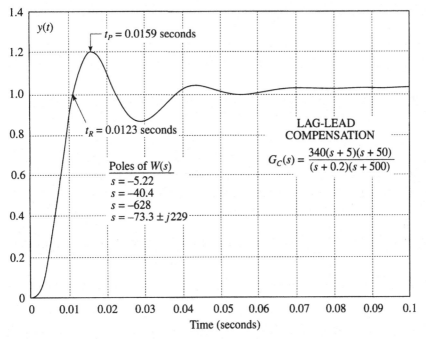

Unit step response figure content:

$t_P = 0.0159$ seconds

$t_R = 0.0123$ seconds

LAG-LEAD
COMPENSATION

$$G_C(s) = \frac{340(s + 5)(s + 50)}{(s + 0.2)(s + 500)}$$

Poles of $W(s)$

$s = -5.22$
$s = -40.4$
$s = -628$
$s = -73.3 \pm j229$

Fig. 10.12 Unit step response of lag-lead–compensated system with overshoot of 20 percent.

A properly scaled root-locus plot for this system is drawn in Fig. 10.11. The left shows a large region of the s plane, and the right side provides details of the plot near the origin. K_C is adjusted by trial and error so that $y(t)$ has exactly 20 percent overshoot. This gives K_C a value of 340, and the resultant $y(t)$ is plotted in Fig. 10.12. The other performance figures for this control law, which is called a *lag-lead* compensator, are:

$K_T = 977.5$ N-m/(V error);
gain margin $= 3.51$; and
$t_R = 0.0123$ s.

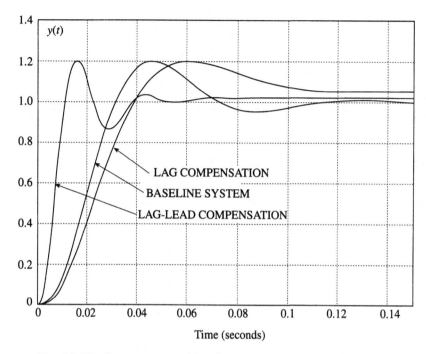

Fig. 10.13   Step responses of baseline system, lag-compensated system, and lag-lead–compensated system compared.

The gain margin for this solution is less than that for the previous two cases, but the speed of response and the torque gain exceed the design specifications. The transient responses for the three cases are superimposed in Fig. 10.13. Note that the long-tailed response in the lag-lead case is noticeably improved over that in the lag case.

The response functions shown in Fig. 10.13 are valid only if the amplifier-motor operates in its linear range. The current output of the amplifier is usually

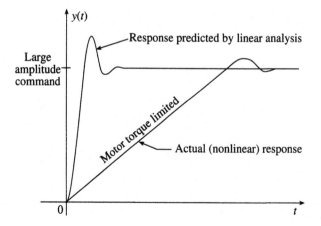

Fig. 10.14   Response of torque-limited system compared to response of linear system.

limited in order to prevent damage to the amplifier and to the motor. A limit of about 5 amperes is typical for the amplifier–motor combination in this example. A practical performance measure of the system is the *linear range* of the servomechanism, defined here as that level of error signal which produces 5 amperes of current in the motor when the motor is at a standstill. When the error exceeds the linear range, the acceleration and velocity capabilities of the motor load are restricted by the current limit. For large inputs, the actual response of the system can be very different from the linear response. The comparison between the two, shown in Fig. 10.14, is typical of a high-performance servomechanism compensated by a lag-lead control law.

The following table summarizes the servomechanism performance for the three control law designs. In each design the overshoot of $y(t)$ in response to a step input is 20 percent of the input command.

Case	$G_C(s)$	K_T	t_R	Gain Margin	Linear Range
Baseline	19.19	22.07	0.0315	4.33	0.208 V
Lag-compensated	$\dfrac{13.5(s+5)}{s+0.2}$	388.1	0.0391	6.23	0.0119 V
Lag-lead–compensated	$\dfrac{340(s+5)(s+50)}{(s+0.2)(s+500)}$	977.5	0.0123	3.51	0.00471 V

The torque gain in both compensated cases can be increased indefinitely simply by shifting the pole at $s = -0.2$ to the origin. This very small change in the root-locus configuration affects the other performance figures only slightly. The compensator then becomes an integrator, permitting it to maintain a nonzero output in the steady state even though the input to the integrator (the system error) is zero. If an integrating compensator is used, the definition of "linear range" should be based upon the amplifier input voltage, not upon the error.

10.3 The PID Controller

Figure 10.15 shows a block diagram of the *proportional-integral-derivative* (PID) device used in the error channel of feedback control systems to provide a control law of the form

$$\frac{Q(s)}{\epsilon(s)} = K_P + \frac{K_I}{s} + K_D s. \tag{10.11}$$

Because the three gains K_P, K_I, and K_D are each adjustable, the available control laws span a significant range. If all three gains are nonzero then the transfer function has a pole at the origin and two finite zeros:

$$\frac{Q(s)}{\epsilon(s)} = \frac{K_D s^2 + K_P s + K_I}{s} = \frac{K(s+Z_1)(s+Z_2)}{s}, \quad \text{where}$$

$$K = K_D, \quad Z_1 + Z_2 = \frac{K_P}{K_D}, \quad \text{and} \quad Z_1 Z_2 = \frac{K_I}{K_D}. \tag{10.12}$$

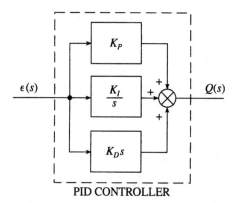

Fig. 10.15 Block diagram for a PID controller.

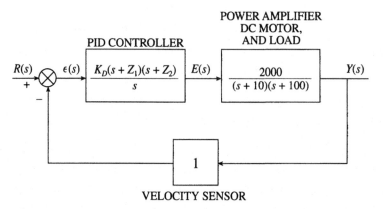

Fig. 10.16 PID controller for a DC motor speed control system.

The two zeros may be real-valued, but if $K_P < 2\sqrt{K_I K_D}$ they will be complex conjugates.

PID controllers are used extensively in industrial processes. A simple example is the DC motor speed control system analyzed in Section 9.4. Figure 10.16 shows a PID controller inserted into the control loop of this system. Figure 10.17 shows two error responses of this system for the two cases discussed in Section 9.4. The control law used in case 1 is a simple proportional controller, and in case 2 the control law is a pure integral controller. The input in this example is $r(t) = E\sigma(t) = 1\sigma(t)$ V. Note that the steady-state error in case 2 is zero, but the response time is about ten times longer than the response in case 1. Root-locus plots for these two cases show clearly why this occurs. With proportional control the system is of second order. The gain $K_P = 2.6$ puts the closed-loop poles at $-55 \pm j56.3$, which corresponds to $\omega_n = 78.7$ rad/s and $\zeta = 0.7$. With integral control, the system is of third order. $K_I = 3$ puts the three closed-loop poles at -100.7 and $-4.67 \pm j6.15$. Because the pole at -100.7 is remote from the complex pair, it has no influence on the settling time. The complex pair corresponds to $\omega_n = 7.72$ rad/s and $\zeta = 0.6$, and since the damping

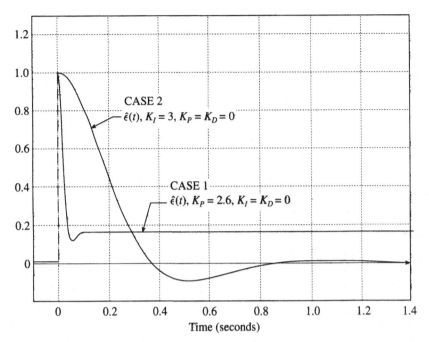

Fig. 10.17 Error response of system in Fig. 10.16 to unit step input: case 1, proportional control only; case 2, integral control only.

ratios of the two systems are comparable, the discrepancies in their response times is closely reflected in the relative values of ω_n.

The PID control law can provide both a fast response and a zero steady-state error in this speed control system if nonzero values for the three gains K_P, K_I, and K_D are invoked. The presence of the integrating property of the controller is assured if K_I is nonzero. Equation 10.12 provides a useful guide to finding satisfactory values for the PID gains, as it relates the three gains to the locations of the two compensating zeros. For example, if these two zeros are chosen to lie at $s = -30 + j5$ and $s = -30 - j5$, the root loci for the closed-loop poles that emanate from the open-loop poles at $s = 0$ and $s = -10$ will be attracted toward these zeros. A suitable choice of loop gain will locate the poles so that they are well-damped, with an ω_n significantly greater than 7.72. In this case the three gain values

$$K_P = 12, \quad K_I = 185, \quad \text{and} \quad K_D = 0.2$$

will yield closed-loop poles at $s = -457$ and $s = -26.5 \pm j10.5$. The complex poles correspond to $\omega_n = 28.5$ and $\zeta = 0.93$. Figure 10.18 compares the error response of the system having these PID gain settings to the response of the original system having simple proportional control.

The process of finding satisfactory settings for the PID gains is often called *tuning*. Because there are three gains, it is necessary to have a systematic procedure for tuning that will converge to a solution. If the dynamics of the controlled process are well identified, as in this example, the two zeros of the controller

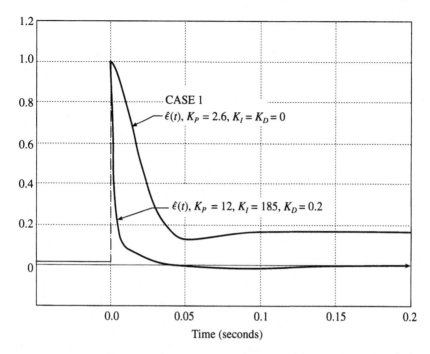

Fig. 10.18 Case 1 of Fig. 10.17 compared to response using full PID control (note difference in time scales).

transfer function may be placed by the designer to achieve a satisfactory root-locus geometry, and the appropriate loop gain may also be realized from the formulas in Eqn. 10.12. If the controlled process is not well identified, an attempt should be made to identify at least the dominant parameters of the process in order to form an approximate transfer function before proceeding with the zero placement and root-locus technique.

10.4 Pole-Zero Cancellation in Series Compensation

In designing a series compensator, the engineer is often tempted simply to place zeros in the compensator where the plant has troublesome poles and to place poles in the compensator where the plant has troublesome zeros, supplementing this pole-zero constellation with a transfer function that will give the desired closed-loop performance. Figure 10.19 illustrates an extreme example of this approach, where the entire plant transfer function is canceled in the compensator by the factor $1/G_P(s)$ and supplemented with a suitable $G(s)$. The closed-loop transfer function is

$$W(s) = \frac{G(s)\left[\dfrac{1}{G_P(s)} \cdot G_P(s)\right]}{1 + G(s)\left[\dfrac{1}{G_P(s)} \cdot G_P(s)\right]} = \frac{G(s)}{1 + G(s)}. \tag{10.13}$$

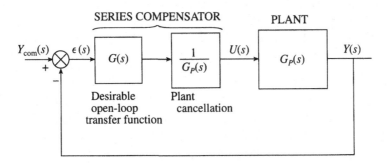

Fig. 10.19 Extreme case of series compensation.

An acceptable $W(s)$ can be formulated from the design specifications, and $G(s)$ can be obtained by inverting Eqn. 10.13, as indicated in Eqn. 8.8:

$$G(s) = \frac{W_{\text{acceptable}}(s)}{1 - W_{\text{acceptable}}(s)}. \tag{10.14}$$

The control law design can be written directly in terms of the plant transfer function and the design specifications, which are represented by $W_{\text{acceptable}}(s)$:

$$\begin{bmatrix} \text{control law} \\ \text{transfer function} \end{bmatrix} = \frac{U(s)}{\epsilon(s)} = \frac{W_{\text{acceptable}}(s)}{1 - W_{\text{acceptable}}(s)} \cdot \frac{1}{G_P(s)}. \tag{10.15}$$

This formula, which is very attractive from a purely mathematical point of view, is often used by novice engineers because it appears to reduce the design procedure to an exercise in algebra that immediately produces an acceptable control law. In practice, the formula usually fails to produce a workable control law for the following reasons.

1. The transfer function $W_{\text{acceptable}}(s)$ pertains only to the zero-state response of the system, $Y_{\text{ZSR}}(s)$, whereas the total response includes components due to initial conditions and disturbance inputs: $Y(s) = Y_{\text{ZSR}}(s) + Y_{\text{ZIR}}(s) + Y_{\text{dist}}(s)$. The control law design must usually satisfy performance specifications on all three components of the response. Both $Y_{\text{ZIR}}(s)$ and $Y_{\text{dist}}(s)$ contain the undesirable modes of the plant, which are canceled in $Y_{\text{ZSR}}(s)$ by the control law. When the plant is temporarily driven into its nonlinear range by a large disturbance or by an excessive control input, these modes will appear in $y(t)$ as the plant resumes linear operation, and the cancellation technique will fail to achieve the result intended by the designer.

2. Usually the values of the physical parameters of the plant are not precisely known. Therefore, the poles and zeros of $G_P(s)$ cannot be precisely determined, and the pole-zero cancellation illustrated in Fig. 10.19 cannot be achieved. The resulting open-loop transfer function will not be the desirable $G(s)$, nor will the closed-loop transfer function be $W_{\text{acceptable}}(s)$. If the range of uncertainties in the physical parameters can be estimated, the control law transfer function can be designed to minimize the deviations in $W(s)$ induced by the uncertainties in $G_P(s)$, thereby providing a degree of robustness to the system.

3. An important instance of item 2 occurs when the plant transfer function is nonminimum phase. If an imperfect attempt is made to cancel either a plant pole or a plant zero in the right-half plane, a short branch of the root locus thereby created in the right-half plane will contribute a right-half–plane pole to $W(s)$.

4. A control law determined by Eqn. 10.15 will probably be more complicated than necessary. The control law should be as simple as possible to minimize engineering difficulties in manufacturing and testing. If the controller is to be a digital computer, an excessively complex control law design will require extravagant software design and testing. This design defect can be equally serious in a continuous time controller such as an electronic network.

5. Usually poles closest to the right-half s plane and lying near the origin in the plant transfer function are the most troublesome ones in feedback control systems. Control law design based strictly on pole-zero cancellation will yield a compensator having zeros near the origin. Such zeros contribute terms of the form $\dot{\epsilon}(t), \ddot{\epsilon}(t), \ldots$ to the output signal of the compensator, $u(t)$. The error signal $\hat{\epsilon}(t)$ frequently has an unwanted component of rapidly fluctuating noise. This undesirable component will be accentuated in the controller. Thus, the "signal to noise ratio" of $u(t)$ is degraded as compared to that of the input $\hat{\epsilon}(t)$, and in some cases the noise will prevent the plant from responding appropriately to the signal component of $u(t)$.

6. Even if the signal-to-noise ratio at $u(t)$ is tolerable, a rapid fluctuation in the signal component of $\hat{\epsilon}(t)$ may cause the signal component of $u(t)$ to increase to a level that drives the plant into its nonlinear operating range. In this case the linear analysis will fail to describe the dynamic response of the plant. The trajectory of $y(t)$ during such transients may resemble that of a linear plant out of control.

Notwithstanding these six precautions against it, pole-zero cancellation is useful in two respects. First, it provides a guide that may help to establish the first trial design for the control law. Second, pole-zero cancellation of some of the plant singularities is permissible in cases where the listed precautions do not apply. For example, if a range of uncertainty of plant pole and zero locations is known, the compensating poles and zeros can be placed to enhance the robustness of the design.

10.5　Parallel Compensation

It is often advantageous to employ a dynamic compensator in the feedback path, as indicated in Fig. 10.20. The influence that $H(s)$ has upon the closed-loop system dynamics may be determined by the root-locus principles set forth in Section 9.5 for the non-unity feedback system. The closed-loop transfer function for this system is

$$\frac{Y(s)}{R(s)} = W(s) = \frac{G(s)}{1+G(s)H(s)} = \frac{N_1(s)D_2(s)}{D_1(s)D_2(s)+N_1(s)N_2(s)}, \tag{10.16}$$

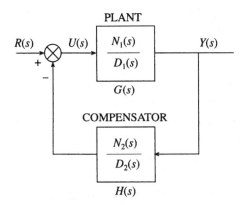

Fig. 10.20 Compensation in feedback path.

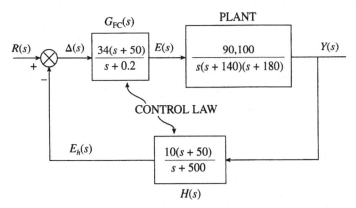

Fig. 10.21 Lag-lead compensation split between forward path and feedback path.

which shows that the product of the two transfer functions $G(s)H(s)$ has the same role in determining the poles of $W(s)$ as the single transfer function $G(s)$ has in the unity feedback case. The zeros of $W(s)$ are the zeros of $G(s)$ and the poles of $H(s)$.

In designing the control law, it is possible to use the feedback path element as part of the compensator. A modification of the lag-lead–compensated system of Section 10.2 provides an interesting example of using both the feedback and feedforward paths for compensation. Figure 10.21 shows half the compensating transfer function as the feedback element $H(s)$ and the other half, $G_{FC}(s)$, in the forward path. Note that the steady-state gain of the feedback element is chosen to be 1 so that the steady-state value of the feedback signal $e_h(t)$ will be equal to the plant output. Also note that $\hat{\Delta}(t)$, the input to the forward-path compensating element, is the system error only when the system is at rest; the dynamic system error, $\hat{\varepsilon}(t) = r(t) - y(t)$, is not available as a physical variable in this system as it is in the system of Fig. 10.4.

We now compare the performance of the system in Fig. 10.21 to that of the system in Fig. 10.4, where $G_C(s) = G_{FC}(s) H(s)$. The closed-loop transfer functions for the two cases are

$$\text{Fig. 10.4:} \qquad W(s) = \frac{(340)(90,100)(s+5)(s+50)}{[\text{denominator}]},$$

$$\text{Fig. 10.21:} \qquad W(s) = \frac{(34)(90,100)(s+5)(s+500)}{[\text{denominator}]}, \qquad \text{where} \qquad (10.17)$$

$$[\text{denominator}] = (s+628)(s+40.43)(s+5.22)[s^2+147s+57,754].$$

Both systems have the same gain margin and the same torque gain in the forward path. The two system transfer functions have the same poles and the same static gain (1), and each has a zero at $s = -5$. The significant difference in the two transfer functions is the location of the second zero, which lies at $s = -50$ in the Fig. 10.4 system and at $s = -500$ in the Fig. 10.21 system. This difference is reflected in the step responses of the two systems (see Fig. 10.22). In the Fig. 10.4 system, the proximity of the zero at $s = -50$ to the pole at $s = -40.43$ causes the residue in $Y(s)$ at that pole to be small, about one fifth of the residue in the Fig. 10.21 system where no zero lies in the vicinity of the pole at $s = -40.43$. The physical reason for the difference in the two responses is that $H(s)$ is a *lead*-type transfer function. Hence rapid fluctuations in $y(t)$ are

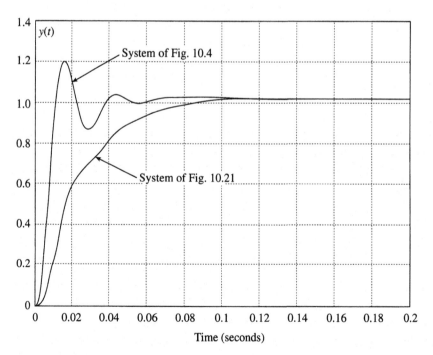

Fig. 10.22 Step response of system using lag-lead compensation in forward path only, compared to system using the same compensation shared between forward and feedback paths.

accentuated in the feedback signal $e_h(t)$, which in turn inhibits (damps) the driving signal to the plant.

Parallel compensation may also be used to alter the dynamic character of the plant when it is not possible to make physical modifications in the plant. An example of this situation is found in the flight-control system of an aircraft.

Consider the longitudinal motion of an airplane that is perturbed slightly from straight and level flight by small displacements of the elevator control surface, $\Delta\hat{\delta}_E(t)$. The equations of motion for small deviations in the pitch attitude of the aircraft $\Delta\hat{\theta}(t)$ are linear differential equations, as derived in Section 2.7. The transfer function relating $\Delta\theta(s)$ to $\Delta\delta_E(s)$ can be derived by taking the Laplace transform of the state-variable model in Eqn. 2.88. A Navion aircraft in cruising flight at low altitude has a transfer function

$$\frac{\Delta\theta(s)}{\Delta\delta_E(s)} = \frac{-11.04(s+1.92)(s+0.052)}{\underbrace{[s^2+0.0356s+0.0456]}_{\text{phugoid}}\underbrace{[s^2+5.02s+13.02]}_{\text{short-period}}}, \quad \text{where}$$

$$\Delta\theta(s) = \mathcal{L}\{\Delta\hat{\theta}(t)\} \text{ and } \Delta\delta_E(s) = \mathcal{L}\{\Delta\hat{\delta}_E(t)\}.$$

(10.18)

The negative sign comes from the sense convention for $\Delta\hat{\theta}(t)$ and $\Delta\hat{\delta}_E(t)$, defined in Fig. 2.14, which shows a positive elevator deflection as being "down" and a positive pitch angle as being "up". For automatic flight control, the force required to move the elevator is provided by an actuator, as shown in Fig. 10.23. The actuator is a servomechanism with a fast response time compared to that of the Navion airframe. The transfer function shown is adequate to represent the influence of the actuator dynamics on the overall system. The input to the actuator is the signal from the controller, $\Delta\hat{\delta}_{E\text{com}}(t)$, which commands the elevator angle.

We examine the transient response of the actuator–airframe system by applying a pulse command to the input of the actuator, as shown in the upper graph in Fig. 10.24. The lower graph shows that the actuator response to this command is approximately a pulse. The airframe response to this elevator input, $\Delta\hat{\theta}(t)$, is plotted in Fig. 10.25. $\Delta\hat{\theta}(t)$ is characterized by two underdamped modes, as shown by the two quadratic factors in the denominator of the transfer function. One of these, the *short-period mode,* represents the well-damped oscillation experienced by the airframe in response to the upward pulse of force at the rear of the airframe due to the momentary deflection of the elevator.

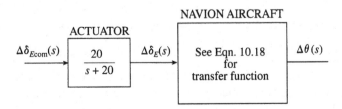

Fig. 10.23 Longitudinal axis control of Navion aircraft, elevator only.

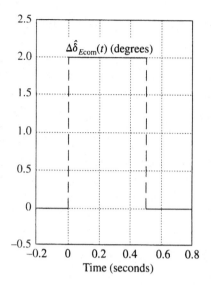

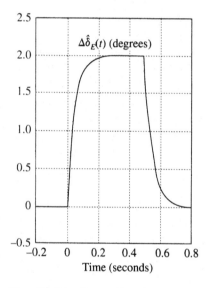

Fig. 10.24 Comparing elevator command signal to elevator motion.

This angular motion is quickly damped by the aerodynamic moments on the airframe, which are essentially opposed to the angular velocity of the airframe. The upper graph in Fig. 10.25, which shows only the first 4 seconds of $\Delta\hat{\theta}(t)$, illustrates this short-lived oscillation. The very lightly damped second mode in $\Delta\hat{\theta}(t)$, the *phugoid mode,* has a comparatively long period, in this case approximately half a minute, as shown in the lower graph in Fig. 10.25. The phugoid mode reflects the vertical motion of the center of mass of the airframe initiated by the upsetting pulse of force at the elevator and reinforced by the change in lift of the wing that accompanies the change in $\Delta\hat{\theta}(t)$. Both the altitude and the airspeed of the aircraft change by small amounts in the same slow oscillatory manner, the airspeed diminishing as the altitude increases, and vice versa. Both

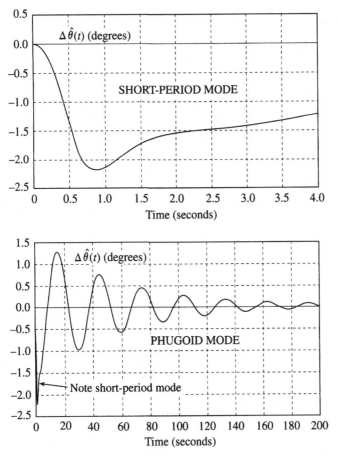

Fig. 10.25 Response of aircraft pitch angle to pulse of elevator command (note difference in time scales between upper and lower graphs).

of these longitudinal modes are characteristic of all aircraft. They may easily be observed in most modern airplanes when the atmosphere is free of turbulence.

In high-performance aircraft, the pronounced phugoid mode presents the autopilot designer with a problem that would not exist if the mode were heavily damped. The analytical reason for the problem is the location of the phugoid poles close to the $j\omega$ axis. Increasing the phugoid damping by physical alterations to the airframe is not an option, since this would destroy the aerodynamic performance of the machine. However, the location of the poles can be altered by installing a feedback loop, as shown in Fig. 10.26. Figure 10.27 is an s-plane plot showing the poles and zeros of $\Delta\theta(s)/\Delta\delta_{E\text{com}}(s)$ and the root loci for constant gains in the control law and feedback path:

$$\begin{bmatrix} \text{control law} \\ \text{transfer function} \end{bmatrix} = K_1, \qquad \begin{bmatrix} \text{attitude sensor} \\ \text{transfer function} \end{bmatrix} = K_2. \qquad (10.19)$$

The s-plane plot is drawn out of scale in order to include all features of the root loci. The product $K_1 K_2$ is made negative so that the flag constant of the loop transfer function will be positive:

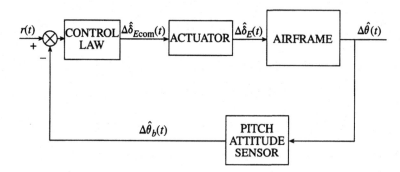

Fig. 10.26 Pitch attitude feedback to damp the phugoid mode.

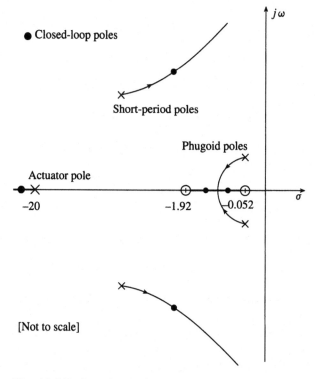

Fig. 10.27 Root-locus plot for constant-gain feedback and constant-gain control law.

$$\begin{bmatrix} \text{loop transfer} \\ \text{function} \end{bmatrix} = \frac{K_1 K_2(-220.8)(s+1.92)(s+0.052)}{\underbrace{(s+20)}_{\text{actuator}}\underbrace{[s^2+0.0356s+0.0456]}_{\text{phugoid}}\underbrace{[s^2+5.02s+13.02]}_{\text{short-period}}}.$$

(10.20)

The two phugoid poles are drawn to the branch of the locus on the real axis between the two zeros, thereby creating an overdamped pair in the closed-loop transfer function. If the product $K_1 K_2$ is set equal to -0.32 and adjusted so that the steady-state gain of the closed-loop system is the same as the steady-state gain of the actuator–airframe combination, namely,

$$K_1 = 1.6 \quad \text{and} \quad K_2 = -0.2, \tag{10.21}$$

then the closed-loop transfer function will be

$$\frac{\Delta\theta(s)}{R(s)} = \underbrace{\frac{-353.3(s+1.92)(s+0.052)}{(s+20.2)}}_{\substack{\text{shifted} \\ \text{actuator}}} \underbrace{(s^2+4.33s+13.94)}_{\text{shifted short-period}} \underbrace{(s+0.29)(s+0.23)}_{\text{damped phugoid}}. \tag{10.22}$$

To compare the response of the closed-loop system to that of the open-loop system, consider $r(t)$ to be a pulse identical to the pulse in the upper graph of Fig. 10.24. The resulting aircraft pitch attitude response is plotted in Fig. 10.28, where the upper graph shows the *shifted short-period mode* and the lower graph shows the overdamped *shifted phugoid mode,* which dies out in one tenth the time of the original phugoid mode shown in Fig. 10.25.

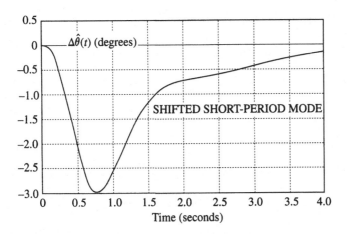

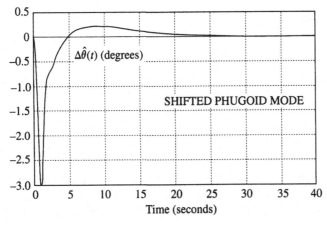

Fig. 10.28 Response of aircraft pitch angle to pulse $r(t)$; compare with Fig. 10.25.

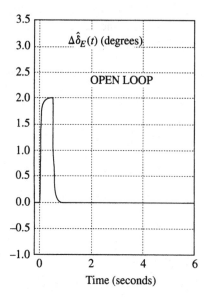

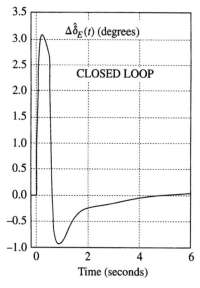

Fig. 10.29 Comparing elevator response in the undamped (open-loop) case to elevator response in the closed-loop case.

This remarkable improvement in phugoid damping is due to the transient motion of the elevator, now driven by $\Delta\hat{\delta}_{E\text{com}}(t)$ and not by $r(t)$. The elevator motion provides the damping forces required to counter the "natural" forces that drove the poorly damped phugoid oscillations when the feedback connection was not employed. The upper graph in Fig. 10.29 compares the elevator movement that occurs in response to the pulse excitation in the open-loop case to the corresponding elevator motion provided by the feedback connection, as shown in the lower graph. A human pilot could perform this same function manually by manipulating the elevators.

The closed-loop system can now be incorporated as a component in a larger automatic flight-control system, such as an altitude control system, with $r(t)$ acting as the elevator input to an airframe having a well-damped phugoid mode.

10.6 Tracking Control

An automatic control system normally operates in one of two fashions:

1. The system is designed to regulate the output $y(t)$ at a fixed level prescribed by a constant input $r(t) = R$ (called the *set point*) in the face of environmental disturbances to the plant.
2. The system is designed to drive the output $y(t)$ to track a changing input $r(t)$ as closely as possible. Such a system is called a *follow-up system* or a *tracking control system*.

A typical tracking command function $r(t)$ is shown in Fig. 10.30. We must somehow relate the performance of our closed-loop system, which up to this point has been expressed primarily in terms of its response to a step function input, to a family of tracking functions that the system will be required to follow in practice. The significance of the step function response as a useful measure of performance for the tracking task is illustrated in Fig. 10.31. Here $r(t)$ is approximated by $\hat{r}_0(t)$, a "staircase" function having discontinuities at instants chosen to maintain an acceptable difference $|r(t) - \hat{r}_0(t)|$ at all times. If the system is driven by $\hat{r}_0(t)$, and if it has a very prompt step response with modest overshoot and zero steady-state error, then the system output $y(t)$ will resemble that sketched in Fig. 10.31. When the system is driven by the less severe command $r(t)$, the response will presumably follow the input command closely enough at all times.

In Section 9.4 it was shown that if the steady-state error in response to a step input is to be zero, an integrating element in the forward path of a unity feedback system is necessary. For example, if the forward-path transfer function is

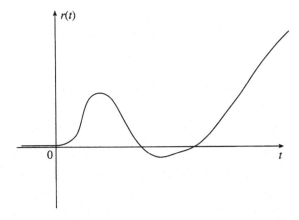

Fig. 10.30 Typical input function $r(t)$ to be tracked.

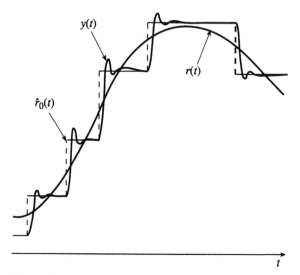

Fig. 10.31 $r(t)$ approximated by a "staircase" function $\hat{r}_0(t)$ and $y(t)$ tracking $\hat{r}_0(t)$.

$$G(s) = \frac{K}{s(s+P)} \tag{10.23}$$

and the input is a step function of A units,

$$r(t) = A\sigma(t) \;\Rightarrow\; R(s) = \frac{A}{s}, \tag{10.24}$$

then the Laplace transform of the error is

$$\epsilon(s) = R(s) \cdot \frac{\epsilon(s)}{R(s)} = \frac{A}{s} \cdot \frac{s(s+P)}{[s^2+Ps+K]}. \tag{10.25}$$

The steady-state value of the error can be found using the final-value theorem:

$$\hat{\epsilon}(t)|_{SS} = \lim_{t \to \infty} \hat{\epsilon}(t) = \lim_{s \to 0} [s\epsilon(s)] = 0, \tag{10.26}$$

provided the system is stable. Since the system will be stable if both P and K are positive, we see that the presence of the integrating element, indicated by the zero at $s = 0$ in the transfer function $\epsilon(s)/R(s)$, is a necessary condition for the steady-state error to be zero.

Figure 10.32 shows an improvement in the approximation to $r(t)$ over the staircase approximation of Fig. 10.31. Here the continuous approximation function $\hat{r}_1(t)$ has discontinuous first derivatives chosen at instants that will maintain an acceptable difference $|r(t) - \hat{r}_1(t)|$. Each segment of $\hat{r}_1(t)$ has a constant slope reminiscent of a *ramp* function. If the control system can be designed so that its steady-state error in response to a ramp function input is zero, then the response of the system to $\hat{r}_1(t)$ could be expected to resemble that shown in Fig. 10.33. Since $\hat{r}_1(t)$ can be a much better approximation to $r(t)$ than is $\hat{r}_0(t)$,

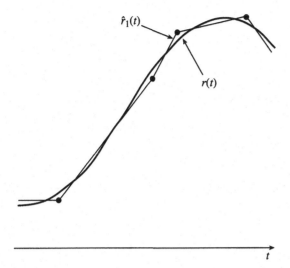

Fig. 10.32 $r(t)$ approximated by $\hat{r}_1(t)$.

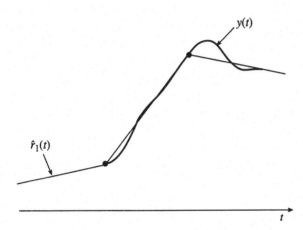

Fig. 10.33 $y(t)$ tracking $\hat{r}_1(t)$.

the response in Fig. 10.33 is superior to the response in Fig. 10.31. We therefore seek a design for the controller that will yield a zero steady-state error in response to a ramp function input,

$$r(t) = At \;\Rightarrow\; R(s) = \frac{A}{s^2}. \tag{10.27}$$

If we use the $G(s)$ given in Eqn. 10.23, the steady-state error will not be zero, as shown by the final-value theorem:

$$\epsilon(s) = R(s) \cdot \frac{\epsilon(s)}{R(s)} = \frac{A}{s^2} \cdot \frac{s(s+P)}{[s^2+Ps+K]} \;\Rightarrow\; \lim_{t \to \infty} \hat{\epsilon}(t) = \lim_{s \to 0}[s\epsilon(s)] = A \cdot \frac{P}{K}. \tag{10.28}$$

If a compensator with an integrator is installed in the error channel of this control system, so that the forward-path transfer function becomes

$$G(s) = \frac{K}{s^2(s+P)},$$ (10.29)

and a ramp function is applied at the input, then the Laplace transform of the error will be

$$\epsilon(s) = \frac{A}{s^2} \cdot \frac{s^2(s+P)}{[s^3+Ps^2+K]}.$$ (10.30)

An attempt to apply the final-value theorem to $\epsilon(s)$ to calculate the steady-state error reveals that

$$\hat{\epsilon}(t)|_{SS} = \lim_{s \to 0} \left\{ s \cdot \frac{A}{s^2} \cdot \frac{s^2(s+P)}{[s^3+Ps^2+K]} \right\} = 0,$$ (10.31)

provided the system is stable.

Therefore, if the steady-state error in response to a ramp input is to be zero, two conditions must be met: (a) the forward-path elements of the unity feedback system must have a *double* integrating property, so that the transfer function $\epsilon(s)/R(s)$ will have two zeros at the origin; and (b) the system must be stable. In the case of Eqn. 10.31 the system is not stable, so the error does not converge to a final value. If the compensating element is augmented to include a zero in its transfer function, as indicated in Fig. 10.34, the system can be stabilized. The closed-loop transfer function relating the input to the error then becomes

$$\frac{\epsilon(s)}{R(s)} = \frac{s^2(s+P)}{s^3+Ps^2+KK_Cs+KK_CZ},$$ (10.32)

and the steady-state error will be zero if the system is stable. Routh's criterion shows the required conditions for stability:

$$KK_C > 0, \quad Z > 0, \quad \text{and} \quad P > Z.$$ (10.33)

The tracking response depicted in Fig. 10.33 can be realized by suitable adjustment of Z and K_C.

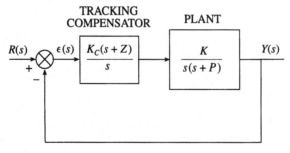

Fig. 10.34 Double-integrator control loop compensated to stabilize feedback system.

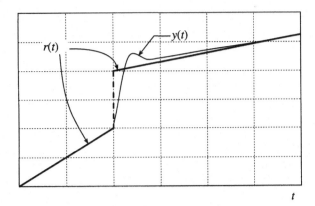

Fig. 10.35 $y(t)$ tracking a discontinuous $r(t)$.

It is interesting to calculate the error when the input is the sum of a step and a ramp:

$$r(t) = A\sigma(t) + Bt \implies R(s) = \frac{As+B}{s^2}. \tag{10.34}$$

For the system shown in Fig. 10.34, the Laplace transform of the error is

$$\epsilon(s) = \frac{As+B}{s} \cdot \frac{s^2(s+P)}{[s^3 + Ps^2 + KK_C s + KK_C Z]}. \tag{10.35}$$

The final-value theorem shows that the steady-state error is zero (provided the system is stable). Therefore, if the step response of the system is sufficiently prompt and without excessive overshoot, it will be able to track (approximately) an input consisting of a series of ramplike segments with discontinuities. Figure 10.35 shows the transient in $y(t)$ that occurs at the junction of two segments of such a command input.

We extend this analysis by considering an input function $r(t) = Ct^2$, which can be termed a *parabolic* input. Its Laplace transform is

$$R(s) = \frac{2C}{s^3}. \tag{10.36}$$

Now, if the unity feedback system has a *triple* integration in the forward path, and the closed-loop system is stable, then the steady-state error will be zero. The output will track the input following the initial transient until the acceleration capability of the plant is reached and the system no longer behaves linearly. Therefore, it follows that the steady-state error in this system will be zero for the composite input

$$r(t) = \underbrace{A\sigma(t)}_{\text{step}} + \underbrace{Bt}_{\text{ramp}} + \underbrace{Ct^2}_{\text{parabolic}}, \tag{10.37}$$

provided the system continues to operate in its linear range.

The pattern established by noting the influence on the steady-state error of one, two, and three integration terms in $G(s)$ can be extended to k integrations and correspondingly to k terms in the power series of $r(t)$. Consider the family of input functions represented by the example function in Fig. 10.30. If this family can be represented by a power series of k terms, the output of the unity feedback system will accurately track the input, except for the transient periods following a discontinuity in $r(t)$, *provided* that:

    (a) $G(s)$ has k (or more) poles at $s = 0$;
    (b) the closed-loop system is stable; and
    (c) the system operates in its linear range at all times.

Because of the stability problem introduced when integrating elements are added in the error channel, these conditions can be met in most practical systems only if k is less than 3 or 4. If $k = 1$ the system is sometimes called a type-1 system; for $k = 2$ it is a type-2 system, and so forth. These designations are useful only for unity feedback single-loop systems. If the system has a dynamic feedback element, or if the integrating elements are not in the forward path of the system, or if multiple feedback loops in the system contain some integrating elements, then the dynamic features of the system pertinent to the tracking problem cannot be summarized by a single "type" number.

An alternate way to analyze the tracking problem is to expand $Y(s)$ in partial fractions and to determine the pole-zero configuration which $Y(s)$ must have so that $y(t)$ will match $r(t)$ following the start-up transient. As an example, consider an nth-order system whose closed-loop transfer function is

$$\frac{Y(s)}{R(s)} = W(s) = \frac{K(s+Z_1)\cdots(s+Z_m)}{(s+Q_1)\cdots(s+Q_n)} \tag{10.38}$$

and whose input is a ramp function $r(t) = Bt$. Then $Y(s)$ is

$$Y(s) = B \cdot \left\{ \frac{K(s+Z_1)\cdots(s+Z_m)}{s^2(s+Q_1)\cdots(s+Q_n)} \right\}, \tag{10.39}$$

which can be expanded in partial fractions as follows, provided $m < n+2$:

$$Y(s) = B \cdot \left[\frac{K_0}{s} + \frac{C_2}{s^2} + \frac{K_1}{s+Q_1} + \cdots + \frac{K_n}{s+Q_n} \right]. \tag{10.40}$$

The inverse transform of $Y(s)$ is

$$y(t) = B \cdot [K_0 + C_2 t + \{\text{transient terms}\}]. \tag{10.41}$$

If all the poles of $W(s)$ lie in the left-half plane, the transient terms die out as time increases, and the steady-state output is

$$y(t)|_{t \to \infty} = B \cdot [K_0 + C_2 t]. \tag{10.42}$$

In order for the steady-state error to be zero, we must have $K_0 = 0$ and $C_2 = 1$. These constants come from the partial-fraction expansion of the term in braces in Eqn. 10.39. K_0 and C_2 are calculated from the formulas given in Eqns. 7.37 and 7.38:

$$C_2 = \frac{K \, \Pi_m \vec{Z}_j}{\Pi_n \vec{Q}_j}\bigg|_{s=0} \quad \text{and} \quad K_0 = C_2 \bigg(\sum_{j=1}^{m} \frac{1}{\vec{Z}_j} - \sum_{j=1}^{n} \frac{1}{\vec{Q}_j} \bigg)_{s=0}. \qquad (10.43)$$

Here C_2 is the static gain of the closed-loop system $W(0)$, which must be 1 – a condition that is satisfied as a matter of course in many systems. Equation 10.43 yields the additional restriction on the pole-zero configuration of $W(s)$,

$$\sum_{j=1}^{m} \frac{1}{\vec{Z}_j}\bigg|_{s=0} = \sum_{j=1}^{n} \frac{1}{\vec{Q}_j}\bigg|_{s=0}, \qquad (10.44)$$

which makes $K_0 = 0$. In a unity feedback system it is necessary for the forward path to contain two integrating elements in order to satisfy Eqn. 10.44. But in a non-unity feedback system it is sometimes possible to satisfy Eqn. 10.44 with a combination of series and parallel compensating elements that do not require integrators.

10.7 Design for Disturbance Rejection

The controlled plant in a feedback system may be subjected to unwanted disturbances, as discussed in Section 8.3 and as illustrated in Fig. 8.7. These disturbances, which drive the plant in the same way as does the controller, induce an unwanted component in $y(t)$. Compensation in the error channel can sometimes minimize the disturbance-induced errors while maintaining the performance of the closed-loop system as it responds to inputs.

To illustrate this aspect of control loop design, let us assume that the servomechanism system described in Fig. 10.2 is subjected to a disturbance on the output shaft. The first step in design is to model the interaction of the disturbance with the control signal. In this example the interaction occurs on the output shaft. The total torque on that shaft is the disturbance torque plus the torque due to the current in the motor. The dynamic relationships between the elements of this example are

$$K_A e(t) = L_a \frac{di(t)}{dt} + R_T i(t) + nK \frac{d\theta_L}{dt}, \quad \text{and}$$

$$\eta nK i(t) + T_D(t) = [J_T] \frac{d^2\theta_L}{dt^2} + [b_T] \frac{d\theta_L}{dt}, \qquad (10.45)$$

where $T_D(t)$ is the disturbance torque, n is the gear ratio, η is the efficiency of the gear train, and $R_T = R_o + R$. K is the machine constant of the motor, and K_A is the amplifier gain. The two coefficients J_T and b_T are the total moment of inertia and the viscous damping factor of the motor–gear–load combination referred to the output shaft:

$$J_T = [J_L + \eta n^2 J_M] \quad \text{and} \quad b_T = [\eta n^2 b]. \qquad (10.46)$$

If the dynamic equations of Eqn. 10.45 are expressed in transfer function form, the interaction of the disturbance torque with the amplifier–motor–gear-load part of the control system can be depicted by the block diagram in Fig. 10.36, which includes the unity feedback connection, the control law block, and the substitution $Y(s) = \mathcal{L}\{\theta_L(t)\}$.

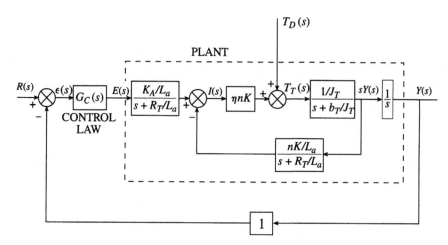

Fig. 10.36 Diagram indicating how a disturbance torque enters the control loop of the electromechanical servomechanism.

The numerical values for the physical parameters of this system are:

$J_M = 2.119 \times 10^{-5}$ kg-m², $\quad K = 0.09674$ N-m/A, $\quad R = 2.108\ \Omega$,

$L_a = 0.01$ H, $\quad b = 0.00475$ N-m/(rad/s), $\quad \eta = 0.88$, $\quad n = 10.5$,

$J_L = 0.002$ kg-m², $\quad K_A = 4$ V/V, $\quad R_o = 1\ \Omega$.

Note that, for $T_D(s) = 0$, the block diagram can be reduced to that in Fig. 10.4. The numerical values for the various physical parameters are rounded slightly and substituted into the plant transfer function blocks, and the control law is given the form

$$G(s) = \frac{K_C(s + Z_1)(s + Z_2)}{(s + P_1)(s + P_2)}. \tag{10.47}$$

The transfer function relating the disturbance torque to the error (for $R(s) = 0$) is

$$\left.\frac{\epsilon(s)}{T_D(s)}\right|_{R(s)=0}$$

$$= -\left[\frac{252(s + 311)(s + P_1)(s + P_2)}{s(s + 140)(s + 180)(s + P_1)(s + P_2) + 90{,}100 K_C(s + Z_1)(s + Z_2)}\right]. \tag{10.48}$$

It has been demonstrated in Section 10.2 that the system can be made stable and sufficiently damped by a suitable selection of K_C, Z_1, Z_2, P_1, and P_2. The steady-state error resulting from a step function of disturbance torque can be determined by inspection of Eqn. 10.48:

$$\hat{\epsilon}(t)|_{SS} = -T\left[\frac{0.87}{G_C(0)}\right] = -T\left[\frac{0.87 P_1 P_2}{K_C Z_1 Z_2}\right]\ \text{rad}, \tag{10.49}$$

where T is the magnitude of the torque in N-m. This calculation shows the desirability of a high static gain for the compensator. If an integration is included

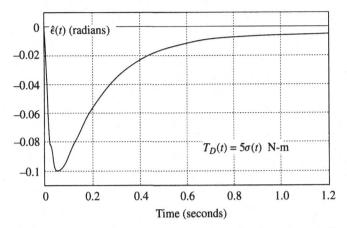

Fig. 10.37 System error following the application of a step disturbance torque.

in the compensator ($P_1 = 0$), the static error due to a constant disturbance torque will be eliminated.

The motor–gear combination in this system can deliver a maximum static torque to the load shaft of approximately 25 N-m. If the maximum expected load torque is 5 N-m, and if the constants given for the lag-lead compensator in Section 10.2 are used, then the maximum static disturbance-torque error will be 0.0051 rad (about 0.3 deg). The transient response of the error to a step input of disturbance torque (see Fig. 10.37) is much slower than the response to a step applied at the input (Fig. 10.12), and the peak error reached during the transient greatly exceeds the static error. It is useful to note that the response shown in Fig. 10.37 may be approximated by using a second-order transfer function in place of that in Eqn. 10.48. This approximation is composed of the poles and zeros of $\epsilon(s)/T_D(s)$ that are close to the origin:

$$\left.\frac{\epsilon(s)}{T_D(s)}\right|_{\text{approx}} = \frac{-1.07(s+0.2)}{(s+5.22)(s+40.4)}. \tag{10.50}$$

The constant -1.07 is chosen to make the steady-state gain of the two transfer functions equal.

10.8 Control of Unstable Plants

A significant problem in control engineering is presented by plants or processes that are unstable. A rocket booster with the driving thrust applied under the center of mass is one example; the cart–inverted pendulum system is another. Some chemical and nuclear processes are unstable, most aircraft are designed to be unstable to enhance their maneuverability, and the kinematics of some short-term navigation problems are represented by unstable transfer functions. These plants and processes may be successfully controlled using feedback and a stabilizing control law, as indicated in Fig. 10.38, but the closed-loop system has a peculiar property not found in systems having stable plants.

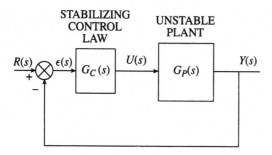

Fig. 10.38 Feedback control of an unstable plant.

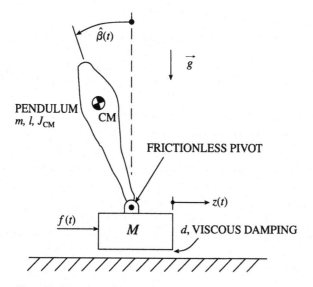

Fig. 10.39 Cart–inverted pendulum system.

We demonstrate this first with the cart–inverted pendulum system shown in Fig. 10.39.

The equations of motion for the cart–*inverted* pendulum system are easily derived from those of the cart–*hanging* pendulum system developed in Section 2.4. It is simply necessary to reverse the sign of the gravitational constant g in the equations of Section 2.4. The following parameter values are typical of a laboratory-size cart–inverted pendulum system:

$$M = 5 \text{ kg}, \quad m = 2 \text{ kg}, \quad g = 9.81 \text{ m/s}^2,$$
$$J_{CM} = 0.8 \text{ kg-m}^2, \quad l = 1 \text{ m}, \quad d = 0 \text{ N/(m/s)}. \tag{10.51}$$

With these values, the transfer function relating the applied force to the pendulum angle is

$$G_p(s) = \frac{\beta(s)}{F(s)} = \frac{-0.128}{(s+2.97)(s-2.97)}. \tag{10.52}$$

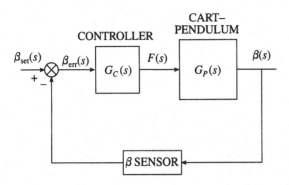

Fig. 10.40 Block diagram for the cart–inverted pendulum system using pendulum angle feedback.

Figure 10.40 shows this transfer function incorporated into the control system. The controller transfer function $G_C(s)$ must be chosen so that the feedback system is stable even though the cart–pendulum itself is unstable. The system can be tested by applying an impulse at the input and observing the transient response $\hat{\beta}(t)$, which must decay toward zero if the system is stable. For simplicity we take the β sensor gain to be 1.

As the first trial design, let

$$G_C(s) = K. \tag{10.53}$$

Figure 10.41 shows a root-locus plot for both positive and negative values of K. Note that the gain constant for the loop transfer function is $-0.128K$. For

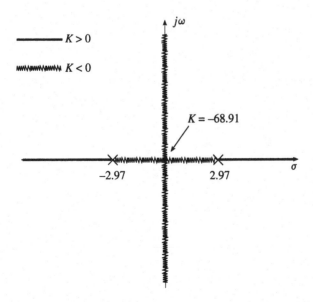

Fig. 10.41 Closed-loop instability for trial design of (simple-constant) compensator.

$K > 0$, one pole of the closed-loop transfer function lies in the right-half plane and one lies in the left-half plane. For $K < 0$, one pole of the closed-loop transfer function is in the right-half plane, one is in the left-half plane for small values of K, and two poles are on the $j\omega$ axis for $K < -68.91$. Poles of the closed-loop transfer function on the $j\omega$ axis represent unacceptable marginal stability. In a practical system, friction and other small nonlinear effects that are neglected in the model would cause instability. The simple controller of Eqn. 10.53 is therefore unsatisfactory for all possible values of K. We require instead a controller transfer function having a pole-zero pattern that will cause the branch of the root locus emanating from the pole at $s = 2.97$ to enter the left-half plane.

As the second trial design, let the controller transfer function be

$$G_C(s) = \frac{K[s^2 + 6s + 10]}{s(s+40)}. \tag{10.54}$$

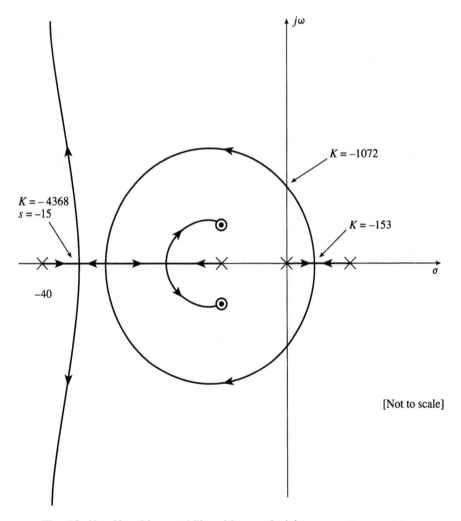

Fig. 10.42 Closed-loop stability with second trial compensator.

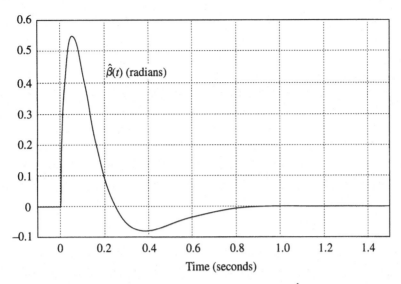

Fig. 10.43 Pendulum angle response to a pulse input $\hat{\beta}_{\text{set}}(t)$.

This design offers an integrating property and a pair of zeros that serve to attract the loci from the right-half plane into the left-half plane. Figure 10.42 shows the interesting root-locus pattern for this system. K must be negative, and it must exceed 1072 in absolute value to ensure stability of the closed-loop system. For $K = -4000$ the gain margin of the system is 3.73, and the four poles of $W(s)$ are located at $-6.70 + j3.48$, $-6.70 - j3.48$, -3.97, and -22.6. If the input to the closed-loop system is an impulse

$$\hat{\beta}_{\text{set}}(t) = 0.05\delta(t), \tag{10.55}$$

then the output $\hat{\beta}(t)$ is the well-damped response shown in Fig. 10.43. The system error is also promptly driven to zero. The compensator defined by Eqn. 10.54 would therefore be considered satisfactory insofar as gain margin, stability, and steady-state error are concerned. Questions of robustness of the design with respect to unmodeled physical properties – such as friction in the pivot and a nonzero damping coefficient – remain to be investigated.

This example reveals the basic problem of controlling an unstable plant, namely, that the root locus always has a branch emanating from the right-half-plane pole of the transfer function. Even when the system is well compensated, as in this example, the stability of the closed-loop system depends on the loop gain remaining *greater* than the critical level. This condition, called *conditional stability,* always exists in systems having an unstable plant. In a control loop with high gain, the linear range is often quite small, as demonstrated in the servomechanism example in Section 10.2. When the error signal exceeds this linear range, the motion of the system cannot be calculated using the linear techniques upon which our stability analysis is based. During the period in which a signal channel is in saturation, the plant operates approximately as it would without feedback, that is, with a significantly lower gain than the linear design requires for stability. For this reason special precautions are taken to prevent saturation in the forward path of systems having an unstable plant. Such precautions are not necessary to maintain the stability of systems having stable plants, provided those systems are not conditionally stable. This problem is explored in Chapter 15.

Control law design for an unstable plant can sometimes be more effective if a second sensor is mounted on the plant. In this example a sensor that measures $\dot{\hat{\beta}}(t)$ could be used to stabilize the plant with an appropriate feedback from that sensor, in order to form an inner loop. An outer loop would then be formed using feedback from the $\hat{\beta}(t)$ sensor, and a second compensator would be designed to perform the desired control function of the complete system.

10.9   Control of Nonminimum Phase Plants

A nonminimum phase plant is one whose transfer function has one or more poles or zeros in the right-half plane. In this section we consider only the control of stable nonminimum phase plants, which are plants whose transfer functions have one or more zeros in the right-half plane. Consider the transfer function of a nonminimum phase plant,

$$G_P(s) = \frac{Y(s)}{U(s)} = \frac{54(s-2)(s+5)}{s(s+3)(s+10)[s^2+6s+18]}. \tag{10.56}$$

Let this plant be controlled by the unity feedback system in the conventional manner (see Fig. 10.44), and let the controller transfer function be a simple gain:

$$G_C(s) = K. \tag{10.57}$$

A root-locus plot for this system, Fig. 10.45, reveals that K must be negative if the closed-loop system is to be stable. The system will be stable for K in the range $0 > K > -1.38$. If K is set at -0.55, the closed-loop pole and zero locations will be as follows.

Poles	Zeros
$-0.729 \pm j0.919$	2
$-4.01 \pm j2.56$	-5
-9.52	

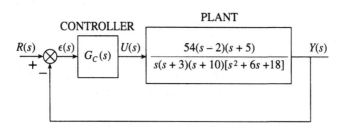

Fig. 10.44   Feedback control of a nonminimum phase plant.

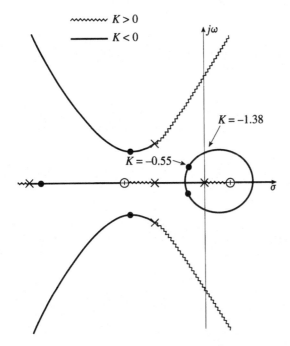

Fig. 10.45 Closed-loop system, stable if K is negative.

The gain margin will be 2.51. When a unit step input is applied, the corresponding response $y(t)$ is that shown in Fig. 10.46. Note that the steady-state error converges to zero and also that the initial acceleration of $y(t)$ is negative, producing a motion that is temporarily opposite to that called for by the input. This behavior is characteristic of the step response of nonminimum phase systems.

The speed of response of the plant can be improved if a controller transfer function having a zero and a pole is used. If the zero is placed just to the left of the plant pole at $s = -3$, the two root loci that come from the poles at $s = 0$ and $s = -3$ will be attracted into the left-half plane, and the two loci that come from the complex poles of the plant transfer function (at $s = -3 \pm j3$) will be the branches which go into the right-half plane. This is shown in Fig. 10.47 for a controller transfer function of

$$G_C(s) = \frac{K(s+3.1)}{s+20}. \tag{10.58}$$

For $K = -4.7$, the poles and zeros of the closed-loop transfer function lie at the following locations.

Poles	Zeros
$-1.2 \pm j1.7$	2
$-2.74 \pm j0.45$	-3.1
-13.1	-5
-18.1	

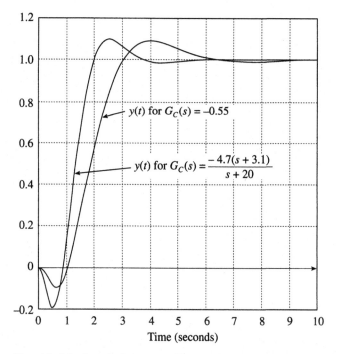

Fig. 10.46 Speed of response improved by lead compensation.

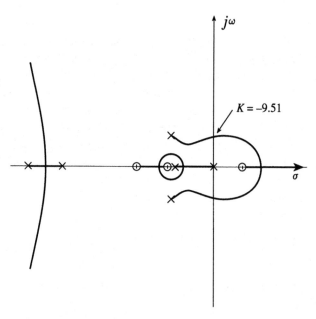

Fig. 10.47 Root-locus plot for lead compensation, $G_C(s)$ in Eqn. 10.58.

The gain margin will be 2.02. The poles at $s = -1.2 \pm j1.7$ in this design are significantly farther out in the left-half plane than are the poles at $s = -0.729 \pm j0.919$ in the first trial design. Since these complex poles dominate the step response of the closed-loop system, that response will be more prompt with the new controller design. The gain constant -4.7 is chosen to give the step response an overshoot approximately the same as that of the first trial design. A comparison between the two responses is made in Fig. 10.46.

This example shows the basic difficulty in controlling a stable but nonminimum phase plant – namely, that the right-half–plane zero attracts the root loci into the right-half plane for relatively low values of loop gain. With proper compensation the control loop can be made to be unconditionally stable. This is a different problem from that encountered in an unstable nonminimum phase plant, such as the cart–inverted pendulum system, where the closed-loop system is necessarily conditionally stable.

10.10 Summary

This chapter has treated design problems restricted to SISO plants having relatively simple design specifications (gain margin, response time, steady-state error, static controller gain, and percent overshoot in response to a step input). A trial-and-error approach is effective in this setting. The first trial using a simple gain as the control law will reveal the basic control problem through a root-locus plot, and it will normally indicate what the second trial design should be. Series compensation or parallel compensation (or both) may be available to the designer; these choices expand the scope of design possibilities. When a satisfactory pole-zero gain configuration has been found for the controller transfer function, hardware for the controller unit must be designed. If the controller unit is an electronic circuit or a mechanical device operating on continuous time signals, the hardware design can be made directly from the desired pole-zero gain configuration.

In many cases the controller unit is a digital signal processor. Because the pole-zero analysis methods in this book pertain to continuous time processes, it is necessary to translate the desired transfer function for the controller into a set of *difference equations* suitable for programming the digital controller. Although this translation process is discussed briefly in Chapter 15, it requires analysis beyond the scope of this book.

The problem of robustness of the controller design with respect to parameter uncertainties in the plant was introduced in Section 10.4, where a warning was issued against using pole-zero cancellation as a primary design tool.

Usually the controlled plant is fixed, and control loop compensation must be designed into the controller. But if post-design testing reveals that the system does not satisfy the performance requirements, an engineering change must be made either in the controller or in the plant. An engineering change proposed for the control law is subject to the same scrutiny and verification procedures as a similar proposal for the plant. In some cases it may be more feasible to make engineering changes in the plant than it is to make engineering changes in the controller. A relatively simple alteration in a structural member of the plant, for example, might shift a troublesome pair of poles of the plant transfer function so that the existing controller will satisfy the system performance specifications.

Some design situations require techniques beyond those suitable for compensation of single-loop systems. These situations arise if the plant has more than one input, or if the performance specifications cannot be interpreted in terms of the pole and zero locations of the closed-loop transfer functions. For example, multi-input plants are usually controlled by more than one feedback loop, each loop being driven by a separate sensor on the plant output. In these cases computer-aided techniques based on modern control theory are used to ensure a harmonious relationship among the various inputs to the plant. Such a harmonious design may be difficult to achieve using the single-loop techniques of this chapter. The design specifications sometimes include an optimization requirement that is also difficult to achieve with elementary compensation techniques. For example, a system might be required to track a given trajectory for a finite time with specifications on some form of an average value of the error, weighted to emphasize the critical periods of the mission. A formal procedure for satisfying such a design specification would be based on analytical techniques such as the calculus of variations. Even the concept of stability, the central theme of classical feedback control theory, may not be relevant to control of a plant during a finite interval.

10.11   Problems

Problem 10.1
Let

$$G(s) = \frac{K}{s(s^2 + as + b)}$$

be the open-loop transfer function for a unity feedback control system.

  (a) Establish conditions on K, a, and b that must prevail to ensure that the closed-loop system will be stable.
  (b) Let $b = 10$. Establish the boundary between closed-loop stability and closed-loop instability on the (K, a) plane.

Problem 10.2
Consider the system in Fig. P10.2(a), where $G(s)$ has no right-half–plane poles or zeros. A unit step input is applied to the system, and the resultant output for $K = 8$ is plotted in Fig. P10.2(b). If K is changed to -8, the output $y(t)$ will also change. Choose the statement which best describes how $y(t)$ will behave when $K = -8$.

  (a) $y(t)$ will be unstable.
  (b) $y(t)$ will be stable but will have pronounced overshoot.
  (c) $y(t)$ could be stable or unstable, depending upon $G(s)$.
  (d) $y(t)$ will be the negative of that shown in Fig. P10.2(b).

Problem 10.3
Let

$$G(s) = \frac{21}{s(s^2 + \beta s + 14)}$$

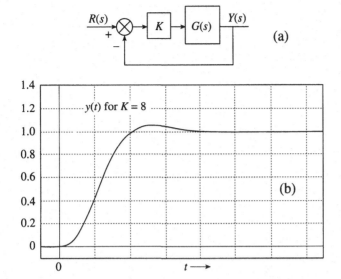

Fig. P10.2

be the open-loop transfer function for a unity negative feedback system. Find the range of β for which the closed-loop system will be stable.

Problem 10.4

Consider the system configuration in Fig. 10.20, where $G(s) = 175/(s+P)^5$ and $H(s) = 4$.

(a) Without the aid of a computer, find the range of values of P for which the closed-loop system will be stable.

(b) Repeat (a) for $H(s) = -4$.

Problem 10.5

Consider the system configuration in Fig. 10.20, where

$$G(s) = \frac{K}{s(s+1)(s+6)} \quad \text{and} \quad H(s) = \frac{2(s+1)}{(s+2)}.$$

Can you find a K for which the closed-loop system will be stable and $Y(s)/R(s)$ will have a pole at $s = -4$? If so, find that K; if not, explain why not.

Problem 10.6

Refer to the system configuration in Fig. 10.21. The plant transfer function is $6/(s+4)$, and $H(s) = 1$. The system input is a ramp function $r(t) = \beta t$.

(a) Find a control law $G_{FC}(s)$ so that the output will have the form $y(t) = \beta t + \{\text{stable transients}\}$, that is, $\lim_{t \to \infty} y(t) = r(t)$.

(b) Determine the robustness of your control law design with respect to small changes in the plant parameters. For example, if the plant gain changes from 6 to 6.1 will your $G_{FC}(s)$ still satisfy the given design requirement?

Problem 10.7

Refer to the system configuration in Fig. 10.21. The plant transfer function is $6/(s+4)$, $G_{FC}(s) = 1$, and $H(s) = K(s+Z)/(s+P)$. The system input is a ramp function, $r(t) = \beta t$.

(a) Find values for K, Z, and P that will produce the same result as in Problem 10.6(a).

(b) Evaluate the robustness of your design with respect to small changes in the plant gain. Compare your design here with the one you found for Problem 10.6.

Problem 10.8

Confirm that Eqn. 10.50 is a reasonable approximation to $\epsilon(s)/T_D(s)$.

Problem 10.9

Investigate the robustness of the controller design in Eqn. 10.54 with respect to a 50% change in the mass of the pendulum.

Problem 10.10

Refer to Problems 4.5 and 6.19. The physical parameters of the motor, amplifier, drive train, cart, and inverted pendulum have the following numerical values.

AMPLIFIER
$K_A = 4$ V/V
$R_o = 1\ \Omega$

DC MOTOR
$K = 0.097$ N-m/A
$R = 2\ \Omega$
$J = 2.1 \times 10^{-5}$ kg-m^2
$b = 3.13 \times 10^{-4}$ N-m/(rad/s)
$L_a = 0.01$ H

CART
Total mass of cart, motor, amplifier, and drive mechanism, including wheels $= M = 6$ kg
Front and rear wheel radius $= 0.05$ m
Belt drive ratio $n = 10$
$J_F = 9.4 \times 10^{-4}$ kg-m^2
$J_R = 1.13 \times 10^{-3}$ kg-m^2

PENDULUM
A slender rod 3 meters long
$m = 15$ kg
$g = 9.81$ m/s^2

The pendulum is to be balanced automatically by the feedback control system diagramed in Fig. P10.10.

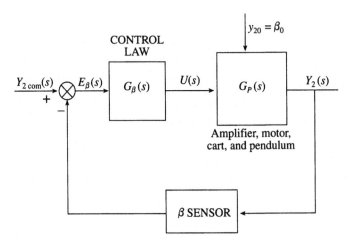

Fig. P10.10

(a) Obtain the transfer function $G_p(s)$ that relates the input voltage of the amplifier to the position of the pendulum. Also obtain the transfer function relating the input voltage of the amplifier to the position of the cart. Note that we employ the notation used in Problem 6.19: $e(t) = u(t)$, $z(t) = y_1(t)$, and $\beta(t) = y_2(t)$.

(b) Let the β sensor have a gain of 1 V/rad. Find a control law $G_\beta(s)$ for which the control loop will be stable, and for which $\beta(t)$ will be driven to zero when the commanded position $y_{2com}(t) = 0$ and the pendulum is initially displaced slightly (0.1 rad) from its equilibrium position. Do not depend on pole-zero cancellation in your design. Find the gain margins for your design.

(c) How will the gain margins of your design be affected if the damping coefficient b is increased to 5×10^{-3} N-m/(rad/s)?

Problem 10.11

Draw the root-locus plots for the three cases of the DC motor speed control example in Section 10.3. Verify the quantitative results presented in the text.

Problem 10.12

Refer to Problem 8.4, in which an initial condition y_0 is incorporated into the system diagram. Let $H(s) = 1$. Write the plant zero state transfer function as

$$G_P(s) = \frac{N_P(s)}{(s+\alpha)D_P(s)}.$$

The factor $(s+\alpha)$ represents a troublesome pole that requires the control law to provide series compensation. The zero-input transfer function of the plant for the initial condition y_0 has the form

$$I_0(s) = \frac{N_0(s)}{(s+\alpha)D_P(s)},$$

where $N_0(s)$ has no $(s+\alpha)$ factor. Assume that the control law is based on the pole-zero cancellation idea and that it therefore has the form

$$G_C(s) = \frac{(s+\alpha)N_C(s)}{D_C(s)},$$

where $D_C(s)$ has no $(s+\alpha)$ factor.

(a) Calculate the two transfer functions $W_R(s)$ and $W_0(s)$ defined in Problem 8.4.

(b) Note that the pole-zero cancellation has had the desired effect on the design for $W_R(s)$, but that the troublesome pole appears as a pole of $W_0(s)$ and therefore a mode $\epsilon^{-\alpha t}$ will appear in $y(t)$ whenever y_0 is nonzero.

(c) Discuss the significance of this analysis if the troublesome pole of the plant is in the right-half plane.

Problem 10.13

Refer to the configuration in Fig. 10.21. Let $H(s) = 1$. The unstable plant has a transfer function

$$\frac{5 \times 10^5}{(s+50)(s+1000)[s^2-(2+\epsilon)s+18]},$$

where ϵ is a small uncertainty with magnitude less than 0.1. Can you find a PID controller design for the control law $G_{FC}(s)$ that will guarantee stability of the closed-loop system for all expected variations in ϵ, and that will also satisfy the following design requirements?

(a) The gain margin must lie between 1/4 and 4.
(b) In response to a step input of $r(t)$, the system error must approach zero with increasing time.

If so, give the PID parameters K_P, K_I, and K_D. Demonstrate that your design is satisfactory. If not, explain why it is not possible to find a solution.

Problem 10.14

The open-loop transfer function for a unity negative feedback system is $K/s(s+40)^2$. The system input is a unit step function.

(a) Find the maximum value for K which will satisfy all three of the following criteria:
  1. gain margin > 4;
  2. overshoot of $y(t) \le 20$ percent; and
  3. time for $y(t)$ to first reach 1 is less than 0.13 s.
(b) Can you repeat (a) if the rise-time specification is changed from 0.13 s to 0.11 s and the overshoot requirement is changed to ≤ 25 percent?

Problem 10.15

Consider the configuration of Fig. 10.20, where

$$G(s) = \frac{K(s+1)}{s(s-2)(s^2+16s+100)} \quad \text{and} \quad H(s) = 1.$$

Note that the closed-loop system will be stable only for $K_{min} < K < K_{max}$. Find K_{min} and K_{max}. Also, find the value of K for which $K/K_{min} = K_{max}/K$.

Problem 10.16

In Section 10.9, a comparison was made between two controller designs for a nonminimum phase plant that provide the same percent overshoot for the closed-loop system. Form a "design figure of merit" that includes the static gain of the controller, the system gain margin, and the response time (the instant $y(t)$ first reaches 1). Use your figure of merit to determine which is the superior design.

Problem 10.17

Refer to the Navion altitude control autopilot system in Problem 9.25. Find a control law $G_{CL}(s) = U(s)/E(s)$ that will satisfy design specifications (a) and (b) for the closed-loop system given in the problem statement.

Problem 10.18

Refer to the Navion longitudinal axis model for Problem 5.12. Show that the transfer function from elevator deflection (deg) to normal acceleration ((ft/s)/s) is approximately

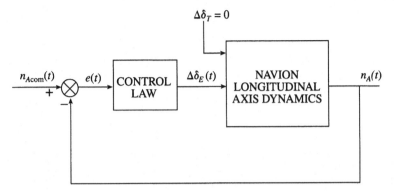

Fig. P10.18

$$\frac{0.48(s)(s+0.02)(s+13)(s-10)}{(s^2+5s+13)(s^2+0.04s+0.05)}.$$

Devise a feedback control law for the system diagramed in Fig. P10.18 so that the output response $n_a(t)$ will be well-damped and the error will approach zero with increasing time in response to a unit step command $n_{acom}(t) = 1\sigma(t)$ (ft/s)/s. We note that this aircraft can sustain a constant normal acceleration for only a limited period at a cruise power setting.

Problem 10.19

The electromagnet–amplifier–sensor system described in Problems 4.13, 5.6, 6.24, and 6.25 is to be used in the stabilizing feedback system diagramed in Fig. P10.19. Use the numerical values for the physical parameters given in the previous problems. Find a control law $E_{in}(s)/E(s)$ such that the closed-loop system will be stable, the gain margin will be adequate, and – if the initial displacement is $z_0 = 0.5$ mm – the control loop will return $z(t)$ to zero. We also require that $|z(t)| \le 0.5$ mm at all times.

Problem 10.20

Refer to Problems 6.29 and 5.13 concerning the ball, rack, and pendulum system. Assume that the torque is developed in a torque motor driven by a power amplifier, as

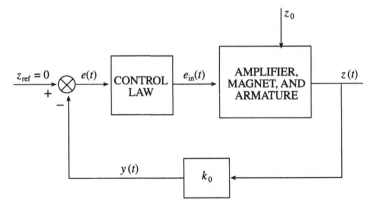

Fig. P10.19

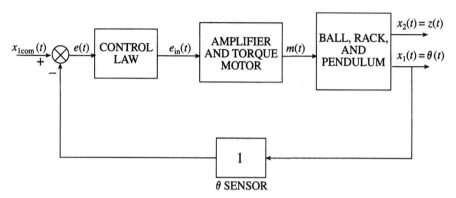

Fig. P10.20

shown in Fig. P10.20. The transfer function of the power amplifier–torque motor combination is $M(s)/E_{in}(s) = 30/(s + 30)$. Use the numerical values for the physical parameters given in Problem 5.13.

(a) Find the transfer functions $X_1(s)/E_{in}(s)$ and $X_2(s)/E_{in}(s)$.

(b) Find the transfer function $X_2(s)/X_1(s)$.

(c) Find a control law $E_{in}(s)/E(s)$ that will stabilize the closed-loop system whose transfer function is $X_1(s)/X_{1com}(s)$. This is the θ control loop.

(d) With the control loop from (c) in operation, will the transfer function $X_2(s)/X_{1com}(s)$ have all its poles in the left-half plane?

Problem 10.21

In Problem 10.20 the ball is uncontrolled. To stabilize the motion of the ball, and to have its equilibrium position at $z = 0$, we add a z control loop to the system in Fig. P10.20, as shown in Fig. P10.21. Devise a control law $X_{1com}(s)/E_2(s)$ to ensure an adequate gain margin of the z control loop.

Problem 10.22

A massive load with moment of inertia J_2 must be driven by DC motor through an elastic shaft, as shown in Fig. P10.22. The amplifier, motor, and gears are characterized by

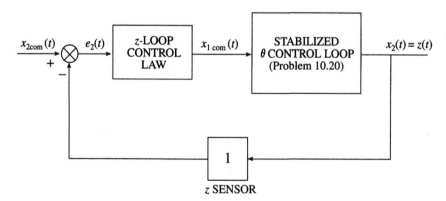

Fig. P10.21

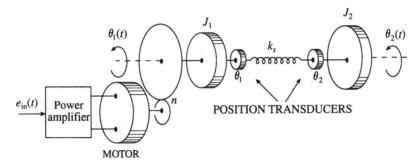

Fig. P10.22

the parameters K_A, R_o, K, R, b, n, and J, following the style established in Chapter 4. The inductance of the motor is negligible. The drive mechanism and load are characterized by J_1, J_2, and k_s, as indicated in Fig. P10.22. Adopt the notation $\theta_1(t) = x_1(t)$ and $\theta_2(t) = x_2(t)$. Obtain the transfer functions $X_2(s)/E_{in}(s)$ and $X_2(s)/X_1(s)$.

Problem 10.23

The amplifier–motor–load combination of Problem 10.22 is instrumented by the two angular position transducers at the ends of the flexible shaft. The output signals are $y_1(t) = k_1 x_1(t)$ and $y_2(t) = k_2 x_2(t)$. The combination is the plant for the feedback system diagramed in Fig. P10.23. The plant parameter values are: $L_a = 0$, $K = 0.05$ N-m/A, $R = 2.4\ \Omega$, $n = 5$, $J = 3 \times 10^{-5}$ kg-m^2, $J_1 = 85 \times 10^{-5}$ kg-m^2, $b = 1.1 \times 10^{-4}$ N-m/(rad/s), $K_A = 4$, $R_o = 1\ \Omega$, $k_s = 2$ N-m/rad, and $J_2 = 60 \times 10^{-5}$ kg-m^2. The feedback sensor gains are $k_1 = k_2 = 1$ V/rad. We are interested primarily in controlling the output shaft position, $x_2(t)$, but we may use a feedback signal from either $x_1(t)$ or $x_2(t)$. *Note:* $\theta_1(t) = x_1(t)$ and $\theta_2(t) = x_2(t)$.

(a) Use the feedback signal $y_1(t)$. The control law transfer function $E_{in}(s)/E(s)$ is a simple gain, K_C. Find the maximum value for K_C that satisfies the following closed-loop performance specification: For a step input $x_{2com}(t) = 0.1\sigma(t)$ V, the overshoot in $x_2(t)$ must not exceed 10 percent of its final value. Plot $x_2(t)$ versus t. Find the gain margin of the system.

(b) Use the feedback signal $y_2(t)$. Repeat (a). Compare the two output responses and the two gain margins.

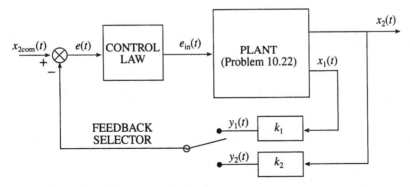

Fig. P10.23

Problem 10.24

Refer to Problem 10.23. We now attempt to improve the performance of the system by designing a dynamic control law $G_C(s) = E_{in}(s)/E(s)$.

(a) Use the feedback signal $y_1(t)$. First, find a $G_C(s)$ that satisfies all of the following performance specifications for the closed-loop system. Then plot $x_2(t)$ versus t and find the gain margin of the system.
   1. For a step input $x_{2com}(t) = 0.1\sigma(t)$ V, the overshoot in $x_2(t)$ must not exceed 10 percent of its final value.
   2. The instant of peak overshoot must be less than 0.2 s following application of the step input.
   3. $x_2(t)$ must be within 0.005 rad of its final value for $t > 0.5$ s.
   4. The static gain of the control law $G_C(0)$ must exceed 1.

(b) Use the feedback signal $y_2(t)$. Use the control law that you found in (a) and the same input step command. Plot $x_2(t)$ versus t. Compare the results with the results found in (a). Discuss the comparison briefly.

Problem 10.25

Refer to Problems 10.23 and 10.24.

(a) Use the feedback signal $y_2(t)$. Find a control law $G_C(s)$ satisfying all four of the design specifications listed in Problem 10.24(a). Plot $x_2(t)$ versus t. Find the gain margin of the system.

(b) Switch to feedback signal $y_1(t)$, and use the $G_C(s)$ from (a). Plot $x_2(t)$ versus t. Find the gain margin of the system. Compare your results with those in (a).

(c) Compare your results in (a) and (b) with the comparable results in Problem 10.24.

Problem 10.26

Let

$$G_P(s) = \frac{K}{s^2(s+3)}$$

in the system diagramed in Fig. P10.26. Find compensating functions $G_C(s)$ and $H(s)$, and specify the gain K that will satisfy both of the following requirements.

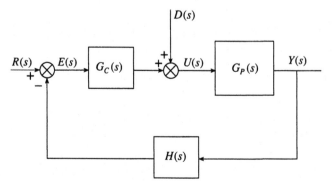

Fig. P10.26

1. When $d(t) = 0$ and $r(t) = \beta t$ (a ramp input), the steady-state value of $e(t)$ must be zero.
2. When $r(t) = 0$ and $d(t) = \beta \sin(4t)$, the steady-state oscillation in $e(t)$ must have an amplitude of less than 0.006β.

Problem 10.27
Refer to the series compensation example in Section 10.2. The load has a moment of inertia of 2×10^{-3} kg-m^2. The lag-lead compensator provides a performance characterized by four quantities: overshoot of 20 percent, gain margin = 3.51, $K_T = 977.5$, and $t_R = 0.0123$. To investigate the robustness of the lag-lead compensator design, calculate how each of these performance measures will be affected if the load moment of inertia is subjected to an uncertainty of 0.7×10^{-3} kg-m^2.

Problem 10.28
Refer to Problem 10.10. We now wish to control the position of the cart, $z(t) = y_1(t)$, by closing an outer loop as indicated in Fig. P10.28. You may obtain the transfer function $G_{21}(s) = Y_1(s)/Y_2(s)$ from the results of Problem 10.10(a). The z sensor has a sensitivity of 1 V/m. Use the closed-loop transfer function for the β loop $Y_2(s)/Y_{2\text{com}}(s)$ that you found in Problem 10.10.

(a) Find a control law transfer function $G_z(s)$ which will provide a stable overall transfer function $Y_1(s)/Z_{\text{com}}(s)$ and which will have a satisfactory gain margin for the outer loop.
(b) Adjust the gain of the inner-loop control law $G_\beta(s)$ until the overall system becomes unstable to determine whether the gain margin of the β loop remains the same as in Problem 10.10.

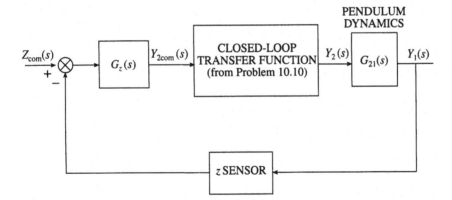

Fig. P10.28

11 Frequency Response Analysis of Linear Systems

11.1 Definition of the Frequency Response for a Linear System

We now consider the experimental set-up for making measurements on a dynamic system (shown in Fig. 11.1) in order to define the frequency response of a linear system. We assume that the system is linear and that it has constant parameters. We also assume, temporarily, that the system is stable. We begin the experiment at time $t = 0$ by applying the input-forcing function $u(t)$ to the system. To define the frequency response, we use the sinusoidal input

$$u(t) = R\sin(\omega t). \tag{11.1}$$

R is called the *amplitude* of $u(t)$, and ω is the *frequency* of $u(t)$ [rad/s]. The output response of the system can be expressed as the sum of a transient response and a steady-state response, as explained in Section 6.2:

$$y(t) = y_{\text{tran}}(t) + y_{\text{SS}}(t). \tag{11.2}$$

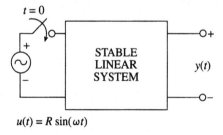

$u(t) = R\sin(\omega t)$

Fig. 11.1 Experimental set-up for measuring the frequency response of a linear system.

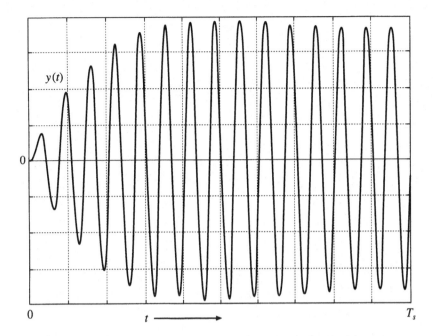

Fig. 11.2   Transient period at the beginning of a frequency response measurement.

Because the system is stable, the transient response will decay toward zero as time increases, becoming insignificant after time T_s. Therefore,

$$y(t)|_{t>T_s} \approx y_{SS}(t). \tag{11.3}$$

Figure 11.2 shows $y(t)$ during the transient period in a typical frequency response test where the initial conditions are zero. (Figure 6.2 shows an example where the initial conditions are not zero.) For tests on electronic amplifiers, T_s may be several microseconds. In servomechanism tests, T_s is often expressed in seconds; in an aircraft maneuver test, T_s will be many seconds; and in chemical manufacturing processes, T_s can be minutes or hours.

The definition of the frequency response is based only on $y_{SS}(t)$. In a linear system $y_{SS}(t)$ will be a sinusoidal function having the same frequency as $u(t)$ but with zero-crossing instants different from those of $u(t)$. Figure 11.3 compares $y_{SS}(t)$ and $u(t)$ in the same test as that in Fig. 11.2. The *period* of the sinusoid is $2\pi/\omega$ s. The zero-crossing instant for the input is t_{u0}, and that for the output is t_{y0}. The analytical expression for $y_{SS}(t)$ is

$$y_{SS}(t) = Y \sin(\omega t + \phi), \tag{11.4}$$

where ϕ is the angle that satisfies the condition

$$\sin(\omega(t_{y0} - t_{u0}) + \phi) = 0, \tag{11.5}$$

which is $\phi = \omega(t_{u0} - t_{y0})$ radians. Expressed in degrees, ϕ is

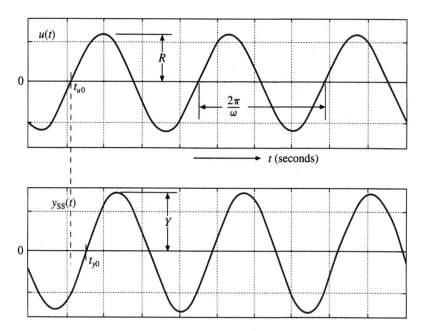

Fig. 11.3   Steady-state oscillations of the input and the output for the frequency response measurement at frequency ω rad/s.

$$\phi = \omega(t_{u0} - t_{y0}) \times \frac{180}{\pi} \text{ deg.} \tag{11.6}$$

Since Fig. 11.3 shows $(t_{y0} - t_{u0})$ to be positive and one eighth the period of the oscillation, we have $\phi = -45°$. When ϕ is negative, as it is in Fig. 11.3, the output is said to *lag* the input because the zero crossing of $y_{ss}(t)$ occurs later than the zero crossing of $u(t)$. In cases where ϕ is positive, the output is said to *lead* the input; in cases where ϕ is zero, the output is said to be *in phase* with the input. In all cases, ϕ is defined as the *phase shift* of the system at the frequency ω:

$$[\text{phase shaft at } \omega] = \phi. \tag{11.7}$$

The amplitude ratio of the response at the frequency ω is defined as follows:

$$[\text{amplitude ratio at } \omega] = M = Y/R. \tag{11.8}$$

In a linear system, the amplitude ratio at any given frequency is independent of the amplitude of the input R. If the test is repeated with a different input amplitude, αR, the amplitude of $y_{ss}(t)$ will be αY, and M will be unchanged. But if the test is repeated with the input at a new *frequency*, the amplitude ratio measured at that new frequency will usually be different from that in the first test. Consequently, frequency response tests on a system must include measurements made at many frequencies and covering a range (or *spectrum*) appropriate to the system. The results of a frequency response test, which take the form of the following table, are often automatically compiled by computer-controlled equipment specially designed for such testing.

Linear System Frequency Response Test Data

ω (rad/s)	R	$Y(\omega)$	$M(\omega)$	$\phi(\omega)$ (deg)
ω_1	R_1	Y_1	M_1	ϕ_1
ω_2	R_2	Y_2	M_2	ϕ_2
$\vdots$	$\vdots$	$\vdots$	$\vdots$	$\vdots$
ω_i	R_i	Y_i	M_i	ϕ_i
$\vdots$	$\vdots$	$\vdots$	$\vdots$	$\vdots$
ω_K	R_K	Y_K	M_K	ϕ_K

The number K of separate runs required to acquire these data may be 10 to 100, depending on how $M(\omega)$ and $\phi(\omega)$ vary with ω. Note that both of the sinusoidal curves shown in Fig. 11.3 for the ith run can be reconstructed from the four numbers ω_i, R_i, Y_i, and ϕ_i, so those four numbers completely describe what occurred on the ith run.

If a great many runs are made, so that a very accurate knowledge of $M(\omega)$ and $\phi(\omega)$ over a wide frequency spectrum is obtained, the experimental data then accurately represent the frequency response of the system, which is defined to be

$$\begin{bmatrix} \text{frequency response of} \\ \text{the linear system} \end{bmatrix} = \{M(\omega), \phi(\omega)\} \quad \text{for } 0 \le \omega < \infty. \tag{11.9}$$

Remark 1 Each run requires the driving frequency to be set to a new value and the transient period to elapse before the measurement of Y and ϕ can be made. In electronic amplifier testing, many hundreds of such runs can be made in a few minutes. However, in slower-acting systems hours of testing may be required for only a few runs. Most aircraft flight hours, for example, are so costly that frequency response tests can be performed only if acceptance tests require demonstration that the system is capable of certain maneuvers – for example, that the phase lag in a given test be less than $10°$ at a driving frequency of 4 rad/s. For this reason, analytical means for calculating the two defining functions $M(\omega)$ and $\phi(\omega)$ are important. Section 11.3 shows how $M(\omega)$ and $\phi(\omega)$ are obtained from the transfer function of the system.

Remark 2 The terms $u(t)$ and $y(t)$ represent physical variables whose values may be expressed in different units. For example, a frequency response test on an electrohydraulic mechanism with $u(t)$ being the input voltage to the amplifier and $y(t)$ the velocity of the ram would result in $M(\omega)$ having units (m/s)/V. It is essential to carry the units along with the $M(\omega)$ data so that someone whose preferred units are (in/s)/mV can use the data properly.

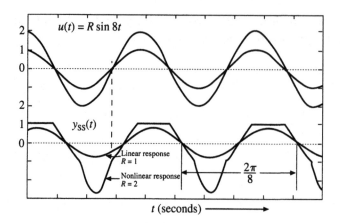

Fig. 11.4   Linear response for $R = 1$, nonlinear response for $R = 2$.

Remark 3   Many physical systems are very nearly linear for small values of the exciting function amplitude R, but are nonlinear if that amplitude becomes large. If the system is operating in its linear range, the waveform of $y_{SS}(t)$ will be sinusoidal at the same frequency as $u(t)$. But if R is large enough to drive the system into its nonlinear range, the waveform of $y_{SS}(t)$ will be distorted. This distortion can take different forms, but in many cases it is slight enough to be overlooked. In reporting experimental frequency response tests in terms of $M(\omega)$ and $\phi(\omega)$, it is important to record the magnitude of Y at each frequency and, if possible, to include a record of the waveform from the tests. Figure 11.4 illustrates the waveform of $y_{SS}(t)$ obtained in a system which is linear for $R = 1$ but is nonlinear for $R = 2$. The response at $\omega = 8$ rad/s shows that $Y(8) \approx 0.8$ and $\phi(8) \approx -39°$ for $R = 1$; for $R = 2$, $y_{SS}(t)$ is periodic at 8 rad/s but the waveform is nonsinusoidal. It would be a mistake to report an "amplitude" for this nonsinusoidal function, and the "phase shift" is similarly undefined. It is important to note the signal level at which the distortion occurs, and it is helpful to record the distorted waveform. A periodic function such as this could be represented as a Fourier series, and certain types of nonlinear analysis are based on the amplitude and phase of the so-called fundamental component of the series. Such analysis is, however, beyond the scope of this book.

Remark 4   The frequency response of an unstable system is well-defined, and it can be determined experimentally. This technique is explained in Section 11.7.

Remark 5   The frequency is sometimes expressed in hertz [Hz], which is also known as cycles per second [cps], and which is denoted by the symbol f; f and ω are related as $\omega = 2\pi f$.

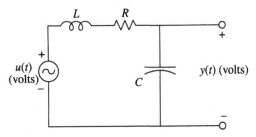

Fig. 11.5   RLC circuit excited by sinusoidal voltage.

11.2   Graphical Representation of Frequency Response Data

We now introduce four ways to plot the amplitude ratio $M(\omega)$ and the phase shift $\phi(\omega)$ in order to display these frequency functions in useful graphical forms; these are the rectangular plot, the polar (Nyquist) plot, the Bode plot, and the log–magnitude phase plot.

We use the frequency response data obtained from tests on the RLC circuit shown in Fig. 11.5 to illustrate these four graphical forms. For this circuit let $L = 1$ H, $C = 1\ \mu$F, and $R = 800\ \Omega$. The input is the voltage $u(t)$, and the output is the voltage $y(t)$:

$$u(t) = 10\sin(\omega t)\ \text{V},$$

$$y(t) = \underbrace{Y(\omega)\sin(\omega t + \phi(\omega))}_{y_{ss}(t)} + y_{tran}(t)\ \text{V}. \tag{11.10}$$

The transient term in $y(t)$ will essentially disappear 20 ms after application of the input, after which the amplitude ratio and phase shift can be measured, recorded, and a new run started. Assume that 25 runs are made and the data from these runs are listed in the following table, which is patterned after the table in Section 11.1.

Frequency Response
of RLC Circuit

ω (rad/s)	$M(\omega)$	$\phi(\omega)$ (deg)	
100	1.007	−4.62	
200	1.028	−9.46	
⋮	⋮	⋮	
700	1.320	−47.7	
825	1.364	−64.1	(peak M)
900	1.343	−75.2	
⋮	⋮	⋮	
2500	0.178	−159.2	

Note that $M(\omega) = Y(\omega)/10$.

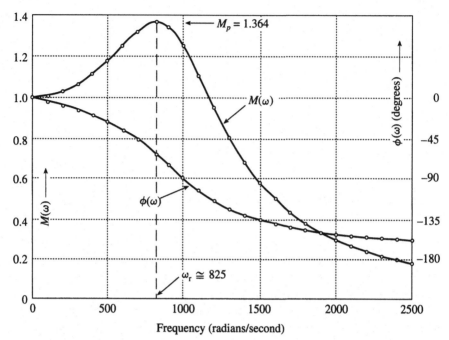

Fig. 11.6 Twenty-five data points for amplitude ratio and phase shift plotted on rectangular coordinates.

Figure 11.6 is the rectangular plot of these 25 data points for $M(\omega)$ and $\phi(\omega)$ versus ω. The abscissa on this graph is the frequency ω and the ordinates are $M(\omega)$ on the left, and $\phi(\omega)$ on the right. The origin of the graph is zero for both abscissa and the left-side ordinate, and the scale along both axes is linear. The 25 experimentally determined data points are connected by smooth curves under the assumption that, if further test runs were made for the frequencies between the data points, the results would likely lie on the curves of interpolation. Physical reasoning gives the points for $\omega = 0$, namely $M(0) = 1$ and $\phi(0) = 0°$. The smooth curves are also extrapolated for $\omega > 2500$ rad/s under the assumption that there is enough experimental evidence to indicate that the asymptotic values of the amplitude ratio and of the phase shift are

$$\lim_{\omega \to \infty} [M(\omega)] = 0 \quad \text{and} \quad \lim_{\omega \to \infty} [\phi(\omega)] = -180°. \tag{11.11}$$

(This assumption is well-founded on the analysis in the following sections of this chapter.) Thus, Fig. 11.6 is the rectangular graphical representation of the frequency response of the RLC circuit as it is defined in Eqn. 11.9.

The tests have shown that the peak value of $M(\omega)$ occurs at $\omega \approx 825$ rad/s, as noted in the data table. This defines the *resonant frequency* for this system, denoted $\omega_r \approx 825$ rad/s, and the peak value of $M(\omega)$ at the resonant frequency is denoted as $M_p = M(\omega_r) = 1.364$.

The polar plot of the amplitude ratio and phase shift data for the RLC circuit is based on representing the amplitude ratio and phase shift pair $\{M(\omega_i), \phi(\omega_i)\}$ for each frequency ω_i by the single complex number

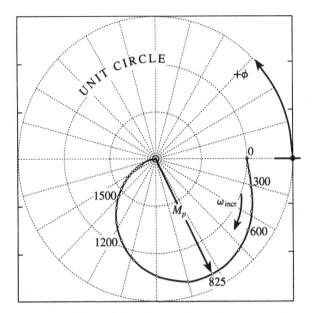

Fig. 11.7   The data of Fig. 11.6 plotted on polar coordinates.

$$M(\omega_i)[\cos(\phi(\omega_i)) + j\sin(\phi(\omega_i))], \tag{11.12}$$

which is expressed in polar form using Euler's formula as

$$M(\omega_i)\epsilon^{j\phi(\omega_i)}. \tag{11.13}$$

Figure 11.7 shows the polar (Nyquist) plot for the RLC circuit test data. The radial coordinate is $M(\omega)$ and the angular coordinate is $\phi(\omega)$, where positive $\phi(\omega)$ is measured counterclockwise from the positive real axis, as indicated, and ω is the parameter along the curve. The polar plot is useful because it combines amplitude ratio and phase shift information in a single parametric curve.

The Bode plot is a useful variation on the rectangular plot that also uses separate curves for $M(\omega)$ and $\phi(\omega)$. In this plot the frequency axis is a logarithmic scale. The amplitude ratio is also plotted on a logarithmic scale, but the phase shift data retain a linear scale. It is convenient to use semilog graph paper for the Bode plot. The amplitude ratio data, converted to decibels [db], are defined as

$$\begin{bmatrix} \text{amplitude ratio} \\ \text{in decibels} \end{bmatrix} = M_{db}(\omega) = 20\log_{10} M(\omega). \tag{11.14}$$

The table in Fig. 11.8 illustrates this conversion. The experimental data on the RLC circuit are graphed on the Bode plot in Fig. 11.9. The experimental data points are connected with smooth curves, and the extrapolations for $\omega \to 0$ and $\omega \to \infty$ are made with dashed lines.

The use of logarithmic coordinates on the Bode plot permits a wide spectrum of frequencies and a wide range of amplitude ratios to be plotted on a single sheet. A minor disadvantage of the logarithmic scales is that the graphs

M	M_{db}
⋮	⋮
0.001	−60
0.01	−40
0.1	−20
0.5	−6.0206
1	0
2	6.0206
10	20
100	40
1000	60
⋮	⋮

Fig. 11.8 Conversion table from amplitude ratio to amplitude ratio in decibels.

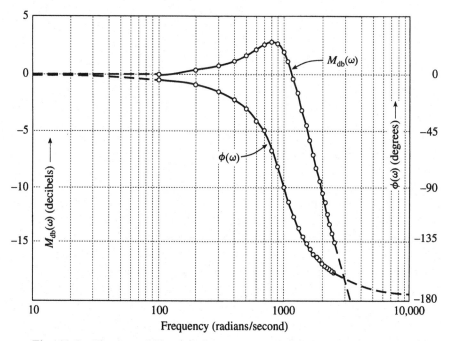

Fig. 11.9 The data of Fig. 11.6 displayed on the Bode chart.

cannot be shown for $\omega = 0$ or for $M = 0$. A major advantage of the Bode plot in control system engineering is that the frequency response of two (or more) cascaded elements can be quickly found from the frequency responses of the individual elements, as indicated in Fig. 11.10. Here the frequency response for system I is the pair of functions $\{M_1(\omega), \phi_1(\omega)\}$ and for system II is the pair $\{M_2(\omega), \phi_2(\omega)\}$. We must be sure that the frequency response for system I was taken while it was driving system II (or a comparable load). Because the output $y(t)$ of system I is the input to system II, the frequency response of the composite system with input $u(t)$ and output $z(t)$ is calculated as

$$M_{total}(\omega) = M_1(\omega) \times M_2(\omega) \quad \text{and} \quad \phi_{total}(\omega) = \phi_1(\omega) + \phi_2(\omega). \tag{11.15}$$

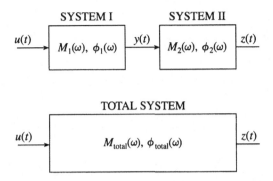

Fig. 11.10 Combining the frequency responses of two cascaded linear systems.

Because the product of two functions corresponds to the sum of the logarithms of the functions, we have

$$M_{\text{db total}}(\omega) = M_{\text{db1}}(\omega) + M_{\text{db2}}(\omega). \tag{11.16}$$

Thus, both the amplitude ratio and phase shift curves for the two systems can be added on the Bode plot to obtain the overall frequency response of the composite system. This addition can usually be performed by inspection, which explains why the Bode plot is popular with engineers who must make decisions by examining data only briefly. The extension of this principle to a composite system with N elements connected in cascade is obvious:

$$M_{\text{db total}}(\omega) = \sum_{i=1}^{N} M_{\text{db}i}(\omega) \quad \text{and} \quad \phi_{\text{total}}(\omega) = \sum_{i=1}^{N} \phi_{\text{db}i}(\omega). \tag{11.17}$$

Examples of this technique are given in Section 11.4.

The log-magnitude phase plot combines the amplitude ratio and phase shift data from the Bode plot into a single curve with frequency as a parameter, just as the Nyquist plot did with the data from the rectangular plot. The ordinate of the log–magnitude phase plot is the logarithmic decibel scale for the amplitude ratio, and the abscissa is a linear scale for the phase shift. The data for the RLC circuit are plotted on this diagram in Fig. 11.11. In Chapter 13 the log–magnitude phase plot is augmented with a second set of coordinates to form the Nichols chart, which is used extensively in analysis and design of single-loop feedback systems.

11.3 Frequency Response Obtained from Transfer Function

We now turn to the analytical method for calculating the frequency response of a linear system whose transfer function $G(s)$ has the usual form, as shown in Fig. 11.12. We assume that all the poles of $G(s)$ lie in the left-half plane and we consider only the zero-state response to a sinusoidal input, $u(t) = A \sin(\omega t)$. This input has the Laplace transform $U(s) = A\omega/(s^2 + \omega^2)$, so that

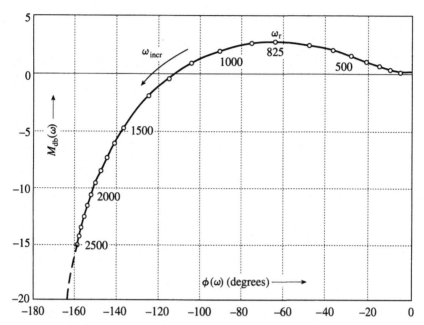

Fig. 11.11 The data of Fig. 11.6 displayed on the log magnitude–phase (Nichols) chart.

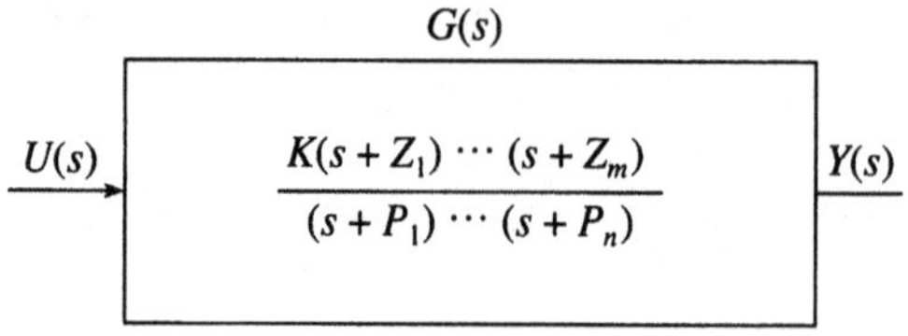

Fig. 11.12 Transfer function of nth-order linear system in standard form.

$$Y(s) = \frac{A\omega K(s+Z_1)(s+Z_2)\cdots(s+Z_m)}{(s^2+\omega^2)(s+P_1)(s+P_2)\cdots(s+P_n)}. \tag{11.18}$$

Expand $Y(s)$ in partial fractions:

$$Y(s) = \underbrace{\frac{K_0}{s+j\omega} + \frac{\bar{K}_0}{s-j\omega}}_{Y_{ss}(s)} + \underbrace{\frac{K_1}{s+P_1} + \frac{K_2}{s+P_2} + \cdots + \frac{K_n}{s+P_n}}_{Y_{tran}(s)}. \tag{11.19}$$

Here K_0 is the residue of $Y(s)$ at $s = -j\omega$, and $\bar{K}_0$ is the residue at $s = j\omega$. $\bar{K}_0$ and K_0 are complex conjugates, and $K_1, K_2, \dots, K_n$ are the residues of $Y(s)$ at the n poles of $G(s)$. All but the first two terms in $Y(s)$ correspond to the transient response, so these need not be calculated for our present purpose of finding only the steady-state component of $y(t)$. We have

$$Y_{SS}(s) = \frac{K_0}{s+j\omega} + \frac{\bar{K}_0}{s-j\omega}. \tag{11.20}$$

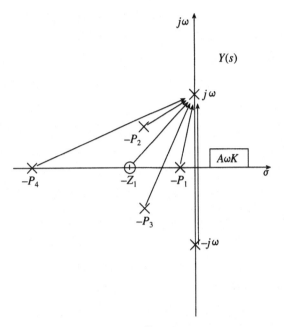

Fig. 11.13 The residue $\bar{K}_0$ calculated by the "arrow" method.

The residue $\bar{K}_0$ can be computed using the "arrow" method introduced in Section 7.2 (see Fig. 11.13):

$$\bar{K}_0 = [A\omega K]\left\{\frac{\Pi_m \vec{Z}_i}{\Pi_{n+1} \vec{P}_j}\right\}_{s=j\omega}, \tag{11.21}$$

where $\Pi_m \vec{Z}_i$ refers to the m zeros of $Y(s)$, and $\Pi_{n+1} \vec{P}_j$ refers to the poles of $Y(s)$, except for the pole at $s = +j\omega$. $\bar{K}_0$ may also be expressed as

$$\bar{K}_0 = \frac{A\omega K}{j2\omega}\left[\frac{(j\omega + Z_1)(j\omega + Z_2)\cdots(j\omega + Z_m)}{(j\omega + P_1)(j\omega + P_2)\cdots(j\omega + P_n)}\right], \tag{11.22}$$

which may also be written as

$$\bar{K}_0 = -j\frac{A}{2}[G(j\omega)] = -j\frac{A}{2}[G(s)]_{s=j\omega} = (\alpha - j\beta), \tag{11.23}$$

so that $Y_{SS}(s)$ can be expressed in the form of pair 16 in the Laplace transform table:

$$Y_{SS}(s) = \frac{\alpha + j\beta}{s + j\omega} + \frac{\alpha - j\beta}{s - j\omega}, \tag{11.24}$$

which has the inverse transform

$$y_{SS}(t) = 2\sqrt{\alpha^2 + \beta^2}\,\sin(\omega t + \phi) = 2|\bar{K}_0|\sin(\omega t + \phi). \tag{11.25}$$

But since $2|\bar{K}_0| = A|G(j\omega)|$, we have

$$y_{SS}(t) = A|G(j\omega)|\sin(\omega t + \phi), \quad \text{where} \quad \phi = \tan^{-1}(\alpha/\beta). \tag{11.26}$$

The phase shift angle ϕ may be calculated by writing $\bar{K}_0$ from Eqn. 11.23 as

$$\alpha - j\beta = -j\frac{A}{2}|G(j\omega)|\epsilon^{j\angle G(j\omega)}, \tag{11.27}$$

which is written equivalently as

$$\beta + j\alpha = \frac{A}{2}|G(j\omega)|\epsilon^{j\angle G(j\omega)} \tag{11.28}$$

and from which we have

$$\phi = \tan^{-1}(\alpha/\beta) = \angle G(j\omega) = \phi(\omega). \tag{11.29}$$

Comparing the input $u(t)$ and the steady-state output $y_{SS}(t)$, we see how the amplitude ratio and the phase shift for the linear system depend on the transfer function:

$$u(t) = A\sin(\omega t),$$
$$y_{SS}(t) = A|G(j\omega)|\sin(\omega t + \angle G(j\omega)). \tag{11.30}$$

This comparison shows these significant facts:

$$\begin{bmatrix} \text{amplitude} \\ \text{ratio} \end{bmatrix} = M(\omega) = |G(j\omega)| \quad \text{and} \quad \begin{bmatrix} \text{phase} \\ \text{shift} \end{bmatrix} = \phi(\omega) = \angle G(j\omega). \tag{11.31}$$

These defining relationships can be used to calculate the frequency response of the RLC circuit in Section 11.1, and to compare the analytical results with the laboratory measurements reported there. The transfer function relating $u(t)$ and $y(t)$ for that circuit is

$$\frac{Y(s)}{U(s)} = G(s) = \frac{1/LC}{s^2 + (R/L)s + 1/LC} = \frac{\omega_n^2}{s^2 + 2\zeta\omega_n s + \omega_n^2}, \quad \text{where}$$
$$\omega_n = \sqrt{1/LC} \quad \text{and} \quad \zeta = (R/2)\sqrt{C/L}. \tag{11.32}$$

Substituting $j\omega$ for s gives

$$G(j\omega) = \frac{\omega_n^2}{[\omega_n^2 - \omega^2] + j2\zeta\omega_n\omega}. \tag{11.33}$$

The amplitude ratio and phase shift are

$$M(\omega) = \frac{\omega_n^2}{[(\omega_n^2 - \omega^2)^2 + 4\zeta^2\omega_n^2\omega^2]^{1/2}} \quad \text{and} \quad \phi(\omega) = -\tan^{-1}\left(\frac{2\zeta\omega_n\omega}{\omega_n^2 - \omega^2}\right). \tag{11.34}$$

When $L = 1$ H, $C = 1$ μF, and $R = 800$ Ω, the damping ratio and undamped natural frequency are $\zeta = 0.4$ and $\omega_n = 1000$ rad/s, respectively. $M(\omega)$ and $\phi(\omega)$ are then

$$M(\omega) = \frac{10^6}{[(10^6 - \omega^2)^2 + 6.4 \times 10^5\omega^2]^{1/2}} \quad \text{and} \quad \phi(\omega) = -\tan^{-1}\left(\frac{800\omega}{10^6 - \omega^2}\right). \tag{11.35}$$

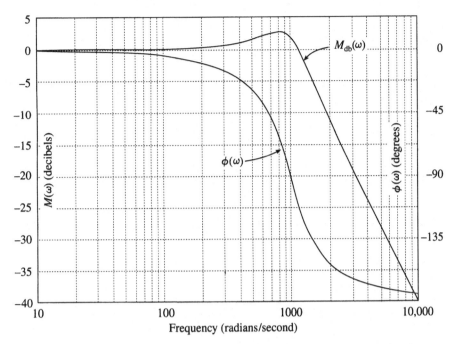

Fig. 11.14 Amplitude ratio and phase shift for RLC circuit; $\omega_n = 1000$ rad/s, $\zeta = 0.4$.

These functions, calculated by computer, are displayed on the Bode chart in Fig. 11.14. The resonant frequency and M_p for this system can be determined analytically by finding the value for ω which makes the derivative of $M(\omega)$ zero. This calculation gives the following results:

$$\omega_r = \begin{cases} \omega_n\sqrt{1-2\zeta^2} & \text{for } 0 < \zeta < \sqrt{2}/2, \\ 0 & \text{for } \zeta > \sqrt{2}/2; \end{cases}$$

$$M_p = \begin{cases} M(\omega_r) = 1/2\zeta\sqrt{1-\zeta^2} & \text{for } 0 < \zeta < \sqrt{2}/2, \\ 1 & \text{for } \zeta > \sqrt{2}/2. \end{cases}$$

(11.36)

For our RLC circuit example, these formulas give the parameter values $\omega_r = 824.62$ rad/s and $M_p = 1.3639$ (or 2.695 db). The phase shift at the resonant frequency is $-64.123°$. These check very closely with the table of laboratory measurements in Section 11.2.

The "arrow" method for calculating $G(j\omega)$ reveals some interesting properties of this second-order system. Figure 11.15 shows the configuration for an underdamped pair of poles. In terms of the "arrow" geometry defined in Section 7.2,

$$G(j\omega) = \frac{\omega_n^2}{\vec{P}_1\vec{P}_2} = \frac{\omega_n^2}{|\vec{P}_1||\vec{P}_2|}\epsilon^{-j(\theta_1+\theta_2)} = |G(j\omega)|\epsilon^{j\angle G(j\omega)}.$$

(11.37)

Because $P_1 = \zeta\omega_n - j\omega_o$ and $P_2 = \zeta\omega_n + j\omega_o$, we have $\vec{P}_1 = \zeta\omega_n - j(\omega_o - \omega)$ and $\vec{P}_2 = \zeta\omega_n + j(\omega_o + \omega)$. We recall that $\omega_o = \omega_n\sqrt{1-\zeta^2}$, which is the frequency of

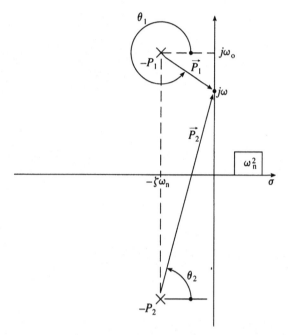

Fig. 11.15   "Arrow" method for calculating $G(j\omega)$ of the RLC circuit.

the transient oscillations occurring in $y(t)$ if the system is excited with a step input, an impulse input, or a nonzero initial condition.

Consider first the phase shift function as ω varies from 0 to ∞. At $\omega = 0$ we have $\angle G(j0) = -(\theta_1 + \theta_2) = -360°$, which is the same as $0°$. As ω increases, both θ_1 and θ_2 increase, so that $\angle G(j\omega)$ decreases. At very high frequencies both θ_1 and θ_2 approach $90°$, so that $\angle G(j\omega) \to -180°$ as $\omega \to \infty$. In Fig. 11.14, the phase shift curve $\phi(\omega)$ shows these tendencies. The curve becomes steeper in the neighborhood of ω_n because the proximity of the upper pole at $s = -P_1$ causes $d\theta_1/d\omega$ to be larger there than in other parts of the spectrum.

Next consider the amplitude ratio function, $|G(j\omega)|$, as ω varies from 0 to ∞. At $\omega = 0$ the two arrows $\vec{P}_1$ and $\vec{P}_2$ have the same length ω_n. Hence $|G(j0)| = 1$. As ω increases from 0, $|\vec{P}_2|$ increases and $|\vec{P}_1|$ decreases, but the fraction of decrease in $|\vec{P}_1|$ for a given small increase in ω is greater than the corresponding fractional increase in $|\vec{P}_2|$; $|G(j\omega)|$ therefore increases with ω in the low-frequency range. As ω approaches the neighborhood of ω_0, $|\vec{P}_1|$ shrinks fractionally very much more than $|\vec{P}_2|$ grows, and $|G(j\omega)|$ reaches its maximum M_p in that neighborhood, at the resonant frequency ω_r. As ω increases beyond the resonant frequency range, both $|\vec{P}_1|$ and $|\vec{P}_2|$ grow, so that $|G(j\omega)|$ continually diminishes with an increase in frequency. At very high frequency, $|G(j\omega)| \to 0$. This characteristic is also exhibited in Fig. 11.14.

The point on the $j\omega$ axis at which resonance occurs can be found geometrically as the point at which the product $|\vec{P}_1||\vec{P}_2|$ is at its minimum, which in Fig. 11.16 is at the intersection of the $j\omega$ axis with a circle centered as $s = -\zeta\omega_n$ and having radius ω_0. This point falls below $j\omega_0$, showing that the three different frequencies with special meanings in this second-order system are:

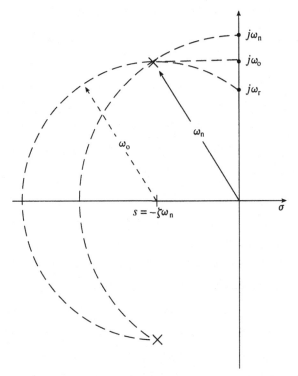

Fig. 11.16 The three characteristic frequencies of a second-order system; ω_n, ω_o, and ω_r defined.

$$
\begin{aligned}
\text{UNDAMPED NATURAL FREQUENCY:} \quad & \omega_n; \\
\text{TRANSIENT OSCILLATION FREQUENCY:} \quad & \omega_o = \omega_n\sqrt{1-\zeta^2}; \\
\text{RESONANT FREQUENCY:} \quad & \omega_r = \omega_n\sqrt{1-2\zeta^2}.
\end{aligned}
\tag{11.38}
$$

Many control systems are well-damped, and these three frequencies are then well-separated in the spectrum. In fact, ω_r becomes zero if the damping ratio exceeds $\sqrt{2}/2$. But in electronic systems, or in structures, systems are often very lightly damped, and these three frequencies are so close together in the spectrum that there is little practical purpose in giving them separate designations.

We note that the expressions for ω_r and M_p in Eqn. 11.36 apply only to a second-order system having a transfer function whose numerator is ω_n^2, so that $M(0) = 1$. A second-order system with a finite zero in the transfer function will have a resonant frequency that depends on ζ, ω_n, and the location of the zero. Similarly, the M_p depends on those three parameters and on the gain constant of the transfer function. In a higher-order system the ω_r and M_p depend on all the poles, all the zeros, and the gain constant of the transfer function.

11.4 Asymptotic Approximation to Bode Plots

As an introduction to a useful technique for approximating the amplitude ratio curve and phase shift curves for a linear system on the Bode plot, we first consider two examples of exact Bode plots.

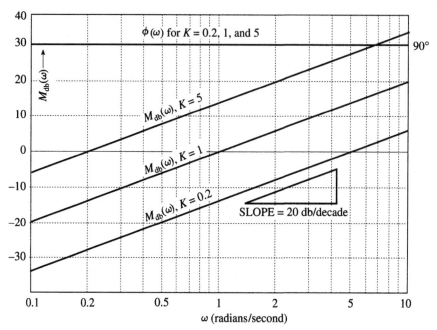

Fig. 11.17   Bode plot for $G(s) = Ks$ for three values of K.

First, consider the system transfer function $G(s) = Ks$, where K is positive. The frequency response is given by

$$M(\omega) = |G(j\omega)| = K\omega,$$
$$\phi(\omega) = \angle G(j\omega) = 90°. \tag{11.39}$$

Note that $M(\omega) = 1$ for $\omega = 1/K$. Figure 11.17 shows the Bode plot for this transfer function for $K = 1/5$, $K = 1$, and $K = 5$. As long as K is positive, the phase shift curve is independent of K, and the amplitude curve depends on K only for its vertical position on the Bode plot. Increasing K by a factor of 5 simply moves the amplitude curve up on the chart by $20\log_{10}(5)$ db (13.98 db), and decreasing K by a factor of 5 moves the curve down by 13.98 db.

To calculate the slope of the amplitude curve on the Bode chart for this example, first make the following substitution. Let $u = \log_{10}\omega$ so that $M_{db}(\omega)$ can be expressed in terms of u:

$$M_{db}(\omega) = 20\log_{10}(K\omega) = 20[\log_{10} K + \log_{10}\omega]; \quad \text{therefore}$$
$$M_{db}(u) = 20[\log_{10} K + u]. \tag{11.40}$$

Next, take the derivative of $M_{db}(u)$ with respect to u:

$$\frac{dM_{db}(u)}{du} = 20 \text{ db/(unit of } u). \tag{11.41}$$

A change of *one unit of u* is a change in ω by a factor of 10, a *decade* on the frequency scale. In this example the slope of the $M_{db}(\omega)$-versus-ω curve is 20 db/decade, and is the same for all values of K. (A change of frequency by a

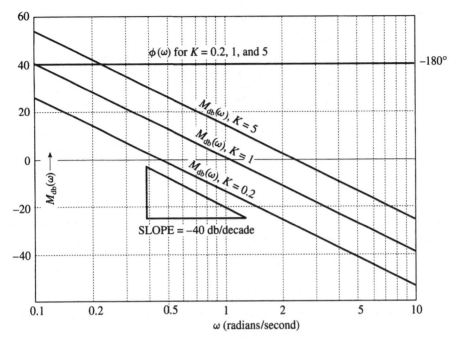

Fig. 11.18 Bode plot for $G(s) = K/s^2$ for three values of K.

factor of 2 on the frequency scale is called an *octave,* so the slope of 20 db/decade is sometimes expressed as 6.0206 db/octave.)

For our second example the transfer function is $G(s) = K/s^2$, where K is positive. The frequency response is given by

$$M(\omega) = |G(j\omega)| = K/\omega^2 \quad \text{and} \quad \phi(\omega) = \angle G(j\omega) = -180°. \tag{11.42}$$

Figure 11.18 shows the Bode plot for this transfer function, using the same values of K as those in the first example. Again, the phase shift is independent of K, provided that K is positive, and the amplitude curve depends on K only for its vertical position on the plot, with a change of K by a factor of 5 causing a vertical shift of 13.98 db. Also, in this second example when $K = 1$, $|G(j1)| = 1$. However, the slope of the $M_{db}(\omega)$ curve is negative, and is calculated as

$$M_{db}(u) = \left[20\log_{10}\left(\frac{K}{\omega^2}\right)\right]_{\omega = 10^u} = 20[\log_{10} K - 2u],$$

$$\frac{dM_{db}(u)}{du} = -40 \text{ db/(unit of } u); \quad \text{so that} \tag{11.43}$$

$$\frac{dM_{db}(\omega)}{d\omega} = -40 \text{ db/decade.}$$

Using these two examples, we can express the exact amplitude and phase shift characteristics for the general case of

$$G(s) = Ks^r \Rightarrow G(j\omega) = K(j\omega)^r. \tag{11.44}$$

If K is positive, then:

(a) the phase shift curve $\phi(\omega)$ is represented on the Bode plot by the constant $\phi(\omega) = r \times 90°$, and

(b) the amplitude ratio curve $M_{db}(\omega)$ is a straight line with slope of $20 \times r$ db/decade having a value of $20 \log_{10}(K)$ db at $\omega = 1$ rad/s.

If r is positive, we say that the transfer function is an rth-order *differentiator;* if r is negative, the transfer function is an rth-order *integrator.*

We now turn to useful techniques for approximating the amplitude and phase characteristics for the general transfer function

$$G(s) = \frac{K(s+Z_1)(s+Z_2)\cdots(s+Z_m)}{(s+P_1)(s+P_2)\cdots(s+P_n)}. \tag{11.45}$$

First, consider the transfer function $G(s) = K(s+Z)$. The amplitude ratio and phase shift for this transfer function are:

$$M(\omega) = |G(j\omega)| = |K|[\omega^2+Z^2]^{1/2},$$
$$\phi(\omega) = \angle G(j\omega) = \tan^{-1}(\omega/Z). \tag{11.46}$$

Now observe the frequency response behavior at very low frequency, that is, for $\omega \ll Z$. Assuming K is positive,

$$M(\omega) \cong KZ \quad \text{and} \quad \phi(\omega) \cong 0°. \tag{11.47}$$

The *low-frequency asymptotes* (LFA) are defined as

$$[\text{LFA}]_M = 20 \log_{10}(KZ) \quad \text{and} \quad [\text{LFA}]_\phi = 0°. \tag{11.48}$$

These are plotted in Fig. 11.19 in the low-frequency range.

Next, consider the frequency response at very high frequencies, for $\omega \gg Z$. In this case $M(\omega) \cong K\omega$ and $\phi(\omega) \cong 90°$, and the high-frequency asymptotes are

$$[\text{HFA}]_M = 20 \log_{10}(K\omega) \quad \text{and} \quad [\text{HFA}]_\phi = 90°. \tag{11.49}$$

The high-frequency asymptote therefore has a slope of 20 db/decade, and passes through the point $(Z, [KZ]_{db})$ on the Bode plot.

For intermediate frequencies around $\omega = Z$, we use Eqn. 11.46 to calculate the amplitude ratio and phase shift, plotting the results in Fig. 11.19. The $M_{db}(\omega)$ curve follows the LFA and the HFA very closely at the extreme frequencies, deviating from the asymptotes by the largest amount at the *break frequency Z*, where it lies 3.0103 db above the asymptotes. One octave below the break frequency, at $\omega = Z/2$, $M_{db}(\omega)$ lies only 0.9691 db above the asymptotes; one octave above the break frequency, at $\omega = 2Z$, $M_{db}(\omega)$ is also 0.9691 db above the asymptote. Therefore, we simply use the asymptotes as a rough approximation to $M_{db}(\omega)$. This approximation may be refined when we remember that the actual curve passes approximately 3 db above the junction of the LFA and the HFA and approximately 1 db above the asymptotes at frequencies an octave above and an octave below the break frequency.

At the break frequency (also called the *corner frequency*), $\phi(\omega)$ is 45°. An approximation to the $\phi(\omega)$ curve in the mid-frequency range is established by

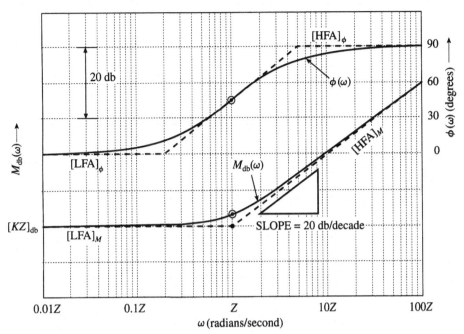

Fig. 11.19 $M_{db}(\omega)$ and $\phi(\omega)$ and their asymptotes on the Bode chart; $G(s) = K(s+Z)$.

calculating the slope of the curve at $\omega = Z$ and using a straight-line tangent to the curve at $\omega = Z$ as the approximation. To calculate the slope on the Bode plot we use the substitution $u = \log_{10}\omega$, as in Eqn. 11.40:

$$\frac{d\phi}{du} = \frac{d\phi}{d\omega} \times \left[\frac{du}{d\omega}\right]^{-1} = \left[\frac{Z^2}{Z^2+\omega^2} \cdot \frac{1}{\omega}\right] \times \left[\frac{\log_{10}\epsilon}{\omega}\right]^{-1}, \quad \text{so we have}$$

$$\left[\frac{d\phi}{du}\right]_{\omega=Z} = \left[\frac{Z^2}{2Z^2}\right] \times \frac{1}{0.43429} = 1.15129 \text{ rad/decade.}$$

(11.50)

The tangent intersects the HFA at $\omega = 4.8189Z$ rad/s and the LFA at $\omega = (1/4.8189)Z$ rad/s. A close approximation to this tangent line, and one that is easy to remember, is a line intersecting the LFA at $\omega = 0.2Z$ rad/s and the HFA at $\omega = 5Z$ rad/s. This straight-line approximation to the $\phi(\omega)$ curve is very good in the immediate neighborhood of $\omega = Z$, but has its maximum error of about 11.7° at the intersection points. A second approximation to $\phi(\omega)$ in the mid-frequency range, also easy to remember, is the line intersecting the asymptotes at $\omega = 0.1Z$ and $\omega = 10Z$. This second approximation suffers larger errors than the first approximation in the neighborhood of $\omega = Z$, but it has a maximum error of only about 5°.

The transfer function $G(s) = K/(s+P)$ is that of a first-order dynamic system, with the frequency response given by

$$M(\omega) = \frac{|K|}{|j\omega+P|} \quad \text{and} \quad \phi(\omega) = -\tan^{-1}\left(\frac{\omega}{P}\right).$$

(11.51)

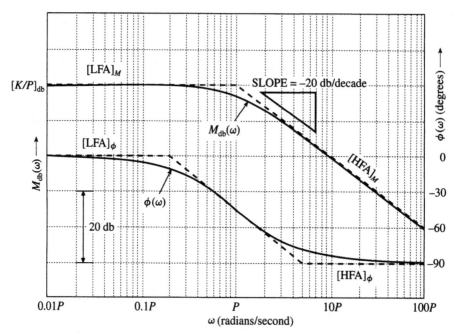

Fig. 11.20 $M_{db}(\omega)$ and $\phi(\omega)$ and their asymptotes on the Bode chart; $G(s) = K/(s+P)$.

The $\phi(\omega)$ curve for this single-pole system is the inverse of that for the single-zero system. Because the first-order term is in the denominator, the $M_{db}(\omega)$ curve breaks downward at the corner frequency $\omega = P$. Both $M_{db}(\omega)$ and $\phi(\omega)$ are plotted in Fig. 11.20, along with the extreme-frequency asymptotes determined using the same techniques as those displayed in Fig. 11.19. The asymptotic approximations to $M_{db}(\omega)$ and $\phi(\omega)$ in the mid-frequency range are also similar to those in Fig. 11.19.

Next we illustrate the effectiveness of the asymptotic approximation by making a quick and accurate sketch of the amplitude and phase characteristics of a system whose transfer function is composed of several first-order terms. The transfer function for this example is

$$G(s) = \frac{5 \times 10^5 (s+1)}{(s+5)(s+10)(s+50)(s+200)}. \tag{11.52}$$

For illustrative purposes, we write this as the product of five terms:

$$G(s) = \left[\frac{(s+1)}{1}\right]\left[\frac{5}{(s+5)}\right]\left[\frac{10}{(s+10)}\right]\left[\frac{50}{(s+50)}\right]\left[\frac{200}{(s+200)}\right]. \tag{11.53}$$

The frequency response $G(j\omega)$ is also the product of the five terms:

$$G(j\omega) = \left[\frac{(j\omega+1)}{1}\right]\left[\frac{5}{(j\omega+5)}\right]\left[\frac{10}{(j\omega+10)}\right]\left[\frac{50}{(j\omega+50)}\right]\left[\frac{200}{(j\omega+200)}\right]. \tag{11.54}$$

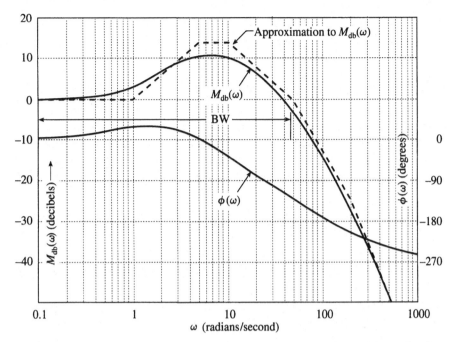

Fig. 11.21 $M_{db}(\omega)$, $\phi(\omega)$, and a straight-line approximation to $M_{db}(\omega)$ for the $G(s)$ in Eqn. 10.52.

The amplitude ratio of the system is the *product* of the five individual amplitude ratios, but when those amplitude ratios are expressed in decibels on the Bode plot, the composite amplitude ratio (in db) is the *sum* of the individual amplitude ratios. Note that each of these five terms has a zero-frequency gain (amplitude ratio) of 1, or 0 db. Since the break frequency of each term is obvious, the asymptotic approximation of the amplitude ratio of each term may be quickly sketched on the Bode plot. Furthermore, because each of the LFAs is the 0-db line, the sum of the five terms is also obtained almost by inspection. Figure 11.21 shows the straight-line approximation to $|G(j\omega)|_{db}$ as the dashed lines. Note a corner in the straight-line approximation at each break frequency. Note also that the approximation breaks positively, adding a slope of 20 db/ decade to the approximation if the break frequency comes from a zero of $G(s)$, and negatively, adding a slope of -20 db/decade, if the break frequency corresponds to a pole of $G(s)$. The actual $|G(j\omega)|_{db}$ curve, denoted as $M_{db}(\omega)$, and the actual phase shift curve $\phi(\omega)$ are also plotted. A straight-line approximation to the phase shift curve can also be obtained by using this technique.

The $M_{db}(\omega)$ curve shows a resonant peak at $\omega = 6.75$ rad/s, where $M_p = 10.46$ db. This resonance is due not to a lightly damped pair of poles in $G(s)$, as in some systems, but rather to the zero in $G(s)$ occurring at a lowest frequency of all the singularities. The *bandwidth* of a control system is the lowest frequency at which the $|G(j\omega)|_{db}$ curve passes through a specified amplitude level, usually 3 db or 6 db below the zero-frequency gain of the system. In this case the bandwidth (BW), defined at the −3-db level, is 48.17 rad/s. The $M_{db}(\omega)$ curve normally diminishes with increasing frequency, as it does in this example, so that the bandwidth can be used as a performance measure for the system: a high bandwidth indicates a fast transient response. In some cases the $M_{db}(\omega)$ curve shows peaks and valleys, crossing the −3-db level at several different frequencies, and in these cases a simple interpretation of bandwidth is less useful as a measure of the dynamic performance of the system.

All transfer functions for linear systems defined by ordinary differential equations having constant, real-valued coefficients can be expressed in the standard form, in which only real-valued coefficients appear:

$$G(s) = \frac{N(s)}{D(s)} = \frac{K[s^m + a_{m-1}s^{m-1} + \cdots + a_1 s + a_0]}{[s^n + b_{n-1}s^{n-1} + \cdots + b_1 s + b_0]}. \tag{11.55}$$

The factors of both $N(s)$ and $D(s)$ are either first-order real polynomials of the form $(s+Z)$ or $(s+P)$, or second-order real polynomials of the form $[s^2 + as + b]$ having complex conjugate roots.

To complete our development of straight-line approximation techniques, we consider the second-order term with complex roots:

$$G(s) = \frac{\omega_n^2}{s^2 + 2\zeta\omega_n s + \omega_n^2} \quad \text{where } \zeta < 1, \tag{11.56}$$

so that the frequency response is

$$|G(j\omega)| = M(\omega) = \left| \frac{\omega_n^2}{(\omega_n^2 - \omega^2) + j2\zeta\omega_n\omega} \right| \quad \text{and}$$

$$\angle G(j\omega) = \phi(\omega) = -\tan^{-1}\left(\frac{2\zeta\omega_n\omega}{(\omega_n^2 - \omega^2)} \right). \tag{11.57}$$

At very low frequency, $\omega \ll \omega_n$, the amplitude ratio is approximately 1 and the phase shift is approximately 0°. The LFAs are

$$[LFA]_M = 0 \text{ db}, \qquad [LFA]_\phi = 0°. \tag{11.58}$$

At very high frequency, $[HFA]_\phi = -180°$ and $[HFA]_M$ intercepts the 0-db line at $\omega = \omega_n$ with a slope of −40 db/decade. These asymptotes for the amplitude ratio are plotted in the upper part of Fig. 11.22 and are labeled A. $M_{db}(\omega)$ curves for $\zeta = 0.2$ and $\zeta = 1$ are also plotted to indicate the type of deviations one might expect between the asymptotes and the actual $M_{db}(\omega)$ curve. The lower part of Fig. 11.22 shows the $\phi(\omega)$ curves for $\zeta = 0.2$ and $\zeta = 1$. A set of amplitude ratio curves and phase shift curves for this transfer function, covering a wide range of damping ratios, is provided in Appendix F.

A typical open-loop transfer function of an electrohydraulic servomechanism will serve as an example of a straight-line approximation of the frequency response of a system having a lightly damped pair of poles. The factored form of

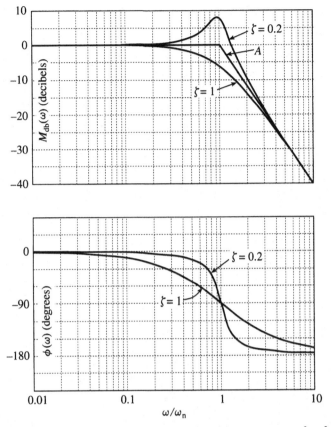

Fig. 11.22 $M_{db}(\omega)$ and $\phi(\omega)$ for the quadratic function $\omega_n^2/(s^2 + 2\zeta\omega_n s + \omega_n^2)$.

$$G(s) = \frac{50(s+4)}{s[s^2+2s+100]} \tag{11.59}$$

makes it convenient to draw the straight-line approximation of the amplitude ratio curve on the Bode plot:

$$G(s) = [2]\left[\frac{s+4}{4}\right]\left[\frac{1}{s}\right]\left[\frac{100}{s^2+2s+100}\right]. \tag{11.60}$$

In Fig. 11.23, the upper graph shows this approximation, along with the calculated $M_{db}(\omega)$ curve. The resonant peak due to the quadratic term accounts for most of the deviation between the approximation and the actual curve in the neighborhood of the resonant frequency. The lower plot shows the phase shift curve.

The frequency response characteristics of any transfer function of the form in Eqn. 11.55 can be quickly approximated on the Bode plot using the asymptotic techniques demonstrated here. Refinements of the straight-line approximations can be made using the corrections illustrated in Figs. 11.19 and 11.20 for the first-order pole and zero factors. The corrections shown in Fig. 11.22 and in Appendix F may be used for the quadratic factors.

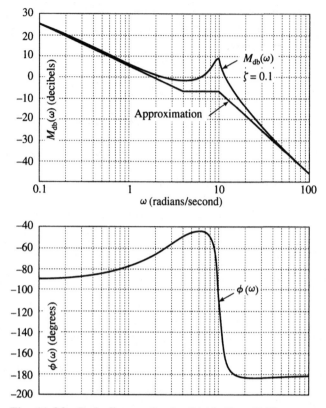

Fig. 11.23   Bode diagram for the $G(s)$ in Eqn. 11.59.

11.5   Frequency Response of Minimum Phase Systems

The distinguishing characteristic of the transfer function of a minimum phase system is that all its poles and zeros lie in the left-half s plane or on the $j\omega$ axis. (Nonminimum phase transfer functions are discussed in Section 11.6.)

If the "arrow" method for calculating $G(j\omega)$ (e.g. Fig. 11.15) is applied to a minimum phase function, it is clear that both the magnitude and argument of each arrow depend on the position of the corresponding pole or zero, as well as on the position of the terminal point $s = j\omega$. Consequently, both the magnitude and argument of the composite of these arrows, $M(\omega)$ and $\phi(\omega)$, depend on all the poles and zeros and also on the frequency. It is therefore reasonable to suppose that, for a given transfer function, an interdependence exists between the $M(\omega)$ and the $\phi(\omega)$ functions. That such an explicit relationship in fact exists in such systems (where all the arrows come from the left side) is shown from the theory of complex variables. Bode and others have explored this relationship for minimum phase transfer functions using that theory, and have developed formulas in the context of electric circuit theory that also pertain to control system analysis.

The frequency response function $G(j\omega)$ may be expressed either in the polar form

$$G(j\omega) = |G(j\omega)|\epsilon^{j\angle G(j\omega)} = M(\omega)\epsilon^{j\phi(\omega)}, \quad \text{where}$$

$$\underbrace{M(\omega)}_{\substack{\text{amplitude} \\ \text{ratio}}} = |G(j\omega)| \quad \text{and} \quad \underbrace{\phi(\omega)}_{\substack{\text{phase} \\ \text{shift}}} = \angle G(j\omega),$$

(11.61)

or in the rectangular form

$$G(j\omega) = \underbrace{\text{Re}[G(j\omega)]}_{\substack{\text{real part} \\ \text{of } G(j\omega)}} + j \underbrace{\text{Im}[G(j\omega)]}_{\substack{\text{imaginary} \\ \text{part of} \\ G(j\omega)}}.$$

(11.62)

In circuit theory it is convenient to work with the rectangular form of $G(j\omega)$, but in control system engineering the polar form of $G(j\omega)$ is more useful because the amplitude and phase functions are related directly to laboratory measurements. Because Bode used the rectangular form, his theorems relate the properties of $\text{Re}[G(j\omega)]$ to the properties of $\text{Im}[G(j\omega)]$, which he designated as *attenuation* and *phase,* respectively. Because the term "attenuation" is used interchangeably with the term "amplitude ratio," incorrect interpretations of Bode's theorems in terms of $M(\omega)$ and $\phi(\omega)$ have appeared in modern literature. Nevertheless, some qualitative features of the relationship between $M(\omega)$ and $\phi(\omega)$ for minimum phase transfer functions can be gleaned from Bode's exact formulas, which relate $\text{Re}[G(j\omega)]$ to $\text{Im}[G(j\omega)]$.

One practical feature of minimum phase functions is that if one specifies the $M(\omega)$ function over the entire spectrum $0 \leq \omega < \infty$ and then constructs a physical system having that $M(\omega)$ characteristic (e.g., an electric network), then one must accept whatever $\phi(\omega)$ function results. Conversely, when specifying the $\phi(\omega)$ function over the entire spectrum, one forgoes any design authority over the resultant $M(\omega)$ function. However, it is possible to prescribe both $M(\omega)$ and $\phi(\omega)$ over limited parts of the spectrum, provided their behavior in the unspecified parts of the spectrum is acceptable. Some examples in Chapter 14 utilize this principle in the design of compensators to improve the performance of servomechanisms.

Another qualitative property that is useful in restricted circumstances comes from the examples which led to Eqn. 11.44. If the amplitude function $M(\omega)$ changes slope only gradually with frequency in a section of the spectrum, then the phase shift $\phi(\omega)$ at a frequency centered in that section will be approximately $(\pi/40) \times S$ rad, where S is the slope of $M(\omega)$ in db/decade at that center frequency. This relationship is exact at all frequencies if $M(\omega)$ has a constant slope, as it has in the examples given in Figs. 11.17 and 11.18. It also approaches exactness at frequencies near zero and at very high frequencies. These extreme-frequency relationships between $M(\omega)$ and $\phi(\omega)$ are often used to advantage in checking experimental data. If the data do not fit this pattern then the system under test may have been operated in its nonlinear range, or perhaps the system has a nonminimum phase transfer function.

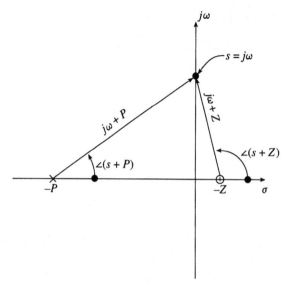

Fig. 11.24 "Arrow" method for calculating $G_{\text{NMP}}(j\omega)$ of Eqn. 11.63.

11.6 Frequency Response of Nonminimum Phase Systems

The amplitude and phase shift properties of nonminimum phase systems are easily studied with the aid of the s-plane "arrow" diagrams introduced in Section 11.3. A simple example is the transfer function

$$G_{\text{NMP}}(s) = \frac{K(s+Z)}{(s+P)},\tag{11.63}$$

where Z is negative. This system is stable if P is positive, but it is nonminimum phase because the zero lies in the right-half s plane. The zero frequency gain $G(0)$ is KZ/P. Figure 11.24 shows a pole-zero plot of this transfer function (for $|Z| < P$), along with the arrows used to calculate $G(j\omega)$. The amplitude ratio and phase shift at frequency ω are given by

$$M(\omega) = \frac{|K||j\omega+Z|}{|j\omega+P|},$$

$$\phi(\omega) = \tan^{-1}\left(\frac{\omega}{Z}\right) - \tan^{-1}\left(\frac{\omega}{P}\right) - 180°,\tag{11.64}$$

where it is assumed that K is negative in order that the zero-frequency gain will be positive. Notice that for very low frequency the phase shift is close to $0°$, but at very high frequency the phase shift approaches $-180°$. It is also clear in this example, since $|Z| < P$, that $\phi(\omega)$ is a monotonically decreasing function of ω. The amplitude ratio for $G_{\text{NMP}}(j\omega)$ is plotted in the upper graph in Fig. 11.25; the phase shift is plotted on the lower graph. Here $Z = -1$, $P = 10$, and $K = -10$.

Now consider a minimum phase transfer function having the same form and the same denominator as that in Eqn. 11.63, but with $Z = 1$ and $K = 10$. Because

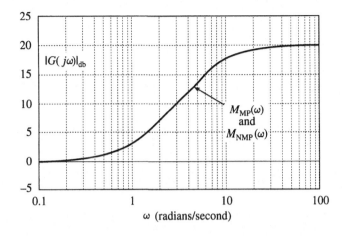

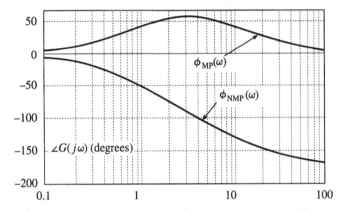

Fig. 11.25 Comparison of phase shift functions between a minimum phase system and a nonminimum phase system with identical amplitude ratio functions.

the magnitude of the $\overline{j\omega + Z}$ arrow will be the same as that for $G_{NMP}(j\omega)$, the amplitude ratio will be identical to that of the nonminimum phase system; however, the argument $\angle(j\omega + Z)$ will be close to 0° at low frequency, increasing monotonically to 90° at high frequency. The phase shift for $G_{MP}(j\omega)$ is plotted in the lower graph of Fig. 11.25 for comparison with that of $G_{NMP}(j\omega)$. This comparison illustrates the origin of the term "nonminimum phase" for systems having zeros in the right-half plane.

It is also interesting to compare the transient responses of these two systems. Let each be driven by a unit step input, so that the output functions are:

$$Y_{MP}(s) = \frac{10(s+1)}{s(s+10)}, \qquad Y_{NMP}(s) = \frac{-10(s-1)}{s(s+10)}. \tag{11.65}$$

The corresponding time responses are:

$$y_{MP}(t) = 1 + 9\epsilon^{-10t}, \qquad y_{NMP}(t) = 1 - 11\epsilon^{-10t}. \tag{11.66}$$

Each has a final value of 1, and each has the same characteristic time constant of 0.1 s, but the two responses are remarkably different, as shown in Fig. 11.26.

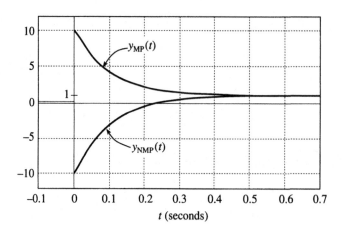

Fig. 11.26   Comparison of transient responses of the two systems of Fig. 11.25.

Consider a second example, which compares a minimum phase system with a nonminimum phase system having these two transfer functions:

$$G_{MP}(s) = \frac{1500(s+1)(s+2)}{(s+3)(s+4)(s+10)[s^2+6s+25]},$$

$$\tag{11.67}$$

$$G_{NMP}(s) = \frac{1500(s-1)(s-2)}{(s+3)(s+4)(s+10)[s^2+6s+25]}.$$

The upper graph in Fig. 11.27 shows the amplitude ratio to be the same for both. But the lower graph shows the phase shift characteristics to be markedly different. Another dramatic difference in the dynamics of the two systems is illustrated in their transient responses to a unit step input, as shown in Fig. 11.28. It is interesting to note that certain significant properties of these two systems are identical:

  both are fifth-order systems having the same characteristic polynomial;
  both have the same zero-frequency gain and same final value in the
    transient response;
  both have two zeros in the transfer function;
  both have the same amplitude ratio function over the entire spectrum
    $0 \le \omega < \infty$.

Yet the dynamic responses are very different. One might be acceptable in a practical application, whereas the other would be unacceptable. This example illustrates one of the principles of control engineering introduced in Chapter 7 – namely, the zeros of a transfer function represent significant dynamic properties of the system. The difference in the phase characteristics (Figs. 11.25 and 11.27) is the manifestation of this principle in the frequency domain.

  Other examples of nonminimum phase transfer functions are provided in Problem 11.17.

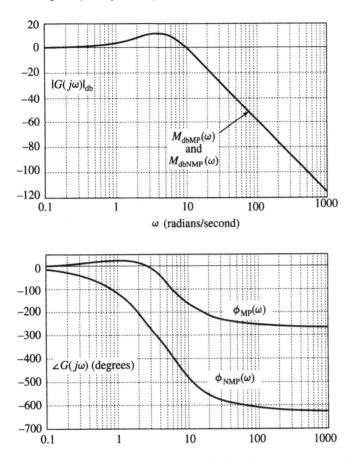

Fig. 11.27 Comparison of phase shift functions between a minimum phase system and a nonminimum phase system with identical amplitude ratio functions (Eqn. 11.67).

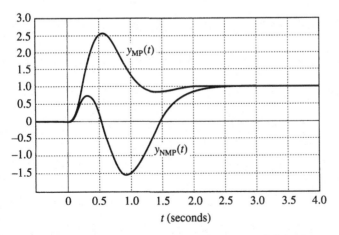

Fig. 11.28 Comparison of transient responses of the two systems of Fig. 11.27.

11.7 Experimental Measurements of the Frequency Response of Unstable Systems

The frequency response of an unstable system is as well-defined as that of a stable system. Furthermore, the frequency response can be measured experimentally in almost the same way as that of a stable system. In Chapter 10, it was shown that an unstable plant can be controlled satisfactorily if a controller of proper design is used in conjunction with feedback from the output of the plant. This simply makes the unstable plant an element of a larger, stable system. The experimental set-up is diagramed in Fig. 11.29. The input to the closed-loop system is driven sinusoidally, $r(t) = A\sin(\omega t)$. Since the system is linear, the variables $\epsilon(t)$, $u(t)$, and $y(t)$ will exhibit responses of the form

$$\epsilon(t) = E\sin(\omega t + \phi_1) + \epsilon_{\text{tran}}(t),$$

$$u(t) = U\sin(\omega t + \phi_2) + u_{\text{tran}}(t), \tag{11.68}$$

$$y(t) = Y\sin(\omega t + \phi_3) + y_{\text{tran}}(t).$$

Because the overall system is stable, all of the transients will decay with time, leaving only the steady-state sinusoidal oscillations

$$\epsilon_{\text{SS}}(t) = E\sin(\omega t + \phi_1),$$

$$u_{\text{SS}}(t) = U\sin(\omega t + \phi_2), \tag{11.69}$$

$$y_{\text{SS}}(t) = Y\sin(\omega t + \phi_3).$$

In this test only $u(t)$ and $y(t)$ are recorded. Note that in the steady state the plant is driven by the sinusoidal input $u_{\text{SS}}(t)$, and responds with the sinusoidal output $y_{\text{SS}}(t)$. The feedback connection provides the stability required for the system to operate in this stable equilibrium condition.

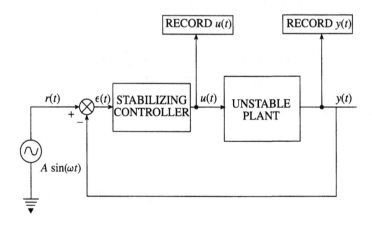

Fig. 11.29 Experimental set-up for measuring the frequency response of an unstable plant.

As an example, consider an unstable plant having the transfer function

$$G_P(s) = \frac{Y(s)}{U(s)} = \frac{-100(s+5)}{(s-1)(s+10)[s^2+12s+50]} \tag{11.70}$$

and a simple stabilizing controller with constant gain

$$G_C(s) = \frac{U(s)}{\epsilon(s)} = -2.4. \tag{11.71}$$

The amplitude of the driving input is kept constant at 1,

$$r(t) = 1\sin(\omega t); \tag{11.72}$$

ω is varied over a range, and the steady-state amplitudes of $y(t)$ and $u(t)$, respectively Y and U, are recorded. The phase shift ϕ_3 of $y(t)$ relative to $r(t)$ is recorded at each test frequency; the phase shift ϕ_2 of $u(t)$ relative to $r(t)$ is also recorded at each test frequency. At each frequency, the amplitude ratio of the plant transfer function is calculated as

$$M_P = 20 \times \log_{10}[Y/U] \text{ db}, \tag{11.73}$$

and the phase shift of the plant transfer function is calculated as

$$\phi_P = \phi_3 - \phi_2. \tag{11.74}$$

The laboratory data and the calculations appear in the following table.

Frequency Response Test on Unstable Plant

ω (rad/s)	Y	ϕ_3 (deg)	U	ϕ_2 (deg)	$[Y/U]_{db}$	$\phi_3-\phi_2$ (deg)
0.1	1.713	−3.518	1.721	−8.426	−0.040	4.91
0.4	1.688	−13.95	1.816	−32.54	−0.635	18.59
0.7	1.638	−24.0	1.993	−53.33	−1.704	29.33
1.0	1.571	−33.48	2.208	−70.32	−2.956	36.84
2.0	1.319	−60.93	2.898	−107.3	−6.837	46.37
4.0	0.907	−107.8	3.701	−145.9	−12.21	38.10
6.0	0.533	−148.3	3.554	−169.1	−16.48	20.80
10	0.169	−191.6	2.800	−181.7	−24.39	−9.90
40	0.0036	−247.5	2.403	−180.2	−56.49	−67.30

The data points for $[Y/U]_{db}$ and $(\phi_3-\phi_2)$ are plotted in Fig. 11.30. A computer-generated plot of the frequency response of the plant transfer function,

$$G_P(j\omega) = \frac{-100(j\omega+5)}{(j\omega-1)(j\omega+10)[(50-\omega^2)+j12\omega]}$$

$$= |G_P(j\omega)|\epsilon^{j\angle G_P(j\omega)}, \tag{11.75}$$

is also plotted, showing the data points lying on the analytical curves.

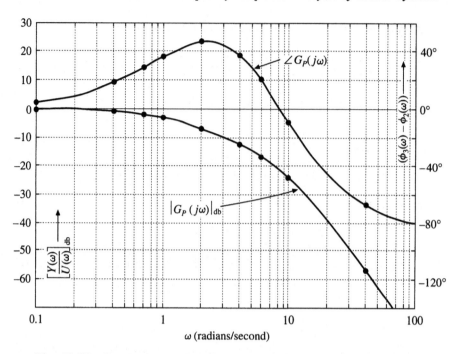

Fig. 11.30 Frequency response of an unstable plant.

11.8 Summary

The frequency response of a dynamic system is defined only for a linear system, and its most common use is for linear systems with constant parameters. It is therefore necessary to take precautions, when making experimental frequency response measurements, to assure that the oscillations do not drive the system into its nonlinear range. Nonsinusoidal waveforms in the output oscillation reveal nonlinear behavior.

The amplitude ratio and phase shift functions of frequency, which comprise the frequency response of a system, may be represented graphically in any one of four forms:

1. rectangular plots display amplitude ratio and phase shift versus frequency on linear scales;
2. Bode plots use a logarithmic scale both for frequency and for the amplitude ratio, and a linear scale for the phase shift;
3. polar (Nyquist) plots depict both amplitude ratio and phase shift in a single curve, along which frequency appears parametrically;
4. the log–magnitude phase plot depicts both amplitude ratio and phase shift on a single curve, along which frequency appears parametrically, with phase shift on a linear scale and with amplitude ratio on a logarithmic scale.

Design specifications stated in frequency response terms often use terminology related to these graphical methods. The break frequency (also called "corner" frequency) and the amplitude ratio slope in db/decade or db/octave on Bode plots are common examples. Peak magnitude (M_p), resonant frequency (ω_r), and bandwidth (BW) are also common terms.

Analytical expressions for the amplitude ratio and phase shift functions for a linear system may be obtained from the transfer function simply by substituting $j\omega$ for s in the transfer function. The s-plane "arrow" diagram of $G(j\omega)$ is a useful tool for evaluating the amplitude ratio and phase shift functions for specific frequencies. Straight-line approximation techniques, coupled with the data on second-order systems in Appendix F, are also effective tools for obtaining a reasonably accurate representation of the amplitude and phase properties from the pole and zeros of the transfer function without resorting to computer calculations. The most practical tool for obtaining precise frequency response plots from transfer functions is the computer.

For a transfer function with m finite zeros and n finite poles, high-frequency behavior (where $\omega \gg$ the highest break frequency) shows the following properties:

$$\left[\frac{dM_{db}(\omega)}{d\omega}\right]_{\omega \to \infty} \cong 20(m-n) \text{ db/decade};$$

$$[\phi(\omega)]_{\omega \to \infty} \cong \begin{cases} (90°)(m-n) & \text{for } K > 0, \\ (90°)(m-n) - 180° & \text{for } K < 0. \end{cases} \tag{11.76}$$

In minimum phase systems, the amplitude and phase functions are intimately related when the entire frequency spectrum is considered. A nonminimum phase system might have the same amplitude ratio function as a minimum phase system but a markedly different phase shift function; Fig. 11.27 is an example of such a case. (It may be necessary to add or subtract 360° to the phase shift property in Eqn. 11.76 for a nonminimum phase system.)

The frequency response of an unstable plant having a transfer function $G(s)$ is found in the same way as that for a stable plant, by substituting $j\omega$ for s to obtain $G(j\omega)$. The phase shift function will exhibit a nonminimum phase characteristic due to the right-half-plane poles in $G(s)$. The amplitude ratio and phase shift functions for an unstable plant may also be determined experimentally. The unstable plant is placed in a feedback loop, which is stabilized by a suitable companion element and driven sinusoidally at a point in the loop different from the input to the plant. The amplitude and phase measurements are made at the input and at the output of the plant, as illustrated in Fig. 11.29.

11.9 Problems

Problem 11.1

The zero-state transfer function of a linear system is

$$\frac{Y(s)}{U(s)} = \frac{5(s+10)}{(s+1)^2(s+50)}.$$

(a) If $u(t) = 3 \sin(6t)$, the output will have the form $y(t) = y_{ss}(t) + y_{tran}(t)$. Find $y_{ss}(t)$.

(b) Find the amplitude ratio $|Y(j\omega)/U(j\omega)|_{db}$ at the frequency 6 rad/s.

Problem 11.2

Let

$$G(s) = \frac{100(s^2 + 6s + 16)}{s^5 + 13s^4 + 99s^3 + 342s^2 + 710s + 195}$$

be the transfer function of a linear system.

(a) Find the DC gain of this system and express it in decibels.

(b) Find the phase shift for $\omega = 0$ and for $\omega \to \infty$.

(c) Is this system stable?

(d) Find the frequency at which the phase shift is $-180°$, and express it in hertz.

Problem 11.3

The transfer function of a second-order system is

$$W(s) = \frac{\omega_n^2}{s^2 + 2\zeta\omega_n s + \omega_n^2}.$$

(a) Prove that the resonant frequency of this system is $\omega_r = \omega_n\sqrt{1 - 2\zeta^2}$.

(b) Prove that the M_p of this system is $1/2\zeta\sqrt{1 - \zeta^2}$, provided that $0 < \zeta < 0.707$.

(c) Find the phase shift at the resonant frequency $\phi(\omega_r)$ in terms of ζ and ω_n.

Problem 11.4

The table in Fig. P11.4 lists the phase shift at several different frequencies for a minimum phase system whose transfer function is $W(s)$. The zero-frequency amplitude ratio of the system is known to be 20 db. There are one or two errors in the table.

ω (rad/s)	Phase Shift (deg)
1	−5.8
10	−46
20	−66
30	−75
50	−86
70	−87
80	−107
90	−127
95	−147
100	−174
110	−219
120	−237
140	−250
200	−260
300	−264
1000	−268

Fig. P11.4

    (a) Find $W(s)$.

    (b) Identify and correct the errors in the table.

Problem 11.5

The frequency response curve (phase shift only) for a minimum phase system is shown in Fig. P11.5. The amplitude ratio is known to be -12 db at $\omega = 1$ rad/s. A unit impulse is applied to the input of the system at $t = 0$. Write the expression for the output response of the system.

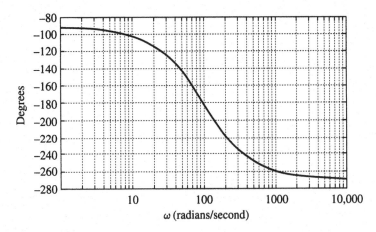

Fig. P11.5

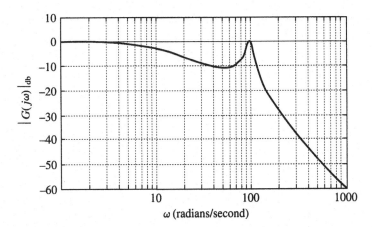

Fig. P11.6

Problem 11.6

A minimum phase system has a transfer function $G(s)$. A Bode plot for the frequency response of this system (amplitude ratio only) is shown in Fig. P11.6. Find $G(s)$.

Problem 11.7

The Bode plot (amplitude ratio only) of the frequency response of a minimum phase system is drawn in Fig. P11.7.

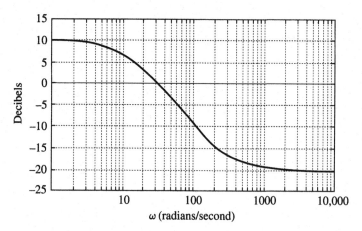

Fig. P11.7

(a) Find the phase shift at $\omega = 100$ rad/s.
(b) A unit step input is applied to this system at $t = 0$. Write the expression for the output response of the system.

Problem 11.8
Refer to Fig. 11.16. Prove the assertions made concerning the locations of the points $j\omega_r$, $j\omega_o$, and $j\omega_n$.

Problem 11.9
Figure P11.9 is the Nyquist plot of the frequency response of a first-order dynamic system. The maximum phase lead of this system, ϕ_{max}, occurs at $\omega = 10$ rad/s. Find ϕ_{max} and the transfer function of the system.

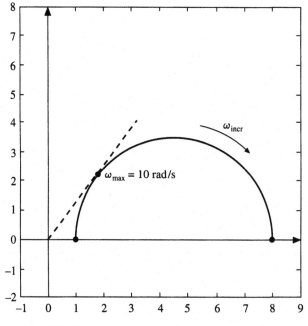

Fig. P11.9

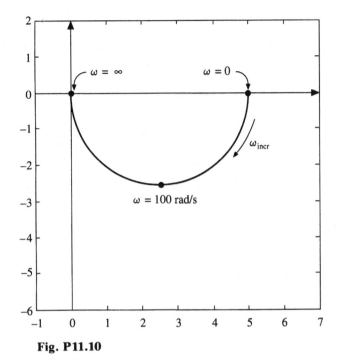

Fig. P11.10

Problem 11.10

The transfer function of a system is $G(s) = K/(s+P)$. Figure P11.10 is the Nyquist plot for $G(j\omega)$. Find K and P. At what frequency is $\angle G(j\omega) = -60°$?

Problem 11.11

Figure P11.11 is the Nyquist plot for the frequency response of a lag-lead device whose transfer function is $K(s+Z)^2/(s+10)(s+P)$.

(a) Find the numerical values for K, Z, and P.
(b) Find the frequency ω_{min}.

Problem 11.12

Select an asymptotic approximation method for the phase shift characteristic of the first-order term $P/(j\omega+P)$, and apply it to the example in Fig. 11.21 to obtain an approximation to the $\phi(\omega)$ curve. Evaluate your approximation technique by calculating the maximum difference between $\phi(\omega)$ and your approximate curve.

Problem 11.13

Let

$$Q(s) = \frac{K(s+10)^3}{(s+2)(s+100)^2(s+200)^2(s+1000)}.$$

Make a straight-line approximation to the Bode plot for $|Q(j\omega)|_{db}$, using the value of K for which $Q(j0) = 1$. Estimate M_p and ω_r from your straight-line approximation to $|Q(j\omega)|_{db}$. Check the accuracy of your estimates by calculating $Q(j\omega)$.

Problem 11.14

How are the two properties (a) and (b) following Eqn. 11.44 altered if K is negative?

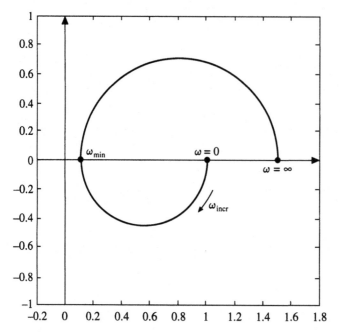

Fig. P11.11

Problem 11.15

(a) Find the poles and zeros of

$$G(s) = \frac{200(s^2 + 3s + 100)}{s^2 + 300s + 20,000}.$$

(b) Make a straight-line approximation to the Bode plot of $|G(j\omega)|_{db}$.

(c) Sketch a smooth approximation to $|G(j\omega)|_{db}$, using your straight-line approximation as a guide. Check your approximation by calculating $G(j\omega)$.

(d) Repeat (b) and (c) for $\angle G(j\omega)$.

Problem 11.16

Let

$$H(s) = \frac{150(\frac{1}{3}s + 1)}{s^4 + 15s^3 + 9s^2 + 145s + 150}.$$

Make straight-line approximations to the Bode plots for both $|H(j\omega)|_{db}$ and $\angle H(j\omega)$. Check your approximations by calculations. Does $H(s)$ represent a minimum phase or nonminimum phase system? Does $H(s)$ represent a stable or an unstable system?

Problem 11.17

Let

$$G(s) = \frac{136(s + Z)}{(s + P)(s + 4)[s^2 + 2s + 17]}.$$

(a) Let $Z = 1$ and $P = 2$.

    1. Is $G(s)$ a minimum phase or a nonminimum phase transfer function?

    2. Find $G(0)$, the zero frequency gain of $G(s)$.

   3. Find $\angle G(j\omega)$ for $\omega = 0$ and $\omega = \infty$.

   4. Find $\angle G(j\omega)$ for $\omega = 2$.

   5. Plot $\angle G(j\omega)$ for $0 \leq \omega < \infty$.

(b) Repeat (a) for $Z = -1$ and $P = 2$.

(c) Repeat (a) for $Z = 1$ and $P = -2$.

(d) Repeat (a) for $Z = -1$ and $P = -2$.

(e) Compare the plots of part 5 for the four cases. Note that $|G(j\omega)|$ is the same for the four cases.

12 Stability Analysis by Nyquist's Criterion

12.1 Introduction

We now describe a method for determining the stability of a feedback system from the frequency response of its open-loop transfer function. This method was developed by Nyquist in the 1920s to explain the way in which the stability of a feedback amplifier depends on its open-loop properties, when those properties are described by amplitude and phase functions. The frequency response view of dynamics in electronic amplifiers is popular because it relates directly to the capability of the devices to transmit audio signals. Nyquist's method for stability determination is less direct than the root-locus method, but it offers the advantage of using frequency response data, which is often the only reliable dynamic information available for a system. It is easily applied to linear systems whose transfer functions are transcendental functions of s, and it is also the basis for some forms of stability analysis in nonlinear systems.

Let us consider the unity feedback system diagramed in Fig. 12.1. This system has an adjustable forward-path gain K. Either the root-locus method or Routh's criterion quickly reveals that the closed-loop system will be stable if

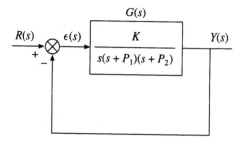

Fig. 12.1 Unity negative feedback system.

and only if the gain lies between 0 and $P_1P_2(P_1+P_2)$. We wish to obtain the same result using Nyquist's method, which requires us to calculate the open-loop frequency response

$$G(j\omega) = \frac{K}{j\omega(j\omega+P_1)(j\omega+P_2)}. \tag{12.1}$$

It is convenient to express $G(j\omega)$ in real and imaginary components:

$$G(j\omega) = \underbrace{\left[\frac{-K(P_1+P_2)}{(P_1^2+\omega^2)(P_2^2+\omega^2)}\right]}_{\text{Re}[G(j\omega)]} + j\underbrace{\left[\frac{K(\omega^2-P_1P_2)}{\omega(P_1^2+\omega^2)(P_2^2+\omega^2)}\right]}_{\text{Im}[G(j\omega)]}. \tag{12.2}$$

We form a Nyquist plot for $G(j\omega)$ by plotting $\text{Im}[G(j\omega)]$ versus $\text{Re}[G(j\omega)]$, with ω as a parameter, as shown in Fig. 12.2. In this example, K, P_1, and P_2 are all positive. In this polar plot of $G(j\omega)$ the arrow from the origin to the point on the curve labeled ω_1 is the complex value of $G(j\omega)$ at the frequency ω_1. Note that, as ω approaches zero, $|G(j\omega)|$ approaches infinity and $\angle G(j\omega)$ approaches $-90°$. Also, $\text{Re}[G(j\omega)]$ approaches the value $-K(P_1+P_2)/P_1^2P_2^2$.

The closed-loop transfer function for the unity feedback system is

$$\frac{Y(s)}{R(s)} = W(s) = \frac{G(s)}{1+G(s)}, \tag{12.3}$$

so the frequency response of the closed-loop system is given by

$$W(j\omega) = \frac{G(j\omega)}{1+G(j\omega)} = |W(j\omega)|\epsilon^{j\angle W(j\omega)}, \tag{12.4}$$

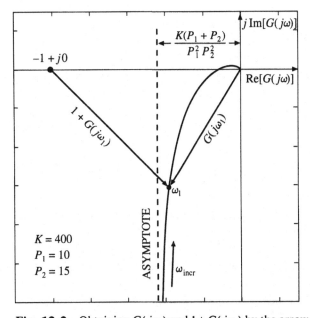

Fig. 12.2 Obtaining $G(j\omega_1)$ and $1+G(j\omega_1)$ by the arrow method.

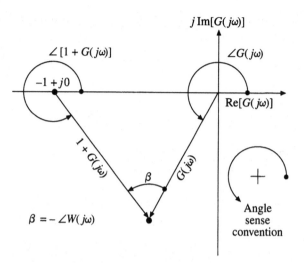

Fig. 12.3   Obtaining $\angle W(j\omega)$ by the arrow method.

where the amplitude ratio and phase shift of $W(j\omega)$ are given by

$$|W(j\omega)| = \frac{|G(j\omega)|}{|1+G(j\omega)|} \quad \text{and}$$

$$\angle W(j\omega) = \angle[G(j\omega)] - \angle[1+G(j\omega)].$$

(12.5)

The complex value of $1+G(j\omega_1)$ is represented by the arrow drawn from the point $-1+j0$ to the point on the $G(j\omega)$ locus labeled ω_1. The angles involved in this analysis are defined in Fig. 12.3 to avoid the ambiguity of sign that can occur when angles between two arrows must be determined. Note that angles are reckoned as positive in the counterclockwise sense. For example, in this diagram $\angle[G(j\omega)]$ is shown as 240° (−120°), and $\angle[1+G(j\omega)]$ is shown as 308° (−52°). It is clear from the triangular configuration that the phase angle of the closed-loop system, $\angle W(j\omega)$, is the negative of angle β, or −68°.

The graphical portrayal of $G(j\omega)$ and $1+G(j\omega)$ in Fig. 12.2 makes it possible to visualize how the closed-loop frequency response varies as ω traverses the entire spectrum from 0 to ∞. For very low frequencies, both the $G(j\omega)$ and the $1+G(j\omega)$ arrows are very long, due to the integrating property of the forward-path element, so the ratio of the absolute values of the two is very close to 1. Because the angle β is very close to zero, $W(j\omega) \cong 1$ at low frequencies. As ω increases toward the mid-frequency range, both $G(j\omega)$ and $1+G(j\omega)$ shrink in absolute value, and because $G(j\omega)$ becomes slightly shorter than $1+G(j\omega)$ the ratio of the absolute values diminishes slightly from its low-frequency value of 1. The angle β becomes larger, indicating that $\angle W(j\omega)$ approaches −90° at mid-range frequencies. At very high frequencies, $|G(j\omega)| \to 0$ and $|1+G(j\omega)| \to 1$, so that $|W(j\omega)| \to 0$. Angle β approaches 270°, so that for high frequencies, $\angle W(j\omega)$ approaches −270°. This behavior of $|W(j\omega)|$ and $\angle W(j\omega)$ is plotted for the curves labeled K_{low} in Fig. 12.4. Note that the amplitude ratio is not represented

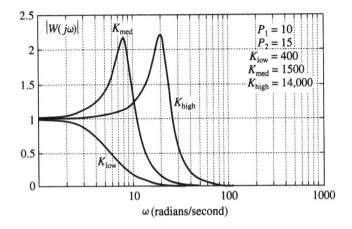

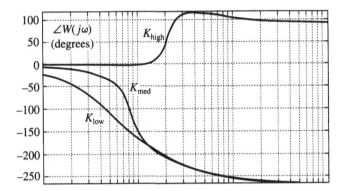

Fig. 12.4 Effect of K on the closed-loop frequency response of the system in Fig. 12.1.

in decibel form on the upper graph. The numerical values used in this example are $P_1 = 10$, $P_2 = 15$, and $K_{low} = 400$.

Next we increase the gain K, observing how this affects the closed-loop frequency response. As long as K remains positive, it does not affect $\angle G(j\omega)$. An increase in K changes the scale of the $G(j\omega)$ curve; it does not affect its shape. Figure 12.5 shows the region in the vicinity of the -1 point when K has a medium value of 1500. Because 1500 is less than $P_1 P_2 (P_1 + P_2)$, the closed-loop system is stable. Note that the ratio of $|G(j\omega)|$ to $|1 + G(j\omega)|$ is greater than 1 in the frequency range in which $G(j\omega)$ is near the -1 point, indicating that $W(j\omega)$ will show a resonant peak in that frequency range. $|W(j\omega)|$ and $\angle W(j\omega)$ are plotted on the curves labeled K_{med} in Fig. 12.4. Note also the abrupt change in the closed-loop phase shift in the frequency range near resonance.

Finally, K is increased to $K_{high} = 14,000$, which is beyond the critical value for stability. The -1 point now lies to the right of the $G(j\omega)$ locus, as shown in Fig. 12.6, where the loci of $G(j\omega)$ for K_{low} and for K_{med} are also shown. Again, in the frequency range in which $G(j\omega)$ is near the -1 point, the ratio of $|G(j\omega)|$ to $|1 + G(j\omega)|$ is greater than 1, indicating a resonance in $W(j\omega)$, as shown in

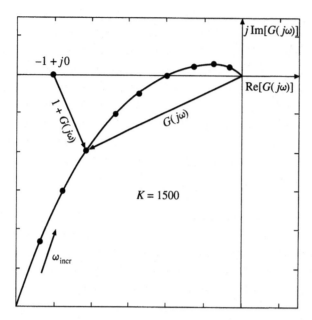

Fig. 12.5 $G(j\omega)$ for $K = 1500$, $P_1 = 10$, and $P_2 = 15$.

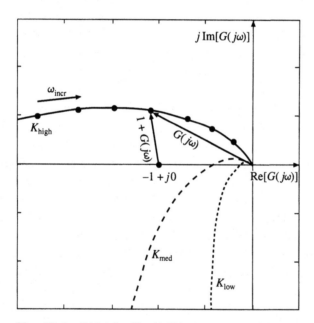

Fig. 12.6 $G(j\omega)$ for $K = 14{,}000$, $P_1 = 10$, and $P_2 = 15$.

the plot of $|W(j\omega)|$ for the K_{high} case in the upper graph of Fig. 12.4. Note that, in the K_{high} case, the resonant frequency is much higher than it is for K_{med}.

It is interesting to trace the progress of the angle β in this case over the entire frequency spectrum, an exercise that explains the closed-loop phase shift curve $\angle W(j\omega)$ in the lower graph of Fig. 12.4. When $K = K_{high}$, the two poles in the right-half plane give $W(s)$ its nonminimum phase character.

The frequency response of the closed-loop system, $W(j\omega)$, describes only the steady-state part of the response of the system to a sinusoidal input. The transient part of that response in an unstable system will grow exponentially, overwhelming the steady-state part, but the *transient instability is not apparent in* $W(j\omega)$ *itself*. The significance of Nyquist's work to control engineering is that it connects the steady-state frequency response $G(j\omega)$ to the stability of $W(s)$.

Nyquist based his stability criterion on a theorem in complex variable theory that had been formulated by Cauchy nearly a hundred years earlier.

12.2 Cauchy's Fundamental Theorem

Cauchy's theorem is effectively stated using graphical representations of complex variables and the corresponding functions of the complex variables. Let $F(s)$ be the ratio of two polynomials. s can be represented on a complex plane, the s plane, having axes σ and $j\omega$. Because $F(s)$ is also complex-valued, it too may be represented on a complex plane, the $F(s)$ plane, having axes $\mathrm{Re}[F(s)]$ and $j\,\mathrm{Im}[F(s)]$.

Both the s plane and the $F(s)$ plane are required in order to graph a function of a complex variable. Figure 12.7 shows such a graphing, where a collection of points in the s plane (chosen to form the letter W) is mapped through the function $F(s) = 2s+1$ onto the $F(s)$ plane. This mapping process is demonstrated in the following table.

Point	s	$F(s) = 2s+1$
A	$-3+j2$	$-5+j4$
B	$-2+j0$	$-3+j0$
C	$-1+j1$	$-1+j2$
D	$0+j0$	$+1+j0$
E	$+1+j2$	$+3+j4$

For point A we have $F(A) = 2[-3+j2]+1 = -5+j4$, which is plotted on the $F(s)$ plane. Points B, C, D, and E will map into the $F(B)$, $F(C)$, $F(D)$, and

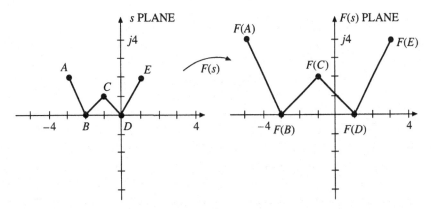

Fig. 12.7   Image on s plane graphed onto $F(s)$ plane, where $F(s) = 2s+1$.

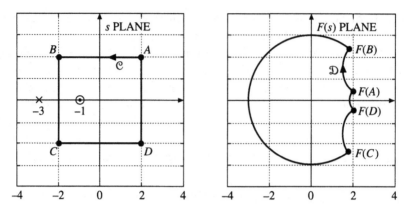

Fig. 12.8 Conformal map of closed contour $\mathcal{C}$ in the s plane into a closed contour $\mathcal{D}$ in the $F(s)$ plane, where $F(s) = 3(s+1)/(s+3)$.

$F(E)$ points indicated in the table and plotted on the $F(s)$ plane. Owing to the nature of the function $F(s)$, points lying on the W curve between these five designated points will map into points on the large W curve between the five corresponding designated points. Note that $F(s)$ simply takes each value of s on the small W curve, multiplies it by 2, and shifts it to the right by 1 unit. Thus the image of the small W in the s plane is a W-shaped curve in the $F(s)$ plane, twice the size of the small W and shifted one unit to the right. An image in the $F(s)$ plane of a collection of points forming a contour in the s plane is said to be a *conformal map* of the s-plane image through the function $F(s)$.

If the mapping function has several poles and zeros, then the mapping of a simple contour like the W in the s plane will be a very much more complicated image in the $F(s)$ plane, but the mapping principle is the same.

Cauchy's theorem is based on the conformal mapping of a closed contour $\mathcal{C}$ in the s plane into a closed contour $\mathcal{D}$ in the $F(s)$ plane. As an example, let the contour $\mathcal{C}$ be the square shown in the left side of Fig. 12.8, and let the mapping function be

$$F(s) = \frac{3(s+1)}{s+3}. \tag{12.6}$$

The pole and zero of $F(s)$ are shown on the s plane, along with contour $\mathcal{C}$. The square contour is mapped through $F(s)$ onto the $F(s)$ plane, as shown in the graph on the right side of Fig. 12.8. We establish a *sense* of the contour $\mathcal{C}$ as the path along $\mathcal{C}$, beginning at point A and traversing $\mathcal{C}$ in the counterclockwise direction in sequence to B then to C then to D and finally back to A. This is defined as a *positive* sense, in keeping with standard graphical notation. The arrow on contour $\mathcal{C}$ between A and B indicates the sense of $\mathcal{C}$, and the arrow between $F(A)$ and $F(B)$ corresponds to the arrow on contour $\mathcal{C}$. In this example, contour $\mathcal{D}$ also has a positive sense. It is interesting to note in this example of conformal mapping that the portions of $\mathcal{D}$ between the corner points are arcs of circles, and that tangents to the adjacent arcs at the corner points make the same angles with one another (in this case, 90°), as the angles made by the adjacent sides of $\mathcal{C}$ at the corners of that contour. The theory of functions of a complex variable explains these facts, but we need not be concerned with that theory to state and use Cauchy's fundamental theorem:

> A closed contour $\mathcal{C}$, with a positive sense in the s plane, which encloses P poles and Z zeros of $F(s)$ and which does not pass through any poles or zeros of $F(s)$, will map through $F(s)$ into a closed contour $\mathcal{D}$ in the $F(s)$ plane which encircles the origin of the $F(s)$ plane N times, where $N = Z - P$.

The encirclements of the origin may be determined in the following way: Draw a line from the origin of the $F(s)$ plane to a point on $\mathcal{D}$. Allow the point to move along $\mathcal{D}$ in the direction of the sense of $\mathcal{D}$, which is established by the conformal mapping. As the point makes a complete circuit of $\mathcal{D}$ and returns to its initial position, the line connecting the origin to this test point will have experienced a net rotation of $2N\pi$ radians in the positive sense, where N is the number of encirclements. N is positive if the net rotation is in the counterclockwise direction; otherwise, it is negative.

Cauchy's theorem is illustrated in Fig. 12.8. Contour $\mathcal{C}$ encloses one zero and no pole of $F(s)$, so $Z = 1$, $P = 0$, and the theorem yields $N = +1$, which means that $\mathcal{D}$ encircles the origin of the $F(s)$ plane once in the positive sense.

It is easy to understand why Cauchy's theorem holds. Consider Fig. 12.8. $F(s)$ may be represented in polar form as

$$F(s) = |F(s)|\epsilon^{j\angle F(s)}, \quad \text{where}$$

$$|F(s)| = \frac{3|s+1|}{|s+3|} \quad \text{and} \quad \angle F(s) = \angle(s+1) - \angle(s+3).$$

(12.7)

Now, $\angle F(s)$ is the angle between the line drawn from the origin of the $F(s)$ plane to a test point on $\mathcal{D}$ and the positive real axis. To determine the number of encirclements of the origin we need only observe how $\angle F(s)$ changes as the test point moves along $\mathcal{D}$ in the sense established by the conformal map.

Consider Fig. 12.8. As a test point in the s plane moves along the path from A to B to C to D and back to A, the net contribution to $\angle F(s)$ due to the term $\angle(s+3)$ is 0° because the pole at $s = -3$ lies *outside* $\mathcal{C}$. $\angle(s+3)$ at point A is 21.8°. It increases to 63.4° at B, decreases to $-63.4°$ at C, increases to $-21.8°$ at D, and increases to 21.8° at A, the net accumulation being 0°. When the zero at $s = -1$ lies inside $\mathcal{C}$, the term $\angle(s+1)$ makes a net contribution of $+2\pi$ rad to $\angle F(s)$. (At A, $\angle(s+1)$ is $+33.7°$; it increases to 116.6° at B, continuing to increase as the test point moves to C to D, and back to A, at which point $\angle(s+1)$ is 383.7°, or just 2π rad more than it was at its initial position.)

Figure 12.9 illustrates Cauchy's theorem once again. Here the contour $\mathcal{C}$ encloses only one pole, so $P = 1$ and $Z = 0$. The mapping function (Eqn. 12.6) produces the conformal map $\mathcal{D}$ shown in the right-hand graph. The theorem requires $N = Z - P = -1$, indicating that $\mathcal{D}$ encircles the origin of the $F(s)$ plane once in the clockwise sense.

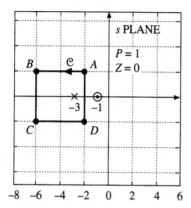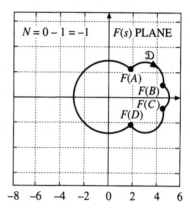

Fig. 12.9　Conformal map of $\mathcal{C}$ in the s plane into $\mathcal{D}$ in the $F(s)$ plane, where $F(s) = 3(s+1)/(s+3)$.

12.3　Nyquist's Stability Criterion

Nyquist's stability criterion is a direct application of Cauchy's theorem. In a unity feedback system, the relation between the open-loop transfer function and the closed-loop transfer function is the familiar equation

$$W(s) = \frac{G(s)}{1+G(s)},\qquad(12.8)$$

and the stability question is: How many poles of $W(s)$ lie in the right-half s plane? An equivalent question is: How many zeros of $1+G(s)$ lie in the right-half s plane? If the answer is one or more, the system is unstable. To apply Cauchy's theorem to this question we introduce the notation

$$1+G(s) = F(s),\qquad(12.9)$$

so that stability is determined by finding the number of zeros of $F(s)$ in the right-half s plane. An obvious but important fact in the following analysis is that a pole of $G(s)$ is also a pole of $F(s)$. We assume that we know how many poles of $G(s)$ lie in the right-half s plane, and we denote that number as P.

We now establish a contour $\mathcal{C}$ enclosing the entire right-half s plane. $\mathcal{C}$ consists of the $j\omega$ axis and a semicircle of infinite radius having the positive sense indicated in Fig. 12.10. Since $\mathcal{C}$ encloses P poles of $F(s)$, it remains only to find Z in order to determine the stability of the closed-loop system. We do this using the frequency response of the open-loop elements, $G(j\omega)$.

Consider the example

$$G(s) = \frac{K}{(s+1)(s+2)(s+3)},$$

$$G(j\omega) = \frac{K}{(j\omega+1)(j\omega+2)(j\omega+3)}.\qquad(12.10)$$

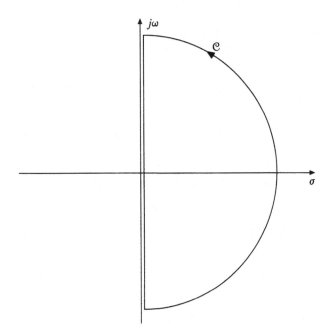

Fig. 12.10 The $\mathcal{C}$ contour for Nyquist's criterion encloses the entire right-half *s* plane.

Here $P = 0$, since no poles of $G(s)$ exist in the right-half *s* plane. The frequency response of the open-loop system, as it might be measured experimentally, is $G(j\omega)$ for frequency ranging from 0 to ∞. But we wish to map the entire contour $\mathcal{C}$, including the negative $j\omega$ axis, onto the $G(s)$ plane. $G(j\omega)$ for $0 > \omega > -\infty$ is simply the complex conjugate of $G(j\omega)$ for $0 < \omega < \infty$. Figure 12.11 shows $G(j\omega)$ for the entire frequency spectrum, with the negative frequency part shown as the dashed contour. Since $G(j\omega)$ is the conformal map of $\mathcal{C}$, it is labeled $\mathcal{D}$ in the figure. Note that the real axis of the $F(s)$ plane coincides with that of the $G(s)$ plane, and the imaginary axis of the $F(s)$ plane is one unit to the left of that of the $G(s)$ plane. The origin of the $F(s)$ plane is therefore the point $-1 + j0$ of the $G(s)$ plane. This is called *the -1 point*. Since the contour $\mathcal{D}$, called the *Nyquist plot,* does not encircle the origin of the $F(s)$ plane in this diagram, Cauchy's theorem (which we now call Nyquist's criterion) gives the answer to the stability question:

$N = Z - P$,
P is known to be 0, and
N is seen to be 0 from the frequency response locus $G(j\omega)$; thus

$$Z = 0 + 0 = 0 = \begin{bmatrix} \text{number of zeros of } F(s) \\ \text{in the right-half } s \text{ plane} \end{bmatrix} = \begin{bmatrix} \text{number of poles of } W(s) \\ \text{in the right-half } s \text{ plane} \end{bmatrix}.$$

The system is stable for the value of K used to plot $G(j\omega)$ in the Nyquist diagram, which is $K = 25$. If K is increased to 75, only the amplitude of $G(j\omega)$ will

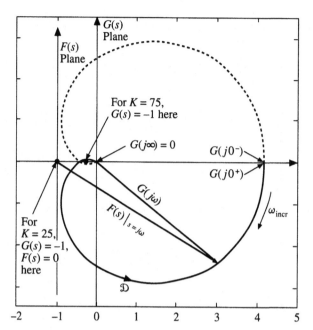

Fig. 12.11 $G(j\omega)$ for $+\infty > \omega > -\infty$, $G(s)$ in Eqn. 12.10.

be changed, and the new contour $\mathfrak{D}$ can be displayed by the Nyquist diagram in Fig. 12.11 simply by changing the scale on the axes of the $G(s)$ plane. On the diagram the -1 point is shifted toward the origin by the appropriate ratio. Now the contour $\mathfrak{D}$ encircles the -1 point twice in the positive sense, and Nyquist's criterion gives the answer

$$N = Z - P, \quad \text{so}$$

$$Z = N + P = 2 + 0 = 2 = \begin{bmatrix} \text{number of poles of } W(s) \\ \text{in the right-half } s \text{ plane} \end{bmatrix},$$

which shows that the closed-loop system is unstable for $K = 75$. A careful study of Fig. 12.11 will show that the value of K between 25 and 75 at which the system passes from being stable to being unstable is $K = 60$. This can also be determined from a root-locus plot, or from Routh's criterion.

A second example illustrates the application of Nyquist's criterion to systems in which $G(s)$ has one or more poles on the $j\omega$ axis. For this exercise we take a typical servomechanism transfer function

$$G(s) = \frac{K}{s(s+2)(s+3)}. \tag{12.11}$$

A difficulty in drawing the contour $\mathcal{C}$ (Fig. 12.10) arises because the pole of $G(s)$ at the origin lies on that contour, a condition not permitted by Cauchy's theorem. Let us therefore modify the contour $\mathcal{C}$ very slightly by drawing a tiny semicircular detour around the origin, as illustrated in Fig. 12.12. Now the conformal map $G(j\omega)$ for $\infty > \omega > -\infty$ is that shown in Fig. 12.13. Note that the

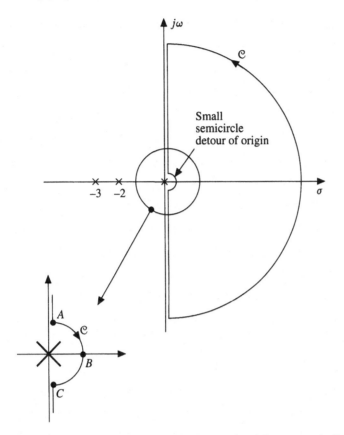

Fig. 12.12 A small detour around the pole of $G(s)$ at $s = 0$, $G(s)$ in Eqn. 12.11.

"infinite" semicircular part of contour $\mathcal{C}$ for $\omega \to \infty$ maps into the point at the origin of the $G(s)$ plane. Since this point is well away from the -1 point, the radius of the "infinite" semicircle portion of $\mathcal{C}$ is unimportant, provided that $\mathcal{C}$ maps into a contour $\mathcal{D}$ lying in a region of the $G(s)$ plane in which no ambiguity occurs when encirclements of the -1 point are counted.

The small semicircular detour in $\mathcal{C}$ around the pole at the origin denoted by arc ABC must be treated carefully. Because point A is a very low value of ω, the conformal map of point A (denoted $G(A)$) in Fig. 12.13 has a very large absolute value, and the graph must be exaggerated to show $G(A)$. Contour $\mathcal{C}$ along the tiny arc ABC has a clockwise sense, and since it is very close to the pole of $G(s)$ at the origin but not near the other poles of $G(s)$, the change in $\angle G(s)$ corresponding to the change in s from A to B to C is $+180°$. $|G(s)|$ remains very large but virtually constant for all points along this tiny detour. Therefore, the conformal map $\mathcal{D}$ corresponding to the detour is the semicircular contour having a very large radius and the positive sense denoted as $G(A)$, $G(B)$, $G(C)$ in Fig. 12.13. This gives $G(s)$ the required change of $+180°$. Again, the radius of this very large semicircular arc in $\mathcal{D}$ is not important, as long as it introduces no ambiguity in the encirclements of the -1 point by $\mathcal{D}$. In the

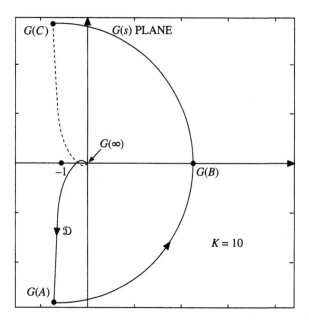

Fig. 12.13 Conformal map of the $\mathcal{C}$ contour of Fig. 12.12 into the $\mathcal{D}$ contour for the $G(s)$ in Eqn. 12.11.

case shown here, $K = 10$ and $N = 0$. The tiny detour has excluded the pole at the origin from the interior of $\mathcal{C}$ so that $P = 0$, and the stability result is

$$Z = \begin{bmatrix} \text{number of poles of } W(s) \\ \text{in the right-half } s \text{ plane} \end{bmatrix} = N + P = 0 + 0 = 0.$$

The system is therefore stable for $K = 10$.

When contour $\mathcal{C}$ is made to detour around the pole at the origin, the detour can be drawn to *include* the pole in the interior of $\mathcal{C}$, as shown in Fig. 12.14. Now $P = 1$, and the large semicircle in the $\mathcal{D}$ contour will have a clockwise sense, as shown in Fig. 12.15. $\mathcal{D}$ now encircles the -1 point once in the negative sense, so $N = -1$, and the Nyquist formula still yields the right answer to the stability question:

$$Z = \begin{bmatrix} \text{number of poles of } W(s) \\ \text{in the right-half } s \text{ plane} \end{bmatrix} = N + P = -1 + 1 = 0, \quad K = 10.$$

If K is increased to 40, the -1 point will lie inside the $\mathcal{D}$ contour. Figure 12.16 is drawn out of scale to clarify the geometry of the $\mathcal{D}$ contour. Now the closed-loop system will be unstable, as indicated by the Nyquist formula:

$$Z = \begin{bmatrix} \text{number of poles of } W(s) \\ \text{in the right-half } s \text{ plane} \end{bmatrix} = N + P = 1 + 1 = 2, \quad K = 40.$$

We now list the steps to be followed in applying the Nyquist stability criterion to negative feedback systems having a loop transfer function $G(s)$.

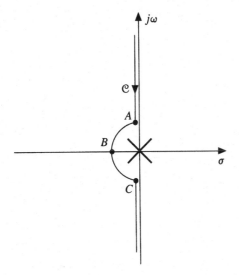

Fig. 12.14 An alternate detour around the pole of $G(s)$ at $s = 0$, $G(s)$ in Eqn. 12.11.

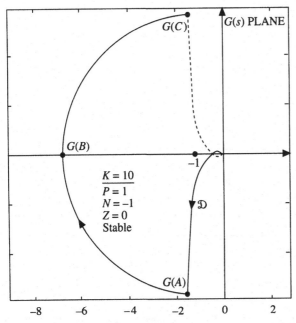

Fig. 12.15 Conformal map of the $\mathcal{C}$ contour, using the alternate detour of Fig. 12.14, into the $\mathcal{D}$ contour for the $G(s)$ in Eqn. 12.11.

1. Determine whether $G(s)$ has poles on the $j\omega$ axis or in the right-half plane. This is obvious if $G(s)$ is available in factored form. If the information on $G(s)$ is experimental frequency response or transient response data, this task requires careful interpretation of the data. If no poles exist on the $j\omega$ axis,

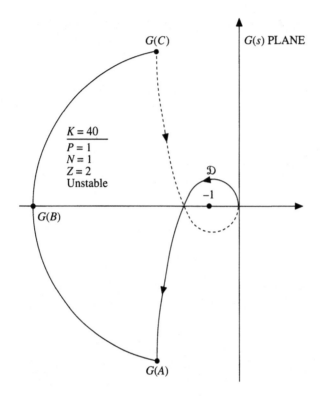

Fig. 12.16   Illustration of instability for $K = 40$; alternate contour in $\mathcal{C}$ used.

the contour $\mathcal{C}$ will simply be the $j\omega$ axis, and the conformal map $G(\mathcal{C})$ will be the frequency response $G(j\omega)$ for $+\infty > \omega > -\infty$. Plot $G(j\omega)$ on the $G(s)$ plane. If poles exist on the $j\omega$ axis, the contour $\mathcal{C}$ must include tiny detours around them as indicated in Fig. 12.12 or 12.14. Plot the conformal map $G(\mathcal{C})$ on the $G(s)$ plane, carefully evaluating the conformal map of the detours.

2. Count the number of poles of $G(s)$ lying inside $\mathcal{C}$. This number is P.

3. Count the number of times $G(\mathcal{C})$ encircles the -1 point in the positive sense. This number N can be positive or negative.

4. Calculate Z using the Nyquist formula $Z = N + P$; Z is the number of poles of the closed-loop transfer function in the right-half plane. If Z is positive, the closed-loop system is unstable. If Z is zero, the system is stable. If Z is negative, a mistake has been made in determining P or N.

Two more examples further illustrate the application of Nyquist's criterion. Consider first a system whose plant has a double integration (see Fig. 12.17). Let the controller transfer function be 1 and the plant gain be 10, so that

$$G(s) = \frac{10}{s^2(s+10)}. \tag{12.12}$$

Use a detour in contour $\mathcal{C}$ that excludes the poles at the origin from the interior of $\mathcal{C}$ (Fig. 12.12). The conformal map $\mathcal{D}$ shown in Fig. 12.18 is exaggerated for

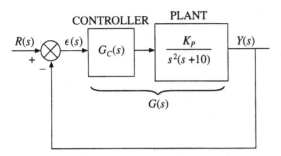

Fig. 12.17   Unity negative feedback system with doubly integrating plant.

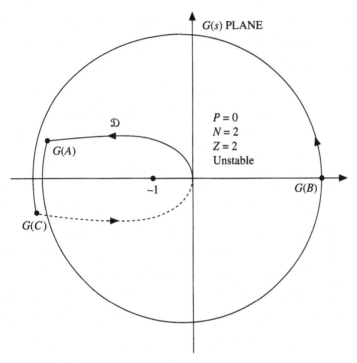

Fig. 12.18   System of Fig. 12.17, unstable for $G_C(s) = K$.

clarity. $Z = 2$ indicates that two poles of $Y(s)/R(s)$ exist in the right-half plane. In order to stabilize this system we introduce some dynamics in the controller. When the controller transfer function is combined with that of the plant, the composite transfer function is

$$G(s) = \frac{K(s+2)}{s^2(s+10)(s+30)}. \tag{12.13}$$

If K is set at 2000, the conformal map $\mathfrak{D}$ has the appearance shown in Fig. 12.19. The presence of the zero at $s = -2$ in $G_C(s)$ has reshaped the $G(j\omega)$ locus in the vicinity of the -1 point, so that $N = 0$ and $Z = 0$. The closed-loop system

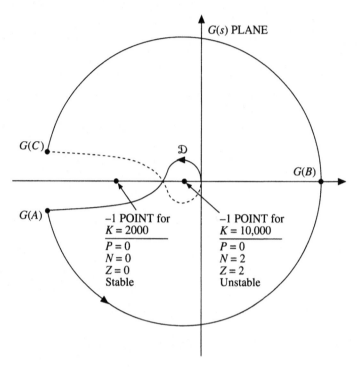

Fig. 12.19   System of Fig. 12.17, stabilized for $G(s) = K(s+2)/[(s^2(s+10)(s+30)]$.

is therefore stable. If K is increased to 10,000, the -1 point will be inside the lobe of $\mathfrak{D}$, creating two encirclements of the -1 point in the positive sense by $\mathfrak{D}$. Therefore, $Z = 2+0 = 2$, and the closed-loop system is unstable.

As a second example, consider a system whose open-loop elements are unstable:

$$G(s) = \frac{K(s+2)}{s(s-10)(s+20)}. \tag{12.14}$$

Since a pole exists at the origin, a tiny detour in contour $\mathcal{C}$ is required. In this example we will use the detour in Fig. 12.14 so that two poles of $G(s)$ lie inside $\mathcal{C}$. The contour $\mathfrak{D}$ is shown in Fig. 12.20. For $K = 100$, the -1 point lies outside $\mathfrak{D}$, so there is no encirclement of that point. Therefore, $Z = 2+0 = 2$, and the closed-loop system is unstable. If the gain is increased to 500, the -1 point will lie inside the lobe of $\mathfrak{D}$, and $\mathfrak{D}$ will encircle that point twice in the negative sense. Therefore, $Z = -2+2 = 0$, and the system is stable.

12.4 Summary

The Nyquist criterion is a method for assessing the stability of a closed-loop system from the frequency response characteristics of its open-loop elements. If $G(s)$ is known in factored form, the root-locus method or Routh's criterion

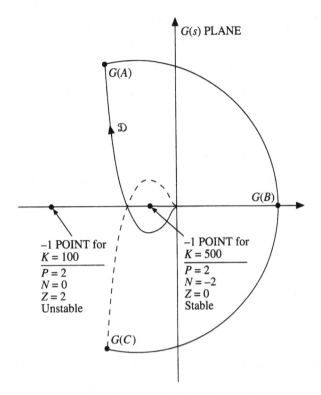

Fig. 12.20 Nyquist criterion applied to a unity negative feedback system with an unstable forward-path element $G(s)$ in Eqn. 12.14 (alternate detour of Fig. 12.14 used).

can be used for the same purpose, and the engineer who has the aid of computer analysis finds these three methods to be equally convenient. But when the open-loop transfer function includes a transcendental function of s (a type we consider in Chapter 13), Routh's method cannot be used. The root-locus method must be modified, but Nyquist's method may be applied directly.

In many cases when $G(s)$ is not known in analytical form, experimental frequency response data on the open-loop elements may be available. This occurs when the internal physical details of the plant are too numerous or too uncertain to be modeled analytically, yet the input–output dynamics are such that the frequency response of the plant can be reliably measured. In these cases the stability of the closed-loop system can often be assessed by inspection, using Nyquist's criterion, whereas the root-locus method or Routh's criterion would require excessive analysis of the available data.

In non-unity feedback systems, where $W(s) = G(s)/[1 + G(s)H(s)]$, Nyquist's method must be applied to the loop transfer function $G(s)H(s)$ to determine closed-loop stability in the same way that $G(s)$ is applied in the unity feedback case.

In some branches of engineering where feedback techniques are employed, the open-loop elements are always stable, making $P = 0$ in Nyquist's equation.

Consequently, Nyquist's formula then reduces to $N = Z$. This abbreviation leads to a common, but incorrect, belief that Nyquist's criterion may be summarized in the statement "the system will be unstable if the frequency response locus $G(j\omega)$ encircles the -1 point." The equally incorrect statement "the system will be unstable if there is a frequency ω_X for which $G(j\omega_X) = -\beta + j0$ where $\beta > 1$" may also be traced to this source.

Unstable open-loop elements are often found in control system engineering. In this case Nyquist's criterion requires the $G(j\omega)$ locus (with the proper representation of any detours in the contour $\mathcal{C}$) to *encircle* the -1 point the proper number of times, and in the proper sense, to ensure a stable system.

12.5 Problems

Problem 12.1
Let

$$G(s) = \frac{K}{s(s+20)(s+40)}$$

be the forward-path transfer function of a unity negative feedback system. Using the s-plane contour given in Fig. 12.14, determine the N, Z, and P terms of the Nyquist equation for each of the following cases, and determine the stability of the closed-loop system for each case.

(a) $K = 40{,}000$.
(b) $K = 50{,}000$.
(c) $K = -40{,}000$.

Problem 12.2
Let

$$G(s) = \frac{K(s^2+4s+9)}{s(s+8)(s^2-2s+9)}$$

be the forward-path transfer function of a unity negative feedback system. Using the s-plane contour given in Fig. 12.14, determine the N, Z, and P terms of the Nyquist equation for each of the following cases. Determine the stability of the closed-loop system for each case.

(a) $K = 40$.
(b) $K = 100$.
(c) $K = -4$.
(d) $K = -40$.

Problem 12.3
In the formal procedure of applying the Nyquist stability criterion, why is it unnecessary for the $\mathcal{C}$ contour to detour around *zeros* of $G(s)$ lying on the $j\omega$ axis?

Problem 12.4
Let

$$G(s) = \frac{K(s-1)}{s(s+5)(s+6)(s+7)}$$

be the forward-path transfer function of a unity negative feedback system.

(a) Make a Nyquist plane plot of $G(j\omega)$, and determine the limits on K for which the closed-loop system will be stable. Identify your N, Z, and P terms for Nyquist's equation in your analysis.

(b) Find a value of K to ensure that the output response of the closed-loop system will be stable and positive if the input is a positive unit step.

Problem 12.5

Repeat Problem 12.4 if the *feedback* is positive.

Problem 12.6

Let

$$G(s) = \frac{K(s^2+2.6s+10)}{s(s-1)(s+1)(s+10)(s+20)}$$

be the forward-path transfer function for a unity negative feedback system.

(a) Determine the limits on K for closed-loop system stability, using Nyquist's method with the $\mathcal{C}$-contour detour given in Fig. 12.12.

(b) Repeat (a) using the $\mathcal{C}$-contour detour given in Fig. 12.14.

(c) Verify your limits on K, using the root-locus method and Routh's criterion.

Problem 12.7

Refer to the non-unity feedback diagram in Fig. P12.7. The control law transfer function is $G_C(s) = K_1$, and the feedback element has the transfer function $H(s) = K_2$. The unstable plant dynamics are described by its transfer function

$$\frac{Y(s)}{U(s)} = G_P(s) = \frac{1}{(s-1)(s+5)(s+8)}.$$

(a) Find values for K_1 and K_2 that satisfy the following design specifications:
  1. The closed-loop system must be stable, as shown by a Nyquist plot, and have a gain margin of at least 3.0.
  2. For $r(t) = \sigma(t)$ the output response $y(t)$ must have a final value of 1 and an overshoot of approximately 10 percent.

(b) With the design values for K_1 and K_2 determined in (a), find the closed-loop frequency response $W(j\omega) = Y(j\omega)/R(j\omega)$ and determine its M_p and ω_r.

(c) Does your design satisfy the stated specifications in a robust manner? That is, if your design value for K_1 should somehow be reduced or increased by 10 percent, would the specifications on stability margin and response to a

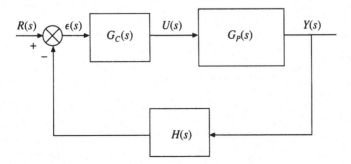

Fig. P12.7

step input be met? If not, what changes in $G_C(s)$ or $H(s)$ would you pro-
pose to improve the robustness of the control loop design?

Problem 12.8

In the system configuration of Fig. P12.7 we have $G_C(s) = K_1$ and $H(s) = K_2$. The plant
is known to be of minimum phase, and its frequency response $G_P(j\omega) = Y(j\omega)/U(j\omega)$ is
plotted on the Bode chart in Fig. P12.8 for $K_1 = K_2 = 1$. Use Nyquist's criterion to determine whether or not the closed-loop system will be stable. If it is stable, find the gain
margin. If it is unstable, determine whether changing K_1 or K_2 could stabilize the system.

Problem 12.9

Provide an example of a stable unity negative feedback system whose open-loop transfer function $G(s)$ is of minimum phase and when there is a frequency ω_X for which
$G(j\omega_X) = -2 + j0$.

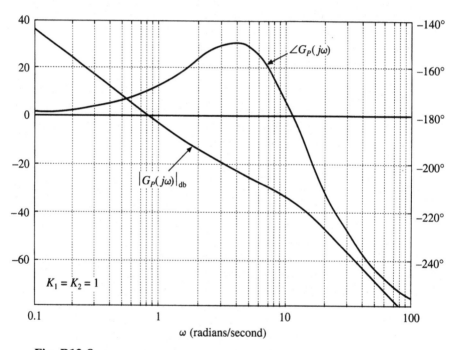

Fig. P12.8

13 Dynamic Analysis of Feedback Systems by Frequency Response Methods

13.1 Introduction

We now apply the frequency response analysis of Chapters 11 and 12 to single-loop feedback systems in order to determine the frequency response performance of the closed-loop system from the frequency response of the open-loop elements. We employ a combination of Nyquist plots, Bode plots, and Nichols diagrams to display both the stability margins and the closed-loop performance quantities – resonant frequency, peak magnitude, and bandwidth. We also analyze systems whose plants have transfer functions that are nonminimum phase, including unstable plants. The influence of time-delay or dead-time components on the stability of feedback systems is also demonstrated in this chapter.

The searchlight servomechanism analyzed in Chapter 9 by the root-locus method is diagramed in Fig. 13.1. The amplifier gain K_1 is variable over positive values. In Fig. 13.2, the open-loop frequency response $Y(j\omega)/\epsilon(j\omega)$ is plotted on the Nyquist diagram for the amplifier gain set at 200. The -1 point for this value of amplifier gain lies outside the $G(j\omega)$ locus, so that locus does not

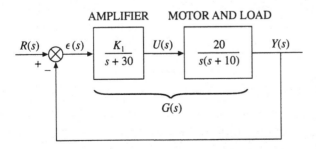

Fig. 13.1 Block diagram for searchlight servomechanism.

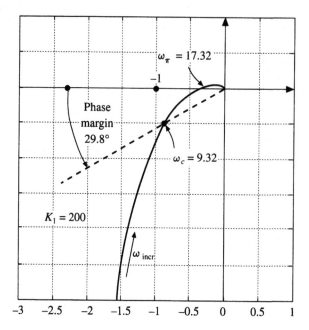

Fig. 13.2   Open-loop frequency response $Y(j\omega)/\epsilon(j\omega)$ with $K_1 = 200$.

enclose the -1 point. Since there are no poles of $G(s)$ in the right-half plane, the Nyquist equation

$$Z = N + P$$

$$= 0 + 0 = 0 \tag{13.1}$$

indicates that the system is stable. We define two special frequencies for this case, ω_c and ω_π. ω_c is the frequency at which the magnitude of $G(j\omega)$ is unity, and ω_π is the frequency at which the argument of $G(j\omega)$ is $-180°$:

$$|G(j\omega_c)| = 1 \quad \text{and} \quad \angle G(j\omega_\pi) = -180°. \tag{13.2}$$

When $K_1 = 200$, ω_c is 9.32 rad/s and ω_π is 17.32 rad/s. The amplitude of $G(j\omega)$ at $\omega = \omega_\pi$ and the argument of $G(j\omega)$ at $\omega = \omega_c$ are:

$$|G(j\omega_\pi)| = 0.3334 \quad \text{and} \quad \angle G(j\omega_c) = -150.2°. \tag{13.3}$$

The *gain margin* of the system, as defined in Section 9.3, is the factor by which the gain constant of the open-loop system (here, $20K_1$) must be increased from its operational value (here, 4000) to make the system unstable. Nyquist's criterion shows that factor to be

$$\left[\begin{array}{c} \text{gain} \\ \text{margin} \end{array} \right] = \frac{1}{|G(j\omega_\pi)|}, \tag{13.4}$$

which in this case is 3. If the loop gain is increased by a factor of 3^+, then the -1 point will lie inside the $G(j\omega)$ locus and the system will be unstable.

The *phase margin,* which is a second indicator of the proximity of the $G(j\omega)$ locus to the -1 point, is defined at the frequency ω_c, as shown in Fig. 13.2:

$$[\text{phase margin}] = \angle G(j\omega_c) - 180°. \tag{13.5}$$

In a stable system the phase margin is normally expressed as a positive number. In this example, when $K_1 = 200$, $\angle G(j\omega_c)$ is expressed as $+209.8°$ (the equivalent of $-150.2°$). This makes the phase margin $29.8°$.

The gain margin (GM) and the phase margin (PM), referred to collectively as *stability margins*, define the two points on the Nyquist diagram at which the $G(j\omega)$ locus intersects the unit circle (for $\omega = \omega_c$) and the negative real axis (for $\omega = \omega_\pi$). In low-order systems where $G(s)$ has neither poles nor zeros close to the $j\omega$ axis, the GM and PM together provide a reliable measure of the *relative stability* of the system. For example, in a low-order system where the GM is known to be 4 and the PM to be $45°$, it is reasonable to assume that the $G(j\omega)$ locus will pass far enough from the -1 point to avoid causing a high resonant peak in $W(j\omega)$. But in high-order systems where $G(s)$ has complex poles and zeros, the $G(j\omega)$ locus can pass relatively close to the -1 point even though the GM is 4 and the PM is $45°$. Figure 13.3 illustrates this case. Here the $G(j\omega)$ locus passes quite close to the -1 point in the frequency range between ω_c and ω_π, so the GM and PM combination is not a reliable indication of the relative stability of the closed-loop system.

In most operating feedback control systems, the gain margin is found in the range from 2 to 6. The phase margin usually lies in the range from $30°$ to

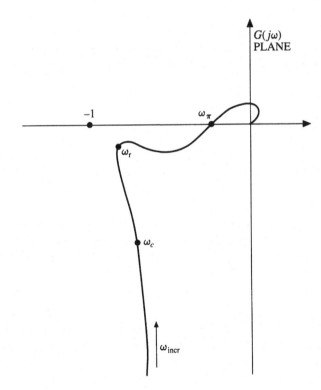

Fig. 13.3 Case in which gain margin and phase margin do not indicate proximity of the $G(j\omega)$ locus to the -1 point.

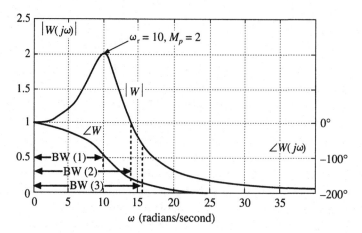

Fig. 13.4   Rectangular plot of closed-loop frequency response for $K_1 = 200$, defining ω_r, M_p, and BW.

80°. In a unity feedback system the relationship of the open-loop frequency response $G(j\omega)$ to the closed-loop frequency response $W(j\omega)$ is $W(j\omega) = G(j\omega)/[1+G(j\omega)]$. In our present example,

$$W(j\omega) = \frac{G(j\omega)}{1+G(j\omega)} = \frac{20K_1}{j\omega(j\omega+10)(j\omega+30)+20K_1}. \qquad (13.6)$$

Figure 13.4 is a rectangular plot of the amplitude ratio and phase shift of $W(j\omega)$ for $K_1 = 200$. It is clear from the plot of $G(j\omega)$ in Fig. 13.2 and from the analytical expression for $W(j\omega)$ in Eqn. 13.6 that the *extreme-frequency* behavior of the closed-loop frequency response is as follows.

For $\omega \to 0$	For $\omega \to \infty$
$\lvert W(j\omega)\rvert \to 1$	$\lvert W(j\omega)\rvert \to 0$
$\angle W(j\omega) \to 0$	$\angle W(j\omega) \to -270°$

$$(13.7)$$

Three additional performance measures for this system, based on the closed-loop frequency response, are defined in Fig. 13.4. These are the *resonant frequency* ω_r, the *peak magnitude* M_p, and the *bandwidth* BW.

The *resonant frequency* is the frequency at which $\lvert W(j\omega)\rvert$ is a maximum. In this example, $\omega_r = 10$ rad/s. The *peak magnitude* is the value of $\lvert W(j\omega)\rvert$ at the resonant frequency; in this example, M_p is 2. The *bandwidth* is defined in various ways, three of which are illustrated in Fig. 13.4. BW(1) is the frequency at which the phase shift of the closed-loop system reaches $-90°$, which in this system is 10 rad/s. BW(2) is the highest frequency for which $\lvert W(j\omega)\rvert = 1$, in this case 13.9 rad/s. BW(3) is the highest frequency for which $\lvert W(j\omega)\rvert = \sqrt{2}/2$, in this case 15.5 rad/s. A statement of the bandwidth of a system, without the definition used, can be misleading.

When the zero-frequency value of $\lvert W(j\omega)\rvert$ is not 1, the M_p is usually defined as the ratio of the *peak value* of $\lvert W(j\omega)\rvert$ to the *zero-frequency value* of $\lvert W(j\omega)\rvert$.

13.2 Stability Margins Determined on Bode Diagrams

The Bode diagram in Fig. 13.5 displays the amplitude and phase data appearing on the Nyquist plot in Fig. 13.2. The PM of 29.8° is determined at frequency ω_c as shown, and the GM, expressed as 9.54 db, is determined at frequency ω_π. The Bode diagram may be used to determine the relationship of the variable gain K_1 to the PM and the GM. Neither the phase shift curve nor the shape of the amplitude curve depends on K_1; a change in K_1 affects only the vertical position of the amplitude curve on the diagram. If K_1 is increased by a factor of 2, for example, the amplitude at each frequency is increased by 6.02 db. This change may be simply incorporated into the diagram by shifting the entire amplitude curve upward by 6.02 db, or by shifting the 0-db line downward by that amount. Note in this system that as K_1 is increased, ω_c also increases toward frequency ω_π. ω_c and ω_π will coincide when the stability limit is reached, in this case when K_1 reaches 600.

In Section 11.5 we discussed the relationship between the amplitude ratio curve and the phase shift at a given frequency, $\phi(\omega_i)$, for a minimum phase system. If the slope of the amplitude ratio curve changes only gradually in the frequency range surrounding ω_i, then the relationship can be approximated by the simple formula

$$\phi(\omega_i) \cong \frac{\pi}{40} \cdot \frac{dM}{d\omega}\bigg|_{\omega_i} \text{ rad,} \tag{13.8}$$

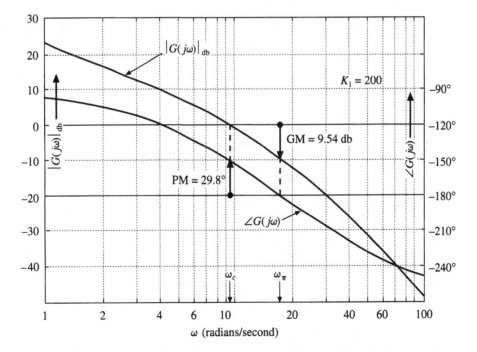

Fig. 13.5 Bode diagram of open-loop frequency response $Y(j\omega)/\epsilon(j\omega)$ with $K_1 = 200$.

where $dM/d\omega$ is expressed in db/decade. The amplitude ratio curve for $G(j\omega)$ in Fig. 13.5 shows a gradual change of slope over the entire spectrum. Therefore, it is possible to estimate the phase shift at ω_c and at ω_π by measuring the slope $d|G(j\omega)|/d\omega$ at each of these frequencies and applying Eqn. 13.8. A graphical measurement on an enlarged version of Fig. 13.5 shows this result.

ω_i	$d\|G(j\omega)\|/d\omega$ (measured)	$\phi(\omega_i)$ (Eqn. 13.8)
$\omega_c = 9.32$ rad/s	−31.7 db/dec	−143°
$\omega_\pi = 17.32$ rad/s	−39.5 db/dec	−178°

In some cases only amplitude ratio data is available. Because the frequency ω_c is evident, an approximation to the PM can be made by this measurement on the $|G(j\omega)|$ curve. (In this example the estimated PM is 37°.) The frequency at which $d|G(j\omega)|/d\omega$ is −40 db/decade can be estimated graphically and taken as the approximation to ω_π. For a minimum phase feedback system whose open-loop amplitude ratio changes slope gradually in the neighborhood of ω_c, as in this example, the engineer may quickly assess the question of stability by examining the slope of $|G(j\omega)|$ at $\omega = \omega_c$. If that slope is shallower than −40 db/decade, the system will probably be stable.

In some systems an ambiguity in the definition of the phase margin occurs. A common case of this sort is the unity feedback system with the open-loop transfer function

$$G(s) = \frac{K}{s(s+0.5)[s^2+0.1s+1]}. \tag{13.9}$$

A Bode plot (see Fig. 13.6) of $G(j\omega)$ for $K = 0.12$ shows a pronounced resonant peak near $\omega = 1$ rad/s. This peak, combined with the phase lag and attenuation caused by both the integration and the low frequency pole, gives us three separate frequencies at which $|G(j\omega)| = 1$. A Nyquist plot of the same $G(j\omega)$ in Fig. 13.7 gives another view of this case.

The following table lists the amplitude and phase shift of $G(j\omega)$ at the five frequencies marked on the Nyquist plot.

ω (rad/s)	$\|G(j\omega)\|$	$\angle G(j\omega)$
$\omega_{c1} = 0.23$	1	−116°
$\omega_\pi = 0.913$	0.664	−180°
$\omega_{c2} = 0.97$	1	−211.7°
$\omega_{peak} = 0.993$	1.084	−235.2°
$\omega_{c3} = 1.013$	1	−258.2°

The formal definition of phase margin applies at three frequencies.

ω_c	Phase Margin
ω_{c1}	+64°
ω_{c2}	−31.7°
ω_{c3}	−78.2°

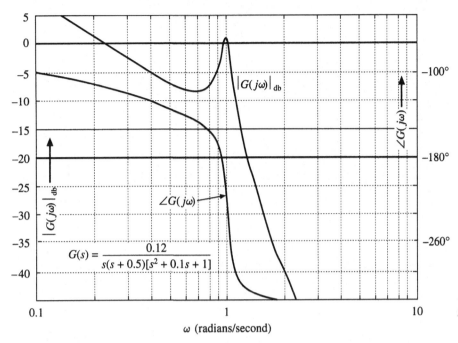

Fig. 13.6 Bode diagram illustrating an ambiguity in the definition of phase margin; Eqn. 13.9 for $K = 0.12$.

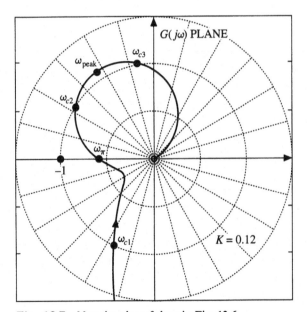

Fig. 13.7 Nyquist plot of data in Fig. 13.6.

In this case the proximity of the $G(j\omega)$ locus to the -1 point should be indicated by the two phase margin values $+64°$ and $-31.7°$, *and* by the gain margin of $1/0.664$.

The sharp resonant peak in $G(j\omega)$ and its proximity to the -1 point is reflected in the closed-loop frequency response $W(j\omega)$. Figure 13.8 shows $W(j\omega)$

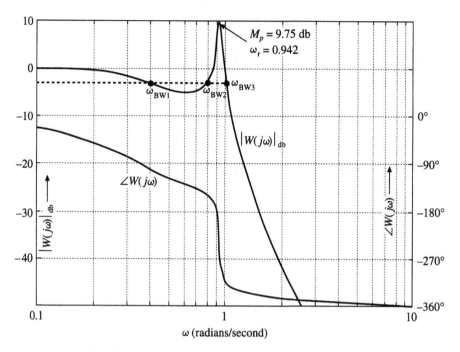

Fig. 13.8   Closed-loop frequency response illustrating an ambiguity in the definition of bandwidth.

to have a sharp resonant peak at $\omega_r = 0.942$ rad/s. The M_p of 9.75 db would be considered excessive in most control applications. Here, if the bandwidth is measured at a frequency for which $|W(j\omega)| = \sqrt{2}/2$ (−3.01 db), the following three possibilities exist: $\omega_{BW1} = 0.426$ rad/s, $\omega_{BW2} = 0.800$ rad/s, and $\omega_{BW3} = 1.021$ rad/s. In this example all three should be included in specifying the bandwidth, which is intended to be an indirect measure of how closely the output of the closed-loop system tracks the input command. The high resonant peak lying between ω_{BW2} and ω_{BW3} indicates that the tracking ability of this system is seriously impaired.

Now consider a system in which the feedback path includes dynamic elements whose transfer function is $H(s)$, such as that defined in Fig. 8.8. The gain margin and phase margin must be defined from the complete open-loop frequency response function $G(j\omega)H(j\omega)$. On the Nyquist plane, the relationship between the $G(j\omega)H(j\omega)$ locus and the −1 point is indicative of the stability of the closed-loop system, just as it is in the unity feedback case, but the ratio

$$\frac{G(j\omega)H(j\omega)}{1+G(j\omega)H(j\omega)} \tag{13.10}$$

is $B(j\omega)/R(j\omega)$ and is *not* the closed-loop frequency response relating the system input to its output. Rather, the closed-loop frequency response $Y(j\omega)/R(j\omega)$ is

$$W(j\omega) = \frac{Y(j\omega)}{R(j\omega)} = \frac{G(j\omega)H(j\omega)}{1+G(j\omega)H(j\omega)} \times \frac{1}{H(j\omega)}. \tag{13.11}$$

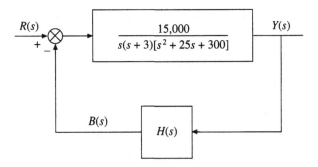

Fig. 13.9 Control system with dynamic feedback elements.

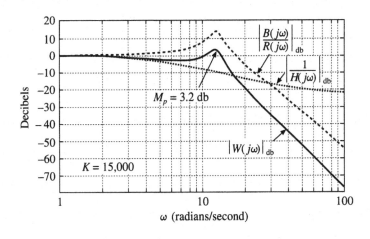

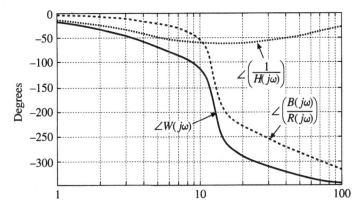

Fig. 13.10 Amplitude ratio and phase shift data for $B(j\omega)/R(j\omega)$, $1/H(j\omega)$, and $Y(j\omega)/R(j\omega)$ for system in Fig. 13.9.

The closed-loop resonant frequency, the peak magnitude, and the bandwidth must be defined in terms of $W(j\omega)$.

As an example, consider the non-unity feedback system diagramed in Fig. 13.9, where $H(s) = [15(s+4)/(s+60)]$. Figure 13.10 is a Bode plot for the closed-loop frequency response $B(j\omega)/R(j\omega)$, which exhibits a high resonant peak

(13.2 db) at $\omega = 12.4$ rad/s. Since $H(j\omega)$ is a lead-type function, it has a rising amplitude ratio curve and its inverse $1/H(j\omega)$ will have a falling amplitude ratio curve, as shown in the Bode plot. The closed-loop frequency response between input and output, $W(j\omega) = Y(j\omega)/R(j\omega)$, will therefore have a smaller amplitude ratio at all frequencies than does $B(j\omega)/R(j\omega)$. This gives the input–output characteristic a modest resonant peak of 3.2 db at $\omega = 12.3$ rad/s. The lower diagram in Fig. 13.10 shows the relationship between the phase shift characteristics of the three transfer functions. It is interesting to note that the gain margin of this control loop is only 1.4 and the phase margin is only 16°. These narrow margins are reflected in the high resonant peak of $B(j\omega)/R(j\omega)$. However, the input–output frequency response resembles that of a unity feedback system having much higher stability margins. If the $H(s)$ in this system is replaced by 1, the forward-path gain would need to be reduced from 15,000 to 2700 to maintain a closed-loop M_p of 3.2 db.

13.3   Closed-Loop M_p and ω_r Determined on the Nichols Chart

Consider the geometry of the $G(j\omega)$ plane in Fig. 13.11. The point labeled ω lies on the $G(j\omega)$ locus of the open-loop frequency response of a feedback system. (The entire locus is not required for the immediate purpose of defining M circles.) At point ω, the ratio

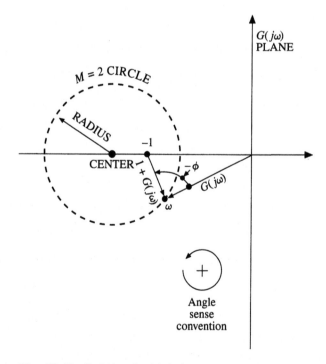

Fig. 13.11 Defining the M circle.

$$\frac{|G(j\omega)|}{|1+G(j\omega)|} \tag{13.12}$$

is denoted as M, which in this sketch is 2. We recognize that this ratio is the amplitude ratio of the closed-loop frequency response of a unity feedback system. Now an infinite number of points in the $G(j\omega)$ plane, not necessarily lying on a frequency response locus, have the same geometric property as point ω – namely, that the lengths of two arrows drawn to the point from the origin and from the -1 point will have the same ratio M as for point ω. These points form a circle, called an *M circle,* in the $G(j\omega)$ plane centered on the real axis at

$$\text{center} = \frac{M^2}{1-M^2}, \tag{13.13}$$

with

$$\text{radius} = \left| \frac{M}{M^2-1} \right|. \tag{13.14}$$

For $M = 2$, the center is at the point $-4/3 + j0$ and the radius is $2/3$. Figure 13.12 shows several M circles for various values of M, as noted.

The angle labeled $-\phi$ in Fig. 13.11, which appears to be approximately $+90°$, is the negative of the argument of $W(j\omega)$. This geometry was also shown in Fig. 12.3. We now have

$$\phi = \angle W(j\omega) = \angle G(j\omega) - \angle[1+G(j\omega)]. \tag{13.15}$$

We recognize that ϕ is the phase shift of the closed-loop system. Once again, an infinite number of points in the $G(j\omega)$ plane, not necessarily lying on a

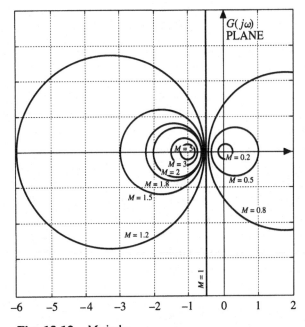

Fig. 13.12 *M* circles.

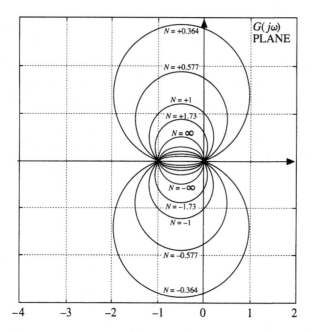

Fig. 13.13 *N* circles.

frequency response locus, have the same geometric property as point ω – namely, that the angle between two arrows drawn to the point from the origin and from the -1 point will be the same as that for point ω. For a given value of ϕ these points lie on a circle, called an *N circle*. The radius and center of the *N* circle are given by

$$\begin{bmatrix} \text{center of} \\ N \text{ circle} \end{bmatrix} = -\frac{1}{2} + j\frac{1}{2N} \quad \text{and} \quad \begin{bmatrix} \text{radius of} \\ N \text{ circle} \end{bmatrix} = \frac{\sqrt{N^2+1}}{|2N|}, \quad \text{where}$$

$$N = \tan \phi.$$

$$(13.16)$$

Figure 13.13 shows *N* circles for several values of *N*, as indicated. Note, for example, that for $\phi = -30°$ the *N* value is -0.577, but that ϕ is $-30°$ only on that portion of the $N = -0.577$ circle lying in the lower half of the $G(j\omega)$ plane. In the upper half plane, the $N = -0.577$ circle corresponds to $\phi = -210°$. A symmetrical situation exists for the $N = +0.577$ circle: the upper half–plane portion corresponds to $\phi = -330°$ and the lower portion to $\phi = -150°$.

The significance of the *M* and *N* circles is that they form a set of coordinates for the closed-loop frequency response superimposed on the set used for the open-loop response. Once the $G(j\omega)$ locus is drawn, the $W(j\omega)$ amplitude and phase can be read from the *M* and *N* circles. As an example, consider the third-order unity feedback system shown in Fig. 12.1, in which

$$G(s) = \frac{K}{s(s+10)(s+15)}.$$

$$(13.17)$$

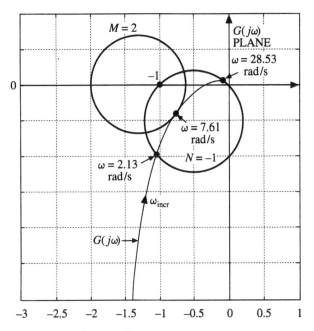

Fig. 13.14   Nyquist plot for $G(j\omega) = 1405/[j\omega(j\omega+10)(j\omega+15)]$.

Figure 13.14 shows the $G(j\omega)$ locus on the Nyquist plane for $K = 1405$. The $M = 2$ circle and the $N = -1$ circle (for $\phi = -45°$ and $-225°$) are also plotted. The $G(j\omega)$ locus is tangent to the $M = 2$ circle at frequency $\omega = 7.61$ rad/s, which is the resonant frequency of the closed-loop system. The $G(j\omega)$ locus cuts the $N = -1$ circle in two places, at frequencies 2.13 and 28.53 rad/s. The closed-loop phase shift is therefore $-45°$ at $\omega = 2.13$ rad/s (the intersection point in the lower half plane) and $-225°$ for $\omega = 28.53$ rad/s (the intersection point in the upper half plane). If several M circles and N circles like those in Figs. 13.12 and 13.13 were superimposed on Fig. 13.14, the closed-loop amplitude and phase could be determined at several frequencies in the range where the $G(j\omega)$ locus is near the -1 point.

However, it is not convenient in design work to use a $G(j\omega)$ plane having several M circles and N circles superimposed upon it. Often it is necessary to try several values for the loop gain to arrive at a satisfactory value. If this trial-and-error process were to be attempted graphically using the $G(j\omega)$ plane, it would be necessary to redraw the $G(j\omega)$ locus for each trial value for the loop gain.

Nichols devised a modification of the $G(j\omega)$-plane plot that enhanced its convenience in relating the open-loop frequency response to the closed-loop frequency response. This modification converts the $G(j\omega)$ plane, which has polar coordinates, into a plot (called the *Nichols plot*) having rectangular coordinates, with $\angle G(j\omega)$ as the abscissa and $|G(j\omega)|_{db}$ as the ordinate. This modification can be considered to be a conformal map of the $G(j\omega)$ plane into the

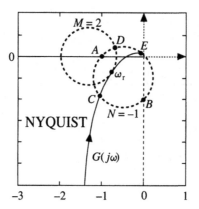

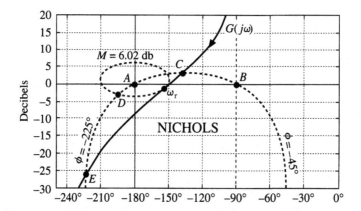

Fig. 13.15   Mapping of the Nyquist plane into the Nichols plane.

Nichols plane through a mapping function that makes the appropriate changes in geometry. Figure 13.15 illustrates this conformal mapping of the Nyquist plot of Fig. 13.14 onto the Nichols plot. Figure 13.16 is a Nichols plot for $\angle G(j\omega)$ in the range from $-250°$ to $0°$, with several M and N contours superimposed. Note that the M contours are labeled in the decibel equivalent of M (which is $|W(j\omega)|_{db}$ in a unity feedback system) and that the N contours are labeled directly in $\angle W(j\omega)$.

The gain margin and phase margin are obvious from a Nichols plot of $G(j\omega)$. Figure 13.17 shows the $G(j\omega)$ locus from Fig. 13.14 on the Nichols plot. For $K = 1405$, the gain margin of 8.53 db and the phase margin of 30.2° are plainly visible in the middle curve. Further, it is possible to determine how both of these stability margins will be affected by changes in K from the value 1405 by sliding the $G(j\omega)$ locus vertically on the Nichols plot, moving up for an increase in the gain and down for a decrease. For example, if K is increased by 2 db to 1769, then the gain margin is reduced by 2 db and the phase margin is reduced to 22.8°, as shown in the upper curve. Sliding the curve down by 2 db from its middle position, so that K is reduced to 1116, increases the gain margin to 10.53 db and the phase margin to 37.6°, as shown in the lower curve. Before

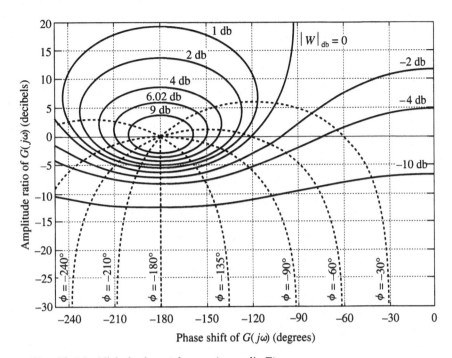

Fig. 13.16   Nichols chart (also see Appendix F).

the advent of computer-aided design, the $G(j\omega)$ locus would be plotted on a transparent overlay of a master copy of the Nichols chart to facilitate the manipulations indicated in Fig. 13.17. A master copy of the Nichols chart is included in Appendix F.

Figure 13.18 (upper curve) is the Nichols chart depicting the frequency response data given on the Bode diagram in Fig. 13.6. The closed-loop resonant peak of 9.75 db ($M_{pCL} = 3.07$) occurs at $\omega_r = 0.942$ rad/s. The resonant peak can be reduced by lowering the gain K. If, for example, an M_{pCL} of 1.5 (3.52 db) is acceptable, then K can be reduced by 2.67 db to 0.0882. This lowers the $G(j\omega)$ locus on the chart so that the peak of the $G(j\omega)$ curve touches the contour $M = 3.52$ db at $\omega = 0.956$ rad/s. The stability margins are also improved by this reduction in K, as evidenced by comparing the two curves in the figure.

13.4   Control of Nonminimum Phase Plants

Consider the unity feedback system in Fig. 13.19. Although the plant is stable, it has a nonminimum phase transfer function owing to the right-half-plane zero in $G_P(s)$ at $s = 1$. The Nyquist plot of $G_P(j\omega)$ is displayed in Fig. 13.20. Assume that the control law is of the simplest possible form, $G_C(s) = K$. Apply Nyquist's criterion to determine how the stability of the system depends on K. Figure 13.21 shows the locus of the loop transfer function $KG_P(j\omega)$ for the entire frequency spectrum $\infty < \omega < -\infty$, using $K = 1$. Nyquist's test for this system requires a small detour in the $\mathcal{C}$ contour around the origin of the s

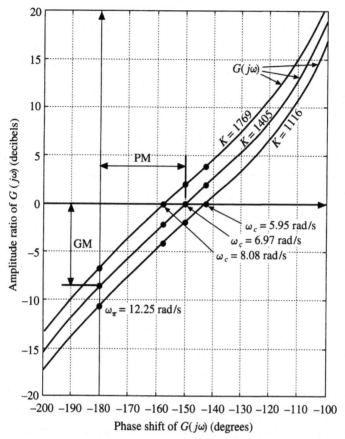

Fig. 13.17 Nichols chart plot of $G(j\omega) = K/[j\omega(j\omega + 10)(j\omega + 15)]$, with K as indicated.

plane. In this example, that small detour is taken to exclude the pole of $G_P(s)$ at the origin, so that the P in Nyquist's equation will be zero. The conformal map of the small detour is the large semicircular arc having a positive sense, so that the $\mathfrak{D}$ contour encircles the -1 point once in the positive sense. Nyquist's test reveals that the closed-loop system is unstable, having one pole of $W(s)$ in the right-half plane:

$$Z = N + P,$$

$$1 = 1 + 0.$$

(13.18)

Furthermore, it is clear from Fig. 13.21 that the $\mathfrak{D}$ contour will encircle the -1 point for all positive values of K. Accordingly, we investigate the possibility that the system might be stabilized by using negative K. Figure 13.22 shows the $\mathfrak{D}$ contour for $K = -1$. The large semicircular arc has a positive sense, as in the previous case. At the crossover frequency ω_c, the magnitude of $KG_P(j\omega)$ is 1.125, so the -1 point lies inside the $\mathfrak{D}$ contour (as shown), which encircles the -1 point twice in the positive sense. Nyquist's test reveals that the closed-loop system is unstable, having two poles of $W(s)$ in the right-half plane:

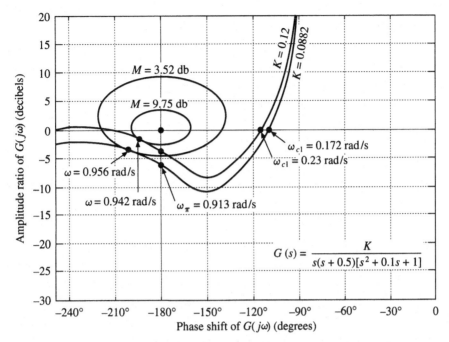

Fig. 13.18   Nichols chart plot of data in Fig. 13.6.

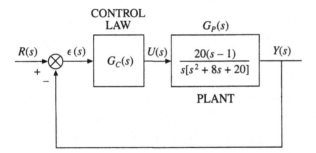

Fig. 13.19   Control system with a nonminimum phase plant.

$$Z = N + P,$$
$$2 = 2 + 0.$$ (13.19)

However, in this case the system can be stabilized by reducing the magnitude of K from 1 to less than $1/1.125$. Hence we conclude that, for the simple control law K, the system will be stable if K lies in the range $-1/1.125 < K < 0$.

We next consider the control of an unstable nonminimum phase plant. Figure 13.23 is a block diagram for an inverted pendulum–cart feedback system that has been stabilized with a compensator designed by the root-locus method in Section 10.8. We first show that the closed-loop system is stable in spite of the instability of the open-loop system. We apply Nyquist's criterion by plotting the frequency response of the loop transfer function on the $G(j\omega)$ plane:

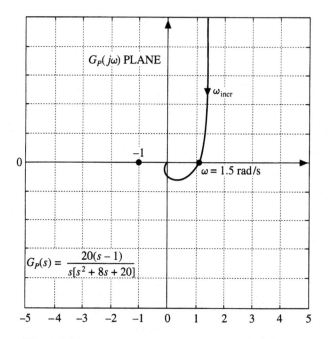

Fig. 13.20 Nyquist plot of nonminimum phase $G_P(j\omega)$.

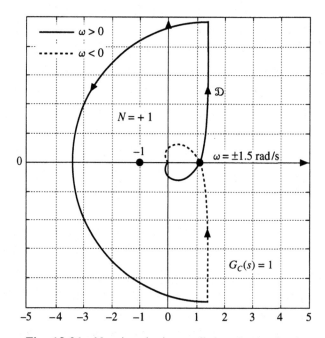

Fig. 13.21 Nyquist criterion applied to $G_C(j\omega)G_P(j\omega) = 20(j\omega-1)/j\omega[(j\omega)^2+8j\omega+20]$.

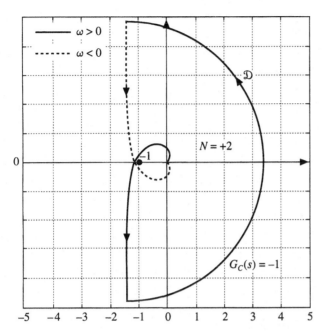

Fig. 13.22   Nyquist criterion applied to $G_C(j\omega)G_P(j\omega) =$
$-20(j\omega - 1)/j\omega[(j\omega)^2 + 8j\omega + 20]$.

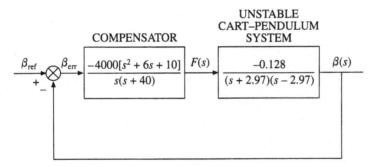

Fig. 13.23   Unstable cart-pendulum system combined with stabilizing
compensator.

$$\begin{bmatrix} \text{Loop} \\ \text{transfer function} \end{bmatrix} = G(s) = \frac{\beta(s)}{\beta_{\text{err}}(s)} = \frac{512[s^2 + 6s + 10]}{s(s+40)(s+2.97)(s-2.97)}; \quad (13.20)$$

$G(j\omega)$ for the entire spectrum $-\infty < \omega < \infty$ is drawn in Fig. 13.24. The small
detour in the $\mathcal{C}$ contour around the origin of the s plane is also required in this
example. It is taken to exclude the pole of $G(s)$ at the origin, so the $\mathcal{C}$ contour
encloses one pole of $G(s)$. The P in Nyquist's criterion is therefore 1. Figure
13.24 shows the corresponding portion of the $\mathcal{D}$ contour as the large semicircular arc (exaggerated in the figure) having a positive sense. For the compensator
gain set at -4000, the $\mathcal{D}$ contour encircles the -1 point once in the negative
sense, and the Nyquist equation yields the result

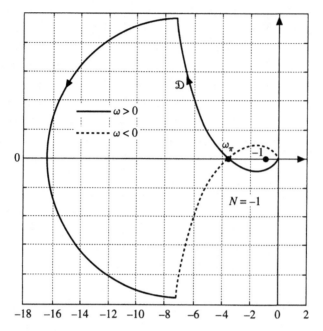

Fig. 13.24 Nyquist criterion applied to $G(j\omega) = 512[(j\omega)^2 + 6j\omega + 10]/j\omega(j\omega + 40)(j\omega + 2.97)(j\omega - 2.97)$.

$$Z = N + P,$$
$$0 = -1 + 1;$$

(13.21)

hence, the closed-loop system is stable.

If the loop gain is increased beyond 512, then the number of encirclements remains unchanged at −1 and the closed-loop system remains stable. However, if the loop gain is reduced below 138, then the −1 point will fall to the left and outside the lobe formed by the $G(j\omega)$ locus, the number of encirclements will change to +1, and the Nyquist equation will reveal that two poles of $W(s)$ lie in the right-half plane. The gain margin is now *less* than 1, a circumstance called *conditional stability* and one that prevails in any feedback system where the open-loop elements are unstable. In most conditionally stable systems the loop gain has both upper and lower limits to its allowable range, owing to phase lag in $G(j\omega)$ exceeding 180° at frequencies higher than ω_c.

Figure 13.25 shows the Nichols plot of $G(j\omega)$ near the $(-180°, 0\text{-db})$ point. This plot shows the following performance parameters for this system for $K = 512$:

$\omega_\pi = 3.43$ rad/s;
$\omega_c = 12.19$ rad/s;
gain margin = −11.4 db (conditional stability);
phase margin = 45.2°;
closed-loop $M_p = 1.64$ (4.28 db);
$\omega_r = 6.73$ rad/s;
bandwidth ($\angle W(j\omega) = -90°$) = 17.2 rad/s.

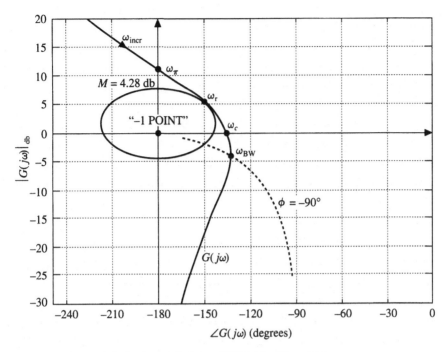

Fig. 13.25 Nichols plot of $G(j\omega) = 512[(j\omega)^2 + 6j\omega + 10]/ j\omega(j\omega + 40)(j\omega + 2.97)(j\omega - 2.97)$.

13.5 Time-Delay Elements in the Control Loop

Figure 13.26 describes an element whose dynamics produce a *time delay* (*dead time* or *transportation lag*). Here the input is a transient function $q(t)$, beginning at $t = 0$. The output $z(t)$ is zero or "dead" during the first t seconds following the application of the input, but it then responds so as to reproduce the input transient, except for a change in amplitude characterized by the gain K_{DT}. Note that this phenomenon differs from that manifested by a *time-lag* element, a device which responds without delay and which produces some distortion of the input waveform. Physical systems described by ordinary differential equations give rise to the time-lag type of response, whereas the time-delay response comes from systems described by partial differential equations (such as transmission lines) or from systems described by difference equations (such as computers or tape-transport mechanisms).

The transfer function for the time-delay device illustrated in Fig. 13.26 is easily derived from the equation that describes the input–output relationship:

$$z(t) = K_{DT}q(t - \tau). \tag{13.22}$$

Take the Laplace transform of Eqn. 13.22,

$$Z(s) = K_{DT}\mathcal{L}\{q(t - \tau)\}. \tag{13.23}$$

Because $q(t - \tau)$ is the delayed version of $q(t)$, we use pair 36 to perform the transformation indicated in Eqn. 13.23,

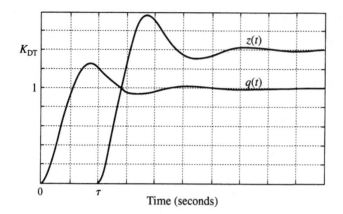

Fig. 13.26 Dynamics of a time-delay element.

$$Z(s) = K_{DT} \epsilon^{-\tau s} Q(s), \tag{13.24}$$

from which the transfer function of the time-delay element is

$$\frac{Z(s)}{Q(s)} = K_{DT} \epsilon^{-\tau s}. \tag{13.25}$$

This is a *transcendental function* of s, not a rational function as we have become accustomed to seeing in previous chapters. $K_{DT} \epsilon^{-\tau s}$ has no poles or zeros in the finite s plane, but its influence on the stability of a closed-loop system can still be assessed, using either a frequency response analysis or a version of the root-locus method termed the *phase-angle–locus* method. Here we use the frequency response approach.

The frequency response of the time delay element is calculated in the same way as that for a system having an algebraic transfer function, namely by substituting $j\omega$ for s:

$$\frac{Z(j\omega)}{Q(j\omega)} = K_{DT} \epsilon^{-j\omega\tau}. \tag{13.26}$$

The amplitude ratio and phase shift of the time delay element are, respectively,

$$\left| \frac{Z(j\omega)}{Q(j\omega)} \right| = K_{DT} \quad \text{and} \quad \angle \left(\frac{Z(j\omega)}{Q(j\omega)} \right) = -\omega\tau, \tag{13.27}$$

provided K_{DT} is positive. A polar plot of this frequency response characteristic is a circle of radius K_{DT}, as shown in Fig. 13.27. The phase lag increases indefinitely as the frequency increases.

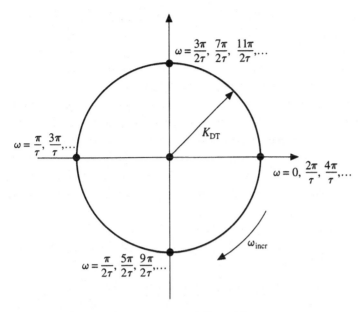

Fig. 13.27 Frequency response of $K_{DT}\,\epsilon^{-j\omega\tau}$.

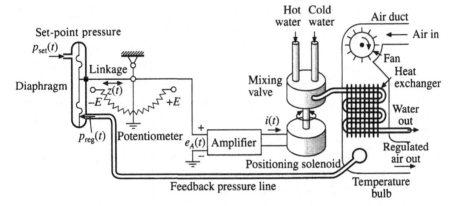

Fig. 13.28 Schematic diagram of an air-temperature regulating system.

The air-temperature regulating system shown in Fig. 13.28 is an application in which a time-delay element is significant. This system controls the temperature of the air flowing out of the duct in accordance with the input set-point pressure $p_{set}(t)$ on the left side of the diaphragm. The controlled temperature directly affects the feedback pressure $p_{reg}(t)$. If this pressure is different from the set point, a force will exist on the spring-restrained diaphragm, causing it to move away from the center position and moving the sensing potentiometer away from its zero-error position. The input voltage at the amplifier, $e_A(t)$, is therefore proportional to the difference in the two pressures and is also proportional to the temperature error in the system, provided that the temperature bulb is properly calibrated. The amplifier generates a current $i(t)$ in the winding of the positioning solenoid so that the position of the mixing valve $q(t)$ is also proportional to the temperature error. If the regulated temperature is exactly at its proper level, then $e_A(t)$ is zero and the mixing valve shaft is centered, delivering the proper temperature of water to the heat exchanger and allowing the system to be at rest. But if the regulated temperature is disturbed by a change in the hot-water temperature or by a variation in the heat-exchanger efficiency, an error will occur. The amplifier will drive the mixing valve toward the HOT position if the temperature error is positive, or toward the COLD position if the error is negative. Because a change in the desired temperature set point will also cause a temperature error, this system will automatically respond both to unwanted disturbances in the heat-exchange mechanisms and to changes in the set-point command.

This qualitative analysis of the feedback system is useful for determining how to connect the diaphragm sensor voltage source, the polarity of the amplifier input and output terminals, and the proper HOT and COLD connections to the plumbing. However, it does not indicate how the system will respond *dynamically* to upsets in the temperature error due to set-point changes or disturbances.

Because none of the individual components in the system can respond instantly to a stimulus, it is necessary to determine the transfer function of each element and to establish a mathematical model for analysis of the closed-loop system. We assume that the system is designed to regulate air temperature over a limited range, that the hot and cold water supplies may have variable temperatures around a nominal value within a known range, and that the center position of the mixing valve will provide the nominal temperature of the water mixture required by the heat exchanger. We then establish the block diagram (see Fig. 13.29) relating the *deviations* of the dynamic variables from their nominal values, and we determine a transfer function for each block using a combination of analysis and dynamic measurements on the individual components. In making such measurements we load the component exactly as it will be loaded during the operation of the system. This important aspect of control engineering is emphasized in Chapter 4.

Let the regulated air temperature be $t_A(t)$. A step increase in this temperature will cause the feedback pressure $p_{\text{reg}}(t)$ to rise abruptly, reaching its new level in a fraction of a second. We neglect this small time lag and take the transfer function of the temperature bulb to be

$$\frac{\Delta P_{\text{reg}}(s)}{\Delta T_A(s)} = K_B = 990 \text{ Pa/}^\circ\text{C}. \tag{13.28}$$

The diaphragm–potentiometer–amplifier–solenoid part of the system is tested with the solenoid connected to the mixing valve and with water flowing through the valve. The tests, which are conducted using the identification techniques of Chapter 7, reveal that this part of the system can be represented accurately by the transfer function

$$\frac{\Delta I(s)}{\Delta P(s)} = \frac{K_1}{(s+1)(s+10)}. \tag{13.29}$$

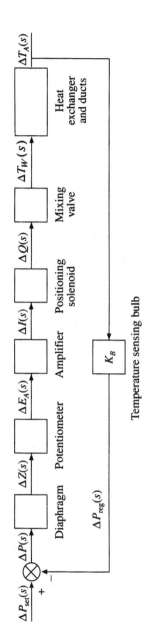

Fig. 13.29 Block diagram for analysis of small changes in air-temperature set point.

The gain K_1 [A/(Pa-s^2)] includes the amplifier gain, which is adjustable. (Note that the steady-state gain of this transfer function $K_1/(1 \times 10)$ has units of A/Pa.) If the tests include dynamic measurements of the solenoid position $q(t)$ and of the water temperature $t_W(t)$ at the output of the mixing valve, then the transfer function of the solenoid–valve combination can also be established. Assume it has the form

$$\frac{\Delta T_W(s)}{\Delta I(s)} = \frac{K_2(s+0.3)}{(s+0.2)(s+0.5)},$$ (13.30)

where $K_2 = 14°$C/(A-s). The steady-state gain of this transfer function is $42°$C/A.

The time delay in this system is due to the transport time of the water flowing from the mixing valve to the heat exchanger. If this transport delay is 8 s, the transfer function relating the water temperature at the output of the mixing valve to the water temperature at the heat exchanger is

$$\frac{\Delta T_{HE}(s)}{\Delta T_W(s)} = \epsilon^{-8s}.$$ (13.31)

To complete the quantitative dynamic description of the block diagram of Fig. 13.29, let us assume that the heat transfer between the water temperature in the heat exchanger and the controlled air temperature is

$$\frac{\Delta T_A(s)}{\Delta T_{HE}(s)} = \frac{0.5}{(s+0.6)}.$$ (13.32)

The upper diagram in Fig. 13.30 is the quantitative description of the air-temperature control system obtained from tests on the system components. The lower diagram, a re-arrangement of the upper diagram, is suitable for determining a range of operating values for the adjustable gain K_1.

The frequency response of the open-loop elements is given by substituting $j\omega$ for s in the transfer function shown in Fig. 13.30:

$$\frac{\Delta T_A(j\omega)}{\Delta T_E(j\omega)} = \frac{6,930 K_1(j\omega+0.3)\epsilon^{-j8\omega}}{(j\omega+0.2)(j\omega+0.5)(j\omega+0.6)(j\omega+1)(j\omega+10)}.$$ (13.33)

The frequency response of this transfer function for $K_1 = 0.00036$ is portrayed on the Nyquist plot in Fig. 13.31. The closed-loop system will be stable for this value of K_1, but the gain margin, measured at $\omega_\pi = 0.234$ rad/s, is only 1.187. The phase margin, measured at $\omega_c = 0.161$ rad/s, is $52.4°$. The spiral frequency response locus for frequencies beyond 0.3 rad/s shows the influence of the time-delay term on the phase shift.

If K_1 is increased slightly beyond the stability limit of 0.000427, then the closed-loop system will be unstable and the open-loop frequency response plot, for ω running from $+\infty$ to $-\infty$, will encircle the -1 point twice, indicating that there will be two poles of the closed-loop transfer function in the right-half s plane. If K_1 is increased slightly beyond 0.00144, the -1 point will lie inside the second ring of the frequency response plot and the number of encirclements will be four, indicating that four poles of the closed-loop transfer function lie

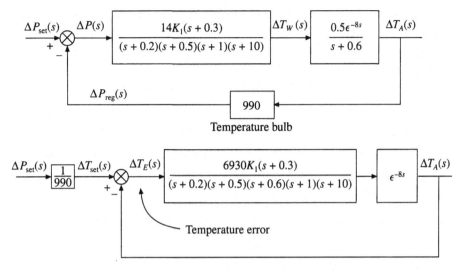

Temperature bulb

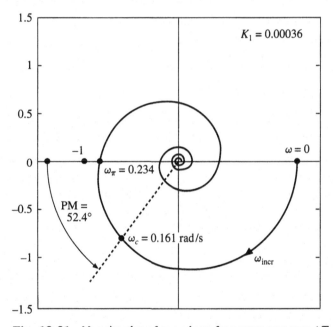

Temperature error

Fig. 13.30 Block diagram reduced to unity feedback configuration.

$K_1 = 0.00036$

$\omega_\pi = 0.234$

$\omega = 0$

PM = 52.4°

$\omega_c = 0.161$ rad/s

ω_{incr}

Fig. 13.31 Nyquist plot of open-loop frequency response $\Delta T_A(j\omega)/\Delta T_E(j\omega)$.

in the right-half plane. It is clear that as K_1 is increased indefinitely, the number of encirclements of the −1 point increases by twos, to six, then to eight, then ten, and so on. The origin of this seemingly infinite supply of poles lies in the transcendental nature of the closed-loop transfer function

$$\frac{\Delta T_A(s)}{\Delta P_{set}(s)} = \frac{(6930/990)K_1(s+0.3)\epsilon^{-8s}}{(s+0.2)(s+0.5)(s+0.6)(s+1)(s+10)+6930K_1(s+0.3)\epsilon^{-8s}}.$$

$$(13.34)$$

Because of the term ϵ^{-8s}, the denominator cannot be factored as a finite-dimensional polynomial.

It is interesting to note that an approximation to the open-loop transfer function $\Delta T_A(s)/\Delta T_E(s)$ can be formed to provide roughly the correct stability margins. We write the open-loop transfer function in factored form as

$$\frac{\Delta T_A(s)}{\Delta T_E(s)} = 3465K_1 \left[\frac{0.2\epsilon^{-8s}}{s+0.2} \right] \left\{ \frac{10(s+0.3)}{(s+0.5)(s+0.6)(s+1)(s+10)} \right\}. \tag{13.35}$$

The denominator of the term in brackets corresponds to the slowest time constant of the transfer function. The term in braces, considered by itself, is the transfer function of a stable system having a steady-state gain of 1 and a fairly fast response, compared to a system whose transfer function is the term in brackets. If we replace the term in braces by 1, we have the approximation

$$\frac{\Delta T_A(s)}{\Delta T_E(s)} \approx \frac{693K_1\epsilon^{-8s}}{s+0.2}. \tag{13.36}$$

In a closed-loop system which uses this approximation as the open-loop transfer function and which has K_1 set at 0.00036, the gain margin, the phase margin, ω_c, and ω_π are roughly comparable to those quantities obtained using the complete open-loop transfer function in Eqn. 13.35.

13.6 Summary

We can determine the stability of a closed-loop system by plotting the open-loop frequency response $G(j\omega)H(j\omega)$ for $-\infty < \omega < +\infty$ and then applying Nyquist's criterion. Although this process is carried out most conveniently using computer-based analysis programs, when computer aid is not immediately available the engineer can approximate $G(j\omega)H(j\omega)$ on a Bode plot, thereby avoiding the intensive calculations required for precise evaluation of that function. It is frequently useful to supplement this Nyquist–Bode procedure with the root-locus techniques of Chapter 9.

The gain margin and the phase margin are apparent either from the Nyquist plot or from the Bode plot. These stability margins are influenced by changes in the open-loop gain, and those changes can be easily assessed from the Bode plot.

The closed-loop performance of the feedback system is reflected in the amplitude and phase properties of $W(j\omega)$. If the amplitude ratio varies smoothly with frequency, showing only a single peak, then the following quantities are expressive of closed-loop performance:

steady-state gain, $W(0)$;
resonant frequency, ω_r;
peak magnitude, M_p; and
bandwidth.

These quantities may be determined from a plot of $G(j\omega)H(j\omega)$ on the Nichols chart; changes caused by variations in loop gain will also be apparent. For this reason, Nichols chart displays are frequently used in computer-aided analysis and design of automatic feedback control systems.

13.7 Problems

Problem 13.1
In the system diagram of Fig. P12.7, let $G_C(s) = 1$, $G_P(s) = K/s(s+10)$, and $H(s) = 1$. When $r(t) = A \sin(10t)$, the system error is $e(t) = \sqrt{2} A \sin(10t + 45°) + \{$stable transients$\}$. Find K and the steady-state component of $y(t)$.

Problem 13.2
In the system diagram of Fig. P12.7, let $G_C(s) = 1$, $G_P(s) = K/s(s^2 + 6s + 100)$, and $H(s) = 1$. K is set so that the closed-loop system M_p is 1.5.

   (a) Find the numerical value of K.
   (b) Find the gain margin and the phase margin of the control loop.
   (c) Find the closed-loop resonant frequency.
   (d) Find the system bandwidth, the highest frequency for which $|W(j\omega)| = -3$ db.

Problem 13.3
In the diagram of Fig. P12.7, let

$$G_P(s) = \frac{K(s+1)}{s(s-2)(s^2 + 16s + 100)},$$

$G_C(s) = 1$, and $H(s) = 1$. Find the value of K for which the closed-loop system will be stable with the maximum possible phase margin. State the phase margin and the gain margin obtained by your K.

Problem 13.4
In a unity negative feedback system, the open-loop transfer function is

$$G(s) = \frac{K}{s(s+40)^2}.$$

   (a) Find the maximum value of K for which all three of the following requirements are satisfied: (i) $M_p \leq 3$ db; (ii) phase margin $\geq 45°$; and (iii) gain margin ≥ 4. *Hint:* Use the Nichols chart.
   (b) What is the closed-loop system resonant frequency and bandwidth (at the -3-db point)?
   (c) A step input is applied to the closed-loop system at $t = 0$. Find the instant at which the peak overshoot of the output occurs. Find the percent overshoot.

Problem 13.5
Show that when the M circles for a unity feedback system are plotted on the $1/G(j\omega)$ plane, they are concentric circles about the -1 point, and that the N "circles" are radial lines through that point.

Problem 13.6
Let $W(s) = G(s)/[1+G(s)H(s)]$ for a nonunity feedback system. Show that the M_p and the ω_r of the closed-loop system can be read directly from a polar chart on which $H(j\omega)$ is added to $1/G(j\omega)$.

Problem 13.7
Refer to Problem 12.8. If $H(s) = 1$ and $G_C(s) = K$, for which value of K is the phase margin at its maximum possible value? What is that maximum value for the phase margin?

Problem 13.8
A process control plant is diagramed in Fig. P13.8. In a calibration test, the process gain K is adjusted so that the control loop is on the verge of instability. A step change in $r(t)$ causes a prolonged oscillation of $y(t)$ with a period of 6 seconds per cycle. Find the delay time τ. Find the maximum value of K if the phase margin of the loop is to be better than 40°. What is the gain margin at that value for K?

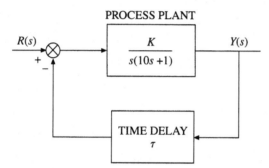

Fig. P13.8

Problem 13.9
Let

$$G(s) = \frac{40}{s(s+P)(s^2+4s+16)}$$

in a unity feedback system, with phase margin of 63°. Find the value of P, the gain margin, and the bandwidth at the -3-db point.

Problem 13.10
Find the phase margin of the control loop and the bandwidth of the closed-loop system $Z(j\omega)/Z_{\text{ref}}(j\omega)$ (at the -3-db point) for your design in Problem 10.19.

Problem 13.11
Consider the system in Problem 13.10. Increase the loop gain until the closed-loop M_p reaches 2 db. Determine the gain margin, the phase margin, the closed-loop resonant frequency, and the closed-loop bandwidth.

Problem 13.12
Refer to the discussion of the PID controller in Section 10.3. The frequency response of a PID controller (amplitude ratio only) is plotted in Fig. P13.12. Find K_P, K_I, K_D, and the frequency at which the phase shift is $+60°$.

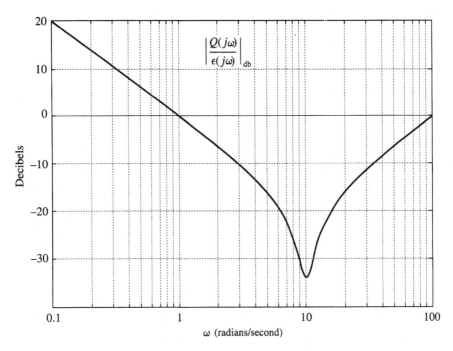

Fig. P13.12

Problem 13.13
A unit step input is applied to the closed-loop system whose frequency response is plotted in Fig. 13.8. Calculate and plot the zero-state time response of this system.

Problem 13.14
Verify the conclusion in Eqn. 13.21, first by the root-locus method and then by Routh's criterion.

Problem 13.15
Refer to the Nichols chart in Fig. 13.25, for which the gain constant of $G(j\omega)$ is 512.

 (a) How must the gain be changed to provide the maximum phase margin for this system?
 (b) How much can you increase the gain from 512 and still retain the closed-loop M_p of 4.28 db? What will the phase margin and the resonant frequency be in this case?

Problem 13.16
Let

$$G(s) = \frac{K(s+Z)}{s(s+10)(s^2+10s+40)}$$

be the open-loop transfer function for a unity positive feedback system. Find values, if any, for K and Z that satisfy the following criteria:

 (a) The closed-loop system must be stable with a phase margin of at least 40°.
 (b) The zero-frequency gain of the closed-loop system must be 1.

If you are unable to find a combination $\{K, Z\}$ to satisfy both (a) and (b), can you find one to satisfy only (a)?

Problem 13.17

In the system configuration of Fig. 8.8, let

$$G(s) = \frac{K}{s^2(s+10)} \quad \text{and} \quad H(s) = \frac{10(s+1)}{(s+10)}.$$

(a) Find the value of K for which the phase margin will be at a maximum. What is the gain margin for this value of K?

(b) Find the maximum value which K can have if the closed-loop M_p, $|Y(j\omega)|/R(j\omega)|_{max}$, is to be 1.414 or less.

(c) For your K in part (b), find the gain margin, the phase margin, the closed-loop resonant frequency, and the closed-loop bandwidth (at the $-40°$ point).

14 Design of Feedback Systems by Frequency Response Methods

14.1 Introduction

The setting for our design problem is defined in Fig. 14.1. We begin the design process by assuming that we know the frequency response characteristics of the fixed plant $G_P(j\omega)$. The design goal is to identify the frequency response characteristics of the control laws $G_C(j\omega)$ and $H(j\omega)$, so that specified performance requirements are satisfied. These requirements pertain both to the open-loop frequency response $G_C(j\omega)G_P(j\omega)H(j\omega)$ and to the closed-loop frequency response

$$W(j\omega) = \frac{Y(j\omega)}{R(j\omega)} = \frac{G_C(j\omega)G_P(j\omega)}{1+G_C(j\omega)G_P(j\omega)H(j\omega)}. \tag{14.1}$$

The term $G_C(j\omega)G_P(j\omega)H(j\omega)$ must often satisfy a specified level of gain at very low frequencies, and it must also provide adequate gain margin and phase

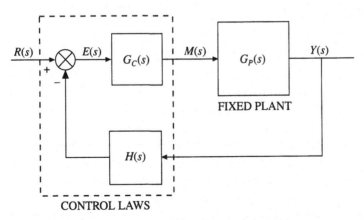

CONTROL LAWS

Fig. 14.1 System configuration for the design problem.

405

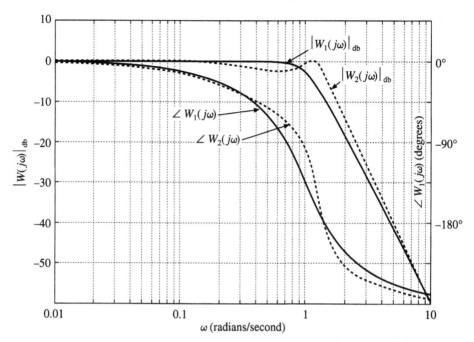

Fig. 14.2   Amplitude and phase characteristics for two closed-loop systems.

margin. Also, $W(j\omega)$ must meet specifications on its amplitude ratio and phase shift, for example, with respect to zero-frequency gain, resonant frequency, peak amplitude ratio, and bandwidth. The design specifications are such that dynamic elements are normally required in the control laws, and we must exercise the technique of series compensation (to find $G_C(s)$) or of parallel compensation (to find $H(s)$).

To obtain an image of desirable closed-loop frequency response characteristics, consider two different systems identified as $W_1(j\omega)$ (system 1) and $W_2(j\omega)$ (system 2). The amplitude and phase characteristics of these two systems are compared in Fig. 14.2. The zero-frequency gain of each system is 0 db, a characteristic that is very common in automatic control systems. The high-frequency amplitude ratios are also equal. The mid-frequency amplitude ratio of system 1 changes slope monotonically and gradually with frequency as compared to that of system 2, whose amplitude ratio has two mild slope reversals in the mid-frequency range. However, both $|W_1(j\omega)|_{db}$ and $|W_2(j\omega)|_{db}$ have maximum values of 0 db. The bandwidth of system 2 (measured at the -3-db point) is 1.39 rad/s, whereas that of system 1 is 1 rad/s. The phase shift characteristics of both systems are nearly identical at extremely low and extremely high frequencies, and are roughly comparable in the mid-frequency range.

The dynamic differences between system 1 and system 2 may also be viewed in the time domain. Figure 14.3 compares the zero-state response of the two systems when each is subjected to a unit step input. These two time responses are similar during the first 3 s of the transient, but the rise time of system 2 is slightly less than that of system 1. The settling time for $y_1(t)$ might be designated

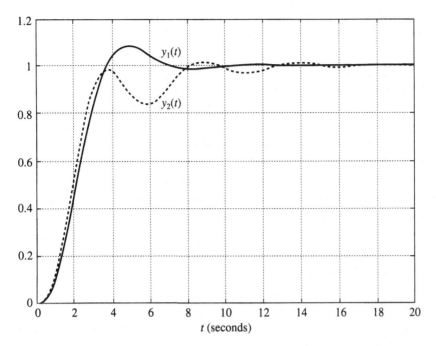

Fig. 14.3 Step responses of the two systems of Fig. 14.2.

as 8 s, whereas that for $y_2(t)$ might be 12 s. The two responses are comparable when $t > 13$ s, but $y_2(t)$ oscillates during the interval 3 s $< t <$ 12 s, indicating that its acceleration $\ddot{y}_2(t)$ reaches significantly higher peaks than does $\ddot{y}_1(t)$.

The comparison of $W_1(j\omega)$ with $W_2(j\omega)$ and the corresponding comparison of $y_1(t)$ with $y_2(t)$ provide a qualitative basis for establishing design specifications for the frequency response of the closed-loop system. The bandwidth of $W(j\omega)$ and the rise time of the step response are inversely related. In system 1, the bandwidth is 1 rad/s and the rise time is approximately 3.3 s; the product of the two is

$$[\text{BW rad/s}] \times [t_R \text{ s}] \approx 3.3. \tag{14.2}$$

The same product for system 2 is approximately 4.25. Here the rise time is taken to be the instant the step response first reaches 90 percent of its final value.

In many applications it is desirable to minimize unnecessary levels of acceleration of the controlled variable in response to abrupt changes in the input signal. Therefore, the frequency response for $W(j\omega)$ is often specified to resemble that of system 1; that is, $|W(j\omega)|$ should vary smoothly with frequency. A slight peak in $|W(j\omega)|$ in the mid-frequency range may be permitted, provided there is only one reversal in the slope of $|W(j\omega)|$ in the mid-frequency range.

Bandwidth design specifications are also dictated by the intended application of the system. In many cases $W(s)$ is simply a component in a larger system, and must conform to the dynamic requirements of that larger system.

14.2 Series Compensation

Consider the third-order electromechanical servomechanism described in Section 10.2 and diagramed in Fig. 14.4. To establish a baseline design we choose a nondynamic control law

$$G_C(s) = K_C,$$ (14.3)

and we require the closed-loop amplitude ratio to show an M_p of 2 db or less. The Nichols chart plot of the open-loop frequency response $Y(j\omega)/E(j\omega)$ for $K_C = 20.1$, shown in Fig. 14.5, reveals the following values for the system performance parameters:

$\omega_c = 62.1$ rad/s, $\omega_\pi = 158.8$ rad/s, $\omega_r = 65.2$ rad/s,

phase margin $= 47°$,   gain margin $= 13$ db, $M_p = 2$ db,

$$\left[\begin{array}{c} \text{low-frequency} \\ \text{open-loop gain} \end{array} \right] \triangleq \left[\frac{Y(j\omega)}{E(j\omega)} \right]_{\omega = 1 \text{ rad/s}} = 37.13 \text{ db}, \quad \text{bandwidth} = 111 \text{ rad/s}.$$

The open-loop frequency response is also plotted on the Bode chart in Fig. 14.6. Here the gain margin, the phase margin, the low-frequency gain, ω_c, and ω_r are identified. The amplitude ratio of the closed-loop frequency response, $|W(j\omega)|_{db}$, is also plotted in Fig. 14.6. The closed-loop amplitude ratio has a smoothly varying form, qualitatively similar to that of system 1 in Fig. 14.2, and with a bandwidth of 111 rad/s. One might accordingly expect to see a time-domain transient response resembling that of system 1 but with slightly more overshoot and a rise time of roughly 0.030 s.

The plots in Fig. 14.6 illustrate a dynamic feature commonly found in unity feedback control systems. In the expression relating the open-loop frequency response to the closed-loop frequency response,

$$W(j\omega) = \frac{G(j\omega)}{1 + G(j\omega)},$$ (14.4)

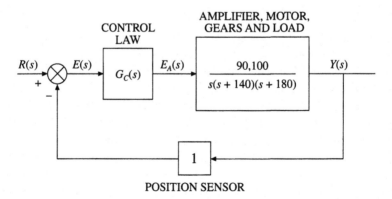

Fig. 14.4 Servomechanism from Chapter 10.

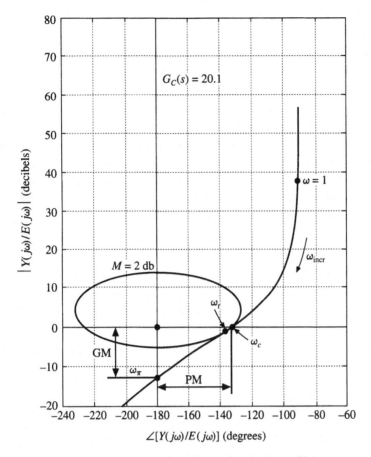

Fig. 14.5 Nichols plot of $Y(j\omega)/E(j\omega)$ for $G_C(j\omega) = 20.1$.

it is clear that if $G(j\omega)$ is very much larger than 1, as it is in this case at low frequency, then $W(j\omega)$ will be very close to 1, as it is for frequencies below 20 rad/s. Further, if $G(j\omega)$ is very much smaller than 1, as it is at frequencies exceeding 200 rad/s in this example, then $W(j\omega)$ will be very close to $G(j\omega)$. Thus, in cases where $G(j\omega)$ resembles this example at the extreme frequencies, from $G(j\omega)$ one can accurately estimate $W(j\omega)$ at both ends of the frequency spectrum.

The low-frequency open-loop gain is an important performance measure because it is directly related to the steady-state torque gain of this system (see Section 10.2). For this example we will measure the low-frequency open-loop gain at $\omega = 1$ rad/s. Let us assume that the low-frequency gain of 37.13 db (obtained in the baseline design) is unsatisfactory, and that an increase of at least 12 db is required. Let us also assume that the closed-loop frequency response should resemble that of the baseline system.

To achieve this revised design objective, we must find a compensator whose amplitude ratio at $\omega = 1$ rad/s is at least 12 db above the 26.06 db provided by

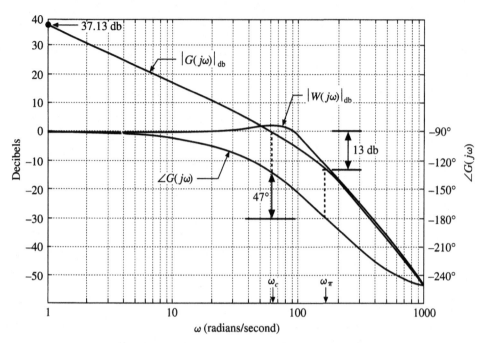

Fig. 14.6   Bode diagrams for the open-loop amplitude ratio and phase shift, and for the closed-loop amplitude ratio, for $G_C(j\omega) = 20.1$.

$K_C = 20.1$ in the baseline system. The amplitude and phase characteristics of the new compensator, when combined with those of the plant, must provide an open-loop frequency locus that is tangent to the contour for $M = 2$ db on the Nichols chart at approximately the same frequency as that of the baseline system, as shown in Fig. 14.5. The compensator must therefore have an amplitude characteristic that decreases with frequency in the range 1 rad/s $< \omega <$ 60 rad/s. The simplest such compensator has the form

$$G_C(s) = \frac{E_A(s)}{E(s)} = \frac{K_C(s+Z)}{s+P},\tag{14.5}$$

where the break frequency P must be lower than the break frequency Z. To provide the required amplitude characteristics, Z should be about a decade below the frequency at which the composite locus $G(j\omega) = Y(j\omega)/E(j\omega)$ touches the contour for $M = 2$ db. P must be sufficiently lower than Z to provide the 12-db improvement in the required low-frequency gain.

Several trials show a satisfactory solution to be

$$G_C(s) = \frac{17.05(s+5)}{s+0.2},\tag{14.6}$$

which is characteristic of a lag compensator. The composite locus $G(j\omega)$ is plotted in Fig. 14.7, along with that of the baseline system. The phase lag introduced by the compensator occurs at low frequencies, where the $G(j\omega)$ locus is well above the -1 point. Therefore, the large phase lag at low frequencies that

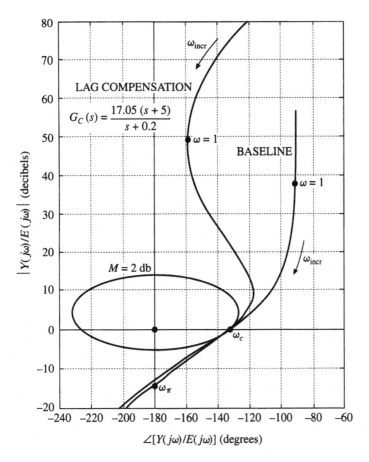

Fig. 14.7 Frequency response of the baseline system compared to that of the lag-compensated system.

is introduced into the control loop by the compensator does not diminish the stability margins. The increase in low-frequency gain over the baseline system satisfies the required 12 db, and the phase margin and gain margin are comparable to those of the baseline system. The bandwidth of the closed-loop system is lower (99.3 rad/s) than that of the baseline system. The performance parameters for this design are:

$\omega_c = 54.6$ rad/s, $\omega_\pi = 151$ rad/s, $\omega_r = 54.5$ rad/s,

phase margin $= 46.8°$, gain margin $= 13.9$ db, $M_p = 2$ db,

$$\begin{bmatrix} \text{low-frequency} \\ \text{open-loop gain} \end{bmatrix} \triangleq \left[\frac{Y(j\omega)}{E(j\omega)} \right]_{\omega=1\,\text{rad/s}} = 49.68 \text{ db}, \quad \text{bandwidth} = 99.3 \text{ rad/s}.$$

The improvement in low-frequency gain provided by the lag compensator is remarkable, and it comes with relatively little expense, since the compensating circuits can be designed as part of the input stage of the power amplifier or, if the control law is realized as a digital computer, by a modest addition to the software. But the new design sacrifices some bandwidth as compared to the baseline system.

An improvement in bandwidth can be achieved if a compensator having a transfer function

$$G_C(s) = \frac{K_C(s+Z_1)(s+Z_2)}{(s+P_1)(s+P_2)} \tag{14.7}$$

is designed. This control law may also be realized with little additional expense over the baseline design. The choice of the five parameters of the new control law, K_C, Z_1, Z_2, P_1, and P_2, is made not only to retain the low-frequency nature of the lag compensator (i.e., high-amplitude gain at low frequency, with the attendant phase-angle lag occurring where it will not influence the stability margins) but also to provide phase-angle lead in the mid-frequency range. The leading phase angle is accompanied by a decrease in amplitude gain at mid-frequency; this decrease is used to shape the composite locus $Y(j\omega)/E(j\omega)$ in the vicinity of the -1 point, thus guaranteeing satisfactory stability margins. This strategy gives $G_C(j\omega)$ a lag-lead characteristic.

We begin the search for a satisfactory set of these parameters by choosing Z_1 and P_1 to be the same as the P and Z used in the lag compensator design. We then choose values for Z_2 and P_2 for their influence on the mid-frequency segment of the composite locus, which must be shaped approximately like the locus of the baseline system in the critical region where the phase shift of $Y(j\omega)/E(j\omega)$ is between $-120°$ and $-180°$. The gain K_C influences the vertical position of the $Y(j\omega)/E(j\omega)$ locus on the Nichols chart, but not its shape. We then "tune" the five parameters around their initial approximate values to arrive at a satisfactory design. Computer-aided calculations are essential in tuning a system in this fashion. A satisfactory solution for this example is $K_C = 250$, $Z_1 = 10$, $Z_2 = 70$, $P_1 = 0.2$, and $P_2 = 500$. The stability margins for this design are very close to those of the baseline system, and the low-frequency gain and the bandwidth are remarkably superior to both the baseline design and the lag-compensated design. The Nichols chart for $Y(j\omega)/E(j\omega)$ showing this comparison appears in Fig. 14.8. The performance parameters for the closed-loop system are:

$\omega_c = 149.1$ rad/s, $\omega_\pi = 356.9$ rad/s, $\omega_r = 171.7$ rad/s,

phase margin $= 48.1°$,   gain margin $= 12.25$ db, $M_p = 2$ db,

$$\left[\begin{array}{l} \text{low-frequency} \\ \text{open-loop gain} \end{array}\right] \triangleq \left[\frac{Y(j\omega)}{E(j\omega)}\right]_{\omega=1\,\text{rad/s}} = 61.82 \text{ db}, \quad \text{bandwidth} = 265 \text{ rad/s}.$$

Figure 14.9 shows the qualitative similarity of the frequency response characteristics of the three closed-loop systems. This implies that the time-domain responses of the three closed-loop systems should also bear a qualitative similarity to one another.

As a second example of series compensation, consider the design of an automatic pilot intended to regulate the altitude of an aircraft in cruising flight,

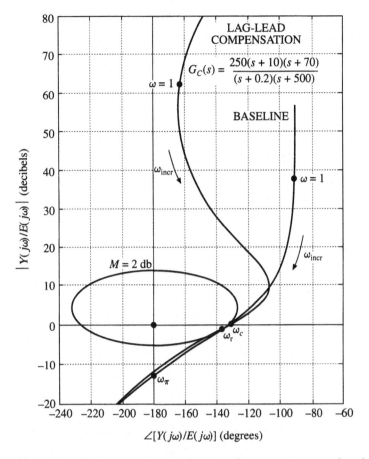

Fig. 14.8 Frequency response of the baseline system compared to that of the lag-lead–compensated system.

as diagramed in Fig. 14.10. The altitude of the aircraft is $h_0 + \Delta h(t)$, where h_0 is the nominal cruising altitude. The throttle remains fixed at its cruise setting, and only the elevator command $r(t)$ is used to control the altitude. The Navion aircraft, whose pitch-axis dynamics were discussed in Section 2.7, is used as the model here. When the phugoid mode is damped, using the scheme described in Section 10.5, the transfer function relating the incremental elevator command to the incremental altitude is

$$\frac{\Delta H(s)}{R(s)} = \frac{900(s+13)(s+0.017)(s-10)}{s(s+20.2)(s+0.3)(s+0.22)[s^2+4.3s+14]} \quad \text{ft/rad.} \quad (14.8)$$

The numerical values in $\Delta H(s)/R(s)$ have been rounded slightly for convenience. Note that this is a nonminimum phase transfer function. It shows that the steady-state response of the aircraft to an elevator command of 1° (a "down" elevator deflection) is a descent rate of 1.86 ft/s.

In this example we seek a control law which will provide a closed-loop system that is stable and very well damped. For a sudden step altitude command

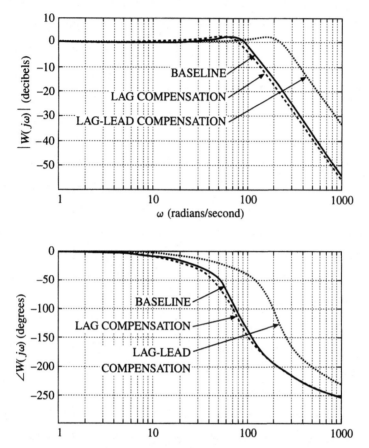

Fig. 14.9 Comparing the closed-loop frequency responses of the baseline system, the lag-compensated system, and the lag-lead–compensated system.

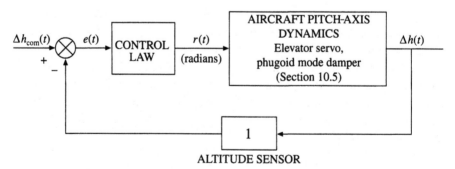

Fig. 14.10 Altitude control system for the Navion airplane.

of 100 feet, the altitude response $\Delta h(t)$ should have a rise time of approximately 15 s, an overshoot of less than 2 ft, and a zero steady-state altitude error. Using system 1 in Section 14.1 as a guide, we set a tentative specification for our closed-loop bandwidth of 0.2 rad/s and the desired M_p at 0 db.

As a baseline design, we choose the simplest control law

$$G_C(s) = \frac{R(s)}{E(s)} = K_C. \tag{14.9}$$

We determine whether it is possible to achieve the desired performance by a suitable choice of K_C. We draw a Nyquist plot for the pitch-axis dynamics defined by the transfer function in Eqn. 14.8. This plot is shown for $K_C = 1$ in Fig. 14.11, where the scale is exaggerated for clarity. To apply Nyquist's stability criterion to this system we draw the $j\omega$-axis contour, using the small semicircle technique of Chapter 12 to exclude the pole of $\Delta H(s)/R(s)$ at the origin from within the contour $\mathcal{C}$. The large semicircular part of the contour $\mathcal{D}$ is then in the position shown. Because $\Delta H(s)/R(s)$ has no poles inside contour $\mathcal{C}$ (the right-half plane), the P in Nyquist's formula equals zero. The N in Nyquist's formula is 3, because the $\mathcal{D}$ contour encircles the -1 point three times in the

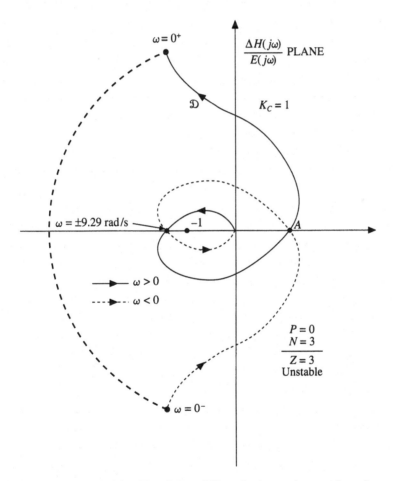

Fig. 14.11 Applying Nyquist's stability criterion to the open-loop frequency response $\Delta H(j\omega)/\Delta E(j\omega)$ for $G_C(j\omega) = 1$.

positive sense. Consequently, the closed-loop system will be unstable because $\Delta H(s)/\Delta H_{com}(s)$ has three poles in the right-half plane. If K_C is increased beyond 1, the size of $\Delta H(j\omega)/E(j\omega)$ will change but not its shape. The position of the -1 point in Fig. 14.11 will move to the right, approaching the origin as K_C approaches infinity. The system will remain unstable for $1 < K_C < \infty$.

If K_C is decreased to a value between 0 and 0.807, the -1 point in Fig. 14.11 will lie between the inner and outer rings of the $\Delta H(j\omega)/E(j\omega)$ locus. The locus will encircle the -1 point once in the positive sense, and since P is 0 as before, Nyquist's formula reveals that the closed-loop system will still be unstable, with a single pole of the closed-loop transfer function, $\Delta H(s)/\Delta H_{com}(s)$, in the right-half plane.

When negative values for K_C are considered, the Nyquist locus for $\Delta H(j\omega)/E(j\omega)$ has exactly the same shape as that in Fig. 14.11, but the negative sign in K_C introduces a phase shift of 180° at all frequencies. Consequently, the $\Delta H(j\omega)/E(j\omega)$ locus is that in Fig. 14.12, where again the scale has been exaggerated for

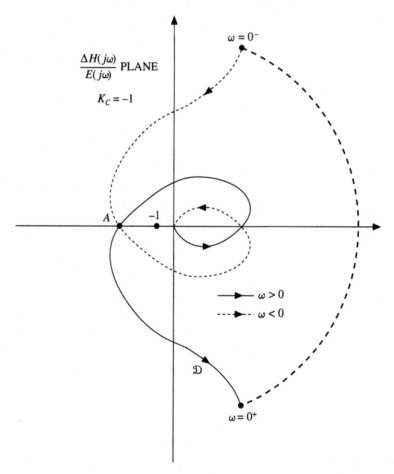

Fig. 14.12 Applying Nyquist's stability criterion to the open-loop frequency response $\Delta H(j\omega)/\Delta E(j\omega)$ for $G_C(j\omega) = -1$.

clarity. Point A then lies at $-325.1+j0$. When $K_C = -1$, the two encirclements of the -1 point by the $\mathcal{D}$ locus indicate that two poles of $\Delta H(s)/\Delta H_{com}(s)$ lie in the right-half plane. But when $0 > K_C > -0.003076$, the -1 point lies to the left of point A. There are no encirclements, and because $P = 0$, the closed-loop system will be stable.

To find the most suitable value for K_C in this range, draw the open-loop frequency response locus $\Delta H(j\omega)/E(j\omega)$ on the Nichols chart and adjust K_C until this locus is tangent to the contour for $M = 0$ db. The appropriate value $K_C = -0.000472$ then produces the Nichols plot displayed in Fig. 14.13. Although the closed-loop frequency response will have the desired $M_p = 1$, it will resemble that of system 2 in Fig. 14.2 because of the abrupt phase lead of $\Delta H(j\omega)/E(j\omega)$ occurring in the frequency range $0.0016 < \omega < 0.36$ rad/s. In order for our closed-loop response to resemble that of system 1 in Fig. 14.2, the $\Delta H(j\omega)/E(j\omega)$ locus must lie close to the contour for $M = 0$ db at low frequency. It must depart from it gradually, without reversing direction on the Nichols chart, cross the contour for $M = -3$ db at the desired bandwidth frequency (approximately 0.2 rad/s), and continue to diminish at higher frequencies. It is therefore

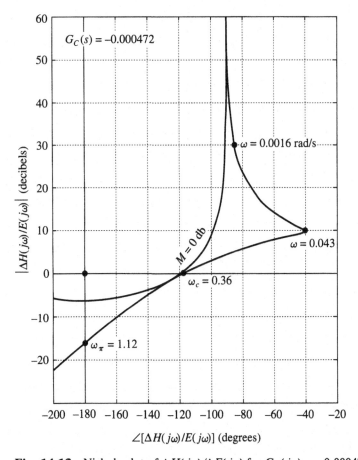

Fig. 14.13 Nichols plot of $\Delta H(j\omega)/\Delta E(j\omega)$ for $G_C(j\omega) = -0.000472$.

necessary to introduce a dynamic element in the control law, so that the composite frequency response

$$\frac{\Delta H(j\omega)}{E(j\omega)} = \underbrace{\left[\frac{R(j\omega)}{E(j\omega)}\right]}_{\substack{\text{control} \\ \text{law}}} \times \underbrace{\left[\frac{\Delta H(j\omega)}{R(j\omega)}\right]}_{\substack{\text{pitch-axis} \\ \text{dynamics}}}$$

(14.10)

will have the desired shape and position on the Nichols chart.

The excessive phase lead of approximately 55 deg that occurs at $\omega = 0.043$ rad/s can be removed by a control law having a comparable lagging phase in that frequency range. We have seen from the analysis in Chapter 11 that a compensator transfer function having a single pole and a single zero will have a frequency response whose phase shift is nearly zero at both low and high frequencies; it will approach as much as 70° or 80° at intermediate frequencies, leading or lagging, depending on the choice of the pole and zero values. For this case a control law of the form

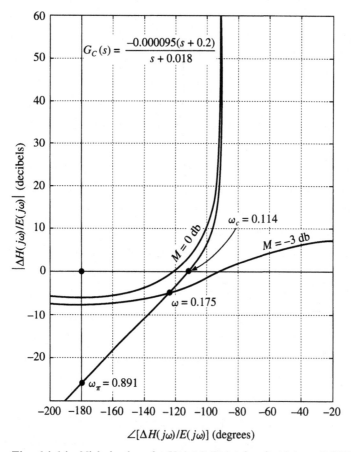

Fig. 14.14   Nichols plot of $\Delta H(j\omega)/\Delta E(j\omega)$ for $G_C(j\omega) = -0.000095(j\omega + 0.2)/ (j\omega + 0.018)$.

$$G_C(s) = \frac{K_C(s+Z)}{s+P} = \frac{K_C(s+0.2)}{s+0.02}, \tag{14.11}$$

which offers a maximum phase lag of 55 deg at $\omega = 0.063$ rad/s, is a suitable choice for a first trial design. A satisfactory final design for the control law is found by adjusting P and Z around these nominal values and by setting K_C to position the $\Delta H(j\omega)/E(j\omega)$ locus on the Nichols chart. Computer-aided calculations are essential in this iterative process. A final design for the control law that satisfies the specifications for the altitude control system is

$$G_C(s) = \frac{-0.000095(s+0.2)}{s+0.018}. \tag{14.12}$$

The resulting Nichols plot is shown in Fig. 14.14. The closed-loop performance parameters achieved in this design are:

$$\omega_c = 0.114 \text{ rad/s}, \quad \omega_\pi = 0.891 \text{ rad/s}, \quad \omega_{BW} = 0.175 \text{ rad/s},$$

phase margin $= 69.6°$, gain margin $= 26.42$ db, $M_p = 0$ db.

Further, it is clear from the Nichols chart that the closed-loop frequency response will resemble that of system 1 in Fig. 14.3, so the time-domain response of the altitude control system $\Delta h(t)$ to a step-function command should have approximately the specified rise time.

14.3 Parallel Compensation

An alternate method for compensating the third-order servomechanism of Section 14.2 is to place a dynamic element in series with the position sensor in the feedback path, as shown in Fig. 14.15. Here the forward-path element is a constant gain of 90, and we accomplish control loop compensation by choosing a suitable $H(s)$. If we assume that the gain of the feedback element at $\omega = 1$ rad/s will be close to 1, the low-frequency open-loop gain $|E_b(j\omega)/E(j\omega)|_{\omega=1}$

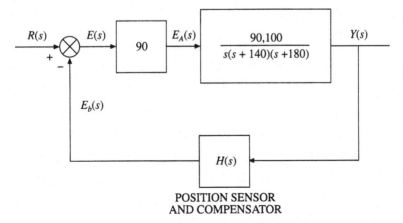

Fig. 14.15 Parallel compensation of the electromechanical servomechanism.

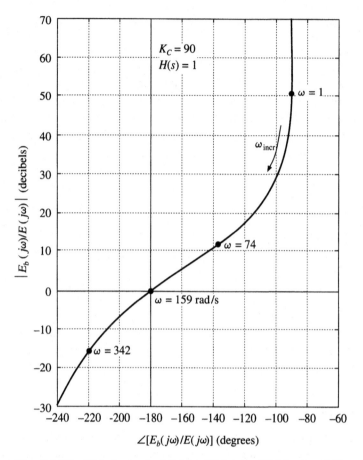

Fig. 14.16   Nichols chart of $E_b(j\omega)/E(j\omega)$ for $H(j\omega) = 1$ and $K_C = 90$.

will be 50.1 db, a gain satisfying the same design specification on this system that the series lag compensator satisfied in Section 14.2.

If we attempt to use a simple gain in the feedback path $H(s) = 1$, the closed-loop system will be unstable, as shown by the Nichols plot of $E_b(j\omega)/E(j\omega)$ in Fig. 14.16. Here the open-loop locus passes very close to the -1 point, so both ω_c and ω_π are close to 159 rad/s. The feedback element must therefore possess dynamics having a frequency characteristic $H(j\omega)$ which will reshape the locus of Fig. 14.16 to avoid the -1 point and which will also provide adequate stability margins.

We must also note that the closed-loop contours on the Nichols chart now pertain to $E_b(j\omega)/R(j\omega)$, and not to $Y(j\omega)/R(j\omega)$. Therefore $H(j\omega)$ must also be such that the ratio $Y(j\omega)/R(j\omega)$ will have acceptable amplitude and phase properties. This ratio is derived from the relationship

$$\frac{Y(j\omega)}{R(j\omega)} = \left\{ \frac{E_b(j\omega)/E(j\omega)}{1 + E_b(j\omega)/E(j\omega)} \right\} \times \left[\frac{1}{H(j\omega)} \right], \tag{14.13}$$

where the term in braces is obtained from the closed-loop contours of the Nichols chart. A third restriction on $H(j\omega)$ is that its steady-state gain, $H(j0)$, must be 1 if the steady-state output is to match the steady-state input.

The simplest compensator that might satisfy these requirements has the form

$$H(s) = \frac{(P/Z)(s+Z)}{s+P}. \tag{14.14}$$

Figure 14.16 shows that $H(j\omega)$ must provide about 60° of phase lead in the approximate frequency range $70 \le \omega \le 350$ rad/s. A series of trial values for Z and P converges to a satisfactory solution, $Z = 130$ and $P = 1300$. The open-loop frequency response plotted on the Nichols chart in Fig. 14.17 shows that the system has a gain margin of 13.4 db and a phase margin of 32.2°. The open-loop locus is tangent to the contour of $M = 5.12$ db at $\omega = 225$ rad/s. Thus $|E_b(j\omega)/R(j\omega)|_{db}$ has a resonant peak at 225 rad/s, as shown in Fig. 14.18. To find the closed-loop response $|Y(j\omega)/R(j\omega)|_{db}$ we add the curve $|1/H(j\omega)|_{db}$ to

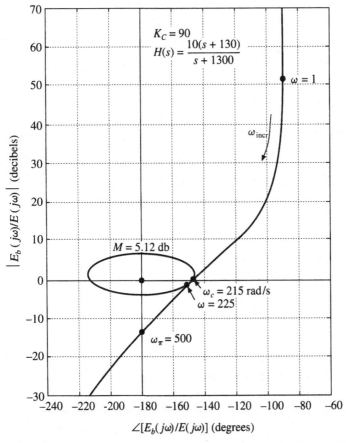

Fig. 14.17   Nichols chart of $E_b(j\omega)/E(j\omega)$ for $H(j\omega) = 10(j\omega+130)/(j\omega+1300)$ and $K_C = 90$.

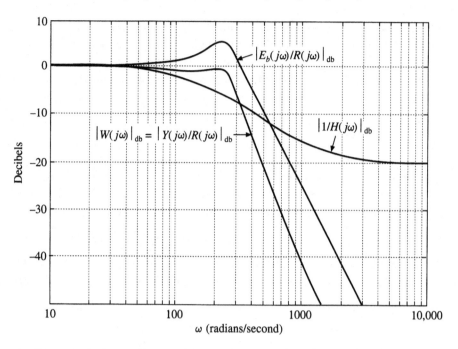

Fig. 14.18 $|W(j\omega)|_{db}$ obtained from the sum of $|E_b(j\omega)/R(j\omega)|_{db}$ and $|1/H(j\omega)|_{db}$ on the Bode chart.

$|E_b(j\omega)/R(j\omega)|_{db}$, as indicated. The resultant $|W(j\omega)|_{db}$ has a shape that is a compromise between those of the two systems shown in Fig. 14.2. The slope reversals of $W(j\omega)$ in this system are less severe than those in system 2; therefore, the time-domain response will exhibit less severe peaks of acceleration than does $y_2(t)$ in Fig. 14.3. The performance parameters for this design are:

$\omega_c = 215$ rad/s, $\omega_\pi = 500$ rad/s,

phase margin $= 32.3°$,   gain margin $= 13.4$ db, $M_p = 0$ db,

$$\left[\begin{array}{l} \text{low-frequency} \\ \text{open-loop gain} \end{array}\right] \triangleq \left[\frac{E_b(j\omega)}{E(j\omega)}\right]_{\omega=1\,\text{rad/s}} = 50.15 \text{ db},$$

bandwidth $\triangleq Y(j\omega)/R(j\omega) = 263$ rad/s.

14.4 Robustness

In designing the compensators for the electromechanical servomechanism in Sections 14.2 and 14.3, we assumed that the physical parameters of the amplifier, motor, gears, and load were known and fixed. In many practical cases the numerical values of these parameters are not precisely known, and after the system is placed in service the parameters may change owing to wear or variations in the operating environment. A system is said to be *robust* if its performance remains satisfactory in the face of such uncertainties and changes.

The original work on feedback amplifiers in the 1920s was motivated precisely by the question of robustness, and for many years designers settled that question simply by providing high loop gain and sufficient gain and phase margins, using compensation where necessary. But as feedback control has been applied to dynamic processes involving electromechanical, aerodynamic, heat-transfer, and chemical phenomena, robustness has re-appeared as an important practical consideration in control system design.

Considerable research effort is being expended on the difficult problem of analytically representing the effect of uncertainties and changes in physical parameters in order to quantitatively assess the robustness of a control system. However, in some cases only a single parameter changes over a known range of values, and in such a case it is possible to calculate the robustness of the system with respect to the changes in this parameter, if those changes occur slowly. As an example, consider the electromechanical servomechanism in Section 14.3 when compensated with a dynamic element in the feedback path, as shown in Fig. 14.15.

The numerical values given in the transfer function for the amplifier–motor–gears–load combination $Y(s)/E_A(s)$ in Fig. 14.4 come from the physical description of this combination given in Section 10.2. One of the parameters is the moment of inertia of the load J_L, which is 0.002 kg-m^2. If we consider this to be an uncertain parameter whose actual value is $J_L = 0.002\Delta$ kg-m^2, the transfer function may be approximately expressed in

$$\frac{Y(s)}{E_A(s)} = \frac{180{,}200/(1+\Delta)}{s[s^2+320s+50{,}400/(1+\Delta)]}, \quad \text{where } \Delta = \frac{J_L}{0.002}. \tag{14.15}$$

Note that for $\Delta = 1$ this transfer function is the same as that in Fig. 14.4. The simple form of this transfer function makes it relatively easy to assess the robustness of this system with respect to variations in J_L. The Nichols chart of Fig. 14.19 is calculated for Δ ranging from 0.25 to 3, representing a variation in J_L of more than one order of magnitude, from 0.0005 to 0.006 kg-m^2. Because the boundaries of the shaded region cross near the contour for $M = 5.12$ db, the shaded region does not quite contain all of the possible loci for all of the values of J_L between 0.0005 and 0.006 kg-m^2. Nevertheless, the shaded region provides a useful representation of the robustness of this system with respect to a wide variation in the design value of J_L. Figure 14.20 shows $|Y(j\omega)/R(j\omega)|_{\text{db}}$ for the two extreme values of J_L considered here. The resultant variations in the performance parameters would be:

gain margin, 9.5 db to 18.8 db;
phase margin, 27° to 32.8°; and
closed-loop bandwidth, 350 rad/s to 179 rad/s.

14.5 Summary

The frequency response method of control loop compensation is an alternate approach to the root-locus method of design, which is the subject of Chapter

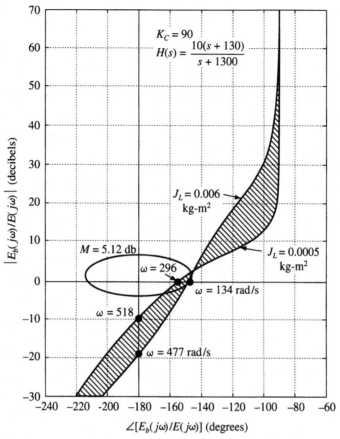

Fig. 14.19 Robustness of the stability margins with respect to uncertainties in J_L.

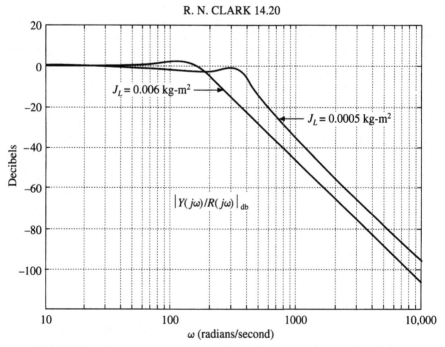

Fig. 14.20 Comparison of the closed-loop frequency response for the two extreme values of J_L.

10. The two methods share a common analytical foundation, and they are equally amenable to computer-aided trial-and-error design procedures. Many automatic control engineers prefer one method over the other. However, because engineering data and specifications appear routinely in both terminologies, engineers must be capable of communicating in either phraseology.

14.6 Problems

Problem 14.1
Let $H(s) = 1$ and $G_P(s) = 250/s(s+4)(s+6)$ in the system configuration of Fig. 14.1. Find a compensating transfer function $G_C(s)$ that will satisfy the following requirements:

(a) The open-loop gain at $\omega = 0.5$ rad/s must be at least 19 db.
(b) The closed-loop M_p must be less than 3 db.
(c) The closed-loop phase lag for $0 \le \omega \le 4.5$ rad/s must be less than 40°.

State the gain margin, the phase margin, the closed-loop resonant frequency, and the bandwidth (at the −3-db point) obtained by your compensator design.

Problem 14.2
Figure P14.2 is the block diagram for an electrohydraulic servomechanism. An undesirable AC voltage at 30 Hz is induced into the feedback sensor by vibration in the driven member. A noise filter is used to attenuate this AC voltage to prevent it from saturating the amplifier. A series compensator must also be found that will satisfy the following performance specifications:

(a) The static gain of the closed-loop system $Y(0)/R(0)$ must be 1.
(b) The static gain of the compensator $G_C(0)$ must be at least 1.
(c) The amplifier gain K_A must be at least 4000.
(d) The closed-loop M_p must be between 1 db and 4 db.

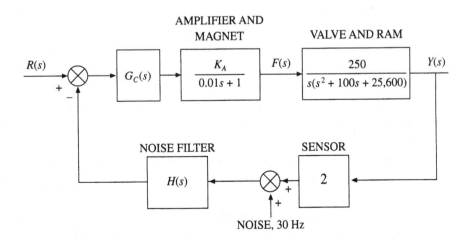

Fig. P14.2

(e) The closed-loop bandwidth (at the −3-db point) must be at least 60 rad/s.

(f) The attenuation of the noise filter must be at least 40 db at the 30-Hz frequency.

Find an $H(s)$ and a $G_C(s)$ that will satisfy these specifications. State the gain margin and the phase margin for your design.

Problem 14.3

(a) Verify the assertion made following Eqn. 14.8 concerning the steady-state descent rate of the Navion aircraft.

(b) If the elevator is given a 1° "up" deflection, calculate the incremental power required of the engine to sustain the resultant climb rate if the airspeed is to remain fixed. The aircraft mass is 85.4 slug. Assume a propellor efficiency of 85%.

Problem 14.4

It is asserted in Section 14.2 that the time-domain responses of the three systems whose frequency response characteristics are plotted in Fig. 14.9 should bear a qualitative similarity to one another. Check this claim by calculating the response of each of these systems to a unit step input.

Problem 14.5

Figure P14.5 is the block diagram for an attitude control system for a rigid spacecraft. Find a gain K and a feedback compensator $H(s)$ that will meet the following design specifications:

(a) closed-loop $M_p = 0$ db;

(b) closed-loop bandwidth (at the −3-db point) must be at least 2 rad/s;

(c) phase margin must be at least 50°;

(d) the ratio $|H(j\infty)|/|H(j0)|$ must be less than 15.

Problem 14.6

At the end of Section 14.2, the following claim is made: For the control law given in Eqn. 14.12, $\Delta h(t)$ will have a rise time of approximately 15 s in response to a step input command to the altitude control system. Investigate that claim.

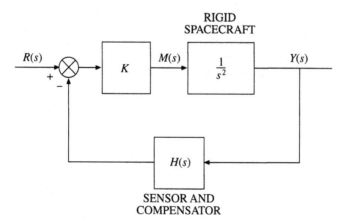

Fig. P14.5

Problem 14.7

At the end of Section 14.3, a speculation is made concerning the time-domain response of the servomechanism if the feedback compensation is $H(s) = 10(s+130)/(s+1300)$. Investigate that speculation.

Problem 14.8

Refer to Problem 10.19. Let $k_0 = 10$ V/m. Neglect the inductance of the magnet coils and the output resistance of the amplifier, so that $\Delta i(t) = 0.25K_A e_{in}(t)$ A. Take the transfer function of the magnet and armature to be $Z(s)/\Delta I(s) = 23.5/(s^2 + 21s - 9790)$. Confirm the following statement: A control law $E_{in}(s)/E(s) = (s+120)/(s+1200)$ and an amplifier gain $K_A = 16{,}000$ will provide a stable and well-damped closed-loop system having a bandwidth of approximately 1000 rad/s, a gain margin of approximately -20 db, and a phase margin of approximately $52°$. Note that the open-loop transfer function is nonminimum phase but the closed-loop transfer function is minimum phase. Note also that, because the system is conditionally stable, the gain margin is less than 1.

Problem 14.9

Continue with Problem 14.8. Find a new control law $E_{in}(s)/E(s)$ that will improve the performance of the system so that it will satisfy the following specifications:

(a) the closed-loop system will be stable and well-damped, and it will have a bandwidth of 5000 rad/s;

(b) for a step input at $z_{ref}(t)$, the steady-state error $e(t)$ will be zero.

Problem 14.10

Consider the system diagramed in Fig. P14.10. Find the PID gains K_P, K_I, and K_D which will permit the system to follow a ramp input with a zero steady-state error and which will also satisfy the following criteria:

(a) the open-loop gain $|Y(j\omega)/E(j\omega)|_{db}$ must be at least 40 db at $\omega = 0.1$ rad/s;

(b) the closed-loop M_p must be less than 3 db;

(c) the gain margin must be at least 12 db;

(d) the phase margin must be at least $40°$.

Problem 14.11

Refer to Problem 2.10. Adopt the notation $\Delta\theta(t) = x_3(t)$ and $\Delta\delta_E(t) = u_1(t)$. Verify that the transfer function relating the elevator deflection in degrees to the pitch angle in degrees is approximately

$$\frac{X_3(s)}{U_1(s)} = \frac{-11(s+1.9)(s+0.06)}{(s^2+5s+13)(s^2+0.04s+0.05)}.$$

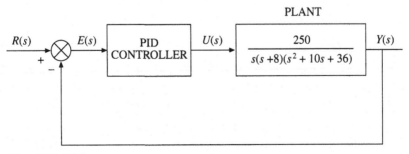

Fig. P14.10

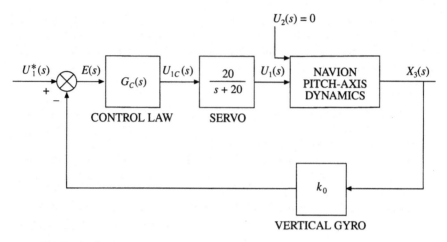

Fig. P14.12

Problem 14.12

We now attempt to damp the phugoid mode of the Navion longitudinal motion by using feedback to the elevator from the pitch angle, as sensed by a vertical gyroscope. Figure P14.12 is a diagram of the feedback system. The throttle setting is held fixed, so that $\Delta\delta_T(t) = u_2(t)$ is zero. The elevator is actuated by a servomechanism whose transfer function is shown in the diagram. Find a control law $G_C(s)$ and vertical gyro gain k_0 which will satisfy the following performance specifications for this control loop:

(a) the steady-state closed-loop gain $X_3(0)/U_1^*(0)$ must be -2;
(b) the phase margin must be at least $50°$;
(c) the gain margin must be at least 12 db;
(d) the closed-loop bandwidth must be at least 5 rad/s, measured at the $+3$-db point, which is 3 db below the zero-frequency gain.

Problem 14.13

Refer to Problem 10.10. Use the numerical values for the amplifier, motor, cart, and pendulum given in that problem. Maintain the notation $e(t) = u(t)$, $z(t) = y_1(t)$, and $\beta(t) = y_2(t)$. The gain of the β sensor is 1.

(a) Obtain the transfer functions $Y_1(s)/U(s)$ and $Y_2(s)/U(s)$.
(b) Consider the inner loop. Find a compensator transfer function $G_\beta(s)$ for which the inner loop has adequate gain and phase margins and a reasonable closed-loop M_p. Do not use pole-zero cancellation in your design for $G_\beta(s)$.

Problem 14.14

Refer to Problem 10.28. Use the numerical values for the amplifier, motor, cart, and pendulum given in that problem. Maintain the notation $e(t) = u(t)$, $z(t) = y_1(t)$, and $\beta(t) = y_2(t)$. Obtain the transfer function $Y_1(s)/Y_2(s)$. The gain of the z sensor is 1. Use the inner-loop compensator which you designed in Problem 14.13.

(a) Consider the outer loop. Find a compensator transfer function $G_z(s)$ for which the outer loop has adequate gain and phase margins and a reasonable closed-loop M_p. Do not use pole-zero cancellation in your design for $G_z(s)$.

(b) Find the bandwidth of the closed-loop transfer function $Y_1(j\omega)/Z_{com}(j\omega)$.

(c) Adjust the gain constant of the inner-loop compensator $G_\beta(s)$ until the overall system becomes unstable. Does this adjustment reveal the same gain margin on the inner-loop gain as that which you found in Problem 14.13(b)?

15 Advanced Topics

15.1 Introduction

The previous chapters and exercise problems have offered the reader a thorough grounding in the principles of classical control theory, analysis, and design – the basic requirements for beginning work as a control engineer. Further progress in that field requires the study of advanced topics, several of which are introduced briefly in this chapter.

15.2 MIMO Linear System Analysis

We first return to Chapter 5, which summarizes the modeling principles for linear multi-input–multi-output (MIMO) systems. The standard form for the state-variable model is

$$\text{DYNAMICS:} \quad \dot{x}(t) = Ax(t) + Bu(t);$$
$$\text{OUTPUT:} \quad y(t) = Cx(t) + Du(t). \tag{15.1}$$

We also return to Chapter 6, in which the solution $x(t)$ for the initial-value problem is given as

$$x(t) = [\epsilon^{At}]x_0 + \int_0^t [\epsilon^{A(t-\tau)}]Bu(\tau)\,d\tau. \tag{15.2}$$

The $n \times n$ state transition matrix, $[\epsilon^{At}]$, is defined as an infinite sum of $n \times n$ matrices

$$[\epsilon^{At}] \triangleq I_n + At + \frac{1}{2!}[At]^2 + \frac{1}{3!}[At]^3 + \frac{1}{4!}[At]^4 + \cdots. \tag{15.3}$$

Because t is a scalar, the term $[At]^k$ may be written as $A^k t^k$, where A^k is an $n \times n$ matrix whose elements are constants. I_n is the $n \times n$ identity matrix, whose

430

elements are all zero except those on the principal diagonal, all of which are 1. The time derivative of $[\epsilon^{At}]$ is easily calculated from its definition as follows:

$$\frac{d}{dt}[\epsilon^{At}] = A + A^2 t + \frac{1}{2!}A^3 t^2 + \frac{1}{3!}A^4 t^3 + \cdots$$

$$= A\left[I_n + At + \frac{1}{2!}A^2 t^2 + \frac{1}{3!}A^3 t^3 + \cdots\right]$$

$$= A[\epsilon^{At}]. \tag{15.4}$$

Consider the homogeneous version of the dynamic equation, along with the initial condition of the state vector:

$$\dot{x}(t) = Ax(t), \quad x(0) = x_0. \tag{15.5}$$

It is clear from Eqn. 15.4 that the vector-valued time function $x(t) = [\epsilon^{At}]x_0$ is the unique solution to the initial-value problem defined in Eqn. 15.5, because $x(t)$ satisfies both the differential equation and the initial condition. An exercise in calculus also shows that the $x(t)$ in Eqn. 15.2 is the general solution to the dynamic equation $\dot{x}(t) = Ax(t) + Bu(t)$.

We next derive the general solution to the dynamic equation using the Laplace transform. We note that a linear operation on a matrix composed of time-variable elements is accomplished simply by performing the operation on each of the individual elements. The Laplace transform of the dynamic equation (a linear operation) yields

$$sX(s) - x_0 = AX(s) + BU(s), \tag{15.6}$$

where s is the scalar Laplace variable, $X(s) = \mathcal{L}\{x(t)\}$, and $U(s) = \mathcal{L}\{u(t)\}$. Eqn. 15.6 may be arranged into the form

$$[sI_n - A]X(s) = x_0 + BU(s). \tag{15.7}$$

We must introduce the $n \times n$ identity matrix I_n in order to maintain consistent dimensions for the terms in the matrix $[sI_n - A]$. If n were 1, $X(s)$ could be found simply by dividing both sides of Eqn. 15.7 by the term $[sI_n - A]$, which in that case would be a scalar, not a matrix. But for $n > 1$, the re-arrangement of Eqn. 15.7 must be accomplished by multiplying both sides by the inverse of the matrix $[sI_n - A]$. The inverse is defined only for a square matrix whose determinant is nonzero. Such matrices are said to be *nonsingular*. Because each diagonal element of $[sI_n - A]$ involves the variable s, the determinant of the matrix is an nth-order polynomial in s. Consequently, the determinant is nonzero, and the inverse exists for almost all s, the only exceptions being the n roots of the polynomial. The inverse of $[sI_n - A]$ is denoted as $[sI_n - A]^{-1}$, and since the product of a matrix with its inverse is an identity matrix, Eqn. 15.7 may be written equivalently as

$$X(s) = [sI_n - A]^{-1}x_0 + [sI_n - A]^{-1}BU(s) \quad \text{for almost all } s. \tag{15.8}$$

The inverse of $[sI_n - A]$ is calculated as $(1/\Delta(s))(\text{adj}[sI_n - A])$, where $\Delta(s)$ is the determinant of $[sI_n - A]$. The $n \times n$ matrix $\text{adj}[sI_n - A]$ is the *adjoint* matrix to $[sI_n - A]$, the elements of which are polynomials of degree $n-1$. We call this adjoint matrix $N_0(s)$, and we call the element in its ith row and jth column $n_{0ij}(s)$. $X(s)$ may now be written as

$$X(s) = G_0(s) \; x_0 \; + G_u(s) U(s), \quad \text{where}$$

$$
\begin{array}{ccccc}
\downarrow & \downarrow & \downarrow & \downarrow & \downarrow \\
n \times 1 & n \times n & n \times 1 & n \times p & p \times 1
\end{array}
$$

$$G_0(s) = \frac{N_0(s)}{\Delta(s)} \quad \text{and} \quad G_u(s) = \frac{N_0(s)B}{\Delta(s)}. \tag{15.9}$$

Note that the element in the ith row and jth column of $G_0(s)$ is the rational function $n_{0ij}(s)/\Delta(s)$, whose numerator is of lower degree than its denominator. $G_0(s)$ is often expressed as $G_0(s) = \{g_{0ij}(s)\} = \{n_{0ij}(s)/\Delta(s)\}$, which is read, "$G_0(s)$ is the matrix having $g_{0ij}(s)$ as the element in its ith row and jth column." Using the same terminology, we can write $G_u(s)$ as

$$G_u(s) = \{g_{uij}(s)\} = \left\{ \frac{n_{uij}(s)}{\Delta(s)} \right\}, \tag{15.10}$$

where the $n_{uij}(s)$ elements come from the matrix product $N_0(s)B$.

We recognize that Eqn. 15.9 has the familiar form associated with the {input} → {transfer function} → {output} configuration derived in Chapter 6 for the SISO case. But since Eqn. 15.9 is a matrix equation, its block diagram employs broad arrows, as indicated in Fig. 15.1. Here $G_u(s)$ and $G_0(s)$ are transfer function *matrices*. For $x_0 = 0$ the system equation reduces to

$$X(s) = G_u(s) \times U(s) ,$$

$$
\begin{array}{ccc}
n \times 1 & n \times p & p \times 1
\end{array}
$$

which in expanded form is

$$
\underbrace{\begin{bmatrix} X_1(s) \\ X_2(s) \\ \vdots \\ [X_i(s)] \\ \vdots \\ X_n(s) \end{bmatrix}}_{X(s)} = \underbrace{\begin{bmatrix} g_{u11}(s) & g_{u12}(s) & \cdots & g_{u1j}(s) & \cdots & g_{u1p}(s) \\ g_{u21}(s) & g_{u22}(s) & \cdots & g_{u2j}(s) & \cdots & g_{u2p}(s) \\ \vdots & \vdots & \vdots & \vdots & \vdots & \vdots \\ g_{ui1}(s) & g_{ui2}(s) & \cdots & [g_{uij}(s)] & \cdots & g_{uip}(s) \\ \vdots & \vdots & \cdots & \vdots & \cdots & \vdots \\ g_{un1}(s) & g_{un2}(s) & \cdots & g_{unj}(s) & \cdots & g_{unp}(s) \end{bmatrix}}_{G_u(s)} \underbrace{\begin{bmatrix} U_1(s) \\ \vdots \\ [U_j(s)] \\ \vdots \\ U_p(s) \end{bmatrix}}_{U(s)} . \tag{15.11}
$$

The meaning of the elements in the $G_u(s)$ matrix is now clear. If there are p nonzero input variables and the n initial conditions are zero, the ith state variable is

$$X_i(s) = g_{ui1}(s)U_1(s) + \cdots + g_{uip}(s)U_p(s) = \sum_{j=1}^{p} g_{uij}(s)U_j(s). \tag{15.12}$$

If there is a single input $u_j(t)$, then $g_{uij}(s)$ is the transfer function between $U_j(s)$ and $X_i(s)$.

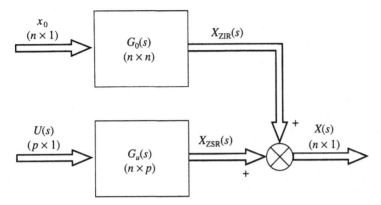

Fig. 15.1   Matrix block diagram for an nth-order linear system relating $X(s)$ to $U(s)$ and to x_0.

We note that the individual transfer functions in both $G_0(s)$ and $G_u(s)$ all have the same denominator, $\Delta(s)$, which is the determinant of $[sI_n - A]$, an nth-order polynomial. The roots of this characteristic polynomial of A are called the *characteristic values* (or the *eigenvalues*) of A. They are also the *poles* of the individual transfer functions, and are the n values of s for which $[sI_n - A]^{-1}$ does not exist.

To obtain the transfer function matrix relating the output $y(t)$ to the stimuli $u(t)$ and x_0, we take the Laplace transform of the output equation and substitute $X(s)$ from Eqn. 15.9. For the simple case where $D = 0$ and $x_0 = 0$, we have only the ZSR response

$$Y(s) = C\left[\frac{1}{\Delta(s)}(\text{adj}[sI_n - A])\right]BU(s). \tag{15.13}$$

$$\underset{m \times 1}{\downarrow} \quad \underset{m \times n}{\downarrow} \qquad \underset{n \times n}{\underbrace{}} \quad \underset{n \times p}{\downarrow}\ \underset{p \times 1}{\downarrow}$$

We now consider the simple but important example portrayed in Fig. 15.2. A SIMO plant employs *full-state feedback,* which means that the plant is instrumented so that each of the multiple outputs is equal to one of the state variables, making $y(t) = x(t)$. We use a third-order plant in this example. The control law is configured so that the feedback signal is a linear combination of the output (state) variables:

$$\underset{1 \times 1}{\underbrace{b(t)}} = \underset{K}{\underbrace{[k_1\ k_2\ k_3]}}\underset{x(t)}{\underbrace{\begin{bmatrix} x_1(t) \\ x_2(t) \\ x_3(t) \end{bmatrix}}}. \tag{15.14}$$

The state-variable model for the plant is

$$A = \begin{bmatrix} 0 & 1 & 0 \\ 0 & 0 & 1 \\ -20 & -4 & 1 \end{bmatrix}, \quad B = \begin{bmatrix} 0 \\ 0 \\ 1 \end{bmatrix}, \quad C = \begin{bmatrix} 1 & 0 & 0 \\ 0 & 1 & 0 \\ 0 & 0 & 1 \end{bmatrix}, \quad D = \begin{bmatrix} 0 \\ 0 \\ 0 \end{bmatrix}. \tag{15.15}$$

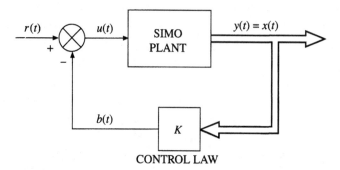

Fig. 15.2   Full-state feedback control of a SIMO plant.

The $[sI_n - A]$ matrix is

$$\begin{bmatrix} s & -1 & 0 \\ 0 & s & -1 \\ 20 & 4 & (s-1) \end{bmatrix}.$$

Its determinant is easily calculated to be $\Delta(s) = s^3 - s^2 + 4s + 20$. Note that the bottom row of the A matrix consists of the negatives of the last three coefficients of $\Delta(s)$, arranged in reverse order. It is also clear that the plant is unstable, since two of the roots of $\Delta(s)$ lie in the right-half plane.

To determine the state-variable model for the closed-loop system, we perform the algebra indicated at the summing point:

$$u(t) = r(t) - Kx(t), \quad \text{which yields}$$

$$\dot{x}(t) = Ax(t) + B[r(t) - Kx(t)], \quad \text{so that} \tag{15.16}$$

$$\dot{x}(t) = [A - BK]x(t) + Br(t).$$

The state distribution maxtrix for the closed-loop system $[A - BK]$ has the same form as A, because the first two rows of B are zero:

$$[A - BK] = \begin{bmatrix} 0 & 1 & 0 \\ 0 & 0 & 1 \\ -(20 + k_1) & -(4 + k_2) & -(k_3 - 1) \end{bmatrix}. \tag{15.17}$$

The characteristic polynomial for the closed-loop system may be written by inspection because $[A - BK]$ has the same *control canonical form* as A:

$$\Delta(s)|_{[A-BK]} = s^3 + (k_3 - 1)s^2 + (4 + k_2)s + (20 + k_1). \tag{15.18}$$

This example illustrates three important characteristics of full-state feedback control systems. First, the poles of the closed-loop system may be placed arbitrarily, simply by choosing the three feedback gains k_1, k_2, k_3 to produce the desired coefficients in the characteristic polynomial. (If we wished to place the three closed-loop poles at $s = -4 + j3$, $s - 4 - j3$, and $s = -5$, for example, we would choose $k_1 = 105$, $k_2 = 61$, and $k_3 = 14$.) Second, this example displays the central role played by the state variables in the dynamics of physical systems and emphasizes the importance of identifying a legitimate set of state variables in a dynamic plant. Third, it provides reliable guidance in the selection of instrumentation for feedback control; sensors that measure state variables directly are the best-suited for this task.

15.3 Discrete Time Systems and Digital Control

The usual configuration of a feedback control system shows a division between signal-processing elements, which consume very little power, and the high-powered elements of the plant. The signal processors perform the summing and compensation tasks of the control law, which are essentially mathematical functions. Figure 15.3 is a diagram for a system in which the control law computations are performed by a digital computer. The plant and sensor are dynamic elements whose behavior is manifested in $u(t)$, $y(t)$, and $b(t)$, which are functions of continuous time. Because the digital computer processes sequences of numbers, not continuous voltages or currents, the connections between the physical plant and the control law must be made through analog-to-digital and digital-to-analog converters, indicated as A/D and D/A in the diagram.

A sample-and-hold device accomplishes the conversion from $b(t)$ to its discrete counterpart $b^*(t)$, as indicated in Fig. 15.4(a). The sample instants t_k, t_{k+1}, t_{k+2}, ... occur at uniform intervals in this example. The digital computer executes its program during the interval between samples, providing outputs at intervals that are normally synchronized with those of the input. The output at time t_k, denoted $u(t_k)$, is the result of computations based on prior input samples, which are temporarily stored in computer memory. Figure 15.4(b) shows the sequence of output samples converted to the continuous time signal $u^*(t)$ (which has discontinuities at the sample instants). This example of a *zero-order hold operation* is commonly used, but other types of "smoothing" of the digital signal between sample intervals are also possible.

The plant and sensor dynamics, defined by differential equations in continuous time, are represented in the conventional state-variable format for an nth-order SISO system as

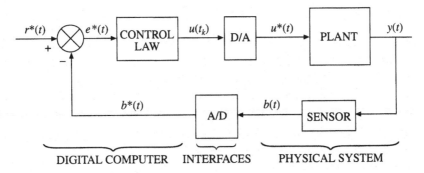

Fig. 15.3   Discrete time feedback control system.

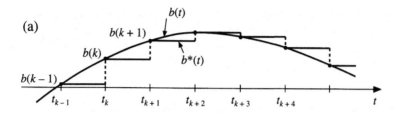

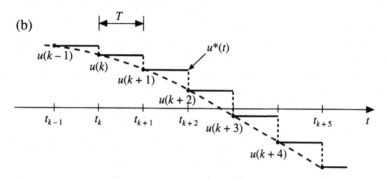

Fig. 15.4 (a) Illustration of zero-order hold operation on $b(t)$. (b) $u^*(t)$, a continuous time function.

$$\dot{x}(t) = Ax(t) + Bu^*(t),$$

$$y(t) = Cx(t) + Du^*(t), \tag{15.19}$$

$$b(t) = Ky(t),$$

where K is the scale factor of the sensor.

Consider the solution of the dynamic equation for the time during the interval of one sample period, from $t = t_k$ to $t = t_{k+1}$. The state vector at time t_k is denoted $x(k)$. During this period, the input $u^*(t)$ is constant at the value $u(k)$, as indicated in Fig. 15.4. The state vector at time t_{k+1} is obtained from Eqn. 15.2 as

$$x(k+1) = \underbrace{\epsilon^{AT}}_{A_P} x(k) + \underbrace{\left[\int_{t_k}^{t_{k+1}} \epsilon^{A(t-\tau)} B\, d\tau\right]}_{B_P} u(k). \tag{15.20}$$

We note that the $n \times n$ matrix A_P and the $n \times 1$ matrix B_P are constant matrices that depend on the sample period T. We also note that A_P is nonsingular because ϵ^{At} is nonsingular for all finite t. Equation 15.20 describes the dynamics of the plant in a form that can relate a sequence of input values $\{u(k), u(k+1), u(k+2), u(k+3), \ldots\}$, denoted $\{u(k)\}$, to a sequence of state vector values $\{x(k), x(k+1), x(k+2), \ldots\}$, denoted $\{x(k)\}$. This is called a *difference equation,* the form that must be used to describe the dynamics of the digital computer–based controller:

PLANT & SENSOR $x(k+1) = A_p x(k) + B_p u(k),$
DYNAMICS:
$$y(k) = Cx(k) + Du(k), \qquad (15.21)$$

$$b(k) = KCx(k) + KDu(k).$$
$$\downarrow \qquad\quad \downarrow$$
$$C_P \qquad\quad D_P$$

The computer operates on the difference between $r^*(t)$ and $b^*(t)$, which we denote as $e^*(t)$, so that its input is the sequence $\{e(k), e(k+1), e(k+2), e(k+3), ...\}$, denoted $\{e(k)\}$:

ERROR DETECTOR: $e(k) = r(k) - b(k).$ $\qquad (15.22)$

Because the control law is normally dynamic in nature, it must be specified as a difference equation having an internal state vector of an appropriate dimension w and the output $u(k)$:

CONTROL LAW $q(k+1) = Fq(k) + Ge(k),$
ALGORITHM: $\qquad\qquad\qquad\qquad\qquad\qquad\qquad (15.23)$
$$u(k) = Hq(k) + Je(k).$$

The matrices F, G, H, and J are selected by the designer to ensure the satisfactory dynamic performance of the closed-loop system.

By combining Eqns. 15.21, 15.22, and 15.23 and then eliminating $b(k)$ and $u(k)$, we obtain the difference equation for the closed-loop system. The total state vector, denoted $p(k)$, consists of the $n \times 1$ plant state vector combined with the $w \times 1$ computer state vector:

$$\underbrace{p(k)}_{(n+w)\times 1} = \begin{bmatrix} x(k) \\ q(k) \end{bmatrix}. \qquad (15.24)$$

The difference equation for the closed-loop system is

$$p(k+1) = F_{\text{CL}} p(k) + G_{\text{CL}} r(k),$$
$$y(k) = H_{\text{CL}} p(k) + J_{\text{CL}} r(k). \qquad (15.25)$$

The system matrices can be partitioned as

$$F_{\text{CL}} = \begin{bmatrix} F_{11} & F_{12} \\ F_{21} & F_{22} \end{bmatrix}, \qquad G_{\text{CL}} = \begin{bmatrix} G_{11} \\ G_{21} \end{bmatrix}, \qquad H_{\text{CL}} = [H_{11} \ H_{12}], \qquad (15.26)$$

where the F_{ij}, G_{ij}, H_{ij} as well as J_{CL} depend on the individual matrices in Eqns. 15.21 and 15.23.

The dynamic nature of a discrete time system is largely disclosed by the solution to the homogeneous form of its difference equation $x(k+1) = Fx(k)$, with initial condition $x(0) = x_0$. This solution is analogous to the solution of the differential equation $\dot{x}(t) = Ax(t)$ with its initial condition $x(0) = x_0$ in the continuous time case.

It is instructive to consider the solution to a first-order difference equation, which can be found simply by calculating the sequence $\{x(k)\}$ beginning with the known initial condition:

$$x(k+1) = f_{11}x(k), \quad x(0) = \beta, \tag{15.27}$$

where f_{11} is a real-valued constant. The solution sequence, starting with $k = 0$, is

$$\{x(k)\} = \{\beta, f_{11}\beta, f_{11}^2\beta, f_{11}^3\beta, \ldots, f_{11}^k\beta, \ldots\} = \{f_{11}^k\}\beta, \tag{15.28}$$

as depicted in Fig. 15.5(a) for f_{11} positive. The sequence is stable for f_{11} between 0 and 1 but is unstable for f_{11} greater than 1. Figure 15.5(b) shows the sequence for f_{11} negative. It is apparent from the difference equation itself that the terms in $\{x(k)\}$ alternate in sign. It is also clear that the boundary between stability and instability occurs for $f_{11} = -1$.

When we consider the solution of the homogeneous nth-order matrix difference equation

$$p(k+1) = F_{CL}\,p(k), \quad p(0) = p_0, \tag{15.29}$$

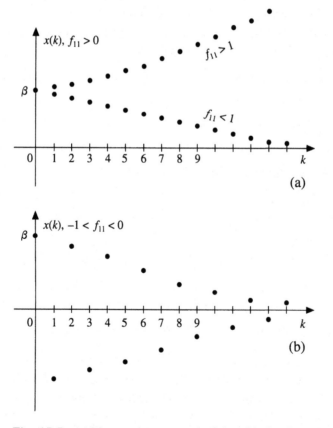

Fig. 15.5   (a) Monotonic sequence $\{x(k)\}$, stable for $f_{11} < 1$ but unstable for $f_{11} > 1$. (b) $\{x(k)\}$ a stable, alternating sequence for $-1 < f_{11} < 0$.

we can observe behavior that is only slightly more complex than that of the first-order case. Assume that F_{CL} is in diagonal form. (If it is not, a new set of states may usually be defined to render it diagonal.) The matrix difference equation has the following structure:

$$\begin{bmatrix} p_1(k+1) \\ p_2(k+1) \\ \vdots \\ p_n(k+1) \end{bmatrix} = \begin{bmatrix} f_{11} & 0 & \cdots & 0 \\ 0 & f_{22} & \cdots & 0 \\ \vdots & \vdots & \vdots & \vdots \\ 0 & 0 & \cdots & f_{nn} \end{bmatrix} \begin{bmatrix} p_1(k) \\ p_2(k) \\ \vdots \\ p_n(k) \end{bmatrix}, \quad p(0) = \begin{bmatrix} p_{10} \\ p_{20} \\ \vdots \\ p_{n0} \end{bmatrix}. \quad (15.30)$$

Now the n state variables are decoupled from one another, so that each row of the matrix equation has the form

$$p_i(k+1) = f_{ii} p_i(k), \quad p_i(0) = p_{i0}, \quad i = 1, 2, \ldots, n, \quad (15.31)$$

which is precisely the form of Eqn. 15.27. Therefore, the solution sequence $\{p_i(k)\}$ has the same form as $\{x(k)\}$, the solution sequence for the first-order equation.

Some elements of the matrix can be complex-valued even though all the parameters of the physical system are real-valued. But since each of the complex-valued f_{ii} will be accompanied by its complex conjugate, the sum of the corresponding terms in the sequence will be real-valued. As an example, let $f_{11} = \alpha + j\beta$ and $f_{22} = \alpha - j\beta$. The initial values $p_1(0)$ and $p_2(0)$ will also be complex conjugates. When we represent f_{11} and f_{22} and the initial values in exponential form and calculate the kth elements in the two sequences, we have

$$\{p_1(k) + p_2(k)\} = 2P_0 M^k \cos(k\phi + \gamma), \quad \text{where}$$
$$p_1(0) = P_0 \epsilon^{j\gamma} \text{ and } f_{11} = M\epsilon^{j\phi}. \quad (15.32)$$

Thus the sum of the two complex-valued sequences is real-valued; it will also be stable, provided that M, the absolute value of f_{11}, is less than 1.

The sequence of state vectors provides the matrix solution to the initial-value problem defined by Eqn. 15.30:

$$\{p(k)\} = \left\{ \begin{bmatrix} p_1(0) \\ p_2(0) \\ \vdots \\ p_n(0) \end{bmatrix}, \begin{bmatrix} p_1(1) \\ p_2(1) \\ \vdots \\ p_n(1) \end{bmatrix}, \begin{bmatrix} p_1(2) \\ p_2(2) \\ \vdots \\ p_n(2) \end{bmatrix}, \ldots, \begin{bmatrix} p_1(k) \\ p_2(k) \\ \vdots \\ p_n(k) \end{bmatrix}, \ldots \right\}. \quad (15.33)$$

The ith state-variable sequence $\{p_i(0), p_i(1), p_i(2), \ldots, p_i(k), \ldots\}$ is called the ith *mode* of the system.

When a matrix is in diagonal form, the elements of the diagonal are the roots of the characteristic polynomial of the matrix. In the continuous time case the characteristic roots of the A matrix in the differential equation are also the poles of each element of the transfer function matrix. Each pole defines the nature of one of the modes of the system; in the continuous time case, a mode will be stable if its defining pole has a negative real part. In the discrete time case, the stability of each mode is also determined by the characteristic polynomial of the F matrix in the difference equation; but now the test for stability is that the magnitude of each of the n roots of that polynomial must be less than 1. In geometric terms these stability conditions are stated as follows.

> Continuous time system: Characteristic values of A must lie in left-half plane.
>
> Discrete time system: Characteristic values of F must lie inside the unit circle.

A necessary task in the design of the control law (see Fig. 15.3) is to choose the F, G, H, and J matrices so that the overall system matrix F_{CL} will have characteristic roots inside the unit circle. This task requires that F_{CL} be expressed as a function of A_p, B_p, C, D, and K. The task is complicated further because A_p and B_p depend on the sampling interval.

The following remarks supplement the short introduction to discrete time system analysis of this section.

Remark 1   The sample period T must be selected both to match the capabilities of the computer and to satisfy the performance requirements of the system. The basic purpose of the control system is to respond to the input $r(t)$ so that the error $e(t)$ remains within limits that are defined in various ways. Therefore, $r(t)$ must be sampled frequently enough to allow its sampled version, $r^*(t)$, to elicit a satisfactory $y(t)$ from the plant. Figure 15.6 shows a typical input signal. It is clear that $r^*(t)$ becomes a better representation of $r(t)$ as

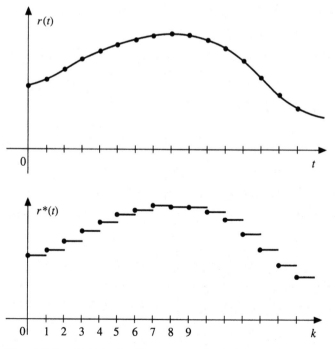

Fig. 15.6 $r^*(t)$ obtained by a zero-order hold operation on $r(t)$.

the sample period becomes shorter. But because the burden on the computer is increased as the sample period is decreased, a design compromise must be reached. The identification of the class of input signals – for example, in terms of its spectral properties – establishes an analytical basis upon which a satisfactory sample period can be determined from the principles of digital signal processing.

The design specifications for the closed-loop response often include a bandwidth requirement. The sample frequency normally turns out to be from eight to twenty times the specified bandwidth.

In some complex control systems each of the several subsystems will have a separate computer operating at a different sample rate. This "multi-rate" system presents synchronization problems that are beyond the rudimentary view of discrete time systems presented in this section.

Remark 2   The *modal analysis* approach to the design of the diagonal F, G, H, and J matrices of the computer algorithm outlined in this section is frequently impractical. A transformation method of analysis suitable to difference equations, called the *z transform,* offers a more convenient approach. This method, analogous to the Laplace transform method for differential equations, produces transfer functions that can be manipulated by simple algebra. Approximation techniques for converting s-domain transfer functions into z-domain transfer functions sometimes permit the design task to be accomplished using a mixture of the two methods.

Remark 3   The A_P matrix ϵ^{AT}, which was derived from the sampled continuous time system, is necessarily a nonsingular matrix. However, an A_P matrix for a digital computer algorithm may be rendered singular by the designer. The control law may therefore be designed to have dynamic properties that are different in principle from those of the plant.

Remark 4   In many systems, the digital computer performs significant control tasks in addition to control law computations. Because the computer can perform logical operations, it is often used to select different modes of behavior in response to remote commands or in response to unforeseen changes in its operating environment. As an example, three different sets of altitude and speed sensing equipment are often carried on commercial and military aircraft. An "air data" set depends on atmospheric pressure, dynamic air pressure, and air temperature. A second set of inertial instruments, aided by computer calculations and occasional radio or optical data corrections, also provides measurements of altitude and airspeed. The third set depends on radio or radar signals. The central control computer is able to select one of these data sets, or some combination thereof, in order to minimize the error in sensed airspeed and altitude.

Computer-based fault-detection schemes can modify the control laws in response to failures in critical components of the plant. Moreover, the computer can often manage off-line recording of the system error and other operating data required for routine maintenance.

15.4   Nonlinear Analysis

Our study in earlier chapters was limited to linear systems that are excited by exponential or sinusoidal functions and whose dynamics are described by ordinary differential equations with constant coefficients. We have seen two significant properties of these systems. First, the solutions consist essentially of linear combinations of exponential and sinusoidal functions; second, the qualitative nature of the response is independent of the magnitude of the stimuli – if the stimuli to the system are doubled, the response is also doubled.

The class of nonlinear systems (consisting of all dynamic systems that are not linear) is less circumscribed than the class of linear systems. Many dynamic phenomena unknown to linear systems, some of them quite spectacular, are observed in the behavior of even simple nonlinear systems. A common property of nonlinear systems is that the qualitative nature of the response is *not* independent of the magnitude of the stimuli. For example, a system may be stable for small magnitudes of stimuli but unstable when excited by larger magnitudes. Even the definition of stability, about which there is no controversy in linear systems, must be refined for nonlinear systems.

We consider only the limit-cycle oscillation here. The limit-cycle oscillation is observed in the zero-input response of some nonlinear feedback systems. A small initial condition produces a stable oscillation, such as a damped sinusoid combined with some decaying exponential terms. Signals in various parts of the system all appear to be of the same family because they all have the same modal content. A slightly larger initial condition produces stable oscillations closely resembling the previous ones, save for amplitudes that are slightly larger.

An even larger initial condition produces stable oscillations exhibiting waveforms that are qualitatively different from those observed in the two previous cases. Further, the waveforms of signals in different parts of the system are noticeably dissimilar. Finally, an even larger initial condition produces an oscillation that neither increases indefinitely nor dies out. The oscillation eventually becomes periodic, and waveforms that may roughly resemble sinusoids in some parts of the system will be markedly different in other parts, even though they have the same period. This steady-state oscillation is a limit cycle. A still larger initial condition may cause the initial part of the response to persist longer, but the response will eventually settle to an oscillation having the same amplitude, the same shape, and the same period as in the previous case.

Because the limit-cycle oscillation is periodic, it is possible in some cases to predict its amplitude and period from a quasilinear analysis. Such a case is illustrated in Fig. 15.7. The limiting amplifier has the nonlinear input-output characteristic shown in Fig. 15.8. The device is incapable of producing an output $u(t)$ greater than α or less than $-\alpha$. The analytical description of the limiting amplifier is

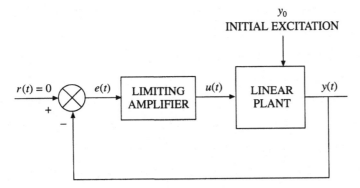

Fig. 15.7 Control system with a nonlinear (limiting) amplifier.

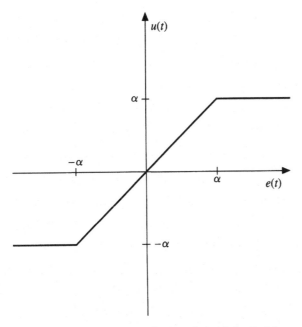

Fig. 15.8 Input–output characteristic of the limiting amplifier.

$$u(t) = \begin{cases} e(t) & \text{for } |e(t)| \le \alpha, \\ \alpha & \text{for } |e(t)| > \alpha, \\ -\alpha & \text{for } |e(t)| < -\alpha. \end{cases} \tag{15.34}$$

The device is linear with a gain of 1 if $e(t)$ does not exceed α in magnitude, so for small signals the control loop behaves according to linear theory.

Figure 15.9 illustrates the nonlinear character of the device for

$$e(t) = E_o \sin(\omega_1 t).$$

In this example $E_o = 1.3\alpha$. Figure 15.10 shows $u(t)$ in detail when $e(t)$ is a pure sinusoidal input and $E_o > \alpha$. In this case $u(t)$ is periodic, having the same period as $e(t)$, which is $2\pi/\omega_1$ s. The term $u(t)$ may be expressed in the following Fourier series:

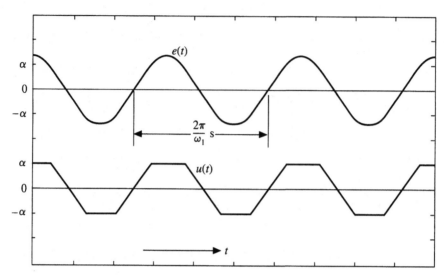

Fig. 15.9 $u(t)$ when $e(t)$ exceeds the limits of the amplifier.

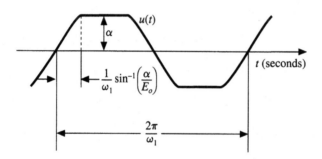

Fig. 15.10 $u(t)$ as it depends on E_o and α.

$$u(t) = B_1 \sin(\omega_1 t) + B_2 \sin(2\omega_1 t) + B_3 \sin(3\omega_1 t) + \cdots. \tag{15.35}$$

We note that the coefficient of each term in the series depends on E_o. The fundamental term, B_1, for example, is given by the formula

$$B_1 = \frac{\omega_1}{\pi} \int_{-\pi/\omega_1}^{\pi/\omega_1} u(t) \sin(\omega_1 t)\, dt, \tag{15.36}$$

which, when the geometry in Fig. 15.10 is taken into account, comes to

$$B_1 = \frac{2}{\pi}[E_o\beta + \alpha \cos(\beta)], \quad \text{where } \beta = \sin^{-1}(\alpha/E_o) \text{ rad} \tag{15.37}$$

when $E_o \geq \alpha$. Note that $B_1 = E_o$ when $E_o < \alpha$. The ratio $[B_1/E_o]$ is called the *describing function* of the limiting amplifier, denoted as $\mathrm{DF}(E_o)$. In the linear range of operation, the describing function is also the transfer function of the device. Thus the describing function for the limiting amplifier is

$$\mathrm{DF}(E_o) = \frac{2}{\pi}[\beta + (\alpha/E_o) \cos(\beta)], \quad \text{where } \beta = \sin^{-1}(\alpha/E_o) \text{ rad}. \tag{15.38}$$

Figure 15.11 is a plot of the describing function as it depends on E_o/α.

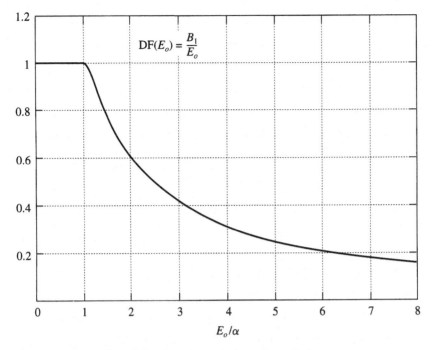

Fig. 15.11 Describing function as it depends on E_o/α.

Because the plant is linear, and because it is excited by a sum of sinusoidal terms, its steady state response $y(t)$ will also be a sum of sinusoidal terms, each corresponding to a term of the input $u(t)$:

$$y(t) = B_1|G(j\omega_1)| \sin(\omega_1 t + \phi_1) + B_2|G(j2\omega_1)| \sin(2\omega_1 t + \phi_2) + \cdots, \quad (15.39)$$

where $G(s)$ is the transfer function of the plant and $\phi_k = \angle G(jk\omega)$ for $k = 1, 2, \ldots$. In many control systems the linear plant has a *low-pass* frequency response characteristic. In this case, the higher harmonic terms in $y(t)$ will be somewhat smaller than the fundamental term, and the following approximation will be reasonable:

$$y(t) \cong B_1|G(j\omega_1)| \sin(\omega_1 t + \phi_1). \quad (15.40)$$

This may be expressed in terms of the describing function as

$$y(t) \cong \underbrace{E_o \times [\mathrm{DF}(E_o)]}_{B_1} \times |G(j\omega_1)| \sin(\omega_1 t + \phi_1). \quad (15.41)$$

Assume that $r(t) = 0$ and that the control loop is excited by an initial condition of the plant, as indicated in Fig. 15.7. Assume further that the control loop is stable in its linear regime of operation, so that any oscillations induced by a small initial excitation will die away with time. But if the initial excitation is large enough to drive the limiting amplifier into saturation then the induced oscillations may either die away or they may persist, eventually settling into a limit cycle. Because $e(t) = -y(t)$ at all times, the following approximation holds after the system has settled into a limit cycle at frequency ω_1:

$$E(j\omega_1) = -Y(j\omega_1) = -G(j\omega_1) \times DF(E_o) \times E(j\omega_1), \tag{15.42}$$

which requires that

$$G(j\omega_1) = -\frac{1}{DF(E_o)}. \tag{15.43}$$

To determine whether a combination $\{E_o, \omega_1\}$ exists that will satisfy Eqn. 15.43, plot both $G(j\omega)$ and $-[1/DF(E_o)]$ on the Nyquist plane, and look for an intersection of the two plots.

For example, consider a linear plant with the transfer function

$$G(s) = \frac{2 \times 10^6(s^2 + 6s + 80)}{s(s + 50)(s + 1000)(s^2 + 2s + 17)}. \tag{15.44}$$

Figure 15.12 shows a Nyquist plot of $G(j\omega)$ (not to scale). The locations of the points labeled A, B, and C on the negative real axis and the frequencies at which they occur on the $G(j\omega)$ locus are as follows.

Point	Coordinate	ω (rad/s)	
A	$-61.8 + j0$	4.48	(15.45)
B	$-3.99 + j0$	8.61	
C	$-0.0415 + j0$	214	

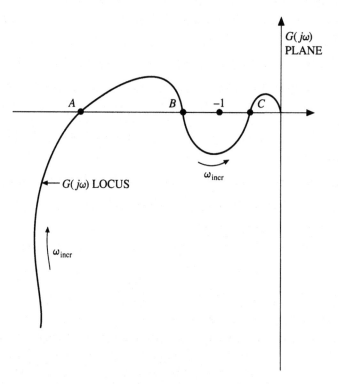

Fig. 15.12   Nyquist plot of $G(j\omega) = (2 \times 10^6[(j\omega)^2 + 6(j\omega) + 80])/$ $((j\omega)(j\omega + 50)(j\omega + 1000)[(j\omega)^2 + 2(j\omega) + 17])$ (not drawn to scale).

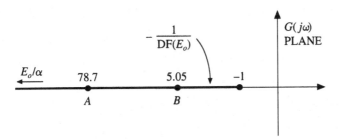

Fig. 15.13 $-1/DF(E_o)$ drawn on Nyquist plane.

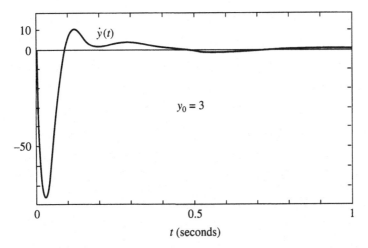

Fig. 15.14   Linear response of $\dot{y}(t)$ to initial condition $y_0 = 3$ when $r(t) = 0$.

The real-valued $-[1/DF(E_o)]$ appears on the Nyquist plane in Fig. 15.13. The intersections of $-[1/DF(E_o)]$ and the $G(j\omega)$ locus are points A and B, which satisfy the condition for a limit cycle.

We study the zero-input response of this system by obtaining its response to a nonzero initial condition y_0. Because the features of significance in this study are more apparent from $\dot{y}(t)$ than they are from $y(t)$, we first look at $\dot{y}(t)$ in response to the initial condition $y_0 = 3$. Figure 15.14 shows this response; $u(t)$ does not reach the nonlinear amplifier limits in this case.

Setting the amplifier limits at ± 1, we repeat the zero-input response experiment with y_0 large enough to drive the amplifier into saturation during the response. The results obtained by a nonlinear simulation of the system are displayed in Fig. 15.15. With $y_0 = 12$ the oscillation dies out in about 4 s. But if y_0 is increased slightly to 12.5 the oscillation builds up, approaching the limit cycle after approximately 10 s. Although the waveform of $\dot{y}(t)$ is obviously nonsinusoidal during the limit cycle, $y(t)$ (the integral of $\dot{y}(t)$) has a waveform that is very close to sinusoidal; its amplitude is approximately 77 and its frequency is approximately 4.57 rad/s. Therefore, $e(t)$ has this same amplitude and frequency. This result closely matches the prediction of {78.7, 4.48} for the limit cycle parameters at point A, as indicated in Figs. 15.12 and 15.13.

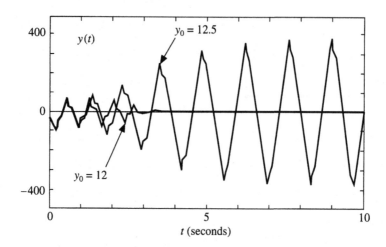

Fig. 15.15 Limit-cycle oscillation resulting from $y_0 = 12.5$ but not from $y_0 = 12.0$.

The limit cycle corresponding to point B is unstable. In principle, a sustained oscillation could exist at an amplitude of approximately 5.05 and frequency of approximately 8.61 rad/s, but any slight perturbation would cause the oscillations to increase in amplitude and to diminish in frequency, and finally to reach the stable limit cycle defined by point A.

15.5   Control over a Finite Interval

Stability and final value are basic ideas in the study of linear system dynamics. When a system is excited by a step input or by nonzero initial conditions, we apply these ideas to the behavior of the system output over the interval $0 \le t < \infty$. Our results have practical significance in those systems that will be in use for prolonged periods. But the operating life of some controlled systems is brief; the interval between the initial time and the final time being comparable to the time constants of the system modes. The performance specifications may thus require one or more of the system state variables to have specific values (with small error tolerances) at the final time. In this case, stability over the infinite interval – which is important in the initial-value problem – is of little practical value. The dynamics problem lies in the category of the *two-point boundary-value problem* (TPBVP).

A simple example of the TPBVP can be made from the spring–mass–damper system in Fig. 15.16. The interval of operation is $0 \le t \le t_F$. At t_0 the position of the mass, measured at its left edge, is zero. The initial velocity $\dot{z}(0)$ is unspecified, but the position of the mass at the final time, $z(t_F)$, is specified. An *initial velocity* must be found to fit the boundary condition specified for the *final position*. The velocity at the final time is not specified. The analytical statement of the TPBVP problem is:

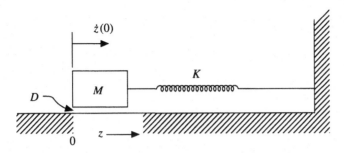

Fig. 15.16 Spring–mass–damper system for the two-point boundary-value problem.

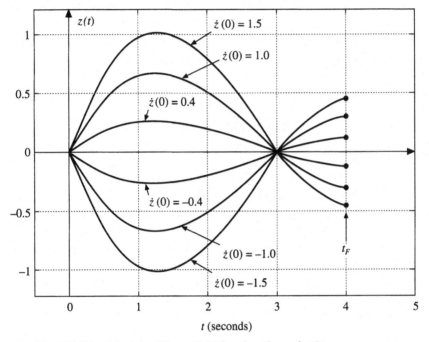

Fig. 15.17 Solution of Eqn. 15.46 for six values of $\dot{z}(0)$.

$$M\ddot{z}(t) + D\dot{z}(t) + Kz(t) = 0, \quad 0 \le t \le t_F$$

$$z(0) = 0; \tag{15.46}$$

for a specified t_F, find $\dot{z}(0)$ so that $z(t_F) = Z_F$.

The solution to the differential equation when $z(0) = 0$ is shown for several values of $\dot{z}(0)$ in Fig. 15.17. In this example $M = 1$, $D = 0.5$, $K = 1.1591$, and $t_F = 4$. We see that this TPBVP has a unique solution for any positive or negative value of the specified terminal value Z_F. For example, if $Z_F = 1.00 \pm 0.01$ is the required value for $z(4)$, then $\dot{z}(0)$ in the range -3.30 to -3.26 will satisfy this terminal specification.

However, if the terminal time should be specified to be 3 for this system, the TPBVP would not have a solution for any nonzero value for Z_F. And if Z_F

should be specified to be zero then there would be no unique solution, since all values for $\dot{z}(0)$ result in $z(3) = 0$. This example illustrates that even in simple linear systems the existence and uniqueness of solutions to a TPBVP depend on the specified boundary conditions. In the initial-value problem no such contingency exists.

15.6   Alternate Control Law Structures

The block diagram of Fig. 8.2 provides a reasonably comprehensive framework upon which to base a scientific study of the automatic control problem. The terms $r(t)$, $u(t)$, $y(t)$, and $b(t)$ can be defined as vectors. The plant, sensor, and control law can be linear or nonlinear, and the controller can be either a continuous time device or a discrete time device with suitable input and output converters. More importantly, the diagram does not constrain the internal structure of the controller as does the succeeding diagram in Fig. 8.3, which dictates that the control law must operate on $e(t)$. In Chapter 8 we limited our attention not only to the configuration of Fig. 8.3, but also to continuous time, linear, SISO blocks. This limitation permitted us to develop classical control theory on a solid scientific basis, to an extent that is useful in practical engineering. It also equipped us to begin an investigation into the more general configuration of Fig. 8.2, although pursuing that investigation will demand knowledge and skills beyond the scope of classical control theory.

Ample evidence from biology (including manual feedback control), from nonlinear system dynamics, and even from multivariable linear system theory indicates the existence of control structures, control mechanisms, and control strategies which fit the nonrestrictive diagram of Fig. 8.2 and which effectively control complex dynamic plants. Many of these plants are nonlinear, time-variable, and unstable, and they cannot be modeled accurately by the normal methods of engineering. The existence of such control mechanisms provides a challenging area for research.

15.7   Summary

The modeling of physical systems using the state-variable matrix format in Chapters 2 through 5, the study of the initial-value problem in Chapter 6, and the modest extension of those ideas in Section 15.2 constitute a useful introduction to modern control theory. But this brief introduction is insufficient to reveal many important and interesting dynamic features of MIMO, constant-coefficient, linear systems. A reasonably complete first course in modern control theory would require a second book of similar length, as well as substantial background in linear algebra and matrix theory. Among the many interesting features to be found in modern control theory are two fundamental properties of linear systems called controllability and observability. These properties are unrelated to stability; they pertain to SISO systems as well as to MIMO systems, and they are not apparent from a study of system dynamics from a purely input–output point of view.

A second course in modern control theory, covering optimal control, would require in addition a background in the theory of differential equations, real analysis, complex variables, the calculus of variations, and the modeling and analysis of random processes. Much of this material has direct application to engineering problems.

A study of variational mechanics and nonlinear system dynamics, of sufficient depth for engineering applications, would also require a large book. Such a study could provide a refreshing change from the emphasis normally placed upon linear analysis in control theory, by revealing the myriad nonlinear dynamic phenomena that cannot exist in linear systems.

Many devices have static nonlinear input–output characteristics similar to the saturating amplifier in Section 15.4, yet even so the devices are dynamic. In such devices both the amplitude and the phase shift of the describing function depend on the amplitude of the input signal, and the curve $-1/DF(E_o)$ will have both real and imaginary parts. The intersections with the $G(j\omega)$ locus may therefore occur at points off the real axis.

Since the mid-twentieth century, the search for alternate control structures for automatic systems has introduced areas of study identified as cybernetics, adaptive control, sliding-mode control, neural networks, fuzzy logic, biomechanics, robotics, fault-tolerant control, and others. This rich field of research will eventually yield new analysis and design procedures to supplement the basic methods of control engineering.

15.8 Problems

Problem 15.1
Given

$$x(t) = \epsilon^{At}x_0 + \int_0^t \epsilon^{A(t-\tau)}Bu(\tau)\,d\tau,$$

show that

(a) $x(0) = x_0$ and
(b) $dx(t)/dt = Ax(t) + Bu(t)$,

and consequently that the assertion made in Section 15.2 regarding the general solution to the dynamic equation is correct.

Problem 15.2
For the MIMO system described by Eqn. 15.1:

(a) Assume that $D = 0$. Find the $m \times n$ and $m \times p$ transfer function matrices $G_0(s)$ and $G_u(s)$ for which $Y(s) = G_0(s)x_0 + G_u(s)U(s)$.
(b) Assume that $D \neq 0$. Find the $m \times n$ and $m \times p$ transfer function matrices $G_0(s)$ and $G_u(s)$ for which $Y(s) = G_0(s)x_0 + G_u(s)U(s)$.

Problem 15.3

(a) Find the four system matrices in Eqn. 15.25 as combinations of the individual matrices in Eqns. 15.21 and 15.23 if D and J are both zero.

(b) Repeat (a) if D and J are both nonzero.

Problem 15.4

(a) Establish a difference equation $x(k+1) = Fx(k) + Gu(k)$ in which F is a nonsingular matrix. Observe the zero-input response sequence $\{x(k)\}$.

(b) Repeat (a) for singular F. Discuss the difference in the results you obtain.

Problem 15.5

Refer to the spring–mass–damper diagram in Fig. 15.16. Assume that the damping force on the mass is of the coulomb variety – namely, constant but opposed to the velocity – so that the equation of motion is nonlinear: $M\ddot{z}(t) + d\,\text{sgn}(\dot{z}(t)) + z(t) = 0$. The positive coefficient d is small but nonzero.

(a) Calculate the response $z(t)$ for the initial conditions $z(0) = Z_0$, $\dot{z}(0) = 0$.

(b) Show that $z(t)$ becomes zero in a finite time which depends upon Z_0.

(c) Show that the "envelope" of the oscillation is a pair of straight lines.

Problem 15.6

For the M, D, and K parameter values given in the example in Section 15.5, the TPBVP has only the trivial solution when $t_F = 3$. Can you find a value for D for which a unique, nontrivial solution will exist for any t_F in the range 3 to 20?

Physical Constants, Units, and Conversion Factors

When you can measure what you are speaking about, and express it in numbers, you know something about it; but when you cannot express it in numbers your knowledge is of a meagre and unsatisfactory kind; it may be the beginning of knowledge, but you have scarcely, in your thoughts, advanced to the stage of science, whatever the matter may be.

William Thomson
(Lord Kelvin) 1824–1907

Control system engineers must work with data which come from many sources and which are expressed in various systems of units. It is desirable to express all the data pertinent to a given problem in a consistent set of units. Some physical constants and conversion factors that are useful for this purpose are given here.

Newton's Law of Gravitation

Newton discovered that a mutual force of attraction, called a *gravitational force*, exists between two bodies by virtue of their masses. The magnitude of the gravitational force is

$$F = G\frac{M_1 M_2}{r^2},$$

where M_1 and M_2 are the masses of the bodies, r is the distance between their centers of mass, and F is the gravitational force acting on each body. G is the *gravitational constant*, having a numerical value considered to lie between 6.6717×10^{-11} and 6.6735×10^{-11} if the masses are in kilograms, the distance in meters, and the force in newtons. It has been found that the gravitational attraction between the earth (M_1) and a mass of 1 kilogram (M_2) located at sea

453

level at 45° latitude is 9.80665 newtons (this number varies less than 0.003%
with longitude). In a different set of units, the force of attraction between the
earth and a mass of 1 slug, at the same location, would be 32.174 pounds.

Consistent Sets of Units

Three sets of consistent units are commonly used in engineering. These are:

1. the foot-pound-second set, in which the slug [slug] is the unit of
    mass, the foot [ft] the unit of length, and the pound [lb] the unit
    of force;
2. the international system (SI), in which the kilogram [kg] is the
    unit of mass, the meter [m] the unit of length, and the newton [N]
    the unit of force; and
3. the centimeter-gram-second (CGS) set, in which the gram [g] is the
    unit of mass, the centimeter [cm] the unit of length, and the dyne
    [dyne] the unit of force.

Time is measured in seconds [s] in each of these sets of units. This second
is determined by the frequency of microwave radiation emitted by a specific
transition of an isotope of the cesium atom, and for many practical purposes is
the same as the mean solar second.

These sets of units are called *consistent* because – in the numerical expression of Newton's second law relating force, mass, and acceleration (in constant-
mass systems) – one force unit is equal to one mass unit multiplied by one acceleration unit. Thus we have:

$$F \quad = \quad M \quad \times \quad a$$

$$1 \text{ pound} = \quad (1 \text{ slug}) \quad \times (1 \text{ (foot/second)/second)},$$

$$1 \text{ newton} = (1 \text{ kilogram}) \times (1 \text{ (meter/second)/second)},$$

$$1 \text{ dyne} \quad = \quad (1 \text{ gram}) \quad \times (1 \text{ (centimeter/second)/second)}.$$

Therefore, a 1-kg mass released from rest at sea level will accelerate toward the
center of the earth at 9.80665 (m/s)/s because the earth, as mentioned previously, exerts a gravitational force of 9.80665 N on the 1-kg mass. The gravitational force on a 2-kg mass at sea level is 19.6133 N, so it too will accelerate
from rest at 9.80665 (m/s)/s. Similarly, any mass released from rest at sea level
will accelerate downward at 9.80665 (m/s)/s. The constant 9.80665 (m/s)/s is
frequently called the *acceleration due to gravity.*

Improper Units and Hybrid Sets of Units

Occasionally one finds an instrument for measuring force calibrated in mass
units (a *gram gauge* is one example) or an instrument for measuring mass calibrated in force units (the household scale calibrated in *pounds,* for example).
Hence engineering data on equipment sometimes appear in improper units.
For example, manufacturers of large electrical machines often give the moment
of inertia of the rotor in "pound-feet^2."

To convert a quantity expressed in improper units to its correct numerical value in a consistent set of units, one need only interpret the correct meaning of the improper unit and then re-express it in its correct unit. For example, an object described as having a "weight" of 192 pounds, if that value has been determined on the earth at sea level and at 45° latitude, has a mass of 5.96755 slugs. Thus, a moment of inertia expressed as 192 lb-ft^2 is, in a consistent set of units, 5.96755 slug-ft^2, which is also 8.0909 kg-m^2.

Sometimes mass will be expressed as the ratio of force to acceleration. For example, a mass of 3 slugs may be called 3 lb/(ft/s^2). Frequently a hybrid set of force and acceleration units is used. An example is: $M_1 = 28.3$ oz/(in/s^2), which converts to $M_1 = 21.225$ slugs.

Force Conversion Factors

1 pound = 4.448216 newtons
1 pound = 16 ounces
1 pound = 1 pound force
1 ounce = 27,801.35 dynes
1 newton = 10^5 dynes

Mass Conversion Factors

1 slug = 14.593885 kilograms
1 kilogram = 1,000 grams
1 pound mass = 0.45359 kilogram

Length Conversion Factors

1 inch = 2.54 centimeters
1 foot = 30.48 centimeters
1 meter = 3.28084 feet
1 meter = 39.37008 inches
1 meter = 10^6 microns
1 micron = 10^4 angstroms

Moment-of-Inertia Conversion Factors

1 slug-foot^2 = 1.3558162 kilogram-meter^2
1 kilogram-meter^2 = 10^7 gram-centimeter^2
1 slug-foot^2 = 13,558,162 gram-centimeter^2

Power Conversion Factors

1 newton-meter/second = 1 watt
1 volt-ampere = 1 watt
1 watt = 1 joule/second
1 horsepower = 550 foot-pounds/second
1 horsepower = 745.6999 watts

Energy Conversion Factors

1 dyne-centimeter = 1 erg
1 joule = 10^7 ergs
1 joule = 1 newton-meter
1 newton-meter = 10^7 dyne-centimeters
1 newton-meter = 141.61211 inch-ounces
1 foot-pound = 1.3558162 newton-meters
1 foot-pound = 192 inch-ounces
1 inch-ounce = 70,615.429 dyne-centimeters

Pressure Conversion Factors

1 pascal = 1 newton/meter^2
1 pound/inch^2 = 6,894.749 pascals

Angle Conversion Factors

1 radian = 57.29577951 degrees
1 degree = 1.111111 grads

Mathematical Constants

π = 3.141592653589
ϵ = 2.718281828459
$\log_{10}(2)$ = 0.301029996

Moments of Inertia of Rigid Bodies

Body of Mass M	Axis	Moment of Inertia
Solid right circular cylinder of radius R	Longitudinal axis	$M\left[\dfrac{R^2}{2}\right]$
Solid right circular cylinder of radius R and length L	Transverse diameter	$M\left[\dfrac{R^2}{4}+\dfrac{L^2}{12}\right]$
Solid right cone with radius of base R	Axis of revolution	$M\left[\dfrac{3}{10}R^2\right]$
Solid sphere of radius R	Any diameter	$M\left[\dfrac{2}{5}R^2\right]$
Uniform slender rod of length L	Transverse axis through center of mass	$M\left[\dfrac{1}{12}L^2\right]$

Unit Abbreviations

Physical Unit	Abbreviation	Physical Unit	Abbreviation
ampere	A	meter	m
centimeter	cm	microfarad	μF
coulomb	C	micron	μm
cycles per second	cps	millimeter	mm
decibel	db	millisecond	ms
degree	deg	millivolt	mV
degree Celsius	°C	minute	min
degree Fahrenheit	°F	newton	N
dyne	dyne	ohm	Ω
erg	erg	ounce	oz
farad	F	pascal	Pa
foot	ft	pound	lb
grad	grad	pounds per square inch	psi
gram	g	pound force	lbf
henry	H	pound mass	lbm
hertz	Hz	radian	rad
horsepower	hp	revolutions per minute	rpm
hour	hr	second	s
inch	in	slug	slug
joule	J	volt	V
kilogram	kg	watt	W
kilowatt	kW	weber	Wb
liter	l		

APPENDIX B
Trigonometric Formulas

Identities

$$\sin(\alpha \pm \beta) = \sin \alpha \cos \beta \pm \cos \alpha \sin \beta$$

$$\cos(\alpha \pm \beta) = \cos \alpha \cos \beta \mp \sin \alpha \sin \beta$$

$$A \sin \alpha + B \cos \alpha = C \sin(\alpha + \theta) = C \cos(\alpha - \delta), \quad \text{where}$$

$$C = [A^2 + B^2]^{1/2}, \quad \theta = \tan^{-1}(B/A), \quad \text{and} \quad \delta = \tan^{-1}(A/B) = 90° - \theta$$

Euler's Formula

$$\epsilon^{j\alpha} = \cos \alpha + j \sin \alpha, \quad \text{where } j^2 = -1$$

Proper Quadrants for Angles

if $\alpha = \tan^{-1}(1/1)$ then $\alpha = 45°$

if $\beta = \tan^{-1}(1/-1)$ then $\beta = 135°$

if $\delta = \tan^{-1}(-1/-1)$ then $\delta = 225°$

if $\phi = \tan^{-1}(-1/1)$ then $\phi = 315°$

In the expression $\tan^{-1}(B/A)$, the sign of B and of A must *both* be known; the sign of (B/A) alone is not sufficient to determine the quadrant in which the angle $\tan^{-1}(B/A)$ lies.

APPENDIX C

The Laplace Transform and Tables

C.1 Positive Time Functions

In solving linear differential equations with constant coefficients, where the independent variable is the continuous variable t (for time in dynamics problems) and the dependent variable represents the response of a dynamic system resulting from excitation of the system by a simple forcing function, we need to deal only with elementary mathematical concepts.

We are concerned mainly with *positive time functions,* that is, functions that are zero for $t < 0$. An example of such a function is:

$$f(t) = \begin{cases} 0 & \text{for } t < 0, \\ \cos(\omega t) & \text{for } t \geq 0. \end{cases} \tag{C.1}$$

Figure C.1 is a graph of this function. Positive time functions are used to represent physical variables such as a sinusoidal voltage which is suddenly applied to an electric circuit at $t = 0$. A very common positive time function is the unit step function, denoted by $\sigma(t - \tau)$ with $\tau \geq 0$, graphed in Fig. C.2. In many cases we will use the notation $f(t)$ to mean the product of $\sigma(t) \times f(t)$, where $\sigma(t)$ is the unit step function occurring at $t = 0$. The positive time function defined in Eqn. C.1 is $f(t) = \sigma(t) \times \cos(\omega t)$, but in many cases it will be denoted simply as $f(t) = \cos(\omega t)$.

The *initial value* of a function $f(t)$ is defined as

$$\begin{bmatrix} \text{initial value} \\ \text{of } f(t) \end{bmatrix} = f(0^+) = \lim_{t \to 0^+} f(t), \tag{C.2}$$

where the limit notation $t \to 0^+$ means that 0 is approached from positive values of t, or "from the right" as it is normally graphed. In the case of Eqn. C.1 the initial value is 1. For the unit step function, the initial value is 0 if $\tau > 0$ but for $\tau = 0$ it is 1; see Fig. C.2.

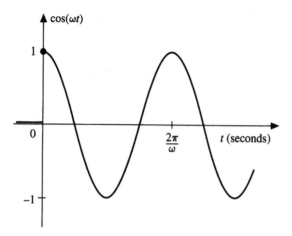

Fig. C.1 Positive time function $\cos(\omega t)$.

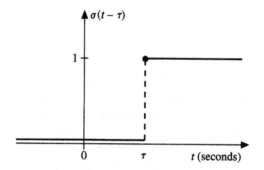

Fig. C.2 Unit step function occurring at $t = \tau$.

The *unit impulse function* $\delta(t)$ is another useful function:

$$\delta(t) = \frac{1}{e}[\sigma(t) - \sigma(t-e)], \tag{C.3}$$

where e is a positive number that is very small. Figure C.3 shows $\delta(t)$, which is used to represent physical phenomena that are very intense but of exceedingly short duration, such as a sharp hammer blow on a massive object. In these cases the duration e is not important, so long as it can be considered to occur instantaneously for the practical purposes at hand. During the brief interval $0 < t < e$ it is the *strength* of the pulse that is important, and not the exact values of the physical phenomenon:

$$\left[\begin{array}{c} \text{strength of the} \\ \text{unit impulse function} \end{array}\right] = \int_0^e \delta(t)\,dt = 1. \tag{C.4}$$

Consider a hammer blow applying the force pulse having the characteristic shown in Fig. C.4. The strength of this blow is approximately 0.02 newton-seconds [N-s]. In analysis this applied force would be characterized as $f(t) = 0.02\delta(t)$ N.

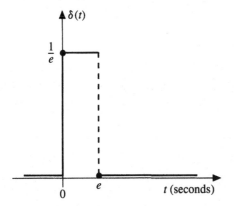

Fig. C.3 Unit input function as defined in this book.

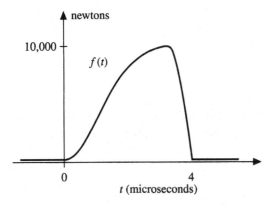

Fig. C.4 $f(t)$ is approximately $0.02 \times \delta(t)$.

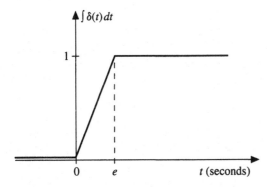

Fig. C.5 The integral of the unit impuluse function is approximately the unit step function.

We note that the integral of $\delta(t)$ is essentially the unit step function, as indicated in Fig. C.5. We define the relationship between $\sigma(t)$ and $\delta(t)$ to be

$$\sigma(t) \triangleq \int_0^t \delta(t) \, dt. \tag{C.5}$$

C.2   The Laplace Transformation

The Laplace transformation is defined only for a limited class of positive time functions, those said to be "sectionally continuous and of exponential order." A function $f(t)$ is of this class provided there is a finite real number α for which

$$\int_0^\infty |f(t)| \epsilon^{-\alpha t} \, dt < \infty. \tag{C.6}$$

We assume here that both $f(t)$ and α are real-valued. We will deal only with time functions satisfying Eqn. C.6, since all the physical phenomena that concern us in this book may be represented by functions from this class. The symbol ϵ is the base of the natural logarithms. The Laplace transform of $f(t)$, denoted $\mathcal{L}\{f(t)\}$, is defined by the integral

$$\mathcal{L}\{f(t)\} = \int_0^\infty f(t)\epsilon^{-st} \, dt, \tag{C.7}$$

where s, called the *Laplace variable,* is simply considered to be a constant for performing the integration indicated in Eqn. C.7. It is worthwhile to note, however, that the exponent of ϵ in the integrand, $-st$, must be dimensionless; if t has the dimension time then s must have the dimension 1/time, which is called *frequency* by engineers. Because the defining integral is a definite integral, the variable of integration t will not appear in the result of the integration, but the Laplace variable s will appear. As an example, let us calculate the Laplace transform of the unit step function $\sigma(t)$:

$$\mathcal{L}\{\sigma(t)\} = \int_0^\infty \sigma(t)\epsilon^{-st} \, dt = \int_0^\infty 1\epsilon^{-st} \, dt = \left[-\frac{1}{s}\epsilon^{-st} \right]_{t=0}^{t=\infty} = \frac{1}{s}. \tag{C.8}$$

We see that the integration has *transformed* our time function $\sigma(t)$ into a function of s. We usually represent this operation by using lower-case letters for the time function and upper-case for the transformed function:

$$\mathcal{L}\{f(t)\} = \int_0^\infty f(t)\epsilon^{-st} \, dt = F(s). \tag{C.9}$$

In the case of the unit step function we would have $\mathcal{L}\{\sigma(t)\} = \Sigma(s) = 1/s$. In cases involving such symbols as β, θ, and ϵ, we evade upper-case notation by using a circumflex to indicate the time function; for example, $\mathcal{L}\{\hat{\theta}(t)\} = \theta(s)$.

Another example shows how the definition of the Laplace transform, Eqn. C.9, is applied directly to calculate the transform of a time function, say $g(t) = \epsilon^{\alpha t}$, with α a constant:

$$\mathcal{L}\{g(t)\} = \int_0^\infty \epsilon^{\alpha t}\epsilon^{-st} \, dt = \int_0^\infty \epsilon^{-(s-\alpha)t} \, dt = \frac{1}{s-\alpha} = G(s). \tag{C.10}$$

A further example, say $h(t) = \cos(\omega t)$, requires more-involved yet still elementary calculus:

$$\mathscr{L}\{h(t)\} = \int_0^\infty \epsilon^{-st}\cos(\omega t)\,dt = \left[\frac{\epsilon^{-st}}{s^2+\omega^2}[-s\cos(\omega t)+\omega\sin(\omega t)]\right]_0^\infty$$

$$= \frac{s}{s^2+\omega^2} = H(s). \tag{C.11}$$

These examples have shown, for the class of time functions used in this book, that the Laplace transformation is an operation that can be performed using elementary calculus. They also illustrate the concept of a *transformation*, a mathematical operation on a function of time that turns it into a function of the Laplace variable s. This may be indicated by the notation: $f(t) \xrightarrow{\mathscr{L}} F(s)$, *and* it is often said that the transformation "takes a function from the *time domain* into the *frequency domain.*" The defining integral makes clear that the units of $f(t)$ are not the same as those of $F(s)$. For example, if $e(t)$ is in units of volts then $E(s) = \mathscr{L}\{e(t)\}$ will be in units of volt-seconds. This simple fact helps an analyst check the dimensional consistency of an equation in the s *domain*.

C.3 Properties of the Laplace Transformation

Several properties of the Laplace transformation that must be understood by the control system analyst are explained briefly here. Further mathematical treatment is readily available in the literature.

Uniqueness and the Inverse Transformation

The integral operation defined in Eqn. C.9 is a one-to-one transformation. This means that if $f_1(t) \xrightarrow{\mathscr{L}} F_1(s)$ then there is no function $f_2(t)$, different from $f_1(t)$, that has $F_1(s)$ as its Laplace transform. This permits us to establish a useful catalog of Laplace transforms of commonly used time functions, called a *Laplace transform table,* which is simply a specialized table of elementary integrals. A table of transforms is given in Section C.5.

Because of the uniqueness property, we can define the *inverse* Laplace transformation of $F_1(s)$ to be $f_1(t)$, represented as: $F_1(s) \xrightarrow{\mathscr{L}^{-1}} f_1(t)$. The inverse transformation is defined as

$$f(t) = \mathscr{L}^{-1}\{F(s)\} = \frac{1}{2\pi j}\int_{c-j\infty}^{c+j\infty} F(s)\epsilon^{st}\,ds, \tag{C.12}$$

where c is a constant that must exceed a certain value determined by $F(s)$, and $j = \sqrt{-1}$. An understanding of how the integration is performed using complex variable theory is not essential for our purposes, since we can determine, directly from the Laplace transform table, any $f(t)$ we require from its corresponding $F(s)$. Alternatively, we can use the partial-fraction expansion method explained in Chapter 7.

Linearity

If $\mathscr{L}\{f(t)\} = F(s)$ then it is clear from the definition, Eqn. C.9, that the Laplace transform has the property of *homogeneity*; that is,

$$\mathcal{L}\{af(t)\} = aF(s), \tag{C.13}$$

where a is a constant. It is likewise clear that the transform has the property of *additivity*; that is, if $\mathcal{L}\{f_1(t)\} = F_1(s)$ and $\mathcal{L}\{f_2(t)\} = F_2(s)$ then the transform of $\{f_1(t) + f_2(t)\}$ is

$$\mathcal{L}\{f_1(t) + f_2(t)\} = F_1(s) + F_2(s). \tag{C.14}$$

It also follows from the definition that the transform is *linear,* since it possesses both the properties of additivity and homogeneity. Therefore, if a and b are constants we have

$$\mathcal{L}\{af_1(t) + bf_2(t)\} = aF_1(s) + bF_2(s). \tag{C.15}$$

Differentiation

If $\mathcal{L}\{f(t)\} = F(s)$, and if $f(0^+)$ is the initial value of $f(t)$, then it can be shown from the definition of the Laplace transform (Eqn. C.7) that the transform of the time derivative of $f(t)$ is

$$\mathcal{L}\left\{\frac{df(t)}{dt}\right\} = sF(s) - f(0^+) \tag{C.16}$$

and further, if $d^n f(t)/dt^n$ is denoted as $f^{(n)}(t)$, we have

$$\mathcal{L}\left\{\frac{df^n(t)}{dt^n}\right\} = s^n F(s) - s^{n-1} f(0^+) - s^{n-2} f^{(1)}(0^+) - \cdots - f^{(n-1)}(0^+). \tag{C.17}$$

As an example, consider the function $f(t) = a + b \cos \omega t$. We then have

$$f^{(1)}(t) = -b\omega \sin \omega t \quad \text{and} \quad f^{(2)}(t) = -b\omega^2 \cos \omega t.$$

Also, $F(s) = a[1/s] + b[s/(s^2 + \omega^2)]$. We apply the formula in Eqn. C.17 to calculate the Laplace transform of $d^3 f(t)/dt^3$:

$$\mathcal{L}\left\{\frac{d^3 f(t)}{dt^3}\right\} = s^3 F(s) - s^2(a+b) - s(0) + b\omega^2 = \frac{b\omega^4}{s^2 + \omega^2}. \tag{C.18}$$

Equation C.18 is easily checked, since $d^3 f(t)/dt^3 = b\omega^3 \sin \omega t$, and the table of Laplace transforms shows $\mathcal{L}\{b\omega^3 \sin \omega t\} = b\omega^3 \mathcal{L}\{\sin \omega t\} = b\omega^3 \{\omega/(s^2 + \omega^2)\}$.

Integration

Denote the indefinite time integral of $f(t)$ as $\int f(t)\, dt = f^{(-1)}(t)$. It can be shown from the definition (Eqn. C.7) that

$$\mathcal{L}\left\{\int f(t)\, dt\right\} = \frac{1}{s}[F(s) + f^{(-1)}(0^+)]. \tag{C.19}$$

For the definite time integral we have

$$\mathcal{L}\left\{\int_0^t f(t)\, dt\right\} = \frac{1}{s} F(s). \tag{C.20}$$

As an example, consider the voltage–charge relationship in a capacitor with current $i(t)$ amperes flowing into it. The relationship between current and charge is

$$q(t) = \int i(t)\, dt = q(0^+) + \int_0^t i(t)\, dt \text{ coulombs,}$$

and the voltage–charge relationship is a simple proportion, where C is the capacitance in farads:

$$e(t) = \frac{1}{C}[q(t)] = \frac{1}{C}\left[q(0^+) + \int_0^t i(t)\, dt \right] \text{ volts,} \tag{C.21}$$

so from Eqn. C.19 we have

$$\mathcal{L}\{e(t)\} = \frac{1}{C}\left[\frac{q(0^+)}{s} + \frac{I(s)}{s} \right] = E(s) = \frac{1}{C}Q(s). \tag{C.22}$$

Delayed Function

Consider a unit step function that does not occur at $t = 0$ but rather is delayed by τ seconds following $t = 0$. The Laplace transform of the delayed step function is calculated as

$$\mathcal{L}\{\sigma(t-\tau)\} = \int_0^\infty \sigma(t-\tau)\epsilon^{-st}\, dt$$

$$= \int_\tau^\infty 1\epsilon^{-st}\, dt = \left[-\frac{1}{s}\epsilon^{-st} \right]_{t=\tau}^{t=\infty} = \epsilon^{-\tau s}\left[\frac{1}{s} \right]. \tag{C.23}$$

This illustrates a general property of the Laplace transform:

$$\mathcal{L}\{f(t-\tau)\} = \epsilon^{-\tau s}F(s). \tag{C.24}$$

Unit Impulse Function

The unit impulse function is defined in Eqn. C.3 as $\delta(t) = (1/e)[\sigma(t) - \sigma(t-e)]$, where e is a small positive number in the neighborhood of 0. Now take the Laplace transform of $\delta(t)$,

$$\mathcal{L}\{\delta(t)\} = \frac{1}{e}\left[\frac{1}{s} - \frac{\epsilon^{-es}}{s} \right] = \frac{1}{es}[1 - \epsilon^{-es}]. \tag{C.25}$$

Expand ϵ^{-es} in its infinite-series representation:

$$\epsilon^{-es} = 1 - es + \frac{(es)^2}{2!} - \frac{(es)^3}{3!} + \cdots,$$

which, for $es \ll 1$, may be approximated as $\epsilon^{-es} \cong 1 - es$. Using this approximation in Eqn. C.25, we have

$$\mathcal{L}\{\delta(t)\} \cong \frac{1}{es}[1 - (1-es)] = \frac{1}{es}[es] = 1. \tag{C.26}$$

When the impulse function is used in this book, we always assume that e is so small that the $\cong$ sign can be replaced by the $=$ sign, so we take as a definition

$$\mathcal{L}\{\delta(t)\} \triangleq 1. \tag{C.27}$$

Exponential Multiplier

Let $h(t) = \cos(\omega t)$ so that $H(s) = s/(s^2 + \omega^2)$. We now calculate the Laplace transform of $\epsilon^{\alpha t}h(t) = \epsilon^{\alpha t}\cos(\omega t)$:

$$\mathcal{L}\{\epsilon^{\alpha t}\cos(\omega t)\} = \mathcal{L}\{\epsilon^{\alpha t}h(t)\} = \int_0^\infty \epsilon^{-st}\epsilon^{\alpha t}\cos(\omega t)\,dt = \frac{(s-\alpha)}{(s-\alpha)^2 + \omega^2}. \tag{C.28}$$

This example illustrates the following general property of the Laplace transform:

$$\mathcal{L}\{\epsilon^{\alpha t}f(t)\} = F(s-\alpha). \tag{C.29}$$

Initial-Value Theorem

This useful theorem permits one to determine the *initial value* $f(0^+)$ of $f(t)$, by inspection of $F(s)$, without taking the inverse Laplace transform of $F(s)$: If $f(t)$ and $df(t)/dt$ each have a Laplace transform, and if $\lim_{s\to\infty}[sF(s)]$ exists, then

$$\lim_{t\to 0^+}[f(t)] = \lim_{s\to\infty}[sF(s)] = f(0^+). \tag{C.30}$$

For example, if $f(t) = A\cos(\omega t)$ so that $F(s) = As/(s^2 + \omega^2)$, the application of Eqn. C.30 to $F(s)$ yields

$$\lim_{s\to\infty}\left[\frac{As^2}{s^2 + \omega^2}\right] = A = f(0^+).$$

As a second example, we attempt to apply this theorem to the unit impulse function. We have $\mathcal{L}\{\delta(t)\} = 1$, so the limit $\lim_{s\to\infty}[s\times 1]$ does not exist and hence the initial-value theorem cannot be applied.

Final-Value Theorem

If $f(t)$ and $df(t)/dt$ both have Laplace transforms, and if $\lim_{t\to\infty}f(t)$ exists, then we have this useful theorem that permits us to determine the *final value* $f(\infty)$ of $f(t)$ by inspection of $F(s)$:

$$f(\infty) = \lim_{t\to\infty}f(t) = \lim_{s\to 0}[sF(s)], \quad \text{provided } \lim_{t\to\infty}f(t) \text{ exists.} \tag{C.31}$$

As an example, let $f(t) = a + b\epsilon^{-\alpha t}$, so that $F(s) = [(a+b)s + a\alpha]/s(s+\alpha)$. Applying the final-value theorem by the formula in Eqn. C.31, we have

$$\lim_{s\to 0}[sF(s)] = a. \tag{C.32}$$

But now we must note that, in this example, $f(t)$ has a final value *only if α is a positive number*. If $\alpha = 0$ then the final value of $f(t)$ is $(a+b)$; if α is negative, the final value of $f(t)$ does not exist because in that case $f(t)$ increases exponentially as $t\to\infty$. In this latter case the conditions on the validity of the final-value theorem are not met, so the formula in Eqn. C.31 cannot be expected to reveal the nature of the unstable $f(t)$, even though $\lim_{s\to 0}[sF(s)]$ exists.

It is therefore necessary to know if $f(t)$ *has* a final value before attempting to find it.

Convolution

Let $f_1(t)$ and $f_2(t)$ be positive time functions having Laplace transforms $F_1(s)$ and $F_2(s)$. The *convolution of $f_1(t)$ with $f_2(t)$*, denoted $f_1(t) * f_2(t)$, is defined to be

$$f_1(t) * f_2(t) = \int_0^t f_1(t-\tau)f_2(\tau)\,d\tau, \tag{C.33}$$

where τ is a dummy variable of integration. The calculation indicated here is accomplished by plotting $f_2(\tau)$ versus τ for $\tau \geq 0$. This plot is identical to that of $f_2(t)$ versus t. Next we plot $f_1(-\tau)$ versus τ by reversing the direction of the independent variable, and we shift the plot of $f_1(-\tau)$ to the right by an amount t, to obtain $f_1(t-\tau)$. We then plot the product $f_1(t-\tau)f_2(\tau)$ and integrate that product from $\tau = 0$ to $\tau = t$ to give the value of $f_1(t) * f_2(t)$ at time t. This process is illustrated in Fig. C.6 for the case $f_1(t) = 2\epsilon^{-t}$ and $f_2(t) = \sigma(t)$, a unit step function. This definition of $f_1(t) * f_2(t)$ leads to the following calculation for this simple example:

$$f_1(t) * f_2(t) = \int_0^t 2\epsilon^{(\tau-t)}(1)\,d\tau = 2\epsilon^{-t}[\epsilon^\tau]_0^t = 2[1-\epsilon^{-t}]. \tag{C.34}$$

The Laplace transform of $f_1(t) * f_2(t)$, calculated from the definition of the Laplace transform, is the product of the individual transforms of $f_1(t)$ and $f_2(t)$:

$$\mathcal{L}\{f_1(t) * f_2(t)\} = F_1(s)F_2(s). \tag{C.35}$$

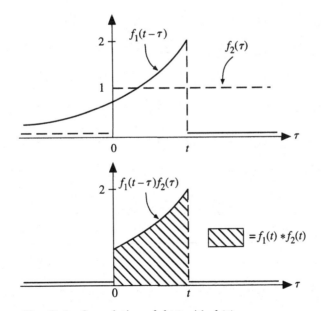

Fig. C.6 Convolution of $f_1(t)$ with $f_2(t)$.

C.5 Laplace Transform Tables

The following table of Laplace transforms and properties of Laplace transforms is compiled only for positive time functions. All transform pairs listed here can be derived from the definition, Eqn. C.7. It is usually possible to convert an $F(s)$ that does not appear here into a sum of simpler functions (each of which does appear in this table) by using the partial-fraction expansion technique described in Chapter 7. The symbol ϵ is used as the base of the natural logarithms in this table, as it is in the body of the text, and should not be mistaken for the infinitesimal notation used in Section C.1.

Pair Number	$F(s)$	$f(t)$
1	1	$\delta(t)$, unit impulse at $t = 0$
2	$\dfrac{1}{s}$	$\sigma(t)$, unit step at $t = 0$
3	$\dfrac{1}{s^2}$	t, unit ramp starting at $t = 0$
4	$\dfrac{n!}{s^{n+1}}$	t^n, $n = 1, 2, 3, \ldots$
5	$\dfrac{1}{s+b}$	ϵ^{-bt}
6	$\dfrac{1}{(s+b)^2}$	$t\epsilon^{-bt}$
7	$\dfrac{n!}{(s+b)^{n+1}}$	$t^n \epsilon^{-bt}$, $n = 1, 2, 3, \ldots$
8	$\dfrac{\omega}{s^2+\omega^2}$	$\sin(\omega t)$
9	$\dfrac{s}{s^2+\omega^2}$	$\cos(\omega t)$
10	$\dfrac{\omega}{(s+b)^2+\omega^2}$	$\epsilon^{-bt}\sin(\omega t)$
11	$\dfrac{s+b}{(s+b)^2+\omega^2}$	$\epsilon^{-bt}\cos(\omega t)$
12	$\dfrac{\omega^2}{s(s^2+\omega^2)}$	$1-\cos(\omega t)$
13	$\dfrac{s+a}{s^2+\omega^2}$	$\dfrac{1}{\omega}(a^2+\omega^2)^{1/2}\sin(\omega t+\psi)$, $\psi = \tan^{-1}\!\left(\dfrac{\omega}{a}\right)$
14	$\dfrac{s+a}{s(s^2+\omega^2)}$	$\dfrac{a}{\omega^2}-\dfrac{(a^2+\omega^2)^{1/2}}{\omega^2}\cos(\omega t+\psi)$, $\psi = \tan^{-1}\!\left(\dfrac{\omega}{a}\right)$
15	$\dfrac{s+a}{(s+b)^2+\omega_0^2}$	$\dfrac{1}{\omega_0}[(a-b)^2+\omega_0^2]^{1/2}\epsilon^{-bt}\sin(\omega_0 t+\psi)$, $\psi = \tan^{-1}\!\left(\dfrac{\omega_0}{a-b}\right)$

Pair Number	$F(s)$	$f(t)$
16	$\dfrac{(\alpha+j\beta)}{(s+b+j\omega_o)} + \dfrac{(\alpha-j\beta)}{(s+b-j\omega_o)}$	$2(\alpha^2+\beta^2)^{1/2}\epsilon^{-bt}\sin(\omega_o t+\psi),\ \ \psi = \tan^{-1}\left(\dfrac{\alpha}{\beta}\right)$
17	$\dfrac{\omega_n^2}{s[(s+b)^2+\omega_o^2]}$	$1+\left(\dfrac{\omega_n}{\omega_o}\right)\epsilon^{-bt}\sin(\omega_o t-\psi),\ \ \psi=\tan^{-1}\left(\dfrac{\omega_o}{-b}\right),\ \ \omega_n^2=b^2+\omega_o^2$
18	$\dfrac{\omega_n^2}{s[s^2+2\zeta\omega_n s+\omega_n^2]}$	$1+\dfrac{1}{\sqrt{1-\zeta^2}}\epsilon^{-\zeta\omega_n t}\sin(\omega_n\sqrt{1-\zeta^2}\,t-\psi),\ \ \psi=\tan^{-1}\left(\dfrac{\sqrt{1-\zeta^2}}{-\zeta}\right),\ \ \zeta<1$

(Notice that pairs 17 and 18 are different forms of the same function, where $b=\zeta\omega_n$ and $\omega_o=\omega_n\sqrt{1-\zeta^2}\,$.)

Pair Number	$F(s)$	$f(t)$
19	$\dfrac{s+a}{s[(s+b)^2+\omega_o^2]}$	$\dfrac{a}{\omega_n^2}+\dfrac{1}{\omega_o\omega_n}[(a-b)^2+\omega_o^2]^{1/2}\epsilon^{-bt}\sin(\omega_o t+\psi),\ \ \psi=\tan^{-1}\left(\dfrac{\omega_o}{a-b}\right)-\tan^{-1}\left(\dfrac{\omega_o}{-b}\right),\ \ \omega_n^2=b^2+\omega_o^2$
20	$\dfrac{1}{s(s+b)}$	$\dfrac{1}{b}(1-\epsilon^{-bt})$
21	$\dfrac{s+a}{s(s+b)}$	$\dfrac{a}{b}+\left(1-\dfrac{a}{b}\right)\epsilon^{-bt}$
22	$\dfrac{1}{s(s+b)^2}$	$\dfrac{1}{b^2}[1-(1+bt)\epsilon^{-bt}]$
23	$\dfrac{s+a}{s(s+b)^2}$	$\dfrac{a}{b^2}+\left[\left(\dfrac{b-a}{b}\right)t-\dfrac{a}{b^2}\right]\epsilon^{-bt}$
24	$\dfrac{s+a}{(s+b)(s+c)}$	$\dfrac{(a-b)\epsilon^{-bt}-(a-c)\epsilon^{-ct}}{(c-b)},\ \ b\neq c$

Pair Number	$F(s)$	$f(t)$
25	$\dfrac{1}{s(s+b)(s+c)}$	$\dfrac{1}{bc} + \dfrac{c\epsilon^{-bt} - b\epsilon^{-ct}}{bc(b-c)}, \quad b \neq c$
26	$\dfrac{s+a}{s(s+b)(s+c)}$	$\dfrac{a}{bc} + \dfrac{(a-b)}{b(b-c)}\epsilon^{-bt} + \dfrac{(a-c)}{c(c-b)}\epsilon^{-ct}, \quad b \neq c$
27	$\dfrac{1}{s(s+c)[(s+b)^2+\omega_o^2]}$	$\dfrac{1}{c\omega_n^2} - \dfrac{\epsilon^{-ct}}{c[(b-c)^2+\omega_o^2]} + \dfrac{\epsilon^{-bt}}{\omega_o\omega_n[(b-c)^2+\omega_o^2]^{1/2}}\sin(\omega_o t - \psi),$ $\psi = \tan^{-1}\left(\dfrac{\omega_o}{-b}\right) + \tan^{-1}\left(\dfrac{\omega_o}{c-b}\right), \quad \omega_n^2 = b^2 + \omega_o^2$
28	$\dfrac{s+a}{s(s+c)[(s+b)^2+\omega_o^2]}$	$\dfrac{a}{c\omega_n^2} + \dfrac{(c-a)\epsilon^{-ct}}{c[(b-c)^2+\omega_o^2]} + \left[\dfrac{(a-b)^2+\omega_o^2}{(b-c)^2+\omega_o^2}\right]^{1/2}\dfrac{\epsilon^{-bt}}{\omega_o\omega_n}\sin(\omega_o t + \psi),$ $\psi = \tan^{-1}\left(\dfrac{\omega_o}{a-b}\right) - \tan^{-1}\left(\dfrac{\omega_o}{c-b}\right) - \tan^{-1}\left(\dfrac{\omega_o}{-b}\right), \quad \omega_n^2 = b^2 + \omega_o^2$
29	$\dfrac{1}{(s+b)(s^2+\omega^2)}$	$\dfrac{\epsilon^{-bt}}{b^2+\omega^2} + \dfrac{\sin(\omega t - \psi)}{\omega[b^2+\omega^2]^{1/2}}, \quad \psi = \tan^{-1}\left(\dfrac{\omega}{b}\right)$
30	$\dfrac{1}{(s+b)(s+c)[s^2+\omega^2]}$	$\dfrac{\epsilon^{-bt}}{(c-b)[\omega^2+b^2]} + \dfrac{\epsilon^{-ct}}{(b-c)[\omega^2+c^2]} + \dfrac{\sin(\omega t + \psi)}{\omega\sqrt{\omega^2+b^2}\sqrt{\omega^2+c^2}}, \quad \psi = 180° + \tan^{-1}\left(\dfrac{\omega}{b}\right) + \tan^{-1}\left(\dfrac{\omega}{c}\right)$

The following pairs define important properties of and operations with the Laplace transform.

Pair Number	$F(s)$	$f(t)$
31 Linearity	$C_1 F_1(s) + C_2 F_2(s) + \cdots$	$C_1 f_1(t) + C_2 f_2(t) + \cdots$
32 Differentiation	$sF(s) - f(0^+)$	$\dfrac{df(t)}{dt} = f'(t)$
33 Multiple differentiation	$s^n F(s) - s^{n-1} f(0^+)$ $- s^{n-2} f^{(1)}(0^+) - \cdots - f^{(n-1)}(0^+)$	$\dfrac{d^n f(t)}{dt^n} = f^{(n)}(t)$
34 Indefinite integral	$\dfrac{1}{s}[F(s) + f^{(-1)}(0^+)]$	$\displaystyle\int f(t)\, dt = f^{(-1)}(t)$
35 Definite integral	$\dfrac{1}{s} F(s)$	$\displaystyle\int_0^t f(t)\, dt = \int_{-\infty}^t f(t)\, dt - f^{(-1)}(0^+)$
36 Delayed function	$\epsilon^{-\tau s} F(s)$	$f(t - \tau), \; t > \tau$
37 Exponential multiplier	$F(s - b)$	$\epsilon^{bt} f(t)$
38 Time-scale change	$\alpha F(\alpha s)$	$f\left(\dfrac{t}{\alpha}\right)$
39 Initial-value theorem	$\displaystyle\lim_{t \to 0^+}[f(t)] = \lim_{s \to \infty}[sF(s)], \;$ provided $\displaystyle\lim_{s \to \infty}[sF(s)]$ exists	
40 Final-value theorem	$\displaystyle\lim_{t \to \infty}[f(t)] = \lim_{s \to 0}[sF(s)], \;$ provided $\displaystyle\lim_{t \to \infty}[f(t)]$ exists (i.e., provided $f(t)$ is a stable function)	

APPENDIX D

Routh's Stability Criterion

In 1877, E. J. Routh presented an analytical technique for determining from the coefficients of a polynomial the number of roots of that polynomial which lie in the right-half plane. Some twenty years later, A. Hurwitz, using an approach different from Routh's, arrived at the same result. Routh's stability criterion (also called the Routh–Hurwitz stability criterion) became an important tool in control engineering before the advent of modern computers. It permitted the engineer to determine the stability of a system from the coefficients of the characteristic polynomial without having to factor the polynomial. Today Routh's criterion is less important, because polynomials of moderately high order can be quickly and quite reliably factored by the computer. Nevertheless, this technique remains useful in identifying which physical parameters have the most significant impact on the stability of low-order systems. It also provides an independent check on the validity of a computer result, and can be applied in situations where a computer is unavailable.

The characteristic polynomial for a linear system can always be expressed as

$$s^n + a_{n-1}s^{n-1} + a_{n-2}s^{n-2} + \cdots + a_1 s + a_0, \tag{D.1}$$

which in factored form is

$$(s+q_1)(s+q_2)\cdots(s+q_n). \tag{D.2}$$

Because in engineering analysis the coefficients a_i are algebraic combinations of the physical parameters of the system, they are real-valued. The n roots of the characteristic polynomial, $-q_1, -q_2, \ldots, -q_n$, are therefore real-valued or, if complex-valued, they occur in complex conjugate pairs. It is also of interest to note two useful relationships between the coefficients a_i and the roots q_i:

$$q_1 + q_2 + \cdots + q_n = a_{n-1},$$

$$q_1 q_2 \cdots q_n = a_0. \tag{D.3}$$

In order to determine how many of the n roots lie in the right-half plane, we form *Routh's array* as follows.

First, arrange the $n+1$ coefficients of the polynomial in two rows.

Next calculate a third row of constants $b_1 b_2 b_3 \cdots$ from the first two rows.

Then calculate a fourth row of constants $c_1 c_2 c_3 \cdots$ from the second and third rows, and continue this pattern until the array terminates at the $(n+1)$th row, which will consist of a single element, r_1.

Routh's array then appears as

$$
\begin{array}{lllll}
\text{Row 1} & 1 & a_{n-2} & a_{n-4} & \cdots & a_1 \\
\text{Row 2} & a_{n-1} & a_{n-3} & a_{n-5} & \cdots & a_0 \\
\text{Row 3} & b_1 & b_2 & b_3 & \cdots \\
\text{Row 4} & c_1 & c_2 & c_3 & \cdots \\
\vdots & \vdots & \vdots & \vdots \\
\text{Row } n+1 & r_1 &
\end{array}
\tag{D.4}
$$

We calculate the elements of the third row by the formula

$$
b_1 = \frac{(a_{n-1})(a_{n-2}) - (1)(a_{n-3})}{a_{n-1}},
$$

$$
b_2 = \frac{(a_{n-1})(a_{n-4}) - (1)(a_{n-5})}{a_{n-1}},
\tag{D.5}
$$

$$
\vdots
$$

We calculate the elements of the fourth row in a similar manner:

$$
c_1 = \frac{(b_1)(a_{n-3}) - (a_{n-1})(b_2)}{b_1},
$$

$$
c_2 = \frac{(b_1)(a_{n-5}) - (a_{n-1})(b_3)}{b_1},
\tag{D.6}
$$

$$
\vdots
$$

The calculation of the rows in the array continues in this pattern until $n+1$ rows exist in the array. Because each of the coefficients $a_0, a_1, a_2, \ldots, a_{n-1}, 1$ is real-valued, all elements of Routh's array will be real-valued.

We now consider the sequence consisting of the $n+1$ elements of the first column of Routh's array, which Routh called *test functions*:

$$
\{1, a_{n-1}, b_1, c_1, \ldots, r_1\}.
\tag{D.7}
$$

We call this ordered set the *Routh sequence*. The number of roots of the polynomial lying in the right half-plane is equal to the number of changes in the algebraic sign of the elements in the Routh sequence.

As an example, consider the fifth-order polynomial

$$
s^5 + 3s^4 + 12s^3 + 34s^2 + 104s + 80.
\tag{D.8}
$$

Routh's array for this example is

Row 1	1	12	104
Row 2	3	34	80
Row 3	2/3	232/3	0
Row 4	−314	80	0
Row 5	77.503	0	
Row 6	80		

$$(D.9)$$

The Routh sequence $\{1, 3, 2/3, -314, 77.503, 80\}$ has two changes of sign, revealing that two roots of the polynomial lie in the right-half plane. Factoring the polynomial shows the roots to lie in the positions $s = -1$, $s = -2+j2$, $s = -2-j2$, $s = 1+j3$, and $s = 1-j3$.

In calculating the elements in the rows of Routh's array, it is permissible to divide all the elements in any row by a positive number without changing the outcome of the sign changes in the Routh sequence. In our example the array could be simplified to:

1	12	104
3	34	80
1	116	0
−1	0.255	0
1	0	
0.255		

$$(D.10)$$

Two special cases can occur in Routh's array for systems having roots in the right-half plane or on the $j\omega$ axis.

Case 1 The first element in one of the rows is zero but the others in that row are nonzero. Calculation of subsequent rows is not possible in this case. However, the Routh sequence may still be calculated simply by replacing the zero element with a small nonzero real number ϵ, continuing the calculation of the subsequent rows to form the Routh sequence, which now will depend on the small ϵ. Terms in the sequence involving ϵ^2 and higher powers of ϵ are neglected. We illustrate this case with the polynomial

$$s^5 + 2s^4 + 3s^3 + 6s^2 + 5s + 5.$$

$$(D.11)$$

Routh's array for this case is

Row 1	1	3	5	
Row 2	2	6	5	
Row 3	0	1		(after multiplying the row by 2/5)

$$(D.12)$$

Now we replace the zero by a small ϵ, which we assume for the moment is positive. Thus we have

Row 3 ϵ 1

Row 4 $(6\epsilon - 2)$ 5ϵ (after multiplying the row by ϵ)

Row 5   1   0   (upon ignoring the ϵ^2 term)

Row 6 5ϵ

(D.13)

Here Routh's sequence is $\{1, 2, \epsilon, -2, 1, 5\epsilon\}$, where the term 6ϵ has been ignored, as compared to -2. Since we assumed $\epsilon > 0$, there are two sign changes in the sequence. Factoring this polynomial gives the roots as approximately -1.8, $-0.5 + j1$, $-0.5 - j1$, $0.4 + j1.4$, and $0.4 - j1.4$. Interestingly, the number of sign changes in the sequence will be the same whether the small ϵ is taken as positive or negative. However, if ϵ is taken to be negative then row 4 cannot be multiplied by ϵ as in (D.13).

Case 2   The polynomial has roots which are mirror images of one another with respect to the $j\omega$ axis, or which lie on the $j\omega$ axis. In this case all elements of a row will be zero, but Routh's array can still be completed using a technique involving an *auxiliary polynomial*. We illustrate this case by an example. Let the characteristic polynomial be

$$s^6 + 5s^5 + 22s^4 + 80s^3 + 196s^2 + 500s + 600. \tag{D.14}$$

The first four rows of Routh's array will be

Row 1   1   22   196   600

Row 2   5   80   500

Row 3   6   96   600

Row 4   0   0

(D.15)

Calculation of row 5 is not possible. Form the auxiliary polynomial using the coefficients of the last nonzero row, which in this example is row 3:

$$6s^4 + 96s^2 + 600. \tag{D.16}$$

The auxiliary polynomial has only even powers of s, and since in this case row 3 has three elements, the auxiliary polynomial is of degree 4. Now take the derivative of the auxiliary polynomial with respect to s:

$$24s^3 + 192s. \tag{D.17}$$

Use the coefficients of this derived polynomial as row 4, expressing the altered version of the complete array as:

Row 1   1   22   196   600

Row 2   5   80   500

Row 3   6   96   600

Row 4   24   192

Row 5   1   12.5   (after dividing the row by 48)

Row 6 -1 (after dividing the row by 108)

Row 7   12.5

(D.18)

The Routh sequence for this case has two sign changes, indicating that two roots lie in the right-half plane. The roots of this polynomial are $-2, -3, -1+j3$, $-1-j3, 1+j3$, and $1-j3$.

If the last term in Routh's array, r_1, happens to be zero, this must be treated as if it were a full row of zeros, and not as a row whose first element is zero. Hence the auxiliary polynomial technique, not the ϵ technique, should be used to determine the proper sign of r_1.

If the characteristic polynomial has roots on the $j\omega$ axis then these roots will also be roots of the auxiliary polynomial. Four possibilities exist, as follows.

Case A All the $j\omega$-axis roots are nonzero and distinct. The Routh sequence will have no change in sign, but the zero-input response of the system will not decay to zero. Insead, it will reach a steady-state oscillation whose magnitude depends on the initial conditions giving rise to the zero-input response. The system is characterized as being *marginally stable*.

Case B At least one pair of repeated roots lies on the $j\omega$ axis. Here the Routh sequence will not show a change of sign, but the zero-input response will be *unstable*, exhibiting an oscillatory mode with increasing amplitude. An example is the polynomial

$$s^5 + s^4 + 8s^3 + 8s^2 + 16s + 16 = (s+1)(s+j2)^2(s-j2)^2. \qquad (D.19)$$

The zero-input response of the system having this characteristic polynomial will exhibit an unstable mode of the form

$$Kt[\sin(2t+\phi)]. \qquad (D.20)$$

It is important to note that Routh's criterion will not reveal this special case of instability.

Case C A single root exists at the origin of the s plane. This case is easily recognized from the characteristic polynomial, since the coefficient a_0 is zero. The zero-input response will converge to a nonzero constant as $t \to \infty$, provided all the other roots lie in the left-half plane. This response is *marginally stable*.

Case D A multiple root of order r exists at the origin of the s plane. This case is also easily recognized from the characteristic polynomial, since the r coefficients $a_0, a_1, ..., a_{r-1}$ are all zero. The zero-input response will be *unstable*, exhibiting modes having the forms:

$$
\begin{aligned}
&K_r t^{(r-1)}, \\
&K_{r-1} t^{(r-2)}, \\
&\quad \vdots \\
&K_2 t, \\
&K_1.
\end{aligned}
\qquad (D.21)
$$

We note that if the characteristic polynomial has one (or more) negative co-efficient then there will be at least one sign change in the Routh sequence, indicating that the system is unstable. We therefore see that a *sufficient* condition for *instability* is that one (or more) of these coefficients be negative, which would be obvious from an inspection of the characteristic polynomial. This inspection will reveal only that there is at least one root in the right-half plane; it will not tell how many such roots exist.

It follows that a *necessary* (but not sufficient) condition for *stability* is that $a_0, a_1, \ldots, a_{n-1}, 1$ all be positive. In this case the stability question must be resolved either by calculating the Routh sequence or by factoring the polynomial. If one or more of the coefficients is zero then Routh's criteria for no right-half-plane roots might be satisfied (some roots on the $j\omega$ axis), but the system will at best be marginally stable. Because of this, the necessary condition on the coefficients of the characteristic polynomial is usually stated as: they must be nonzero and positive.

In many cases most of the physical parameter values of the system are fixed, with only a few being adjustable. In such cases the n conditions for stability obtained from Routh's sequence, namely,

$$
\begin{aligned}
a_{n-1} &> 0, \\
b_1 &> 0, \\
c_1 &> 0, \\
&\ \vdots \\
r_1 &> 0,
\end{aligned}
\tag{D.22}
$$

can be used to determine whether or not the system can be stabilized by adjusting those parameters over their permitted ranges. Because these conditions consist of inequalities, care must be used in making these adjustments. For a third-order polynomial having all positive coefficients, the three conditions may be reduced to the single requirement

$$a_1 a_2 > a_0. \tag{D.23}$$

For a fourth-order polynomial having all positive coefficients, the basic inequality conditions also reduce to a single requirement:

$$a_1(a_2 a_3 - a_1) > a_0(a_3)^2. \tag{D.24}$$

For polynomials of fifth and higher order, the basic inequality conditions reduce to more than one requirement. The establishment of these inequality requirements for a specific design, using the basic inequality conditions derived from Routh's sequence, is an important use of Routh's stability criterion.

APPENDIX E
Normalized Time Response Curves

E.1 Step Responses of Second-Order Systems

A block diagram relating the zero-state response of a second-order system to its input is shown in Fig. E.1. If the input is a step function $u(t) = A\sigma(t)$ then the output may be calculated, as in Section 7.3, to be

$$y(t) = \frac{AK}{\omega_n^2}\left[1 + \frac{\epsilon^{-\zeta(\omega_n t)}}{\sqrt{1-\zeta^2}} \sin((\omega_n t)\sqrt{1-\zeta^2} + \psi)\right],$$

(E.1)

where $\psi = \tan^{-1}\left(\frac{\sqrt{1-\zeta^2}}{-\zeta}\right)$.

Note that the independent variable t appears in $y(t)$ only as a product with the undamped natural frequency $\omega_n t$. $y(t)$ can be normalized with respect to time by introducing the notation $\omega_n t = \tau$, and also with respect to the final value of $y(t)$, AK/ω_n^2, as shown in Section 7.3. We denote this normalized step response as $\hat{y}(\tau)$:

$$\hat{y}(\tau) = \left[1 + \frac{\epsilon^{-\zeta\tau}}{\sqrt{1-\zeta^2}} \sin(\sqrt{1-\zeta^2}\tau + \psi)\right].$$

(E.2)

$\hat{y}(\tau)$ is plotted versus τ in Fig. E.2 for several values of the damping ratio ζ.

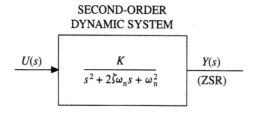

SECOND-ORDER
DYNAMIC SYSTEM

$U(s)$ — $\dfrac{K}{s^2 + 2\zeta\omega_n s + \omega_n^2}$ — $Y(s)$ (ZSR)

Fig. E.1 Second-order system.

478

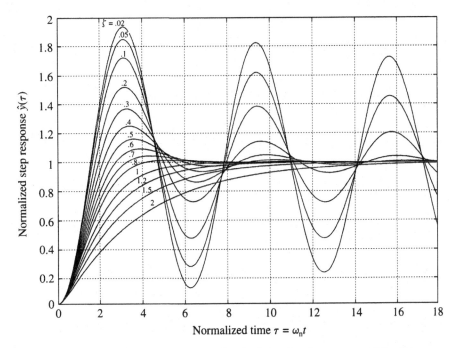

Fig. E.2 Normalized step response for the second-order system of Fig. E.1, zero-state response only.

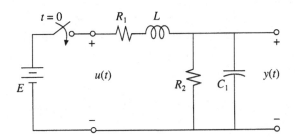

Fig. E.3 Second-order system to be analyzed using the response curves of Fig. E.2.

The curves in Fig. E.2 can be used directly to evaluate the step response of any second-order system that matches the pattern established by Fig. E.1. As an example, consider the RLC circuit shown in Fig. E.3. The capacitor voltage is zero and the inductor current is zero at the time the switch is closed. This situation is represented exactly by the diagram in Fig. E.1, with the parameters defined as follows:

$$A = E, \quad K = \frac{1}{LC_1}, \quad 2\zeta\omega_n = \frac{L + R_1 R_2 C_1}{R_2 L C_1}, \quad \omega_n^2 = \frac{R_1 + R_2}{R_2 L C_1}.$$

If the circuit elements have the following values,

$$E = 12 \text{ V}, \quad L = 1 \text{ H}, \quad R_1 = 15 \ \Omega, \quad R_2 = 1000 \ \Omega, \quad C_1 = 0.001 \text{ F},$$

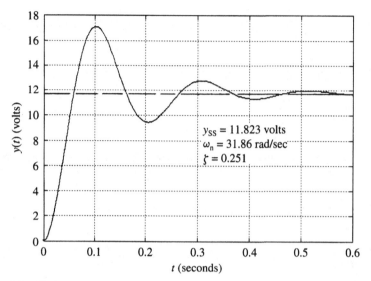

Fig. E.4 $y(t)$ obtained by denormalizing $\hat{y}(\tau)$ from Fig. E.2.

then the transfer function parameters are $\zeta = 0.251$ and $\omega_n = 31.86$ rad/s. The
final value for the output voltage is $y_{SS} = 11.823$ V. The output response $y(t)$
can be approximated in this case by interpolating between the curves for $\zeta = 0.2$
and $\zeta = 0.3$ in Fig. E.2. The amplitude scale is obtained by equating $y_{SS}(t) =$
$11.823\hat{y}_{SS}(\tau)$. The time scale is obtained by equation $t = [\tau/31.86]$ s. $y(t)$ is
plotted in Fig. E.4.

The transfer function for some second-order systems has a single finite zero:

$$\frac{Y(s)}{U(s)} = \frac{K(s+Z)}{s^2 + 2\zeta\omega_n s + \omega_n^2}. \tag{E.3}$$

For $u(t) = A\sigma(t)$, the Laplace transform of the zero-state response is

$$Y(s) = \frac{AK(s+Z)}{s[s^2 + 2\zeta\omega_n s + \omega_n^2]}. \tag{E.4}$$

The inverse transform of $Y(s)$ can be represented by a set of normalized curves
as in the previous case, but now the normalization depends on four parameters: K, Z, ζ, and ω_n. Consequently, the normalization cannot be displayed on
a single graph as in the case of Fig. E.2, where only three parameters take part
in the normalization.

A convenient way to construct the set of normalized step response curves
for this class of second-order systems is to represent Z as a fraction of the real
part of the complex poles of the transfer function:

$$Z = \beta(\zeta\omega_n). \tag{E.5}$$

In this case the final value of $y(t)$ is $AK\beta\zeta/\omega_n$, and the amplitude of the response is normalized to this value. The time scale is normalized with respect
to ω_n, as before: $\tau = \omega_n t$.

Figures E.5 through E.9 show $\hat{y}(\tau)$, the normalized versions of the $y(t)$
whose Laplace transform is given in Eqn. E.4. Each of these figures shows how

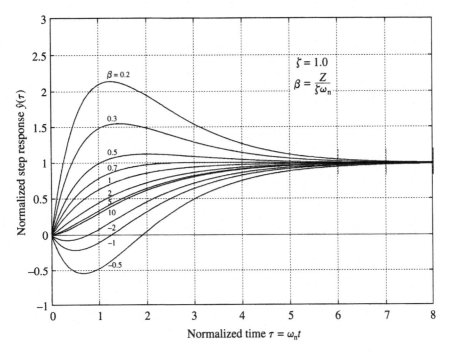

Fig. E.5 Normalized step response $\hat{y}(\tau)$ for the second-order system whose transfer function has a finite zero, $\zeta = 1.0$.

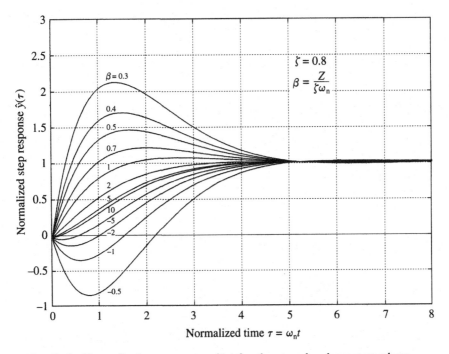

Fig. E.6 Normalized step response $\hat{y}(\tau)$ for the second-order system whose transfer function has a finite zero, $\zeta = 0.8$.

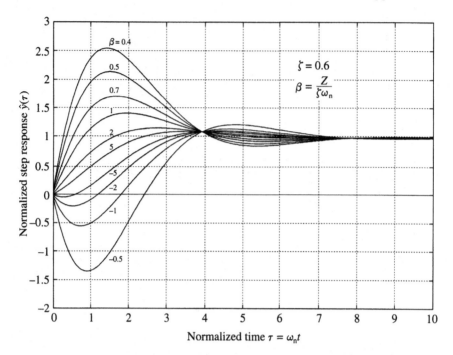

Fig. E.7   Normalized step response $\hat{y}(\tau)$ for the second-order system whose transfer function has a finite zero, $\zeta = 0.6$.

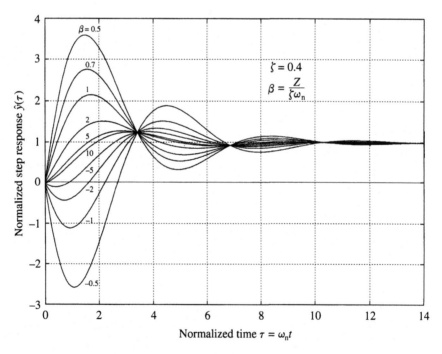

Fig. E.8   Normalized step response $\hat{y}(\tau)$ for the second-order system whose transfer function has a finite zero, $\zeta = 0.4$.

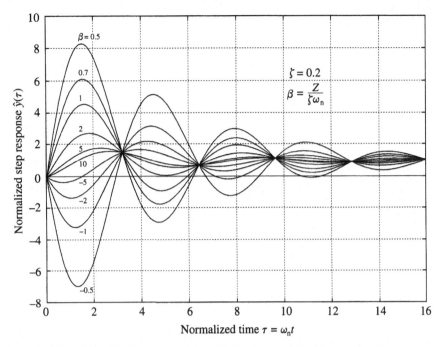

Fig. E.9 Normalized step response $\hat{y}(\tau)$ for the second-order system whose transfer function has a finite zero, $\zeta = 0.2$.

$\hat{y}(\tau)$ depends on the position of the zero of $Y(s)$ relative to the real part of the complex poles, for the damping ratio specified on the figure.

E.2 Step Responses of a Third-Order System

The system analyzed in Section 7.4 is an example of a third-order system whose transfer function has no finite zeros. The general form for such a transfer function, relating the input to the zero-state response, is

$$\frac{Y(s)}{U(s)} = \frac{K}{(s+P)[s^2+2\zeta\omega_n s+\omega_n^2]}. \tag{E.6}$$

If the input is a step of strength A, the Laplace transform of the zero-state response will be

$$Y(s) = \frac{AK}{s(s+P)[s^2+2\zeta\omega_n s+\omega_n^2]}. \tag{E.7}$$

Assuming the system is stable, $y(t)$ will have a final value of $AK/P\omega_n^2$, which establishes the normalization factor for the amplitude of $y(t)$. It is convenient to express P as a fraction of the real part of the complex poles of $Y(s)$:

$$P = \beta(\zeta\omega_n). \tag{E.8}$$

Figures E.10 through E.15 show normalized versions of $y(t)$. Each figure shows how $\hat{y}(\tau)$ depends on the location of the real-valued pole of $Y(s)$ with respect to the real part of the complex poles, for the damping ratio specified on the figure.

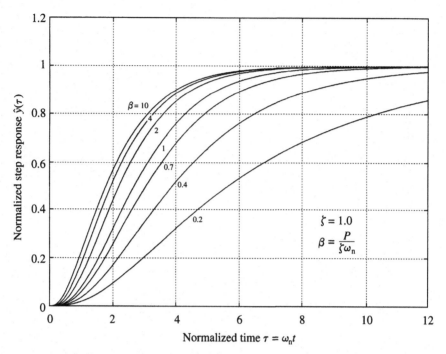

Fig. E.10 Normalized step response $\hat{y}(\tau)$ for the third-order system whose quadratic term has damping ratio $\zeta = 1.0$.

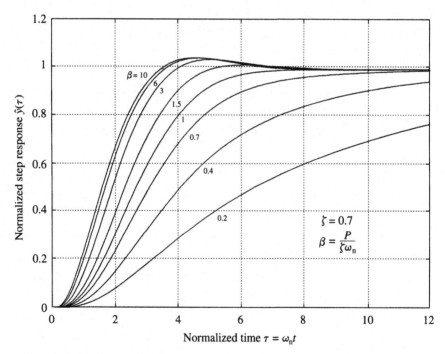

Fig. E.11 Normalized step response $\hat{y}(\tau)$ for the third-order system whose quadratic term has damping ratio $\zeta = 0.7$.

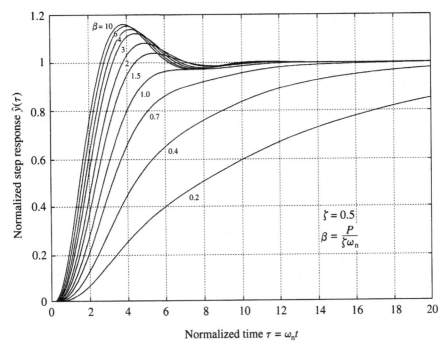

Fig. E.12 Normalized step response $\hat{y}(\tau)$ for the third-order system whose quadratic term has damping ratio $\zeta = 0.5$.

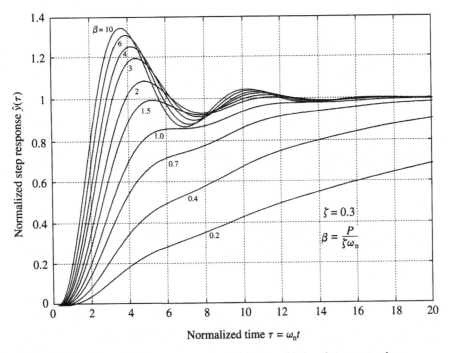

Fig. E.13 Normalized step response $\hat{y}(\tau)$ for the third-order system whose quadratic term has damping ratio $\zeta = 0.3$.

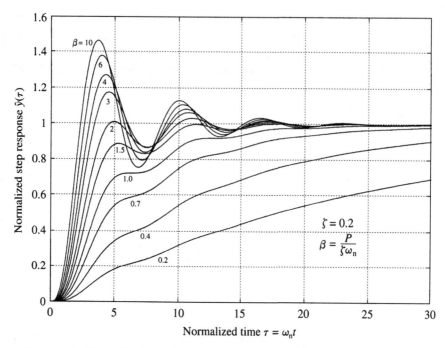

Fig. E.14 Normalized step response $\hat{y}(\tau)$ for the third-order system whose quadratic term has damping ratio $\zeta = 0.2$.

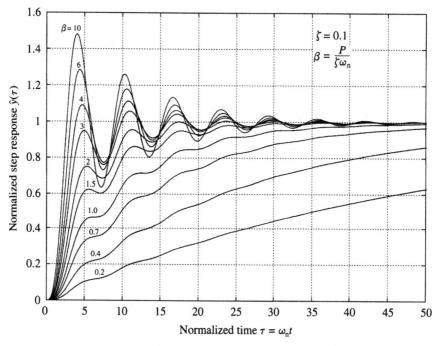

Fig. E.15 Normalized step response $\hat{y}(\tau)$ for the third-order system whose quadratic term has damping ratio $\zeta = 0.1$.

Normalized Frequency Response Curves and Nichols Chart

F.1 A Second-Order System

A common type of second-order system has a transfer function with no finite zeros:

$$G(s) = \frac{K}{s^2 + 2\zeta\omega_n s + \omega_n^2}. \tag{F.1}$$

The frequency response of this system is given by

$$G(j\omega) = \frac{K}{(\omega^2 - \omega_n^2) + j2\zeta\omega_n\omega}. \tag{F.2}$$

We normalize $G(j\omega)$ with respect to ω_n by introducing the notation $u = \omega/\omega_n$, so that Eqn. F.2 becomes

$$G(ju) = \frac{K/\omega_n^2}{(1-u^2) + j(2\zeta)u}, \tag{F.3}$$

where u is the *normalized frequency*. The frequency response function is also normalized with respect to its value at $u = 0$ by multiplying $G(ju)$ by ω_n^2/K to give

$$\hat{G}(ju) = \frac{1}{(1-u^2) + j(2\zeta)u}. \tag{F.4}$$

A straight-line approximation to the amplitude ratio characteristic of $\hat{G}(ju)$ is sketched on a Bode chart in Fig. F.1(a), and an approximation to the phase shift is sketched in Fig. F.1(b). The deviations of $|\hat{G}(ju)|_{db}$ and of $\angle\hat{G}(ju)$ from these straight-line asymptotic approximations depend on the damping ratio ζ.

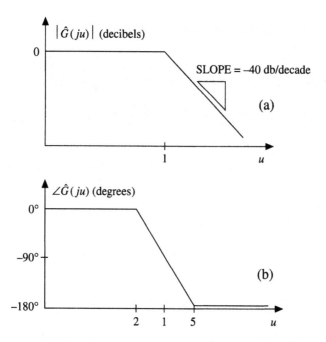

Fig. F.1 (a) Straight-line approximation to $|\hat{G}(ju)|_{db}$ on Bode plot.
(b) Straight-line approximation to $\angle\hat{G}(ju)$ on Bode plot.

Accurate Bode plots of $|\hat{G}(ju)|_{db}$ and $\angle\hat{G}(ju)$ for damping ratios ranging from 0.05 to 2.0 are given in Figs. F.2A and F.2B. These charts may be used directly to obtain accurate amplitude and phase data for second-order systems whose transfer functions match the form given in Eqn. F.1. It is necessary to know the damping ratio ζ, the undamped natural frequency ω_n, and the flag constant K for the specific system under study in order to establish the numeriical scales on the abscissas and ordinates of the amplitude and phase charts.

F.2   Nichols Chart

The Nichols chart is described in Section 13.3, and illustrated in Figs. 13.15 and 13.16. An accurate Nichols chart for $\angle G(j\omega)$ ranging from $-360°$ to $0°$ and $|G(j\omega)|_{db}$ ranging from -30 db to 20 db is drawn as Fig. F.3. The *M con-tours*, those of constant values for $|G(j\omega)/(1+G(j\omega))|_{db}$, are labeled in decibels ranging from -24 db to 12 db. The *N contours*, those of constant values for $\phi = \angle[G(j\omega)/(1+G(j\omega))]$, are labeled in degrees ranging from $-350°$ to $-10°$.

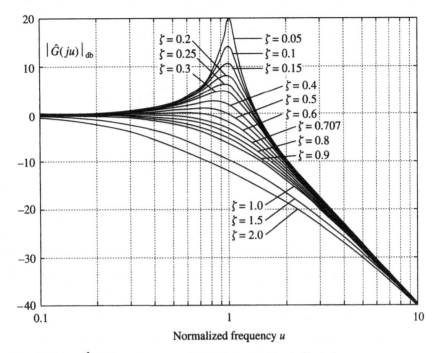

Fig. F.2A $|\hat{G}(ju)|_{db}$ versus u on Bode plot, with ζ as indicated.

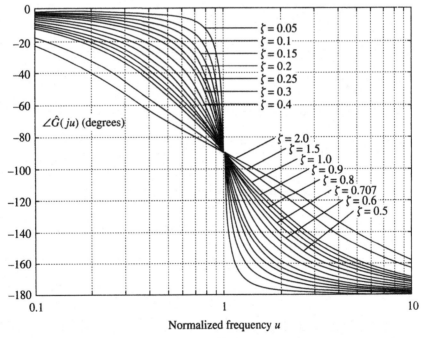

Fig. F.2B $\angle\hat{G}(ju)$ versus u on Bode plot, with ζ as indicated.

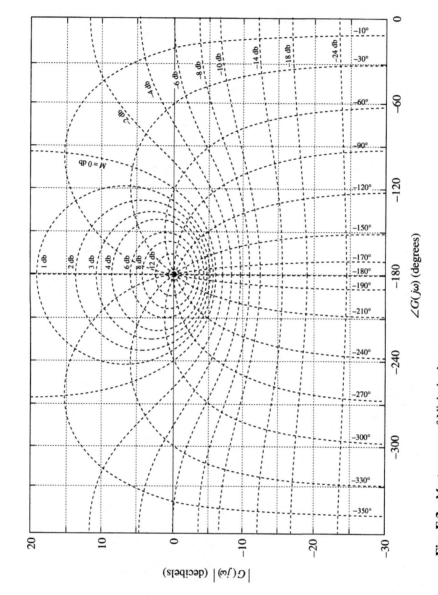

Fig. F.3 Master copy of Nichols chart.

Answers to Problems

Chapter 2

2.1 0.001031 N-m.

2.3 $\begin{bmatrix} x_1 \\ x_2 \end{bmatrix} = \begin{bmatrix} z \\ \dot{z} \end{bmatrix}$, $a_{11} = 0$, $a_{12} = 1$, $a_{21} = -k/M_E$, $a_{22} = -\alpha/M_E$, $b_{11} = 0$, $b_{12} = 1/M_E$,
$C = [1\ 0]$, and $M_E = M + J/r^2$.

2.4 $M = 2$ kg, $\alpha = 4$ N/(m/s), $a_{11} = 0$, $a_{12} = 1$, $b_{11} = 0$, $b_{12} = 0.5$ kg^{-1}, $c_{11} = 1$, $c_{12} = 0$.

2.8 All a_{ij} and b_{ij} are zero except $a_{12} = a_{34} = 1$, $a_{23} = \dfrac{m^2 l^2 g}{[M_E + m]J_{CM} + mM_E l^2}$,

$a_{43} = \dfrac{mgl[M_E + m]}{[M_E + m]J_{CM} + mM_E l^2}$, $b_{21} = \dfrac{J_{CM} + ml^2}{[M_E + m]J_{CM} + mM_E l^2}$, and

$b_{41} = \dfrac{ml}{[M_E + m]J_{CM} + mM_E l^2}$, where $M_E = M + \dfrac{2J_R}{r_R^2}$. All $c_{ij} = 0$ except

$c_{11} = c_{23} = 1$.

2.10 $A = \begin{bmatrix} -0.0451 & 0.0361 & -32.174 & -1.84 \\ -0.3697 & -2.0244 & -0.3369 & 176 \\ 0 & 0 & 0 & 1 \\ 0.0019 & -0.0395 & 0.0017 & -2.9857 \end{bmatrix}$, $B = \begin{bmatrix} 0 & 0.1 \\ -28.17 & 0 \\ 0 & 0 \\ -11.0437 & 0 \end{bmatrix}$,

$C = \begin{bmatrix} 0 & 0 & 1 & 0 \\ 0 & 0 & 0 & 1 \end{bmatrix}$, $D = \begin{bmatrix} 0 & 0 \\ 0 & 0 \end{bmatrix}$.

2.13 $a_{11} = a_{21} = a_{22} = b_{11} = 0$, $a_{12} = 1$, $C = [1\ 0]$, $\eta = 0.86$.

2.17 $A = \begin{bmatrix} 0 & 1 & 0 & 0 \\ a_{21} & 0 & a_{23} & 0 \\ 0 & 0 & 0 & 1 \\ a_{41} & 0 & a_{43} & 0 \end{bmatrix}$, $B = \begin{bmatrix} 0 \\ -\phi \\ 0 \\ \psi \end{bmatrix}$, $C = [0\ 0\ 1\ 0]$, $D = [0]$, where

$a_{21} = \psi M_B g - \phi\Delta$, $a_{23} = -\phi M_B g$, $a_{41} = \psi\Delta - \dfrac{\alpha M_B g\psi}{\beta}$, $a_{43} = \psi M_B g$, and

$$\psi = \frac{\beta}{\beta^2 - \alpha\gamma}, \phi = \frac{\gamma}{\beta^2 - \alpha\gamma}, \Delta = g[M_B h + M_A d], \beta = M_B h + \frac{J_B}{r},$$

$$\alpha = J_A + M_A d^2 + M_B h^2 + J_B, \gamma = M_B + \frac{J_B}{r^2}.$$

Chapter 3

3.1 $a_{11} = \frac{-1}{C_1(R_1 + R_3)}, a_{12} = \frac{-R_3}{C_1(R_1 + R_3)}, a_{21} = \frac{R_3}{L(R_1 + R_3)},$

$a_{22} = \frac{-[R_1 R_2 + R_1 R_3 + R_2 R_3]}{L(R_1 + R_3)}, b_{11} = \frac{R_3}{C_1(R_1 + R_3)}, b_{21} = \frac{R_1 R_3}{L(R_1 + R_3)},$

$c_{11} = \frac{R_3}{R_1 + R_3}, c_{12} = \frac{-R_1 R_3}{R_1 + R_3}, D = \left[\frac{R_1 R_3}{(R_1 + R_3)}\right].$

3.4 $a_{11} = \frac{-R_1}{L}, a_{12} = \frac{-1}{L}, a_{21} = \frac{K+1}{C_1}, a_{22} = \frac{-1}{R_2 C_1}, b_{11} = \frac{1}{L}, b_{21} = 0,$

$C = [0\ 1], D = [0].$

3.6 **(a)** $\ddot{e}_c(t) + \left[\frac{L + R_1 R_2 C_1}{LR_1 C_1}\right]\dot{e}_c(t) + \left[\frac{R_1 + R_2}{LR_1 C_1}\right]e_c(t) = \left[\frac{1}{LC_1}\right]u(t).$

For $x(t) = \begin{bmatrix} i(t) \\ e_c(t) \end{bmatrix}$: $a_{11} = \frac{-R_2}{L}, a_{12} = \frac{-1}{L}, a_{21} = \frac{1}{C_1}, a_{22} = \frac{-1}{R_1 C_1}, b_{11} = \frac{1}{L},$

$b_{21} = 0, C = [0\ 1], D = [0].$

3.7 $R_1 = R_2 = 10^6\ \Omega, R_3 = 0.5 \times 10^6\ \Omega.$

3.9 $G = \dfrac{KR_2 + R_o}{(1 - K)R_1 + (R_o/R_L)(R_1 + R_2 + R_L) + R_2}$. For $K \lll -1, G = -\dfrac{R_2}{R_1}.$

3.13 $363.503 < R_{\text{bal}} < 363.530.$

Chapter 4

4.1 $K = 0.48\ \text{Nm/A}, R = 2\ \Omega, b = 0.0048\ \text{Nm/(rad/s)}.$

4.3 $J = 0.0004\ \text{kg-m}^2, K_A = 3.9\ \text{V/V}, R_o = 1\ \Omega.$

4.4 $a_{12} = a_{21} = a_{22} = a_{32} = b_{21} = b_{31} = 0, a_{23} = 1, a_{11} = \frac{-R_E}{L_a}, a_{13} = \frac{-K_E}{L_a}, a_{31} = \frac{K_E}{M_E},$

$a_{33} = \frac{-b_E}{M_E}, b_{11} = \frac{K_A}{L_a}, C = [0\ 1\ 0], D = [0],$ where $R_E = R + R_o, K_E = \frac{nK}{r},$

$b_E = \frac{n^2 b}{r^2},$ and $M_E = M + \frac{n^2 J + J_F + J_R}{r^2}.$

4.5 $a_{12} = a_{14} = a_{15} = a_{21} = a_{22} = a_{24} = a_{25} = a_{32} = a_{35} = a_{41} = a_{42} = a_{43} = a_{44} = 0;$

$a_{52} = a_{55} = b_{21} = b_{31} = b_{41} = b_{51} = 0; a_{23} = a_{45} = c_{12} = c_{24} = 1; d_{11} = d_{21} = 0;$

$c_{11} = c_{13} = c_{14} = c_{15} = c_{21} = c_{22} = c_{23} = c_{25} = 0; a_{11} = \frac{-R_E}{L_a}, a_{13} = \frac{-K_E}{L_a}, b_{11} = \frac{K_A}{L_a},$

$a_{31} = \frac{K_E}{M_T}, a_{33} = \frac{-b_E}{M_T}, a_{34} = \frac{m^2 l^2 g}{M_T J_0}, a_{51} = \frac{mlK_E}{\Delta}, a_{53} = \frac{-mlb_E}{\Delta},$ and

$a_{54} = \frac{mgl(M_E + m)}{\Delta}.$ Here $R_E = R + R_o, K_E = \frac{nK}{r}, b_E = \frac{n^2 b}{r^2},$

$$M_E = \left[M + \frac{n^2 J + J_F + J_R}{r^2}\right], \Delta = [J_{CM}(M_E + m) + mM_E l^2], J_0 = [J_{CM} + ml^2], \text{ and}$$

$$M_T = \left[M_E + \frac{mJ_{CM}}{J_0}\right].$$

4.9 All a_{ij}, b_{ij}, and c_{ij} are zero except for the following: $a_{13} = a_{24} = c_{12} = c_{21} = 1$,

$$D = \begin{bmatrix} 0 \\ 0 \end{bmatrix}, b_{51} = \frac{1}{L_a}, a_{54} = \frac{-nK}{L_a}, a_{55} = \frac{-R}{L_a}, a_{31} = \frac{-r^2 g M_B \beta}{Q},$$

$$a_{32} = \frac{r^2 g[M_B(J_T + (J_B/r)(r-h)) - M_A d\beta]}{Q}, a_{34} = \frac{r^2 n^2 b\beta}{Q}, a_{35} = \frac{-\beta n K r^2}{Q},$$

$$a_{41} = \frac{gr^2 M_B \sigma}{Q}, a_{42} = \frac{gr^2[M_A d\sigma + (h-r)(M_B J_B)/r^2]}{Q}, a_{44} = \frac{-n^2 b r^2 \sigma}{Q},$$

$$a_{45} = \frac{nK r^2 \sigma}{Q}, \text{ where the state vector is } x = [z \ \theta \ \dot{z} \ \dot{\theta} \ i]^T \text{ and } \beta = [M_B h + J_B/r],$$

$\sigma = [M_B + J_B/r^2]$, $Q = [M_B J_B(h-r)^2 + r^2 M_B J_T + J_B J_T]$, and
$J_T = [J_A + M_A d^2 + n^2 J]$.

4.10 For a state vector $x = [\Delta i \ z \ \dot{z}]^T$: $a_{12} = a_{13} = a_{21} = a_{22} = b_{21} = b_{31} = 0$, $a_{23} = 1$,
$a_{11} = -2500$, $a_{33} = -60$, $b_{11} = 500$, $a_{31} = 1810$, $a_{32} = 3.62 \times 10^5$.

4.13 (a) $i_0 = 0.833$ A. (b) $\left[\frac{2K_M i_0}{g_0^2}\right] \Delta i = M_A \ddot{z} + b\dot{z} - \left[\frac{2K_M i_0^2}{g_0^3}\right] z$, where $K_M = \frac{A\mu_0 N^2}{2}$.

Chapter 5

5.2 For $x_1 = \dot{y}_1 + 2y_1 - 4u_1$, $x_2 = y_1$, and $x_3 = y_2$: $C = \begin{bmatrix} 0 & 1 & 0 \\ 0 & 0 & 1 \end{bmatrix}$, $D = \begin{bmatrix} 0 & 0 \\ 0 & 0 \end{bmatrix}$, and

$$A = \begin{bmatrix} 0 & 3 & 3 \\ 1 & -2 & 0 \\ 0 & -5 & 3 \end{bmatrix}, B = \begin{bmatrix} 0 & 0 \\ 4 & 0 \\ 1 & 1 \end{bmatrix}.$$

5.5 $\ddot{i} + b_1 \dot{i} + b_0 i = a_0 u + a_1 \dot{u}$, where $b_1 = \frac{L + R_1 R_2 C_1}{LC_1 R_1}$, $b_0 = \frac{R_1 + R_2}{LC_1 R_1}$, $a_0 = \frac{1}{LC_1 R_1}$,

and $a_1 = \frac{1}{L}$. Let $x_1 = i - \frac{1}{L}u$ and $x_2 = i$ so that $C = [0 \ 1]$, $D = [0]$, $b_{11} = \frac{-R_2}{L^2}$,

$b_{21} = a_1$, $a_{11} = -b_1$, $a_{12} = -b_0$, $a_{21} = 1$, $a_{22} = 0$.

5.6 For $x = [z \ \dot{z}]^T$: $a_{11} = b_{11} = c_{12} = 0$, $c_{12} = k$, $D = [0]$, $a_{12} = 1$, $a_{22} = \frac{-d}{M_A}$,

$a_{21} = \frac{2K_M i_0^2}{M_A g_0^3}$, $b_{21} = \frac{2K_M i_0}{M_A g_0^2}$, where $K_M = \frac{A\mu_0 N^2}{2}$.

5.9 For $x = Qq$ and $q = Px$, $P = \begin{bmatrix} 3 & -1 \\ -2 & 2 \end{bmatrix}$ and $Q = \begin{bmatrix} 0.5 & 0.25 \\ 0.5 & 0.75 \end{bmatrix}$.

5.11 (b) For $x = Ww$, $W = \begin{bmatrix} 1/R & C_1 \\ 1 & 0 \end{bmatrix}$.

5.12 $C = \begin{bmatrix} 0 & 0 & 1 & 0 \\ 0 & 0 & 0 & 1 \\ 0.3697 & 2.0244 & 0.3369 & 0 \end{bmatrix}$, $D = \begin{bmatrix} 0 & 0 \\ 0 & 0 \\ 28.17 & 0 \end{bmatrix}$.

5.14 If $x = [\theta_1 \ \theta_2 \ \dot{\theta}_1 \ \dot{\theta}_2]^T$: $C = \begin{bmatrix} \Psi_1 & 0 & 0 & 0 \\ 0 & \Psi_2 & 0 & 0 \end{bmatrix}$, $D = \begin{bmatrix} 0 \\ 0 \end{bmatrix}$.

Chapter 6

6.1 (a) $F(s) = \dfrac{3}{s} + \dfrac{5}{s^2} - \dfrac{2\omega_0}{s^2 + \omega_0^2}$. (d) $f(t) = 0.2\epsilon^t + 0.8\epsilon^{-4t}$.

6.2 (b) $b(t) = 6\epsilon^{-t} - 1.5 - 4.5\epsilon^{-2t}$. (d) $f(t) = 2\delta(t) - 2\epsilon^{-2t}$.

6.4 (a) $\dfrac{E_L(s)}{E_{in}(s)} = \dfrac{s}{s + R/L}$. (b) $\dfrac{E_L(s)}{I(s)} = Ls$.

6.6 (b) $z_c(t) = \frac{1}{8}[\epsilon^{-2t} + \epsilon^{-6t}\sin(8t + 53.13°) + 1.118\epsilon^{-6t}\sin(8t - 116.57°)]$.

6.9 $t_1 = 0.2$ s.

6.14 If $u(t) = 0$ then $\dfrac{Y(s)}{e_C(0)} = \dfrac{1 + R_2 C_1 A}{s - A}$, where A is defined in Eqns. 3.32 and 3.33.

6.17 (e) $E(s) = \dfrac{-18(s+5)(s-1)}{(s+1)(s+10)\underbrace{[s^2+9]}_{(s+j3)(s-j3)}}$.

6.18 (b) No, because $c(t)$ is an unstable function and so $\lim_{t\to\infty}(c(t))$ does not exist for $K \neq 0$.

6.20 $\dfrac{Y_1(s)}{U(s)} = \dfrac{3041(s+2.426)(s-2.426)}{s(s-2.435)(s+2.416)(s+34.27)(s+968)}$,

$\dfrac{Y_2(s)}{U(s)} = \dfrac{1825(s)}{(s-2.435)(s+2.416)(s+34.27)(s+968)}$.

6.22 $\dfrac{Y(s)}{E_a(s)} = \dfrac{29{,}590}{(s+9.846)(s+169.5)(s+492.7)}$, $\theta_{LSS} = 1.8$ rad.

6.25 $\dfrac{Y(s)}{E_{in}(s)} = \dfrac{94{,}128}{(s+109.5)(s-89.51)(s+1000)}$.

6.27 $\dfrac{X_1(s)}{U(s)} = \dfrac{(1/J_1)[s^2 + a_1 s + a_0]}{s^4 + b_3 s^3 + b_2 s^2 + b_1 s + b_0}$, where $a_1 = \dfrac{d_2}{J_2}$, $a_0 = \dfrac{k_1 + k_2}{J_2}$,

$b_3 = \dfrac{J_1 d_2 + J_2 d_1}{J_1 J_2}$, $b_2 = \dfrac{J_1(k_1 + k_2) + J_2 k_1 + d_1 d_2}{J_1 J_2}$, $b_1 = \dfrac{k_1(d_1 + d_2) + k_2 d_1}{J_1 J_2}$,

$b_0 = \dfrac{k_1 k_2}{J_1 J_2}$.

6.29 $\dfrac{X_1(s)}{M(s)} = \dfrac{[6.3036 \times 10^{-6}]s^2}{(s+6.85)(s-6.85)(s+j4.9)(s-j4.9)} \dfrac{\text{rad}}{\text{dyne-cm}}$.

Chapter 7

7.1 (a) $Q(s) = \dfrac{K(s-1)}{s(s+1)(s^2+4s+13)}$. (b) $q(0^+) = \dot{q}(0^+) = 0$, $\ddot{q}(0^+) = K$.

(c) $q(t)$ is stable; $q(\infty) = 1$ if $K = -13$.

7.3 (b) All P_i except one must be positive; $b_0 = 0$; and $Ka_0/b_1 = A$.

(d) $n > m + r + 1$.

7.5 $G(s) = \dfrac{-(s-1)}{s+1}$.

7.6 (b) $\dfrac{2}{s} + \dfrac{2}{s+2} - \dfrac{2}{s+1+j\sqrt{3}} - \dfrac{2}{s+1-j\sqrt{3}}$.

7.9 (a) 0.6667 m/s. (b) 0.101 s. (d) 0.201 s.

7.11 $A_R = 5$ cm^2, $k_v = 400$ dynes/(cm/s).

7.13 Unstable for $-50 \le K \le -2$.

7.16 $R(s) = \dfrac{8(s-1)}{s(s^2+s+4)}$.

7.18 (b)(3) $Z_1 < 0$, $Z_2 < -Z_1$, and $KZ_1Z_2 = 75$. Example: $K = -25$, $Z_1 = -3$, $Z_2 = 1$.

7.24 $H(s) = \dfrac{5(s+0.2)}{s(s+0.1)(s^2+0.2s+1)}$.

Chapter 8

8.2 $\dfrac{\epsilon(s)}{R(s)} = \dfrac{M(s)+G_C(s)G_P(s)[M(s)H(s)-1]}{1+G_C(s)G_P(s)H(s)}$.

8.4 $W_0(s) = \dfrac{I_0(s)}{1+G_C(s)G_P(s)H(s)}$.

8.5 (a) $W(s) = \dfrac{136}{(s+4.914)[s^2+9.086s+55.35]}$. (b) Yes.

(c) $G(0) = \infty$, $W(0) = 0.5$. (d) System will be unstable.

8.8 (a) $W(0) = 1$. (b) $W(s) = \dfrac{20(s+6)}{[s^2+9.42s+31.76][s^2+0.5796s+3.778]}$.

(c) $W(s) = \dfrac{20(s+6)}{[s^2+10.54s+39][s^2-0.54s+7.692]}$; system is unstable.

Chapter 9

9.1 (b) Three.

9.2 (a) One. (b) One.

9.4 Three conditions must be met: (1) all $a_i > 0$; (2) $a_3a_4 > a_2$; and
(3) $a_1a_2a_3a_4 + 2a_0a_1a_4 + a_0a_2a_3 > a_0a_3^2a_4 + a_1^2a_4^2 + a_1a_2^2 + a_0^2$.

9.7 (b) $W(s) = \dfrac{5000}{(s+11.31)[s^2+13.9s+84.31][s^2-0.2092s+13.77]}$.

9.9 $P = 1$.

9.10 $G(s) = \dfrac{0.25(s+5)}{s}$.

9.12 $A_0 = 5$ deg, $A_1 = 6.32$ deg, $\phi = -71.5°$, $K_A = 4000$ V/V, $C = 200$ μF.

9.15 $K_{A\text{crit}} = 29.04$ V/V, $k_e = 2 \times 10^5$ dyne/cm.

9.18 (a) Zero. (b) Two. (c) One.

9.20 (c) $K_C = 3.64$.

9.24 $K = 28.03$.

9.25 No.

Chapter 10

10.1 (a) $0 < K < ab$.

10.3 $\beta > 1.5$.

10.7 (a) K, P, and Z must satisfy three conditions: (1) $3KZ > -2P$;
(2) $3K(Z+2) = 2$; (3) $3KZ = P$.

10.10 (a) $\dfrac{Y_1(s)}{U(s)} = \dfrac{679.63(s+2.2147)(s-2.2147)}{s(s+1.99)(s-2.39)(s+12.96)(s+288.5)}$,

$\dfrac{Y_2(s)}{U(s)} = \dfrac{339.81(s)}{(s+1.99)(s-2.39)(s+12.96)(s+288.5)}$.

10.14 (a) $K = 27,800$. (b) No.

10.17 First trial design: $G_{CL}(s) = \dfrac{-80[s^2+0.04s+0.04]}{(s+0.04)(s+3)}$.

10.22 $\dfrac{X_2(s)}{X_1(s)} = \dfrac{k_s/J_2}{s^2+(k_s/J_2)}$, $\dfrac{X_2(s)}{E_{in}(s)} = \dfrac{K_T}{s^4+b_3s^3+b_2s^2+b_1s}$, where

$K_T = \dfrac{nKK_Ak_s}{R_TJ_2J_T}$, $b_3 = \dfrac{R_Tb_T+n^2K^2}{R_TJ_T}$, $b_2 = \dfrac{k_s(J_2+J_T)}{J_2J_T}$, $b_1 = \dfrac{k_s[b_TR_T+n^2K^2]}{R_TJ_2J_T}$,

and $R_T = R+R_o$, $J_T = J_1+n^2J$, $b_T = n^2b$.

10.23 (a) $K_C = 0.475$. (b) $K_C = 0.48$, gain margin $= 14.17$.

Chapter 11

11.1 (a) $y_{SS}(t) = 0.094 \sin(6t - 137.5°)$. (b) -30 db.

11.2 (d) $\omega_{180} = 6.143$ rad/s, so $f_{180} = 0.9777$ Hz.

11.4 (a) $W(s) = \dfrac{10^6}{(s+10)[s^2+20s+10^4]}$. (b) Phase shift at $\omega = 70$ should be $-97°$.

11.7 (b) $3.162 - 3.062\epsilon^{-10t}$.

11.11 $\omega_{min} = 196.4$ rad/s.

11.14 (a) $\phi(\omega)$ will be constant at $(90r - 180)$ deg. (b) Unchanged.

11.17 $G(s)$ is minimum phase for (a) and nonminimum phase for (b), (c), and (d).
For $\omega = 2$, $\angle G(j\omega)$ is (a) $-25.33°$, (b) $27.897°$, (c) $-115.2°$, (d) $-62.1°$.

Chapter 12

12.1 (a) $N = -1$, $P = 1$, $Z = 0$; stable.
(b) $N = 1$, $P = 1$, $Z = 2$; unstable.
(c) $N = 0$, $P = 1$, $Z = 1$; unstable.

12.2 (b) $N = -3$, $P = 3$, $Z = 0$; stable.

12.5 (b) Not possible.

12.7 (a) $K_1 = 40$ and $K_2 = 2$ yield gain margins of 4.55 and 0.5.

12.9 $G(s) = \dfrac{14(s+1)}{(s+2)(s+5)}$, $\omega_X = \sqrt{3}$.

Chapter 13

13.1 $K = 100$, $y_{SS}(t) = A \sin(10t - 90°)$.

13.3 $K = 448$, phase margin = 12.1°, gain margins are 0.57 and 2.

13.7 $K = 14.02$, phase margin = 30.7°.

13.9 $P = 3$, gain margin is 3.63, bandwidth is 1.67 rad/s.

13.12 $K_P = 0.02$, $K_I = 1$, $K_D = 0.01$, phase shift is 60° at $\omega = 11.88$ rad/s.

13.16 **(a)** $Z = -1$, $K = 235$. **(b)** Not possible.

Chapter 14

14.2 $H(s) = \dfrac{0.5[s^2 + 35,400]}{(s+50)(s+708)}$, $G_C(s) = \dfrac{10(s+50)}{(s+500)}$, $K_A = 4050$.

14.3 **(b)** 10.9 horsepower.

14.10 $K_P = 2.397$, $K_I = 1.128$, $K_D = 0.282$.

14.12 $G_C(s) = \dfrac{2.5(s+0.5)}{s}$, $k_0 = -0.5$.

14.13 $G_\beta(s) = \dfrac{110(s+1.7)}{(s-2)}$.

Chapter 15

15.2 **(b)** $G_u(s) = \{C[sI - A]^{-1}B + D\}$, $G_0(s) = \{C[sI - A]^{-1}\}$.

15.3 **(a)** Assume that the dynamic variables have the following dimensions: $x(k)$ is $n \times 1$; $u(k)$ is $p \times 1$; $y(k)$ is $m \times 1$; $b(k)$, $e(k)$, and $r(k)$ are $r \times 1$; and

$q(k)$ is $l \times 1$. Then $p(k)$ is $(n+l) \times 1$ and $F_{CL} = \begin{bmatrix} [A_p] & [B_p H] \\ [-GKC] & [F] \end{bmatrix}$,

$G_{CL} = \begin{bmatrix} [0]_G \\ [G] \end{bmatrix}$, and $H_{CL} = [[C] \ [0]_C]$, where $[0]_G$ is $n \times r$ and $[0]_C$ is $m \times l$.

15.6 $D > 2.16$.

Bibliography

Ackermann, J., *Robust Control, Systems with Uncertain Physical Parameters.* New York: Springer-Verlag, 1993.

Anderson, B. D. O., and Moore, J. B., *Optimal Control.* Englewood Cliffs, NJ: Prentice-Hall, 1990.

Bellman, R., and Kalaba, R., *Mathematical Trends in Control Theory.* New York: Dover, 1964.

Bennett, S., *A History of Control Engineering, 1800–1930.* London: Peregrinus, 1979.

Blakelock, J. H., *Automatic Control of Aircraft and Missiles,* 2nd ed. New York: Wiley, 1991.

Bode, H. W., *Network Analysis and Feedback Amplifier Design.* New York: Van Nostrand, 1945.

Boyd, S. P., and Barratt, C. H., *Linear Controller Design.* Englewood Cliffs, NJ: Prentice-Hall, 1991.

Bryson, A. E., *Control of Spacecraft and Aircraft.* Princeton, NJ: Princeton University Press, 1994.

Buckley, P., *Techniques of Process Control.* New York: Wiley, 1964.

Canfield, E. B., *Electromechanical Control Systems and Devices.* New York: Wiley, 1965.

Cannon, R. H., *Dynamics of Physical Systems.* New York: McGraw-Hill, 1967.

Clark, R. N., *Introduction to Automatic Control Systems.* New York: Wiley, 1962.

D'Azzo, J. J., and Houpis, C. H., *Linear Control System Analysis and Design,* 3rd ed. New York: McGraw-Hill, 1988.

DiStefano, J. J., Stubberud, A. R., and Williams, I. J., *Feedback and Control Systems,* 2nd ed. New York: McGraw-Hill, 1990.

Doebelin, E. O., *Measurement Systems, Application and Design,* 4th ed. New York: McGraw-Hill, 1990.

Dorf, R. C., *Modern Control Systems,* 5th ed. Reading, MA: Addison-Wesley, 1989.

Doyle, J. C., Frances, B. A., and Tannenbaum, A. R., *Feedback Control Theory.* New York: Macmillan, 1992.

Electro-Craft, *DC Motors, Speed Controls, Servo Systems,* 5th ed. Minneapolis, MN: Electro-Craft Corp., 1980.

Evans, W. R. *Control System Dynamics.* New York: McGraw-Hill, 1954.

Fallside, F., *Control System Design by Pole-Zero Assignment.* New York: Academic Press, 1977.

Franklin, G. F., Powell, J. D., and Emami-Naeini, A., *Feedback Control of Dynamic Systems,* 2nd ed. Reading, MA: Addison-Wesley, 1991.

Franklin, G. F., Powell, J. D., and Workman, M. L., *Digital Control of Dynamic Systems,* 2nd ed. Reading, MA: Addison-Wesley, 1990.

Friedland, B., *Control System Design.* New York: McGraw-Hill, 1986.

Gantmacher, F. R., *Applications of the Theory of Matrices.* New York: Wiley, 1959.

Gardner, M. F., and Barnes, J. L., *Transients in Linear Systems.* New York: Wiley, 1942.

Greenwood, D. T., *Principles of Dynamics,* 2nd ed. Englewood Cliffs, NJ: Prentice-Hall, 1988.

Herceg, E. E., *Handbook of Measurement and Control.* Pennsauken, NJ: Schaevitz Engineering, 1976.

Himmelblau, D. M., and Bischoff, K. B., *Process Analysis and Simulation.* New York: Wiley, 1968.

Horowitz, I. M., *Synthesis of Feedback Systems.* New York: Academic Press, 1963.

Hughes, P. C., *Spacecraft Attitude Dynamics.* New York: Wiley, 1986.

Ince, E. L., *Ordinary Differential Equations.* New York: Dover, 1956.

James, H. M., Nichols, N. B., and Phillips, R. S., *Theory of Servomechanisms.* New York: McGraw-Hill, 1947.

Jamshidi, M., and Herget, C. J., *Computer-Aided Control Systems Engineering.* Amsterdam: North-Holland, 1985.

Kassakian, J. G., Schlecht, M. F., and Verghese, G. C., *Principles of Power Electronics.* Reading, MA: Addison-Wesley, 1991.

Kaplan, M. H., *Modern Spacecraft Dynamics and Control.* New York: Wiley, 1976.

Kaplan, W., *Advanced Mathematics for Engineers.* Reading, MA: Addison-Wesley, 1981.

Kuo, B. C., *Automatic Control Systems,* 6th ed. Englewood Cliffs, NJ: Prentice-Hall, 1991.

Laub, A. J., "Numerical Linear Algebra Aspects of Control Design Computations," *IEEE Transactions on Automatic Control* 30 (1985), pp. 97–108.

Leonhard, W., *Control of Electric Drives.* New York: Springer-Verlag, 1985.

LePage, W. R., *Complex Variables and the Laplace Transform for Engineers.* New York: Dover, 1980.

Maciejowski, J. M., *Multivariable Feedback Design.* Reading, MA: Addison-Wesley, 1989.

MATLAB, *The Student Edition of.* Englewood Cliffs, NJ: Prentice-Hall, 1992.

Maxwell, J. C., "On Governors," *Proceedings of the Royal Society, London* 16 (1868), pp. 270–83.

Merritt, H. E., *Hydraulic Control Systems.* New York: Wiley, 1967.

McCloy, D., and Martin, H. R., *The Control of Fluid Power.* New York: Wiley, 1973.

McCormick, B. W., *Aerodynamics, Aeronautics, and Flight Mechanics.* New York: Wiley, 1979.

Nering, E. D., *Linear Algebra and Matrix Theory,* 2nd ed. New York: Wiley, 1970.

Nesline, F. W., and Zarchan, P., "Why Modern Controllers Can Go Unstable in Practice," *AIAA Journal of Guidance, Control, and Dynamics* 7 (1984), pp. 495–500.

Newton, G. C., Gould, L. A., and Kaiser, J. F., *Analytical Design of Linear Feedback Controls.* New York: Wiley, 1957.

Nise, N. S., *Control System Engineering.* Redwood City, CA: Benjamin-Cummings, 1992.

Nyquist, H., "Regeneration Theory," *Bell System Technical Journal* 11 (1932), pp. 126–47.

Ott, H. W., *Noise Reduction Techniques in Electronic Systems,* 2nd ed. New York: Wiley, 1988.

Ralston, A., *A First Course in Numerical Analysis.* New York: McGraw-Hill, 1965.

Rizzoni, G., *Principles and Applications of Electrical Engineering.* Boston: Irwin, 1993.

Roberts, G. E., and Kaufman, H., *Table of Laplace Transforms.* Philadelphia: Saunders, 1966.

Routh, E. J., *Treatise on the Dynamics of a System of Rigid Bodies (Advanced Part),* 6th ed. New York: Dover, 1955.

Skilling, H. H., *Electromechanics.* New York: Wiley, 1962.

Stevens, B. L., and Lewis, F. L., *Aircraft Control and Simulation.* New York: Wiley, 1992.

Strogatz, S. H., *Nonlinear Dynamics and Chaos.* Reading, MA: Addison-Wesley, 1994.

Truxal, J. G., *Automatic Feedback Control System Synthesis.* New York: McGraw-Hill, 1955.

Uspensky, J. V., "Appendix III," in *Theory of Equations.* New York: McGraw-Hill, 1948.

Watton, J., *Fluid Power Systems.* Englewood Cliffs, NJ: Prentice-Hall, 1989.

Wertz, J. R., *Spacecraft Attitude Determination and Control.* Boston: Reidel, 1984.

Index

acceleration, normal, 109
active device, 90
actuator, 189
adaptive control, 451
additivity, 115, 464
aircraft, Navion, 40, 109, 279, 306, 426
airspeed, 27
Albrecht, R. W., 503
altitude, 27, 41
  automatic control of, 258
Ampere's rule, 72, 77
amplifier
  electrohydraulic, 88
  four-quadrant, 81
  "infinite-gain," 57
  push–pull, 85, 88
amplifier, operational (op-amp), 53, 67, 68
  dynamic stability of, 54
  as integrator, 56
  limiting, 442
  as summing amplifier, 55
amplitude ratio, 314
analysis, frequency domain, 107, 116
angle
  of attack, 27
  of departure, 235
  flight path, 27
  pitch, 27
angular momentum, 28
arrow, 143, 146
  calculation of residues, 170, 323
asymptote
  high-frequency, 330
  low-frequency, 330
  of root loci, 229
axis, output, of rate gyroscope, 11

backlash, gear, 21
ball–screw mechanism, 25
bandwidth, 333, 376
baseline design, 264
belt drive, 23, 94
bias point, 99
biomechanics, 451
block diagram
  of feedback control system, 197
  of linear system, 49, 100
Bode formulas, 337
Bode plot, 319
  approximation to, 327, 400
body, rigid, 5
boundary-value problem, two-point, 448, 452
break frequency, 330
breakaway point, 234
brushless motor, 81

calculus of variations, 451
canonical forms, 104
Cauchy's fundamental theorem, 357
CCCS (current controlled current source), 51, 65
CCVS (current controlled voltage source), 51
characteristic equation, 216
  polynomial, 121, 144, 213
circuit, active, 51
  analog, 44
  bridge, 60, 69
  gain, 68
  integrator, 56
  passive, 45
  output, 51
commutation, in DC machine, 75, 81

503

compensator
  lag, 265, 349, 410, 419
  lag-lead, 268, 277, 350, 412
  lead, 278, 348, 421
  parallel, 276, 419
  series, 175, 408
complex function
  argument of, 217
  magnitude of, 217
  rectangular expression, 217
  polar expression, 217
computer-aided calculations, 412
conformal map, 358
continuity of $\angle G(s)$, 223
control inputs
  elevator, 27
  throttle, 27
control law, 189, 197, 242, 405
  algorithm, 437
  structures, 191, 450
control theory
  basic, 1
  classical, 1, 3, 116, 192, 202
  modern, 1, 2, 6, 21, 450
controllability, 450
controlled sources, 51
controller, 175
  design, 192
converters, A/D and D/A, 435
convolution, 467
coordinate system, in state space, 103
corner frequency, 330
cost function, 201
crane, overhead, 19
critical K, 225
current, differential, 85, 89
cybernetics, 451

D matrix, 53, 106
damped sinusoid, 147
damping
  critical, 151
  over-, 151
  ratio ζ, 154, 167, 324
  under-, 153
  viscous, 83
DC machine, 71
  armature-controlled, 72
  back-voltage constant, 75
  circuit model of, 76
  machine constant, 72, 75
  torque constant, 75
dead time, 393
dead zone, 81
decade, of frequency, 328
decibel (db), 319
decomposition, of time function, 172
describing function, 36, 444

design, by trial and error, 201
  hardware realization of, 202
device, active, 90
diagram, block, 49
  device-based, 91
  transfer function, 121, 124
diagram, free-body, 6
differential equation
  coefficients, physical meaning of, 9
  nonlinear, 10, 18
  state-variable form, 2, 6, 10
  vector-matrix, 2, 33
digital control, 435
discrete time systems, 435
disturbance
  input, 191, 198
  random, 26
  rejection, 291
  in spacecraft, 25
disturbance–control interaction, 291
drag, aerodynamic, 25, 32
droop, 261
duality, 132
dynamic order, 121, 123
dynamometer, 93

effect, loading, 44
efficiency
  belt drive, 24
  gear, 23
eigenvalues, 433
electromechanical principles, 71, 72
electromagnets, 83
  force current in, 87
  force displacement in, 86
  push–pull, 85
  tractive, 84
encoder, shaft, 13
  absolute and incremental, resolution of, 59
energy source, of controlled plant and
    actuator, 189
envelope
  of exponential function, 173
  of nonlinear oscillation, 452
equation, difference, 301, 436
error, 193, 198
  steady-state, 242
  unmeasurable, 198
error-actuated system, 193
Euler's formula, 147, 217, 458
Euler's law, 5, 12, 17, 24, 34, 77
Evans, W. R., 214
existence of solutions, 36, 110
expansion, partial-fraction, 125, 130
exponential response, 78

failure detection, 201
Faraday's law, 45, 72

fault detection, 441
fault-tolerant control, 451
feedback, 2
  full-state, 202, 205
  negative, 194
  non-unity, 243
  positive, 256, 258
  velocity, 205
feedback amplifiers, 423
feedback control, 87, 191
  automatic, 192, 252
  manual, 192, 252
figure of merit, 201, 306
filtering, of electrical signals, 62
final value, of $y(t)$, 140
final-value theorem, 134, 140, 466
finite interval, control over, 448
flag constant, 143, 169
flight condition, equilibrium, 29
follow-up system, 285
forces
  aerodynamic, 28
  contact, 7, 16
  damper, 8
  flow in hydraulic valves, 89
  gravity, 7, 16
  moment of, 7, 17
  spring, 8
  thrust, 28
  viscous, 16
Fourier series, 316, 444
fraction
  improper, 123
  proper, 123
  rational, 123
frequency, 462
  of oscillation, ω_o, 155, 327
  resonant, ω_r, 327
  undamped natural, ω_n, 154, 327
frequency response
  of closed-loop system, 353
  definition of, 315
  of minimum phase system, 336
  of nonminimum phase system, 338
  obtained from transfer function, 321
  of unstable plant, 316
frequency response curves, normalized for
    second-order system, 487
friction
  in gears, 21
  in hydraulic servovalve, 90
  in servomotors, 79, 83
full-state feedback, 204, 434
fuzzy logic, 451

gain, voltage, of amplifier, 52
gain margin (GM), 227, 239, 374
gauge, strain, 59

Gauss's theorem, 138
gear ratio, 22
gimbal, 11
gradient
  gravity, 25
  voltage, 59
gravitation, Newton's law of, 453
ground, 47
gyroscope
  pitch attitude, 33
  rate, 11

heat exchanger, 396
hertz (cycles per second), 61
homogeneity, 115, 463
Hooke's law, 8
Hurwitz, A., 472

identity, 142
identity matrix, 431
impulse function, 465
  relation to step function, 461
  strength of, 460
impulse response, 168
inductance, of armature circuit, 76
inductor, 45
inequality conditions, in Routh's criterion, 477
inertia, moment of, 22
inertial space, 15, 34
initial conditions, 15, 115, 158
  in feedback system, 207
  nonzero, 178, 181, 199
initial state, 10
initial value, of $y(t)$, 140
initial values, 115, 158
initial-value problem, 6, 10, 33, 49
  for MIMO system, 115
  solution of, 106, 110, 113
initial-value theorem, 140, 466
inner-loop compensation, 429
input, 19
input vector, 32
instability, static, 87
integral, convolution, 114
integrating factor, 112
integrating property, 241
integration, numerical, 111
integrator, 57, 242
interaction, of devices, 92
interface specifications, 82
inverse transformation, 119, 146

$j = \sqrt{-1}$, 125, 141
Jordan normal form, 104
junction, summing, 121

Kelvin, Lord, 453
kinematics, 16, 27, 92

Kirchhoff's laws, 101
current, 48, 54
voltage, 45

Laplace transformation, 462
properties of, 463
tables of, 468
limit cycle, 205, 442
unstable, 448
linear algebra, 20
linear models, approximate for nonlinear
systems, 35
linear range, 271
linear system, response of, 115
linear transformation, 20
load cell, 62
loading effect, 44, 51, 55, 59, 81, 129
locked-rotor test, 77
log-magnitude phase plot, 321
loop gain, 237
loop transfer function, 244
lost motion, 21
low-frequency loop gain, 409
LVDT (linear variable differential
transformer), 61

M circle, 383
M_p (peak magnitude), 318, 376
machine constant, 72, 75
magnet force constant, 162
margin
gain, 227, 239
phase, 374
marginally stable system, 476
MATLAB®, 4
matrix
adjoint, 431
control canonical form, 434
diagonal, 439
input distribution, 100
inverse, 431
nonsingular, 103, 431
notation, 20
singular, 441
state distribution, 100, 434
state transition, 115, 430
transfer function, 432
transformation, 108
Maxwell, J. C., 192, 202, 214
MBB (make before break), 46
mesh, gear, ideal, 23
micrometeorite, 25
MIMO system, 2, 111, 189, 430, 451
mode, 144
system, 168, 213
models
air-temperature regulator, 395
aircraft, longitudinal axis, 32, 41, 109, 135,
258, 413, 428

amplifier–motor, 82, 83, 94, 96, 116, 128,
135, 136
ball–rack–pendulum, 41, 96, 109, 135, 136,
307
belt-drive system, 23, 41, 94
cart–motor, 94
electric car, 94
electrohydraulic, 236, 256, 334, 425
electrohydraulic amplifier–magnet–valve–
ram, 91, 96, 162, 184
electromagnet–amplifier, 88, 97, 108, 135,
307, 427
gear train, 21
inverted pendulum–cart, 39, 95, 134, 304,
311, 391, 428
pendulum, moving base, 15
rate gyroscope, 11, 41
RC circuit, 52, 65, 67, 111
rigid spacecraft, 25, 426
RL circuit, 46, 131, 132
RLC circuit, 48, 50, 51, 64, 65, 66, 102,
108, 132, 157, 317, 479
servomechanism, position control, 254,
262, 308, 311, 408
servomechanism, searchlight, 193, 210, 373
servomechanism, speed control, 241, 255
spring–mass–damper, 6, 37, 118, 132, 134,
151, 183, 203, 209, 452
two-rotor system, 43, 109, 135
models, linear system, 100
from differential equations, 107
in standard state-variable form, 102, 106
moment of inertia, of DC machine, 81
equivalent, 22
moments
aerodynamic, 29
control, 25
disturbances in spacecraft, 26
unwanted, 25
momentum, linear, 8
motion, planar, 5, 27
multi-loop control, 302
multi-rate digital system, 441

N circle, 384
negative K, 218, 220, 246, 416
neural networks, 451
Newton's laws, 5, 6, 13, 16, 21, 34, 453
NFPG system, 250
Nichols chart, 321, 490
Nichols plot, 386
noise, electrical, 62, 177
nonlinear systems, 442
nonminimum phase system, 161, 179, 248
control of, 298
transient response of, 341
normalized frequency u, 328
Nyquist plot, 319, 361
inverse, 401

Nyquist's criterion, 360
  applied to nonminimum phase plant, 387
  applied to positive feedback system, 371
  applied to unstable plant, 368, 370, 389

observability, 450
octave, of frequency, 329
Ohm's law, 45
optimal control, 201, 451
order, dynamic, 115, 213
oscillation
  limit cycle, 205
  unstable, 240
output
  circuit, 51, 53
  equation, 49, 104, 116
  signals, 33, 50
overshoot, in second-order system, 156, 160

parabolic function, 289
parameter, hidden, 92
partial derivatives
  first-order, 31
  high-order, 30
partial fractions, 119, 140
  expansion, 141
pendulum, inverted, 294
permeability, magnetic, 84
perturbations, in flight condition, 29
PFNG system, 250
phase margin (PM), 374
phase shift, 314
phase-angle–locus method, 394
phugoid mode
  of aircraft, 281
  damped, 283
PID (proportional integral derivative)
  circuit, 67, 184
  controller, 271, 306, 402, 427
  tuning, 273, 427
pitch rate, 28
pole, of rational function, 140
pole-zero cancellation, 175, 261, 274
pole-zero plot, 141
poles, closed-loop, 232
polynomial
  auxiliary, 475
  characteristic, 121, 144
  factoring, 125, 251
  monic, 124
  roots of, 138, 215
position control, 79, 236
positive feedback, 248
positive K, 248
positive time function, 459
potentiometer, 13, 58
principles of control system engineering, 93,
  158, 167, 170, 213, 228, 232, 247, 250,
  297, 301, 315, 345, 378, 425, 433, 434

process, non-integrating, 241
proximity sensor, 108

qualitative analysis, of stability, 213
quasilinear analysis, 442

ram, hydraulic, 88
ramp function, 57, 132
rational function, 138
  improper, 142, 150
  partial-fraction expansion of, 125, 141
  proper, 142
  residues of, 142
  standard form of, 124, 134
relative stability, 375
residues, 142, 168
  calculation of, by arrow method, 143,
  170
  calculation at repeated pole, 150
  relative values of, 168, 174
resistance
  brush, 78
  contact, 59
  equivalent, 165
  input, 49
resolution, transducer, 59
resonant frequency, 318, 376
response
  long-tailed, 267
  normalized, 156
  overdamped, 120
  of torque-limited system, 271
  underdamped, 120
rise time, 264
robotics, 451
robustness, 276, 303, 372, 422
  of design, 201, 205
root loci, 224
root locus
  angle requirement, 218
  branch of zero length, 245
  computer calculations for, 216
  magnitude requirement, 218
  method, 215
  plot, 162
  rules, 251
roots of polynomials, 138, 139
  complex, 119
  equal, 119
  real, 119
rotor
  of large machine, 36
  of rate gyroscope, 12
Routh, E. J., 472
Routh
  array, 473
  criterion, 214, 352, 472
  sequence, 473
  test functions, 473

RVDT (rotary variable differential transformer), 13, 62

$s = \sigma + j\omega$, Laplace variable, 120
sample and hold, 435
sample period, 440
scale factor
  of accelerometer, 105
  of position transducer, 91
  of sensor, 3, 13
second-order system, 150
  damping ratio ζ, 154
  frequency, undamped natural ω_n, 154
    normalized, 156
  frequency, of oscillation ω_o, 155
  step-function input, 151
  step response of, 157, 160
sensor, 58
  airspeed, 33
  equation, 14
  output, 49
series compensation, 408
servomechanism
  position control, 254, 262, 308, 311, 408
  searchlight, 193, 210, 373
  speed control, 241, 255
  theory, 1
set point, 285, 396
settling time, 406
short-period mode
  of aircraft, 281
  shifted, 283
signal
  conditioner, 175, 198
  processing, 62
  shaping, 176
SIMO system, 433
simulators, 201
singularity, of rational function, 141
singularity function, 150
SISO system, 1, 117, 122, 189
sliding-mode control, 451
solution
  homogeneous, 113
  particular, 113
  time-domain, 119
source, controlled, 51
speed, vertical, 27
speed control, 79, 241
  of DC motor, 272
spring constant, 8
stability
  conditional, 297, 392
  conditions for, 140
  in discrete time systems, 440
  of feedback system, 216
stability derivatives, 31
stability margins, 375
  on Bode diagrams, 377

state-variable equations, 6, 10, 20
state-variable model, 14
state variables, 48
  change of, 103
  choice of, 100
  importance of, 434
  model obtained from differential equation, 107
  natural set, 103
state vector, augmented, 41, 95, 97, 109
steady-state gain, 168
steady-state response, 113
step function, 10, 459
summing circuit, 55
superposition, 115
system
  multi-input, 115
  multi-output, 116
  multivariable, 115
  nth-order, 122
  second-order, 150
  third-order, 162
system error, 260
system model, 260
system type number, 290

Taylor series, 30
terminal time control, 449
test functions, 190
theory, communication, 62
third-order system, 162
time constant, 78
time delay, 393, 402
time response curves, normalized for second- and third-order systems, 478
torque
  gain, 263
  limit, 271
  load on DC machine, 77
tracking control, 285
transcendental function, 394
transducers
  electromechanical, 58
  gain, 59
transfer function, 107, 117, 121, 130
  approximation to, 170
  closed-loop, 195, 199
  composite, 126
  forward-path, 194
  including initial conditions, 121, 124
  loop, 199, 244
  open-loop, 194
  poles of, 196, 199
  related to impulse response, 132
  standard form, 134
  zeros of, 196, 199
transform, Laplace, 118
transient, 113
transient response, 130

transmission characteristic, unidirectional, 130
transportation lag, 393
trial and error, 261
trim condition, 30, 99
tuning, of PID controller, 273, 427
two-point boundary-value problem, 36, 110

uncertain parameters, 177
unidirectional drive, 23
unique solution, 102
uniqueness of solutions, 36, 110
units
    abbreviations for, 457
    consistent sets of, 454
    mixture of, 27
unity feedback, 197
unstable device, 87
unstable plant, 93, 178, 252, 256
    control of, 293
    response, 130, 178, 211

valve
    hydraulic, 88
    nozzle–flapper, 88
variational mechanics, 451
VCCS (voltage-controlled current source), 51
VCVS (voltage-controlled voltage source), 51
vector, 20
vector space, 20
voltage gain, 52
voltage gradient, 59

Watt, J., 202

z transform, 441
zero-order hold operation, 435
zeros
    at infinity, 141, 229
    of rational function, 140
ZIR (zero-input response), 110, 113
ZSR (zero-state response), 110, 113